Power Electronics and Power Quality

Power Electronics and Power Quality

Special Issue Editor

José Gabriel Oliveira Pinto

MDPI • Basel • Beijing • Wuhan • Barcelona • Belgrade • Manchester • Tokyo • Cluj • Tianjin

Special Issue Editor
José Gabriel Oliveira Pinto
Universidade do Minho
Portugal

Editorial Office
MDPI
St. Alban-Anlage 66
4052 Basel, Switzerland

This is a reprint of articles from the Special Issue published online in the open access journal *Energies* (ISSN 1996-1073) (available at: https://www.mdpi.com/journal/energies/special_issues/power-electronics-power-quality).

For citation purposes, cite each article independently as indicated on the article page online and as indicated below:

LastName, A.A.; LastName, B.B.; LastName, C.C. Article Title. *Journal Name* **Year**, *Article Number*, Page Range.

ISBN 978-3-03928-358-3 (Pbk)
ISBN 978-3-03928-359-0 (PDF)

Contents

About the Special Issue Editor

José Gabriel Oliveira Pinto (Ph.D.) is an Assistant Professor in the Department of Industrial Electronics of the University of Minho where he teaches graduate and post-graduate courses for the Engineering School. He is an Integrated Researcher for the Group of Energy and Power Electronics (GEPE) of the Centro ALGORITMI of the University of Minho. He received a degree in Electronic Engineering (2001), his MSc in Industrial Electronics (2004) and his Ph.D. in Electronics and Computer Engineering (2012) from the University of Minho. Since 2001, he has an integrated team and more than 15 R&D projects. He is the author or co-author of more than 90 publications in international journals and conferences indexed by SCOPUS and Web of Science. He previously supervised a Ph.D. thesis and 20 MSc dissertations and is currently the supervisor of a new Ph.D. thesis and three MSc dissertations. In addition to the scientific and pedagogical activities, he participated in numerous studies related to the Power Quality in Power Systems for several entities that requested advice from GEPE. His research interests relate to power electronics, power quality, renewable energy and digital control of power electronic converters.

Preface to "Power Electronics and Power Quality"

The ease of transport, the lack of noise, the absence of gas and odor emissions, combined with a reasonable production cost, make electricity an excellent energy and one that is essential for the development of most economic activity sectors. With the development of electronics, namely power semiconductors, it was possible to optimize electrical systems and equipment, in terms of efficiency and controllability, to previously unimaginable levels. However, these technological developments do not only bring advantages. The new systems based on power electronics were responsible for the appearance of disturbances in the power system that caused malfunctions, and even breakdowns, in more sensitive equipment. Therefore, beyond the availability and service continuity, the quality of electric power started to assume a role of extreme relevance. Studies conducted by several international organizations reveal that electrical power quality (PQ) is a fundamental factor for the productivity of companies, and that the economic losses caused by PQ problems are, nowadays, very high. Furthermore, power systems are experiencing a new set of challenges: decentralized generation, integration of renewable energy systems, widespread electric vehicles, and the electrification of railway systems, all of which bring a higher level of importance to PQ and power electronics. In this way, the research and development of methodologies and technologies to understand and mitigate PQ problems is a subject of utmost importance. This book brings together a set of high quality papers reflecting the latest investigations and the new trends in terms of power electronics and PQ.

José Gabriel Oliveira Pinto
Special Issue Editor

Article

Single-Phase Shunt Active Power Filter Based on a 5-Level Converter Topology

José Gabriel Oliveira Pinto [1,*], Rui Macedo [2], Vitor Monteiro [1], Luis Barros [1], Tiago Sousa [1] and João L. Afonso [1]

[1]   Centro ALGORITMI—University of Minho, Campus de Azurém, 4800-058 Guimarães, Portugal; vitor.monteiro@algoritmi.uminho.pt (V.M.); luis.barros@algoritmi.uminho.pt (L.B.); tiago.sousa@algoritmi.uminho.pt (T.S.); jla@dei.uminho.pt (J.L.A.)
[2]   Department of Industrial Electronics, University of Minho, Campus de Azurém, 4800-058 Guimarães, Portugal; a55743@alunos.uminho.pt
*   Correspondence: gpinto@dei.uminho.pt; Tel.: +351-253-510-190

Received: 14 March 2018; Accepted: 16 April 2018; Published: 22 April 2018

Abstract: This paper presents a single-phase Shunt Active Power Filter (SAPF) with a multilevel converter based on an asymmetric full-bridge topology capable of producing five distinct voltage levels. The calculation of the SAPF compensation current is based on the Generalized Theory of Instantaneous Reactive Power (*p-q* theory) modified to work in single-phase installations, complemented by a Phase-Locked Loop algorithm and by a dedicated algorithm to regulate the voltages in the DC-link capacitors. The control of the SAPF uses a closed loop predictive current control, followed by a multilevel Sinusoidal Pulse-Width Modulation technique with two vertical distributed carriers, which were specially conceived to deal with the asymmetric nature of the converter legs. Along the paper, some simulation results are used to show the main characteristics of the 5-level converter and control algorithms, and the hardware topology and control algorithms are described in detail. In order to demonstrate the feasibility and performance of the proposed SAPF based on a 5-level converter, a laboratory prototype was developed and experimental results obtained under diverse conditions of operation, with linear and non-linear loads, are presented and discussed in this paper.

Keywords: Shunt Active Power Filter; digital control; harmonics; multilevel converter; power quality; *p-q* theory

---

1. Introduction

The field of power electronics has experienced a major development in the last decades, allowing for the utilization of electrical loads with higher efficiency for a vast range of appliances. Nonetheless, this development is still present nowadays and it shows a predisposition to continue in the future [1,2]. An overview of the key considerations for an improved utilization of both presently used and upcoming power devices is presented in [3]. Despite its provided benefits, the development of power electronics shifted the paradigm of electrical loads from linear to non-linear, i.e., the current in this type of load presents a waveform different from the supplied voltage, which is due to a production of harmonic currents [4]. Harmonic currents are harmful to the power systems since they contribute to overheating, for instance, in power transmission cables and transformers, and interfere with communication systems. As such, power transmission cables and transformers should be derated due to harmonic currents flowing, which carries additional costs [5,6]. Harmonic currents can be produced by loads such as adjustable speed drives [7], arc furnaces [8], cycloconverters [9], electric vehicle battery chargers [10], lighting equipment, such as compact fluorescent and LED lamps [11], among other domestic appliances. Therefore, power quality, and especially harmonic currents, has deserved significant attention throughout the years.

The compensation of harmonic currents was primarily performed by passive filters. However, these are bulky and are only capable of compensating a limited range of harmonics. Besides, passive filters can produce resonance phenomena when connected to the power grid [12]. On the other hand, Shunt Active Power Filters (SAPFs) are power electronics devices that are capable of compensating harmonic currents, power factor and current imbalances in three-phase systems. Moreover, the compensation capability of a SAPF is dynamic, i.e., it is capable of compensating the aforementioned power quality problems regardless of the loads connected to the power grid. The operation of a SAPF consists in producing the harmonic currents and reactive power required by the connected loads, so that, the power grid currents acquire a sinusoidal waveform in phase with the voltages. Although this operation principle was proposed in [13], the SAPF was initially presented in [14] and, since then, has gained popularity due to the referred advantages over the traditional compensation systems [15]. Furthermore, several control theories for SAPFs have been developed and described in the literature, with the p-q theory [16], the FBD [17], the CPC [18] and the synchronous reference frame d-q theory [19] being among the main methods for obtaining the compensation currents for SAPFs. An extended review concerning control algorithms for SAPFs is presented in [20].

Regarding its structure, a SAPF is a DC-AC power converter, which can be either of the voltage source type or the current source type [21]. Moreover, a DC-AC power converter can be also classified in accordance with the number of levels of the output voltage or current, whereby it is considered a multilevel converter if the number of levels is three or more [22]. Multilevel converters are advantageous for medium voltage applications since they are capable of providing higher output voltages than the voltage that power semiconductors have to withstand. Besides, multilevel converters produce less dv/dt and have a better performance in terms of Total Harmonic Distortion (THD) when compared to two-level converters [23]. There are three classical topologies of multilevel converters, namely the neutral point clamped (or diode clamped) [24], the flying capacitor (or capacitor clamped) [25], and the cascaded multilevel converter [26]. Nevertheless, several topologies have been proposed in the literature, aiming to achieve higher numbers of levels with reduced switch counts [27–31]. Accordingly, control and modulation techniques for multilevel converters have been also widely investigated [32,33].

In this context, this paper presents a single-phase SAPF composed of a 5-level converter topology. The principle of operation of the proposed converter is explained in detail, from the implemented power theory to the applied modulation technique. Simulation results are depicted to analyze the feasibility of the converter. A laboratory prototype was developed and tested in real conditions of operation. The obtained experimental results validate the implemented single-phase 5-level SAPF hardware topology and control algorithms. The main contributions of this paper are: a low digital resource for the control theory implementation, which is based on the concepts introduced by the p-q theory, adapted to work in single-phase installations, and combined with a Phase-Locked Loop (PLL) to allow sinusoidal currents at the source side; a Sinusoidal Pulse-Width Modulation (SPWM) modulation strategy based on the vertical distribution of two triangular carriers that deals with the asymmetries of the 5-level converter legs; a comprehensive description of the hardware prototype development and a step-by-step validation of the control algorithms.

2. Proposed 5-Level Converter Topology

This section introduces a detailed description of the 5-level converter topology, which is depicted in Figure 1. As it can be seen, it is constituted by a three-level Diode-Clamped Multilevel Inverter (DCMLI) leg and by a two-level leg. The DC-link is composed of two capacitors in order to obtain the different voltage levels during both positive and negative half-cycles.

A relevant characteristic of the 5-level converter is the asymmetrical switching frequency of each leg, as well as the nominal voltage that each semiconductor of both legs withstands. The two-level leg (formed by S_1 and $S_1{}'$) is responsible for the polarity of the AC voltage and the switching frequency of this leg is equal to the fundamental frequency of the voltage to be synthetized by the converter, i.e., equal to the frequency of the power grid voltage, in this work. The semiconductors of this two-level leg should be able to support a voltage of v_{DC}, i.e., the sum of the voltage across each capacitor of the DC-link (v_{DC1} and v_{DC2}). On the other hand, the DCMLI leg (formed by S_2, S_2', S_3, and S_3') is controlled in order to establish the voltage level that the converter should produce (v_{DC}, $v_{DC}/2$ or 0) in both positive and negative half-cycles. This leg is switched with a higher frequency in comparison with the switching frequency of the two-level leg and is responsible for defining the converter state in each half-cycle, and, therefore, for defining the waveform of the produced voltage. By increasing the switching frequency, it is possible to improve the quality of the produced voltage, resulting in an improved compensation current. The semiconductors of the DCMLI leg should be able to support a nominal voltage of $v_{DC}/2$, i.e., the voltage of each DC-link capacitor, which is half of the voltage that the semiconductors of the two-level leg should support.

The different states that the 5-level converter can assume during the positive and negative half-cycles are presented in Table 1. This table illustrates that the semiconductors S_1 and S_1' are switched at the fundamental frequency of the produced voltage, and the other semiconductors (S_2, S_2', S_3, and S_3') are switched to establish the different voltage levels.

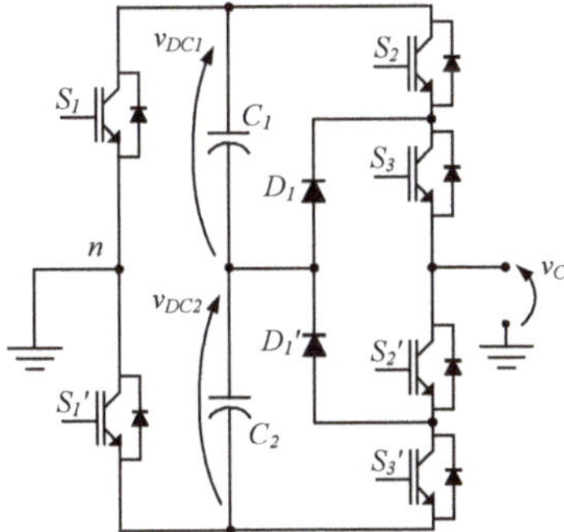

Figure 1. Topology of the 5-level converter used in the Shunt Active Power Filter (SAPF).

Table 1. Voltage levels that the 5-level converter can assume during the positive and negative half-cycles.

Voltage Polarity	S_1	S_1'	S_2	S_2'	S_3	S_3'	v_c
	OFF	ON	ON	OFF	ON	OFF	v_{DC}
$v_C > 0$	OFF	ON	OFF	ON	ON	OFF	$v_{DC}/2$
	OFF	ON	OFF	ON	OFF	ON	0
	ON	OFF	ON	OFF	ON	OFF	0
$v_C < 0$	ON	OFF	OFF	ON	ON	OFF	$-v_{DC}/2$
	ON	OFF	OFF	ON	OFF	ON	$-v_{DC}$

3. Control Algorithms

This section introduces the proposed control algorithms for the SAPF based on a 5-level converter, including a description about the *p-q* theory complemented by a Phase-Locked Loop (PLL) and adapted to work with single-phase installations, the DC-link control, the predictive current control, and the SPWM strategy.

3.1. p-q Theory

The *p-q* theory was initially developed to be applied in the analyses of three-phase three-wire electrical systems and later, it was expanded to three-phase four-wire systems to be usable in unbalanced systems with zero sequence components [34,35]. With some modifications, it is also possible to use the *p-q* theory in the analyses of single-phase systems [36]. The basics of this time-domain theory resides in a transformation of referential from the *a-b-c* coordinates, where the fundamental voltages are shifted by 120° for each other, to the *α-β-0* orthogonal system of coordinates. The change in the coordinates referential is performed by the application of the Clarke transformation to the voltages,

$$
\begin{bmatrix} v_0 \\ v_\alpha \\ v_\beta \end{bmatrix} = \sqrt{\frac{2}{3}} \begin{bmatrix} \frac{1}{\sqrt{2}} & \frac{1}{\sqrt{2}} & \frac{1}{\sqrt{2}} \\ 1 & -\frac{1}{2} & -\frac{1}{2} \\ 0 & \frac{\sqrt{3}}{2} & \frac{\sqrt{3}}{2} \end{bmatrix} \begin{bmatrix} v_a \\ v_b \\ v_c \end{bmatrix},
\tag{1}
$$

and to the currents,

$$
\begin{bmatrix} i_0 \\ i_\alpha \\ i_\beta \end{bmatrix} = \sqrt{\frac{2}{3}} \begin{bmatrix} \frac{1}{\sqrt{2}} & \frac{1}{\sqrt{2}} & \frac{1}{\sqrt{2}} \\ 1 & -\frac{1}{2} & -\frac{1}{2} \\ 0 & \frac{\sqrt{3}}{2} & \frac{\sqrt{3}}{2} \end{bmatrix} \begin{bmatrix} i_a \\ i_b \\ i_c \end{bmatrix},
\tag{2}
$$

of the installation under analysis. After the transformations, it is possible to calculate the instantaneous real power (p), the instantaneous imaginary power (q), and the instantaneous zero sequence power (p_0), applying:

$$
\begin{bmatrix} p_0 \\ p \\ q \end{bmatrix} = \begin{bmatrix} v_0 & 0 & 0 \\ 0 & v_\alpha & v_\beta \\ 0 & -v_\beta & v_\alpha \end{bmatrix} \begin{bmatrix} i_0 \\ i_\alpha \\ i_\beta \end{bmatrix}.
\tag{3}
$$

Each one of the instantaneous powers calculated by Equation (3) can be decomposed in an average value and an oscillating value:

$$
p = \bar{p} + \tilde{p},
\tag{4}
$$

$$
q = \bar{q} + \tilde{q},
\tag{5}
$$

$$
p_0 = \bar{p}_0 + \tilde{p}_0,
\tag{6}
$$

where the average value of the instantaneous real power ($\bar{p}$), is related to the energy that is transferred in a continuous way from the source to the loads and is associated with the active power. The oscillating value of the instantaneous real power ($\tilde{p}$), is related to a parcel of energy that is exchanged between the source and the load. The average and oscillating values of the instantaneous real power ($\bar{p}_0$ and $\tilde{p}_0$) relates to the energy that is transferred and exchanged with the help of the neutral conductor. The instantaneous imaginary power components ($\bar{q}$ and $\tilde{q}$) are associated with the presence of storage elements (capacitors or inductors) in the loads, and are related to energy that is exchanged between the source and loads or between the loads with the help of the source. In three-phase systems, the average value of the instantaneous imaginary power ($\bar{q}$) is exchanged between the three-phases of the system and is not associated with any effective energy transfer from the source of the loads. However, the presence of this component results in an increase in the source currents. The adoption of the *p-q* theory to calculate the compensation currents for a SAPF results from the physical interpretation of the power components described above. The power components that are involved with an effective energy transfer from the source to the load are the average value of the instantaneous real power ($\bar{p}$) and the average value of the instantaneous zero sequence power ($\bar{p}_0$). Therefore, in an installation with a SAPF, these components are the only power components that should be supplied by the source, while all the other components should be provided by the SAPF. However, $\bar{p}_0$ is transferred with the help of the neutral conductor, and so, it contributes to an increase in the current in the neutral conductor, which is an undesired phenomenon. Nevertheless, with a three-phase four-wire SAPF and with an appropriate control algorithm, it is possible to transfer the energy related to the average value of the zero-sequence power, in an equilibrated way by the three phases of the system, and so, eliminating the current in the neutral conductor upstream of SAPF. The calculation of the compensation currents that the SAPF must produce to eliminate the undesired power components, which are supplied by the source, can be accomplished by:

$$\begin{bmatrix} i_{F\alpha}^* \\ i_{F\beta}^* \end{bmatrix} = \frac{1}{v_\alpha^2 + v_\beta^2} \begin{bmatrix} v_\alpha & -v_\beta \\ v_\beta & v_\alpha \end{bmatrix} \begin{bmatrix} -\tilde{p} + \overline{p}_0 \\ -q \end{bmatrix}, \tag{7}$$

$$i_{F0}^* = -i_0, \tag{8}$$

where $i_{F\alpha}^*$, $i_{F\beta}^*$ and i_{F0}^* are the references for the compensation currents of the SAPF in the α-β-0 coordinates. In order to obtain the references for the compensation currents in the a-b-c coordinates, it is used the inverse Clarke transformation, defined as:

$$\begin{bmatrix} i_{Fa}^* \\ i_{Fb}^* \\ i_{Fc}^* \end{bmatrix} = \sqrt{\frac{2}{3}} \begin{bmatrix} \frac{1}{\sqrt{2}} & 1 & 0 \\ \frac{1}{\sqrt{2}} & \frac{1}{2} & \frac{\sqrt{3}}{2} \\ \frac{1}{\sqrt{2}} & \frac{1}{2} & \frac{\sqrt{3}}{2} \end{bmatrix} \begin{bmatrix} i_{F0}^* \\ i_{F\alpha}^* \\ i_{F\beta}^* \end{bmatrix}. \tag{9}$$

According to this procedure, it is possible to apply the p-q theory to separate the desired power components from the undesired ones, and, therefore, develop a control algorithm for a SAPF to eliminate the undesired power components and improve the power quality of the installation. With a balanced and sinusoidal three-phase system, the application of the compensation strategy, based on p-q theory, results in balanced and sinusoidal currents in the source. However, if the voltages are distorted and unbalanced, by the application of the p-q theory, the source currents become neither sinusoidal nor balanced, which is a disadvantage. To overcome this drawback, it is possible to use the positive sequence of the fundamental components of the source voltages instead of the real ones, and so, the compensation results always in sinusoidal and balanced currents in the source, even with distorted and unbalanced voltages. The positive sequence of the fundamental components of the voltages can be obtained by the application of a PLL algorithm. There are interesting PLL algorithms in the literature, for three-phase systems [34] and for single-phase [35].

The application of the p-q theory to single-phase systems requires the emulation of a balanced three-phase system. For that, the voltages and currents in phases b and c are obtained applying a phase shift of $\pm120°$ to the measured voltage and current in phase a. The application of the Clarke transformation to a three-phase balanced system produces two equal α and β components shifted by $90°$ degrees, one from the other. Therefore, it is possible to implement the emulation of a three-phase, balanced system, directly in the α-β-0 coordinates, by applying:

$$\begin{bmatrix} v_\alpha \\ v_\beta \end{bmatrix} = \begin{bmatrix} v_{Sa} \\ v_{Sa}\, e^{-s\frac{3T}{4}} \end{bmatrix}, \tag{10}$$

and

$$\begin{bmatrix} i_\alpha \\ i_\beta \end{bmatrix} = \begin{bmatrix} i_{La} \\ i_{La}\, e^{-s\frac{3T}{4}} \end{bmatrix}, \tag{11}$$

where T is the period of the power grid fundamental frequency, and $e^{-s\frac{3T}{4}}$ corresponds to a time delay of $\frac{3}{4}T$, which corresponds to a phase shift of $+90°$ in the fundamental frequency. It is important to explain that, in a balanced three-phase system, the zero sequence power component does not exist and, therefore, the reference for the compensation current for a single-phase SAPF can be calculated by [36]:

$$i_F^* = \frac{1}{v_\alpha^2 + v_\beta^2} \begin{bmatrix} v_\alpha & -v_\beta \end{bmatrix} \begin{bmatrix} -\overline{p} \\ -q \end{bmatrix}. \tag{12}$$

Also, in single-phase systems it is possible to use PLL algorithms combined with the p-q theory to obtain sinusoidal currents at the source side. In this work, it was used the PLL algorithm described in [35].

3.2. DC-Link Voltage Control

In order to the SAPF produce the compensation current, it is necessary to guarantee that the total DC-link capacitors voltage is greater than the peak value of the source voltage. The regulation of the DC-link capacitors voltage, in steady state, is done by absorbing a small amount of active power from the power grid, which corresponds to the converter losses. During some transients, it can be necessary to absorb or to deliver active power to the power grid, in values which can be greater than the ones involved in steady state. This is required to maintain the capacitors voltage regulated. In this way, it is necessary to include an additional instantaneous real power parcel (p_{reg}) in the calculation of the references for the SAPF compensation currents. The value of p_{reg} is determined by means of a PI controller that eliminates the error between the DC-link voltages and the reference value. It is important to explain that, during the normal operation of the SAPF, the oscillating values of the power components defined by the *p-q* theory are exchanged between the loads and the DC-link capacitors of the SAPF, and so, it is natural that the DC-link capacitors voltage oscillates to compensate these power components. Taking into account this situation, the objective of the DC-link voltage control algorithm is to regulate the average values of the DC-link capacitor voltages (V_{DC1} and V_{DC2}) instead of the instantaneous values (v_{DC1} and v_{DC2}). Therefore, the calculation of the errors is not performed by the difference between the reference voltage and the DC-link voltages, but by the difference between the reference and the average values of the DC-link capacitors voltage. Despite the possibility to use other values, in this work it was used an average value of an entire grid period.

As the DC-link of the SAPF is composed by two capacitors, it is necessary to take into account which capacitor is used in each instant, in order to select the correct value of p_{reg} and add it to the average value of the instantaneous real power ($\bar{p}$) in Equation (12). The identification of the capacitor that is used in each time instant can be done by consulting Table 1, from where it is possible to conclude that capacitor C_1 is used when the converter output voltage is negative and the capacitor C_2 is used when the output voltage is positive. The block diagram of the DC-link voltage control is represented in Figure 2, and, as it is possible to see, two similar PI controllers are used to obtain the value of p_{reg}.

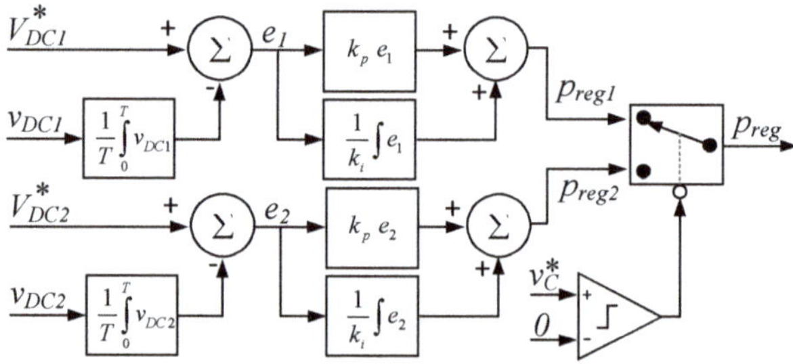

Figure 2. Block diagram of the DC-link voltage control algorithm.

3.3. Predictive Current Control

A predictive strategy was adopted to control the current in the AC side of the 5-level converter. This strategy requires measuring the values of the AC current (i_F) and the source voltage (v_S), which are obtained using a signal acquisition board, as well as the nominal values of the main electrical parameters of the system. Therefore, by analyzing the voltages and currents between the 5-level converter and the power grid, a set of equations can be established and used to predict the next state (i.e., during the time interval of [k, $k + 1$]) that the converter should assume in terms of produced voltage. By using this strategy, it is predictable that the filter current will follow its reference during each sampling time interval. The block diagram of the implemented predictive current control is illustrated in Figure 3.

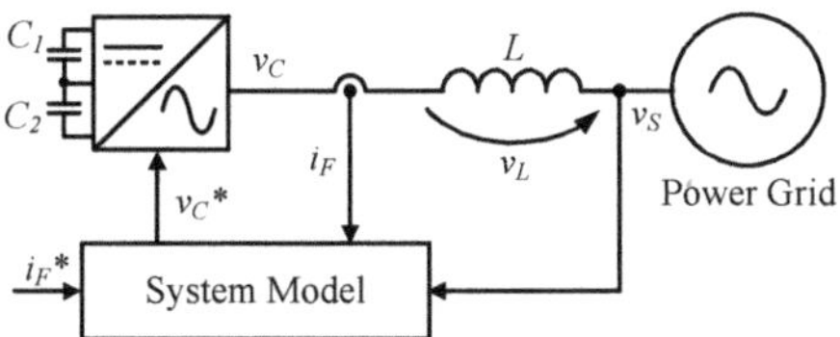

Figure 3. Block diagram of the predictive current control strategy implemented in the 5-level converter SAPF.

The electrical model of this type of system, generally can be described by the source voltage (v_S), by the converter output voltage (v_C), and by the coupling passive filters voltage (v_L), where, in this specific case, an inductor is used as a passive filter. Therefore, by analyzing the voltages in the AC side of the system, it is possible to establish the output voltage of the 5-level converter as a function of the source voltage and the voltage across the inductor, as described by:

$$v_C = v_L + v_S. \tag{13}$$

Substituting the inductance voltage (v_L) by the time derivative of its current multiplied by its nominal value, it is possible to obtain the equation:

$$v_C = R_L \, i_F + L\frac{di_F}{dt} + v_S, \tag{14}$$

where is also considered the voltage across the internal resistance of the inductor. Taking into account that the voltage in the resistive part is residual due to the reduced value of the internal resistance, it can be neglected, resulting in the equation:

$$v_C = L\frac{di_F}{dt} + v_S. \tag{15}$$

As occurs in all the systems with a feedback control, also in this current control strategy an error between the reference signal and the measured current is calculated as:

$$i_{error} = i_F^* - i_F. \tag{16}$$

Taking into account the electrical model of the system, it is possible to substitute the produced current (i_F) established in Equation (14), allowing obtaining the equation:

$$L\frac{di_{error}}{dt} = -v_C + v_S + L\frac{di_F^*}{dt}. \tag{17}$$

Considering a digital implementation and using a high sampling frequency, the variation of the error derivative is almost linear and equal to the increase in the current error (Δi_{Ferror}). Therefore, by selecting a high switching frequency, the resultant current ripple will be reduced and the error will be practically zero, i.e., the produced current will be centered in its reference. The current error can be obtained according to:

$$L\frac{i_{error}}{T_{aq}} = -v_C + v_S + L\frac{\Delta i_F^*}{T_{aq}}. \tag{18}$$

Taking into account the position of the sensors and implemented control algorithm, it is important to note that the voltage established by the converter should produce a current in phase opposition with the one established in Equation (18). In this way, changing the signal to the term i_{error}, it is obtained the control equation defined by:

$$-\left(L\frac{i_{error}}{T_{aq}}\right) = -v_C + v_S + L\frac{\Delta i_F^*}{T_{aq}}. \tag{19}$$

Rearranging the terms in both sides of Equation (19), the final control equation is established according to:

$$v_C = L\frac{i_{error}}{T_{aq}} + v_S + L\frac{\Delta i_F^*}{T_{aq}}. \tag{20}$$

Considering this final control equation, the output voltage is the reference of the voltage (v_c^*) used in the SPWM, i.e., the signal that is compared with the carrier. Taking into account that is used a digital platform to implement the control algorithm, the final control equation in discrete-time is obtained according to:

$$v_C^*[k] = L\frac{i_{error}[k]}{T_{aq}} + v_S[k] + \frac{L}{T_{aq}}(i_F^*[k] - i_F^*[k-1]). \tag{21}$$

Using Equation (21), expanding the terms, a simplified equation is obtained according to:

$$v_C^*[k] = v_S[k] + \frac{L}{T_{aq}}(2\,i_F^*[k] - i_F^*[k-1] - i_F[k]). \tag{22}$$

Taking into account that this is a linear control, it has some characteristics that are similar to a traditional PI controller, however without the necessity to adjust the gains of the controller, which represents a relevant characteristic of the predictive current control strategy. However, the performance of the predictive current controller is strongly affected by the accuracy of the mathematical model of the circuit. In real implementations, the inductance value of the coupling inductor varies as a function of the frequency, current amplitude, and core temperature, adversely affecting the results.

3.4. SPWM Strategy

In order to produce the desired current, a modulation strategy is required to drive the IGBTs accordingly. Multilevel inverters require different SPWM arrangements than those required by two-level inverters, hence more than one triangular carrier is needed. The different carriers can be distributed either vertically or horizontally. Since the studied converter derives from the DCMLI topology, the adopted SPWM strategy is based on Phase Disposition (PD) modulation, i.e., the triangular carriers are vertically distributed and in phase with each other. However, an adaptation was performed in order to reduce the number of required triangular carriers. Hence, instead of using four carriers and a single reference signal, a single carrier and adapted reference signals are used.

Figure 4 shows the adopted strategy and Figure 5 shows the obtained reference signals to generate the IGBT gate commands. As it can be seen in Figure 4, the reference signal is used in two comparisons, one being used for emulating the upper carrier ($Carrier_2$) and the other for the lower carrier ($Carrier_1$). The comparisons between Ref and the two carriers can be accomplished by using a single carrier and two modified reference signals, which are obtained by means of a branch function. The reference signal in the positive $Carrier_1$ level, i.e., when Ref is positive and inside the carrier amplitude, corresponds to the original reference. However, in the negative $Carrier_1$ level, i.e., when Ref is negative and below the carrier amplitude, the reference signal consists in the sum of the original reference with twice the carrier amplitude. This reference signal corresponds to Ref_1 in Figure 5. On the other hand, the reference signal in $Carrier_2$ level in the positive half-cycle, i.e., when Ref is positive and above the carrier amplitude, is obtained by subtracting the original reference and the carrier amplitude. Finally, the reference signal in $Carrier_2$ level in the negative half-cycle, i.e., when Ref is negative and inside the carrier amplitude, is attained by summing the original reference and the carrier amplitude. This signal corresponds to Ref_2 in Figure 5. As it can be seen in Figure 5, a comparison with zero is performed, which corresponds to the polarity of the original reference. With this approach, a single triangular carrier can be used to achieve a 5-level PD modulation, which would require four carriers in the conventional implementation. Besides this advantage, the IGBTs S_1 and S_1' switch at the line frequency (50 Hz), whereby switching losses are reduced.

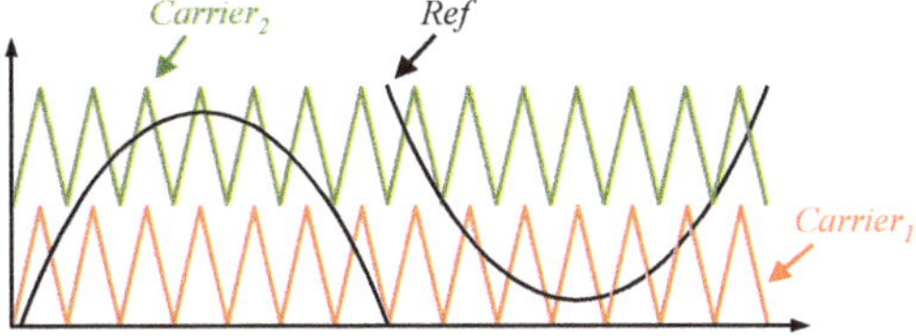

Figure 4. Adapted reference signal used in the modulation strategy.

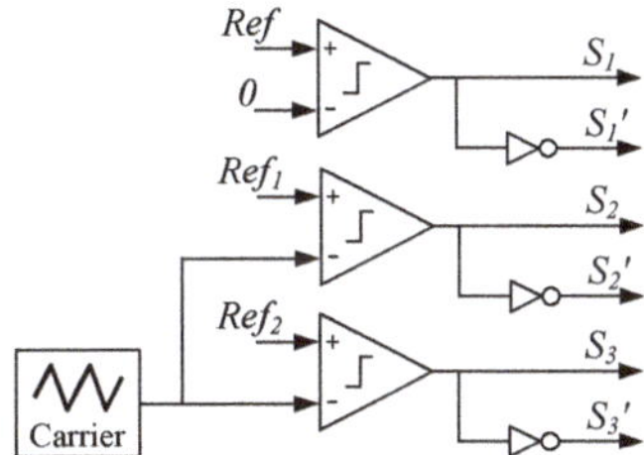

Figure 5. Sinusoidal Pulse-Width Modulation (SPWM) signals generation based on the adapted reference signals.

Figure 6 presents a block diagram that shows an overview of the algorithms used to control the SAPF. As it is visible in the figure, the source voltage (v_S) is used by the PLL algorithm to synchronize the SAPF controller with the fundamental component of the source voltage. Based in the α-β voltages calculated by the PLL algorithm, the DC-link regulation power (p_{reg}) and the load current (i_L), the p-q theory calculates the compensation current ($i_F{}^*$). The compensation current is processed by the predictive current controller to obtain the reference voltage ($v_C{}^*$) of the converter. Finally, the modulation block defines the semiconductors that must be activated in each instant.

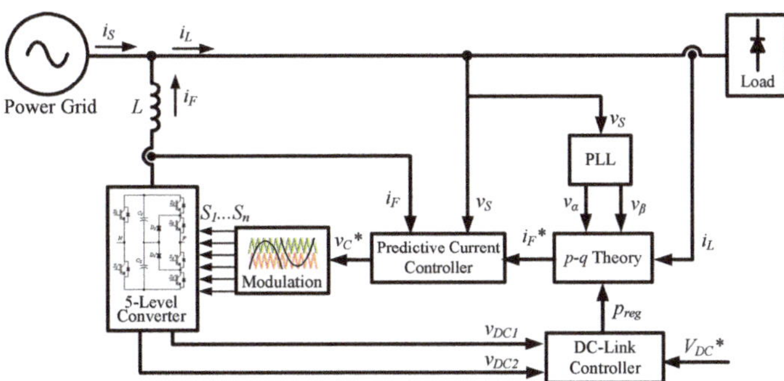

Figure 6. Block diagram of the 5-level converter SAPF control algorithms.

4. Simulation Analysis

This section presents a simulation analysis of the proposed single-phase 5-level SAPF, where the key control functionalities are described, such as the modulation strategy and DC-link voltage control, and both the transient and steady-state operation of the SAPF are analyzed. The simulations were performed using the software PSIM v9.1 from PowerSim, Inc (Rockville, MD, USA).

4.1. SPWM Strategy

Figure 7 shows a simulation result of the implemented modulation strategy, where a sinusoidal signal was used as reference signal. As it can be seen, the reference signal is used in two comparisons, one for emulating the upper carrier (Ref_2) and the other for the lower carrier (Ref_1).

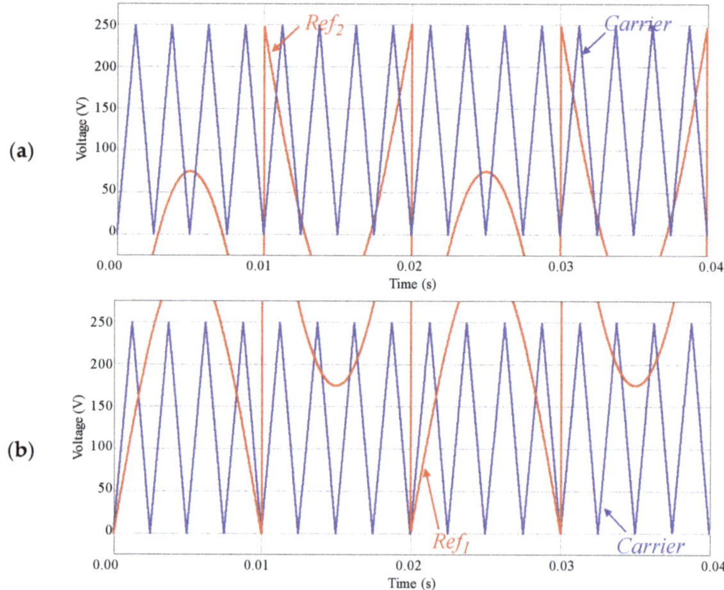

Figure 7. Adapted reference signals used in the modulation strategy: (**a**) Ref_2; (**b**) Ref_1.

In order to assess the generation of the desired voltage levels based on the adapted SPWM strategy, Figure 8 depicts the output voltage produced by the 5-level converter (v_c) and its reference ($v_c{}^*$). As it can be perceived, the converter is capable of producing the distinct 5 voltage levels according to the reference voltage established by the predictive current controller.

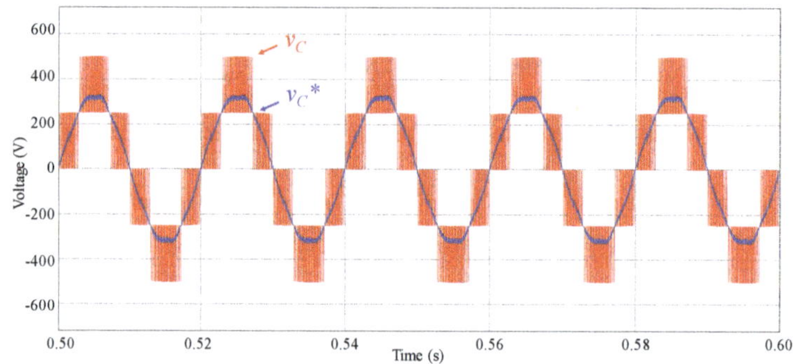

Figure 8. Simulation results of the modified multilevel SPWM modulation technique: Converter reference voltage ($v_c{}^*$) and converter voltage (v_c).

4.2. DC-Link Voltage Control

In order to assure a proper operation for the SAPF, the DC-link voltage must be controlled to an established reference voltage and its average value needs to be as constant as possible. However, since the SAPF is comprised by a voltage source converter and the DC-link capacitors are initially discharger, it cannot be directly connected to the power grid, during the start-up procedure, otherwise, the current absorbed by the DC-link capacitors via the IGBTs antiparallel diodes would damage the power converter. For this purpose, a pre-charge resistor is used, being bypassed when the DC-link voltage reaches a value close to the power grid peak voltage. After the bypass, the DC-link voltage control is activated, whereby the IGBTs are controlled so that the DC-link voltage reaches the established reference value. After this instant, the SAPF compensation is activated.

Figure 9 shows the referred stages of the DC-link voltage control. The first stage is visible between the instants 0 s and 1.1 s, where the pre-charge resistor connected in series with the AC side of the SAPF limits the current, and therefore, the rise of the DC-link voltage. After the instant 1.1 s, the resistor is bypassed, and the DC-link voltage control is activated. For this initial stage, the p-q theory uses only the p_{reg} power component, whereby the operation as SAPF is not performed during this period. Finally, the third stage starts only after the voltage in each DC-link capacitor reaches the established value of 250 V, which occurs at instant 2.7 s. From this instant, the p-q theory is used to calculate the compensation current, including all the power components, whereby the SAPF operation is activated. A slight decrease in the DC-link voltages (v_{DC1} and v_{DC2}) can be noted when the compensation is activated, being shortly after restored by the DC-link control. Besides, a ripple component can be visible in both voltages, which is a result of the power exchange between the SAPF and the power grid. This power exchange is intrinsic to the operation of the SAPF and causes an oscillation with twice the power grid frequency. Therefore, instead of using the instantaneous value of the DC-link voltage in the PI controller, a full-cycle average value is used, so that the DC-link voltage oscillation can be admissible without making the control unstable.

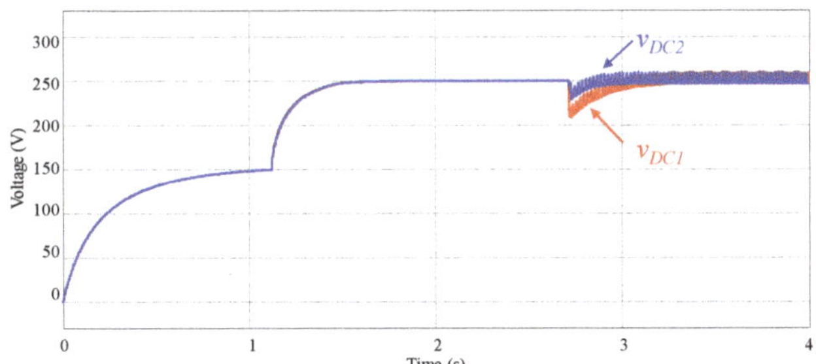

Figure 9. Evolution of the DC-link voltages (v_{DC1} and v_{DC2}) during the start-up of the SAPF.

4.3. SAPF Operation

This section presents several simulation results of the proposed SAPF for operation with different types of electrical loads, namely linear and non-linear loads. Both transient and steady-state conditions are analyzed.

4.3.1. SAPF Operating with Resistive-Inductive (RL) Load

With the advances in power electronics, the number of linear loads has been reducing. However, there is still a considerable number of linear loads, such as, for example, the single-phase AC motors, which are good examples of RL loads. RL loads do not produce harmonic currents if supplied by a sinusoidal voltage, however, they produce reactive power. Since a SAPF is capable of compensating both harmonic currents and reactive power, the SAPF operation with linear loads should be considered.

Therefore, a 30 mH inductor connected in series with a 7.5 Ω resistor was realized. Figure 10 depicts the source voltage (v_S), which is equal to the load voltage, the current drawn by the load (i_L), and the resulting source current in steady-state due to the SAPF operation (i_S). As it can be seen, the displacement power factor is compensated to the unity by the SAPF operation. It is noticeable that the Root Mean Square (RMS) value of the source current decreases due to the improvement in the power factor (PF).

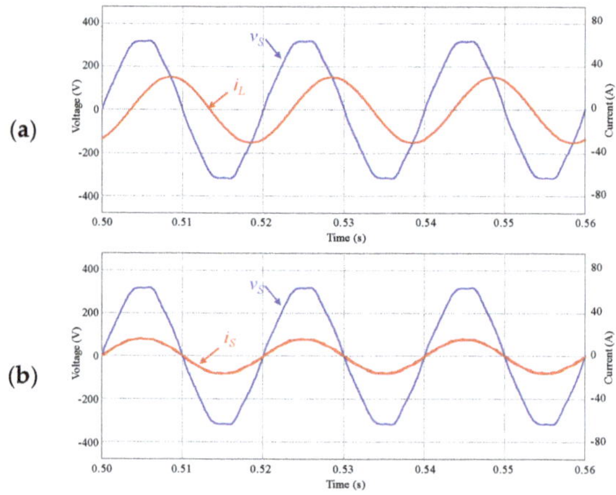

Figure 10. Steady-state analysis of the SAPF operation with linear Resistive-Inductive (RL) Load: (a) Source voltage and load current; (b) Source voltage and source current.

The transient caused by the SAPF starting of compensation, at $t = 0.5$ s, can be depicted in Figure 11. It is noticeable that the source current acquires a sinusoidal waveform in phase with the voltage almost instantly as the SAPF starts to compensate ($t = 0.5$ s), showing a good transient response of the SAPF.

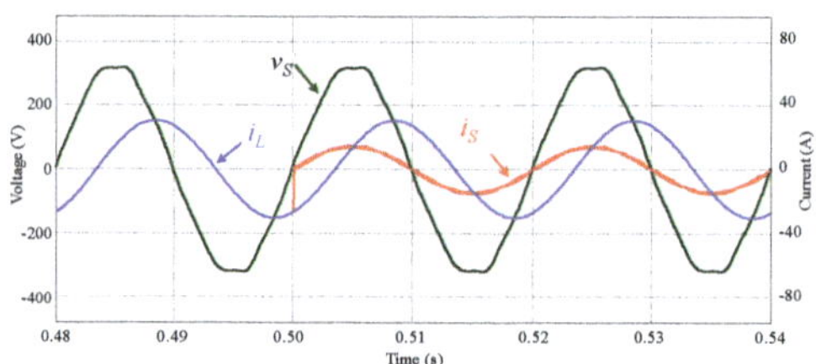

Figure 11. Transient analysis of the SAPF starting of compensation with linear Resistive-Inductive (RL) Load.

4.3.2. SAPF Operating with Diode Rectifier with Resistive-Capacitive (RC) Load

The full-bridge diode rectifier with capacitive output filter and resistive load is one of the most common electrical loads used nowadays, and it is characterized by high level of harmonic current and low Power Factor (PF) due to the presence of harmonics. Real loads that use a diode rectifier with capacitive output filter are very common today as for example: computers, printers, CFL and LED lamps, TV-sets, cellphone chargers, etc. In order to assess the performance of the SAPF towards non-linear loads, the referred load was implemented in the simulation model, being used a 1 mF capacitor and a 16 Ω resistor connected in parallel to the rectifier output, and a 10 mH inductor connected in series with the rectifier input. Figure 12 depicts the load and source voltage (v_S), the current drawn by the load (i_L), and the resulting source current in steady-state due to the SAPF operation (i_S). As it can be seen, the SAPF operation turns the source current sinusoidal and in phase with the source voltage, regardless of the harmonic distortion of the voltage.

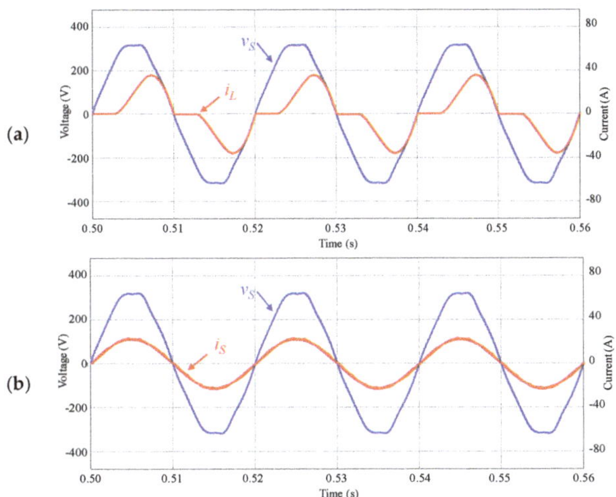

Figure 12. Steady-state analysis of the SAPF operation with non-linear load: (**a**) Source voltage (v_S) and load current (i_L); (**b**) Source voltage (v_S) and source current (i_S).

Figure 13 shows the transient state of the SAPF operation for the same load, where it can be seen the load current (i_L), the source current (i_S), the SAPF reference current ($i_F{}^*$) and the SAPF actual compensation current (i_F). Once again, it is noticeable that the source current acquires a sinusoidal waveform in phase with the source voltage immediately when the SAPF operation begins, showing a good transient response of the SAPF operating with non-linear loads.

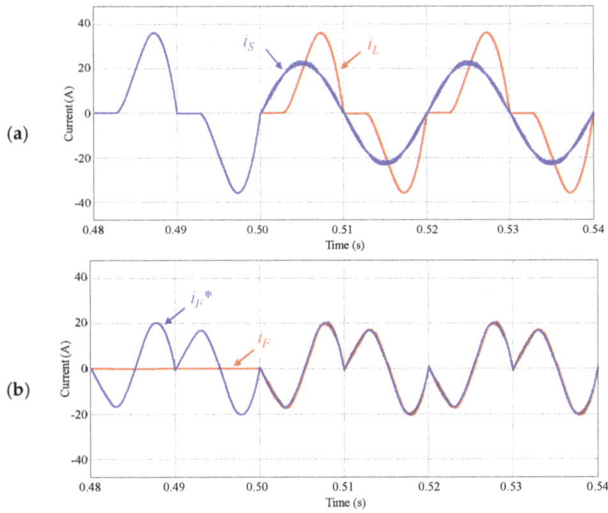

Figure 13. Transient analysis of the SAPF operation with non-linear load: (**a**) Load (i_L) and source currents (i_S); (**b**) SAPF reference ($i_F{}^*$) and actual (i_F) compensation current.

4.3.3. SAPF Dynamic Response towards Load Changing

In order to perform a more insightful analysis of the transient response of the SAPF, its operation towards changes in the connected loads was simulated. Therefore, two non-linear loads were connected to the power grid namely a diode rectifier with output RC load, initially connected, and a diode rectifier with output RL load.

Figure 14 shows the SAPF response towards the connection of the second load at instant $t = 1.5$ s, where it can be seen that the source current presents a sinusoidal waveform even during the transient. The SAPF reference and output currents are also depicted, where an accurate current tracking can be perceived. On the other hand, Figure 15 shows the voltages in each DC-link capacitor. Due to the connection of a load, the DC-link voltage tends to decrease abruptly so that the SAPF can accurately perform the compensation. Despite the initial voltage deviation, the DC-link voltage control compensates the voltage error, whereby the reference value of 250 V is reestablished. This behavior in the DC-Link voltages results from de time delay need to obtain the average values of the power components defined by the *p-q* theory after the load transient. In this work, the determination of the average values is performed by sliding window average of one grid-cycle. This means that after one grid cycle the SAPF operates again in steady state. However, to prevent sudden variations in the source currents, the repositioning of the initial voltage in the DC-link capacitors is performed more slowly.

Figure 16 shows the SAPF response towards the disconnection of the diode rectifier with RL load at instant $t = 1.5$ s. It is noticeable that the source current presents a sinusoidal waveform during the transient and, once again, an accurate current tracking can be noted. The voltages in each DC-link capacitor can be seen in Figure 17. In this case, due to the disconnection of a load, the DC-link voltage increases so that the SAPF performs the compensation during the transient. The DC-link voltage control is also capable of compensating the overvoltage, with the reference value of 250 V being once again restored.

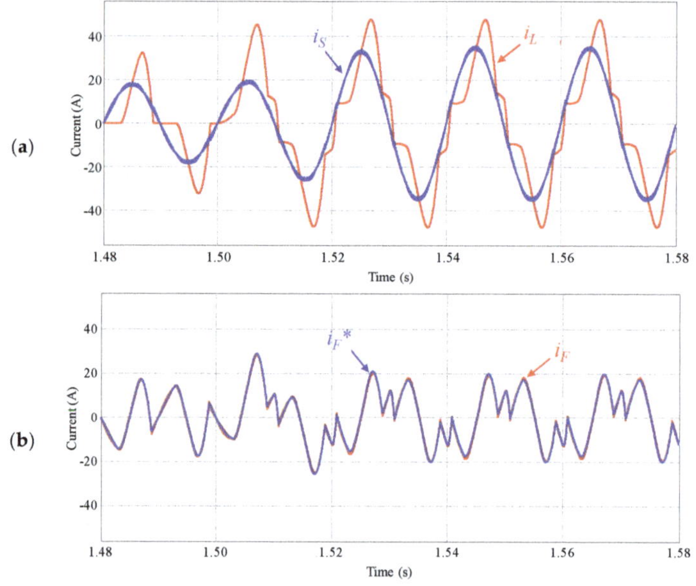

Figure 14. Transient analysis of the SAPF operation with the connection of a non-linear load: (**a**) Load (i_L) and source (i_S) currents; (**b**) SAPF reference ($i_F{}^*$) and actual (i_F) compensation current.

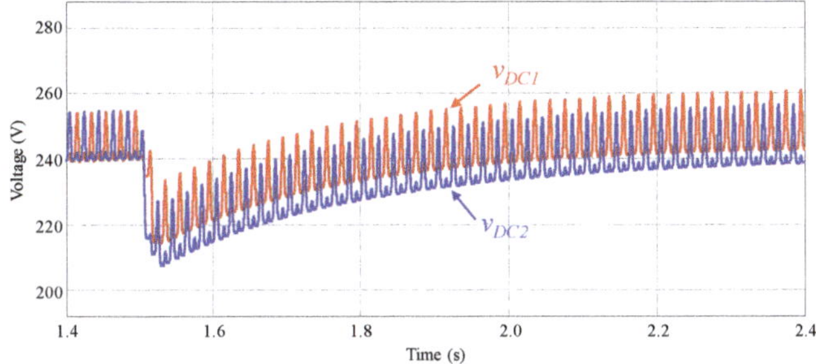

Figure 15. Transient analysis of the SAPF DC-link voltages (v_{DC1} and v_{DC2}) during the connection of a non-linear load.

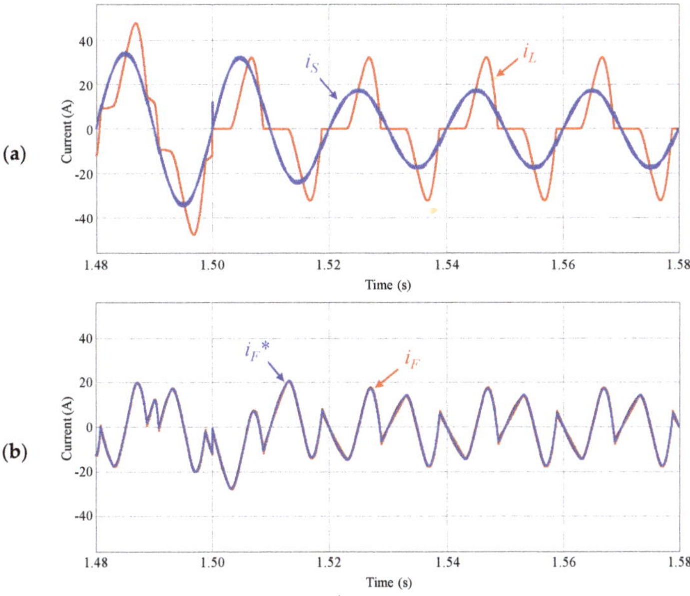

Figure 16. Transient analysis of the SAPF operation with the disconnection of a non-linear load: (**a**) Load (i_L) and source (i_S) currents; (**b**) SAPF reference (i_F^*) and actual (i_F) compensation current.

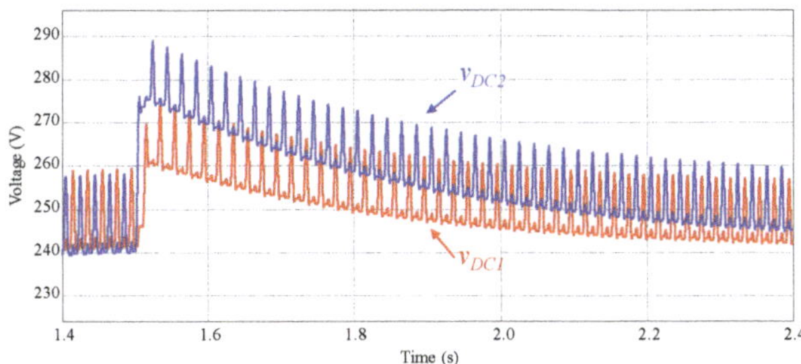

Figure 17. Transient analysis of the SAPF DC-link voltages (v_{DC1} and v_{DC2}) during the disconnection of a non-linear load.

In order to facilitate the comprehension of the benefits provided by the SAPF operation, Table 2 presents a comparison between the RMS, THD and PF of the load and source currents with the SAPF compensation for both the linear (Section 4.3.1) and the nonlinear (Section 4.3.2) loads. As it can be seen, the RMS values decrease in the source current, as well as the THD values. On the other hand, the PF increases practically to unity, as expected with the SAPF operation.

Table 2. Comparison of simulation results of the Root Mean Square (RMS), Total Harmonic Distortion (THD) and power factor (PF) values of the load current and source current with SAPF compensation, for operation with linear and nonlinear loads.

Installation Parameter	Linear Load		Non-linear Load	
	Source	Load	Source	Load
Current (RMS)	10 A	21 A	15.8 A	20.2 A
Current THD (%f)	1.6%	0.5%	2.8%	38%
PF	0.99	0.47	0.99	0.78

5. Implementation of a Laboratorial Prototype of the SAPF

In this section, it is presented the developed hardware, as well as a description about the control algorithms for the proposed 5-level SAPF. In the SAPF it is possible to identify two main hardware circuits: the power circuits and the control circuits, as shown in Figure 18. The power circuits (Figure 18a) include several components, as the DC-link capacitors, the power switching devices, the gate driver boards, and the protection devices. Additionally, in the power circuit, it is also included the coupling inductor and the DC-link pre-charge system. On the other hand, the control circuit (Figure 18b) is constituted by a Digital Signal Controller (DSC), a connection board to access the pins of the DSC, a signal conditioning board to acquire the signals from the sensors, and by a command board. During the experimental tests, it was used a Digital-to-Analog Converter (DAC) board, not necessary for the functionality of the system, but very useful during the process of development and debugging of the control algorithms. Figure 18c shows the structure of the final prototype, constituted by two metallic mounting bases supporting all the hardware. In the first level are fixed all the components of the power circuit, and in the second level all the boards that constitute the control system. The structure and components disposition were sized, taking into account a posterior inclusion in an electric switchboard.

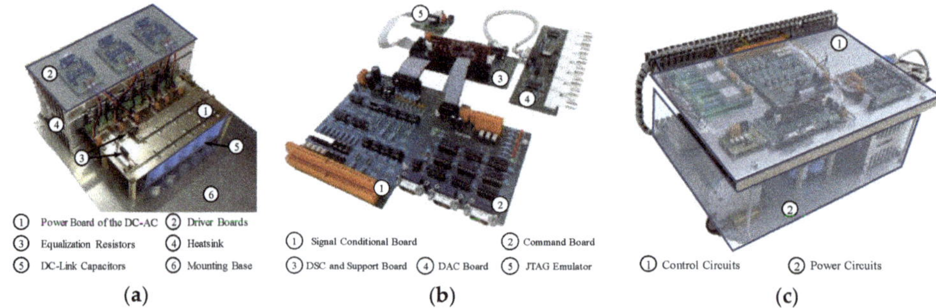

Figure 18. Developed SAPF: (**a**) Power circuits; (**b**) Control circuits; (**c**) Final prototype.

5.1. Power Circuit Hardware

In Figure 18a is presented the power circuit structure, where the 5-level converter was implemented, being possible to identify the heatsink, the power board of the 5-level converter, as well as the capacitors of the DC-link, the gate driver boards and a mounting base, where all the components are fixed. The IGBTs and diodes of the 5-level converter were fixed with the mounting clips that push them against the heatsink.

For each IGBT, a gate protection circuit was used, which consists of a 10 kΩ resistor in parallel with two 16 V zener diodes in series, with common anode. The gate protection is intended to protect the IGBTs gate terminals from voltage surges as well as to prevent unwanted switching. Additionally, in parallel with each complementary pair of IGBTs were placed snubber capacitors with the purpose of protecting the IGBTs from high voltage transients, which can occur when switching inductive loads.

The DC-link is composed of two sets of capacitors, which are placed in series and having the midpoint connected to the central point of the DCMLI. Each of them is charged with 250 V thus obtaining a voltage of 500 V on the DC-link. The use of a split DC-link is necessary to generate the different voltage levels of the converter. To generate these levels, it is possible to use the voltage of a single set of capacitors or the sum of the voltage of the two sets. For the proper operation of the 5-level converter, it is extremely important to maintain the same value of voltage in both sets of capacitors. Thus, in addition to the DC-link voltage balance control algorithm, an equalization resistor is placed in parallel with each set of capacitors, as it can be seen in Figure 18a. Each of the capacitor assemblies contains five electrolytic capacitors connected in parallel, each one with a value of 470 μF, thus obtaining a total of 2350 μF in each set of capacitors. These capacitors withstand a maximum voltage of 400 V.

The coupling of the 5-level converter to the power grid is done through two inductors and one DC-Link pre-charge system. This system consists of a pre-charge resistor in parallel with a bypass switch and a mains circuit breaker. The mains connection circuit breaker is responsible for connecting and disconnecting the SAPF with the power grid. The pre-charge resistor has the effect of limiting the initial current during the charge of the DC-link capacitors. The pre-charge bypass switch is used to create an alternative path for the current after the pre-charge. Whenever the mains circuit breaker is closed, the pre-charge system must be opened. The 5-level converter inductor consists of two windings with mutual coupling, where one is connected to the phase and the other is connected to the neutral. The two windings have an equivalent inductance of 1.6 mH. On the top of the heatsink, as shown in Figure 18a, the gate driver circuit boards are responsible for operating the IGBTs. In order to keep the control system galvanically isolated from the power stage, optical coupling driver circuits were used. Each board is capable of controlling two complementary IGBTs.

5.2. Control Circuit Hardware

Figure 18b shows the boards responsible for the control system, from the acquisition of the voltage and current sensors signals to the generation of the PWM pulses that are applied to the gate driver boards described above. A block diagram of the control system is shown in Figure 19. This system consists of the block of sensors, followed by the block responsible for the signal conditioning and conversion of the analog signals to digital, through an Analog-to-Digital Converter (ADC). This block also contains a comparator circuit, used to detect if there exist values outside of the stipulated limits. The values converted by the ADC are made available to the DSC, where all the control algorithms were implemented. The DAC board connected to the DSC through a serial SPI communication allows a real-time visualization of the variables calculated by the control algorithms. Finally, the control block converts the PWM signals from the 3.3 V Transistor-Transistor Logic (TTL) to the 15 V Complementary Metal-Oxide Semiconductor (CMOS) logic used by the gate driver boards. This block also receives error signals, suspending the switching of all IGBTs in case any error is detected.

The signal conditioning board, previously presented in Figure 18b, is responsible to convert the analog signals into digital signals in order to enable the DSC to process the values acquired from the sensors. Although the DSC used in this project has an internal ADC, it presents some limitations. For this reason, a board with an external ADC was implemented. This ADC, manufactured by MAXIM with the reference MAX1320, can read 8 channels with bipolar input of ±5 V with a resolution of 14 bits. All the digital control algorithms of the system were implemented in the DSC, whereby in this work it is used the TMS320F28335 manufactured by Texas Instruments (Dallas, TX, USA). It is a platform with typical instructions of a Digital Signal Processor and with a set of peripherals typical of a microcontroller. This DSC has a 32-bit processor with a clock frequency up to 150 MHz with native support for floating-point operations, 16 channels of 12-bit ADC and 18 PWM channels. One important characteristic of the PWM module included in the DSC is the possibility to introduce a dead-time between the two complementary IGBTs of the converter leg, avoiding the existence of short circuits during the switching. Figure 20 shows a flowchart with the sequence of processes implemented in the DSC to perform the control system. After the peripheral configurations and variables initialization, the DSC enters in an infinite loop. There is a timer interrupt routine responsible to trigger the ADC start at every 25 µs resulting in a frequency of acquisition of 40 kHz. After the conclusion of the ADC conversion, it is implemented the routine responsible for the PLL algorithm, followed by the DC-link voltage control to calculate the required inputs of the *p-q* theory. The *p-q* theory routine determines the compensation current that is processed by the predictive current routine and, subsequently, by the SPWM strategy to determine the gate signals applied to the DSC PWM peripherals.

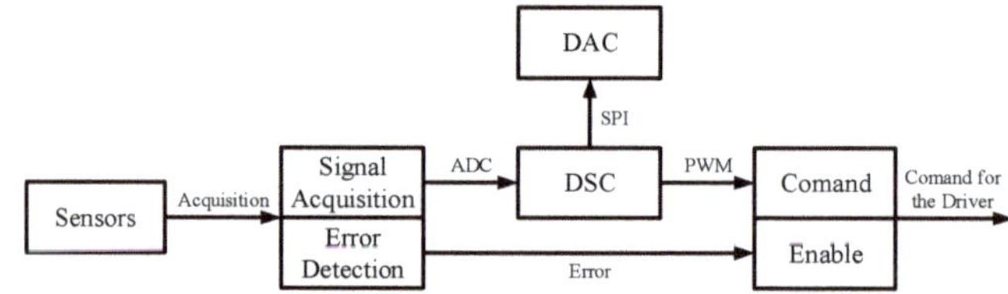

Figure 19. Block diagram of the control system.

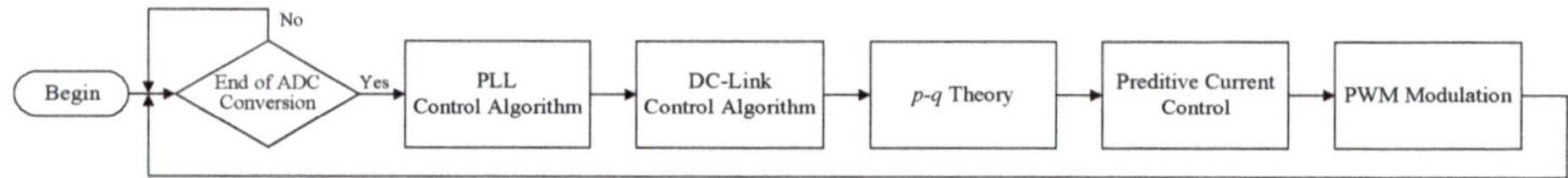

Figure 20. General flowchart of the control system implemented in the DSC.

6. Experimental Results

In this section are presented and discussed the experimental results of the SAPF operating within different conditions. Before the final validation of the power converter operating as SAPF, all the circuits and control algorithms were validated in a step-by-step procedure to prevent damages of the prototype electronic components or the laboratory equipment. For safety reasons, the experimental results were obtained with a 50 V–50 Hz reduced voltage single-phase system. The installation used in the experimental tests was obtained from the low-voltage 230 V–50 Hz electrical grid using a 5.5 kVA step-down transformer with 9:2 voltage ratio.

6.1. Experimental Validation of the PLL Algorithm

The preliminary tests began with the synchronization of the SAPF with the power grid. For this purpose, it was used a PLL algorithm responsible to detect the fundamental component of the source voltage. Figure 21a shows the source voltage (v_S) and the PLL signal (v_{pll}) calculated by the algorithm during the system connection. As it can be seen, the PLL algorithm quickly acquires the phase of v_S, being the amplitude value achieved in a few grid-cycles. After the synchronization transient, the PLL signal (v_{pll}) perfectly matches the fundamental component of v_S as it can be seen in Figure 21b.

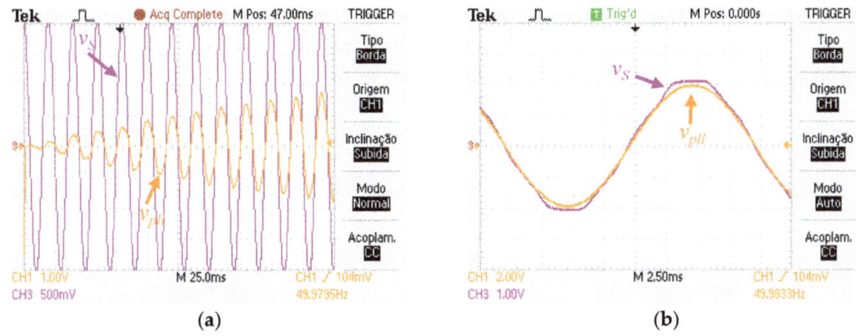

(a)   (b)

Figure 21. Experimental results of the Phase-Locked Loop (PLL) control algorithm: (**a**) PLL synchronization transient; (**b**) PLL in steady state.

6.2. Experimental Validation of the p-q Theory

In this section are presented the experimental results from the *p-q* theory calculation. In the computer simulations, when implementing the *p-q* theory, the source voltage (v_S) and the load current (i_S) were used as the α coordinate, and a delay loop of 270° was implemented to obtain the β coordinate. However, in the experimental implementation, in order to increase the performance and save the DSC resources, it was chosen to keep the β coordinate as the original signal and delay the signals 90° to obtain the α coordinate. Subsequently, it is calculated the compensation current from the β coordinate current (i_β). The experimental implementation of the *p-q* theory algorithm applied to a single-phase installation is presented in Figure 22a shows the source voltage (v_S) and the signals v_α and v_β generated by the PLL algorithm. Figure 22b shows the current (i_L) from a resistive load, as well as the generated signals i_α and i_β. As previously mentioned, the phase shift between i_α and i_β is achieved by the storage of the acquired current values into an array and posterior reading of the values with a 90° delay. In both cases, it is possible to see that the α coordinate has a delay of 90° related with the β coordinate.

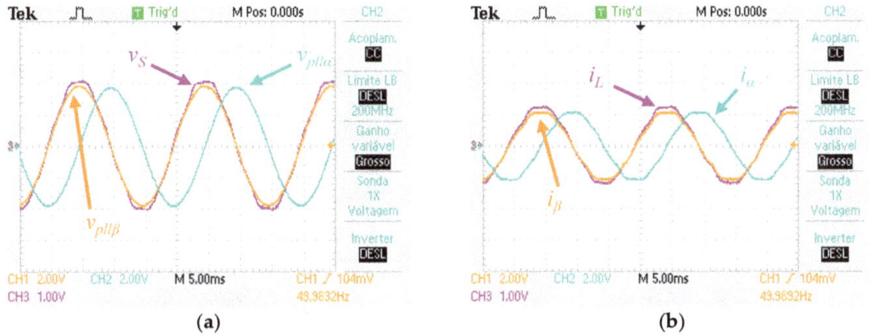

(a)   (b)

Figure 22. Creation of a three-phase virtual balanced system from a single-phase installation for the application of the *p-q* theory: (**a**) Source voltage (v_S) and PLL output signals in the α-β coordinates ($v_{pll\alpha}$ and $v_{pll\beta}$); (**b**) Load currents in the α-β coordinates (i_α and i_β).

In order to test the *p-q* theory algorithms with non-linear loads, it was used a full-bridge diode rectifier with a resistor and a capacitor connected in parallel in the DC-side to calculate the compensation current and the theoretical value for the source current after compensation. Figure 23 shows the load current (i_L), the compensation current calculated from the *p-q* theory (i_F^*), and the theoretical source current after the compensation (i_S^*). It can be seen that the theoretical source current

(i_S*), results from the difference between i_F* with i_L. Additionally, it can be seen that i_S* is perfectly sinusoidal, validating the implemented *p-q* theory control algorithm.

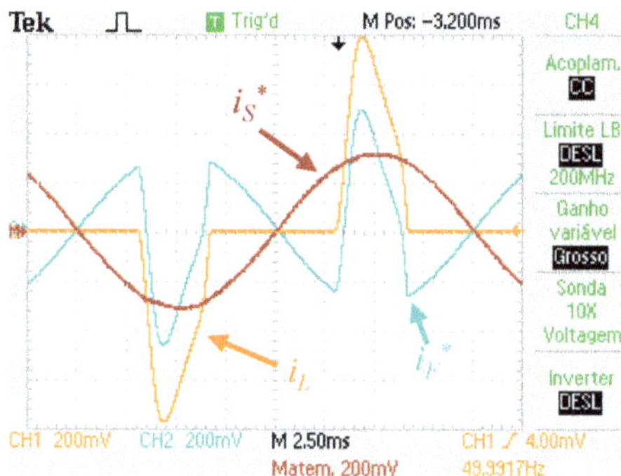

Figure 23. Validation of the *p-q* theory algorithm with a full-bridge diode rectifier: Load current (i_L), compensation current calculated by the *p-q* theory (i_F*), and the theoretical source current after the compensation (i_S*).

6.3. Experimental Validation of the Modulation Technique

With the purpose of validating the modulation technique, it was used a sinusoidal waveform with 50 Hz as the reference to the modulation technique. In Figure 24, it is possible to see the voltage generated by the 5-level converter, as well as the sinusoidal reference. In this figure, the 5-levels generated by the power converter following the reference signal are clearly identifiable. For this experimental test, the 5-level converter was tested in open loop, with a DC power supply connected to the DC-link.

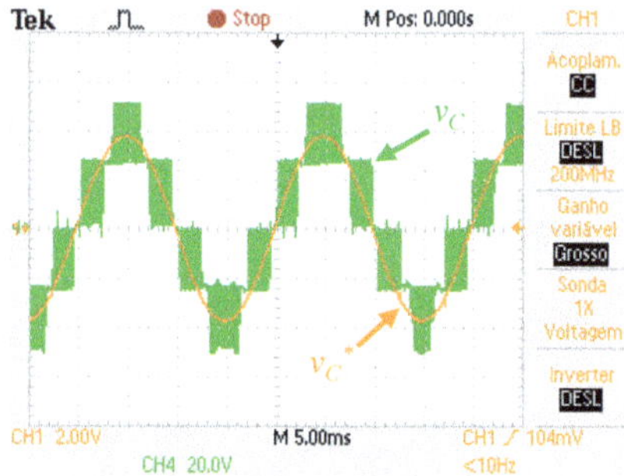

Figure 24. Experimental results of the modified multilevel SPWM modulation technique: Converter reference voltage (v_c*) and converter voltage (v_c).

In order to avoid unnecessary commutations of the two-level leg of the 5-level converter, it was necessary to implement a hysteresis margin around the zero of the reference voltage, otherwise, due to the switching noise in the signals, several zero transitions can occur. To successfully mitigate this problem, it is necessary to consider a hysteresis band slightly higher than the noise.

6.4. Experimental Validation of the Predictive Current Controller

In order to test the predictive current control technique, it was used a resistive load in the output of the SAPF. With that, and once the 5-level converter is not connected to the power grid, the equation of the current control previously referred undergoes a slight change. In this way, the output voltage of the 5-level converter will correspond to the voltage in the output resistive load. During this test, the DC-link of the power converter is fed by two DC power supplies. In Figure 25 are presented the experimental results obtained with the predictive current control producing a sinusoidal current with 2 A of amplitude. Firstly, it was used a reference current with 50 Hz (Figure 25a), and in a second test was used a frequency equal to 100 Hz (Figure 25b). As it is possible to see in the figures, the filter current perfectly follows the references with a low switching ripple.

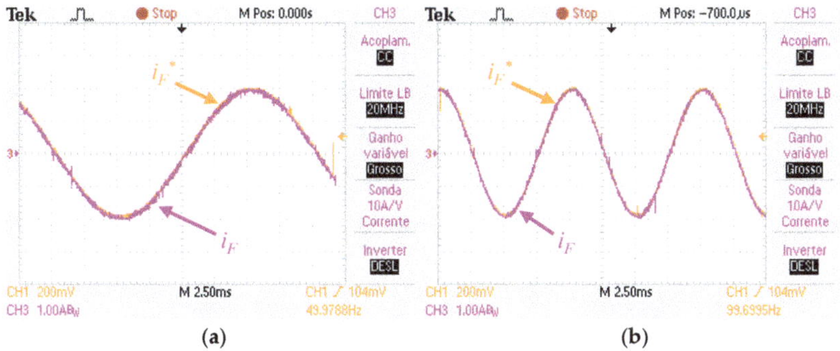

(a)   (b)

Figure 25. Experimental validation of the predictive current control algorithm producing a sinusoidal current with 2 A of amplitude: (**a**) Sinusoidal current with 50 Hz frequency; (**b**) Sinusoidal current with 100 Hz frequency.

6.5. Experimental Validation of the DC-Link Voltage Controller

For a proper operation of the SAPF, the voltage on the DC-link should be slightly higher than the source peak voltage. To prevent a high peak current during the connection of the SAPF with the power grid, the pre-charge system described in the Section 5 needs to be operated correctly. As it can be seen in Figure 26, before the tests, the DC-link capacitors are fully discharged. When the SAPF is connected with the power grid, with the pre-charge system, at the time instant 1, the DC-link voltages start to increase until the peak value of the power grid voltage, approximately 70 V (35 V in each capacitor) and then at instant 2 the bypass contact is activated. A few seconds after, at time instant 3, the DC-link voltage control algorithm was enabled, starting to regulate the DC-link voltage to the reference value of 100 V (50 V in each capacitor). As it can be seen in the figure, the DC-link voltages converge to the reference value without any visible overshoot and maintaining the two voltages balanced, proving that the DC-link voltage control algorithm operates properly.

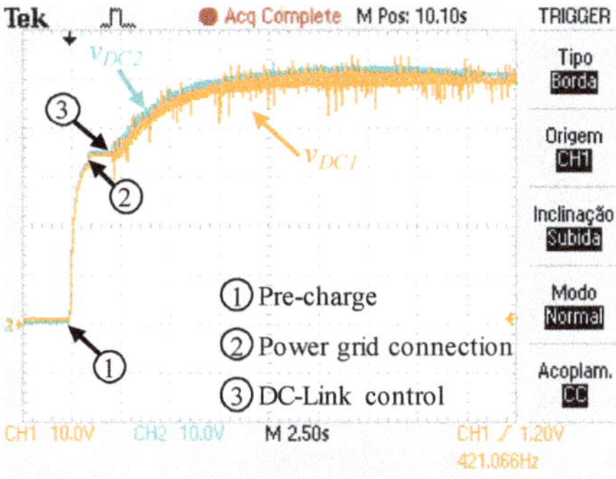

Figure 26. Experimental validation of the DC-link voltages (v_{DC1} and v_{DC2}) control algorithm.

6.6. Experimental Results of the SAPF with RL Load

After validating the DC-link voltage control algorithm, the operation the SAPF was firstly tested with a linear load composed by a resistor in series with an inductor (RL load). The load current (i_L) and the source voltage (v_S) of the single-phase system are shown in Figure 27. In Figure 27a it is possible to see that i_L and v_S have a phase-shift of 42°, resulting in a PF of 0.74. Figure 27b is represented the harmonic spectrum of the load current, where it is possible to verify a THD of 2.5% and an RMS value of 6.4 A.

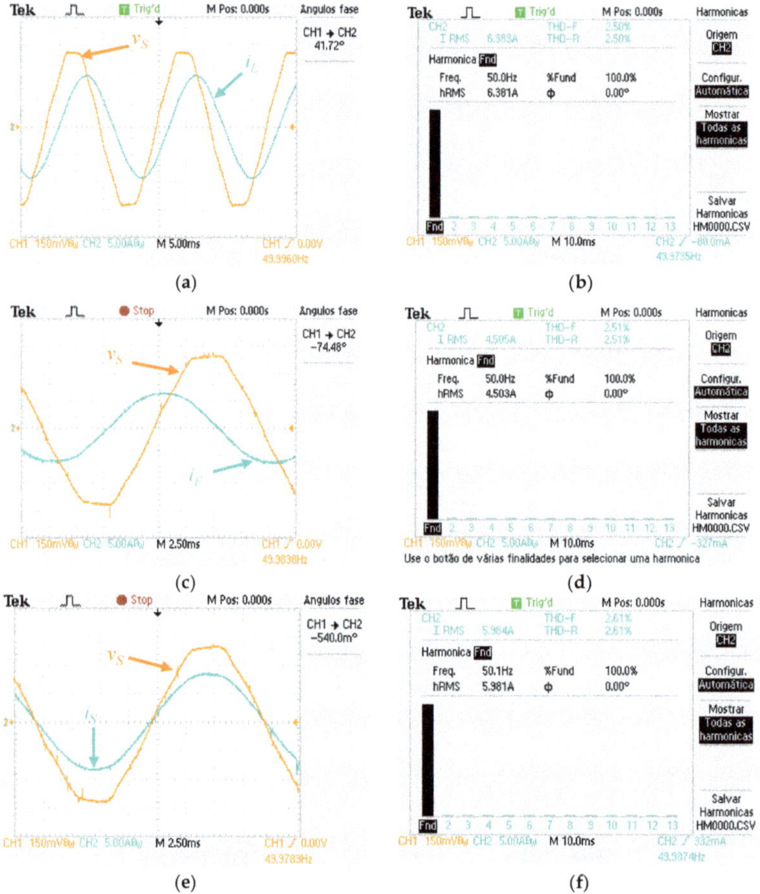

Figure 27. Experimental results of the SAPF with a RL load: (**a**) Source voltage (v_S) and load current, (i_L); (**b**) Harmonic spectrum of the load current; (**c**) Source voltage (v_S) and filter current (i_F); (**d**) Harmonic spectrum of the current of the filter current; (**e**) Source voltage (v_S) and source current (i_S); (**f**) Harmonic spectrum of the source current.

In order to compensate the reactive power of the RL load, the SAPF produces a current with a phase shift of 90° with the source voltage as shown in Figure 27c. After the compensation, the source current (i_S) becomes sinusoidal and in phase with v_S, as visible in Figure 27e. By consulting the Figure 27f, it is confirmed the phase shift of 0° of the fundamental component of the source current in relation to the source voltage, resulting in a PF of 0.99. It is also visible in Figure 27f that the RMS value of the source current is lower than the RMS value of the load current.

In Figure 27, it is visible that the source voltages, which correspond to the actual power grid voltages in the Power Electronics Laboratory, are not sinusoidal, due to distorted voltage drops in the line impedances caused by distorted currents of non-linear loads. Thus, the source voltages present a considerable amount of voltage harmonics. Nevertheless, the source currents after the SAPF compensation are almost perfectly sinusoidal. This result is only possible due to the used PLL, but also proves that the control algorithms are working as intended, allowing sinusoidal currents in the source to be obtainable.

6.7. Experimental Results of the SAPF with a Diode Rectifier with Capacitive Filter

After the validation of operation with a linear load, the SAPF was tested with a non-linear load composed by a full-bridge diode rectifier with a resistor and capacitor connected in parallel in the DC-side. The results obtained during the test are presented in Figure 28. As expected, it is perfectly visible that the load current is distorted (Figure 28a), presenting a THD of 44% and an RMS value of 7.4 A (Figure 28b).

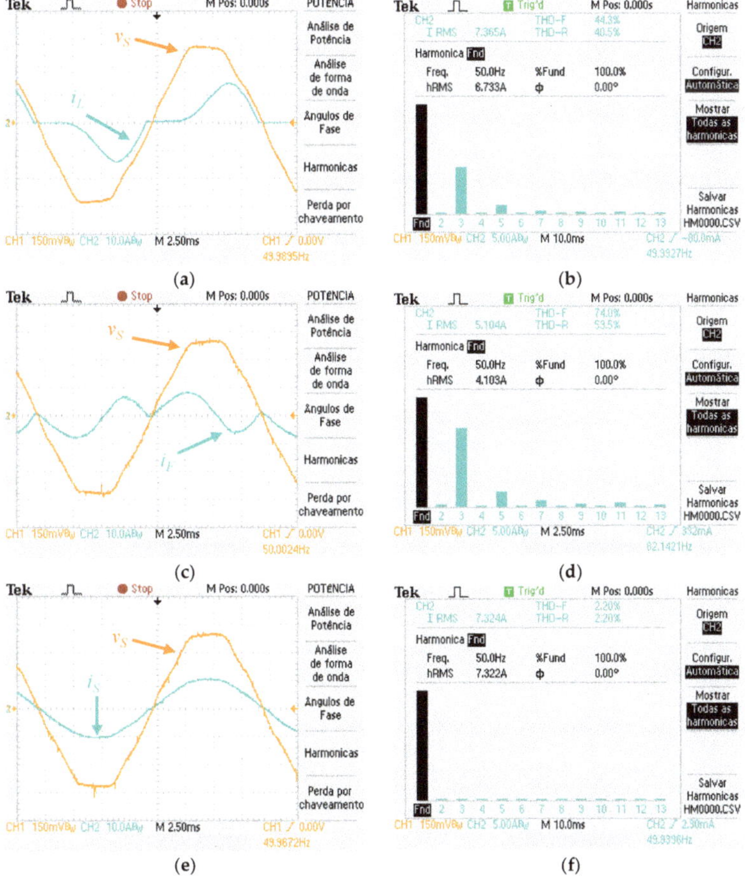

Figure 28. Experimental results of the SAPF with a Resistive-Capacitive (RC) load and a rectifier: (**a**) Source voltage (v_S) and load current (i_L); (**b**) Harmonic spectrum of the load current (i_L); (**c**) Source voltage (v_S) and SAPF current (i_F); (**d**) Harmonic spectrum of the SAPF current; (**e**) Source voltage (v_S) and source current (i_S); (**f**) Harmonic spectrum of the source current (i_S).

In Figure 28c, it is visible the compensation current produced by the SAPF (i_F), as well as the harmonic spectrum in Figure 28d. Once again, and as desired, after the compensation the source current becomes sinusoidal and in phase with source voltage, as shown in Figure 28e. In Figure 28f, it is possible to verify that the THD of the current is reduced to a value of 2.2%. Table 3 presents the main indicators of the experimental results achieved with the SAPF. As it is possible to see, in both test cases, the RMS values of the source current are lower than the RMS values of the load currents. However, the benefits of the SAPF are more visible in terms of current harmonics and reactive power compensation. In both test cases the PF of the installation becomes almost unitary and in the case of the non-linear load, the THD is reduced from 44.3 to 2.2%.

Table 3. Comparison of experimental results of the RMS, THD and PF values of the load current and source current with SAPF compensation, for operation with linear and non-linear loads.

Installation Parameter	Linear Load		Non-linear Load	
	Source	Load	Source	Load
Current (RMS)	6.0 A	6.4 A	7.3 A	7.4 A
Current THD (%f)	2.6%	2.5%	2.2%	44.3%
PF	0.99	0.74	0.99	0.81

7. Conclusions

A single-phase Shunt Active Power Filter (SAPF) based on a 5-level asymmetric full-bridge topology is proposed in this paper. Along the paper are presented the principle of operation of the 5-level power converter, the digital implementation of the p-q theory complemented by a Phase-Locked Loop and adapted to work with single-phase installations, the DC-link voltage control, the predictive current control and the multicarrier Sinusoidal Pulse-Width Modulation strategy. After the theoretical concepts, are described the simulation models and discussed the simulation results. A detailed description of a laboratory prototype implementation, including the sizing and selection of the electronic components is also presented, followed by a comprehensive description of the step-by-step hardware and control algorithms validation. Finally, the experimental results of the SAPF operating with linear and nonlinear loads are presented and discussed.

The simulation and experimental results confirm an excellent performance of the proposed SAPF topology and control algorithms, resulting always in a source current with very low Total Harmonic Distortion and in phase with the source voltage.

Despite the excellent results obtained, due to asymmetries of the full-bridge converter legs, mainly in terms of the switching frequency and maximum withstanding voltage, it may be interesting in a future work to combine the utilization of different semiconductor technologies. For instance, lower withstand voltage and fast switching technology (e.g., MOSFETs), and higher withstand voltage and slower switching technology (e.g., IGBTs) can be combined to implement a 5-level converter, which may result in interesting benefits in terms of efficiency and cost of the proposed solution.

Acknowledgments: This work has been supported by COMPETE: POCI-01-0145-FEDER-007043 and FCT—Fundação para a Ciência e Tecnologia within the Project Scope: UID/CEC/00319/2013. This work is financed by the ERDF—European Regional Development Fund through the Operational Programme for Competitiveness and Internationalisation—COMPETE 2020 Programme, and by National Funds through the Portuguese funding agency, FCT—Fundação para a Ciência e a Tecnologia, within project SAICTPAC/0004/2015-POCI-01-0145-FEDER-016434. Tiago Sousa is supported by the doctoral scholarship SFRH/BD/134353/2017 granted by the Portuguese FCT agency.

Author Contributions: José Gabriel Oliveira Pinto is the mentor of the idea of using the 5-level asymmetric full-bridge topology as a single-phase shunt active filter, participated in the development of the experimental prototype by designing some of the PCBs, and coordinated the organization and writing of the paper. Rui Macedo is the responsible for the implementation of the simulation model, hardware prototype, obtainment of the simulation results and experimental results, and participated in the writing of the paper. Vitor Monteiro participated in the design of the hardware prototype, sizing of electronic components, design of the predictive current control and writing of the paper. Luis Barros participated in the development of the hardware prototype, obtainment of the experimental results and writing of the paper. Tiago Sousa participated in the hardware prototype development, design of the SPWM modulation technique and writing of the paper. João L. Afonso participated in the design of the control algorithms based on the *p-q* theory, in the adaptation of the *p-q* theory to single-phase systems and in the writing of the paper.

Conflicts of Interest: The authors declare no conflict of interest.

References

1. Dugan, R.C.; McGranaghan, M.F.; Santoso, S.; Beaty, H.W. *Electrical Power Systems Quality*, 3rd ed.; McGraw-Hill Education: New York, NY, USA, 2002.
2. Balda, J.C.; Mantooth, A. Power-Semiconductor Devices and Components for New Power Converter Developments: A key enabler for ultrahigh efficiency power electronics. *IEEE Power Electron. Mag.* **2016**, *3*, 53–56. [CrossRef]
3. Hoene, E.; Deboy, G.; Sullivan, C.R.; Hurley, G. Outlook on Developments in Power Devices and Integration: Recent Investigations and Future Requirements. *IEEE Power Electron. Mag.* **2018**, *5*, 28–36. [CrossRef]
4. *IEEE Recommended Practices and Requirements for Harmonic Control in Electrical Power Systems*; 519-1992; IEEE: Piscataway, NJ, USA, 1993; pp. 1–112. [CrossRef]
5. Palmer, J.A.; Degeneff, R.C.; McKernan, T.M.; Halleran, T.M. Pipe-type cable ampacities in the presence of harmonics. *IEEE Trans. Power Deliv.* **1993**, *8*, 1689–1695. [CrossRef]
6. Faiz, J.; Ghazizadeh, M.; Oraee, H. Derating of transformers under non-linear load current and non-sinusoidal voltage—An overview. *IET Electron. Power Appl.* **2015**, *9*, 486–495. [CrossRef]
7. Lee, K.; Carnovale, D.; Young, D.; Ouellette, D.; Zhou, J. System Harmonic Interaction Between DC and AC Adjustable Speed Drives and Cost Effective Mitigation. *IEEE Trans. Ind. Appl.* **2016**, *52*, 3939–3948. [CrossRef]
8. Mendis, S.R.; Gonzalez, D.A. Harmonic and transient overvoltage analyses in arc furnace power systems. *IEEE Trans. Ind. Appl.* **1992**, *28*, 336–342. [CrossRef]
9. Liu, Y.; Heydt, G.T.; Chu, R.F. The Power Quality Impact of Cycloconverter Control Strategies. *IEEE Trans. Power Deliv.* **2005**, *20*, 1711–1718. [CrossRef]
10. Gomez, J.C.; Morcos, M.M. Impact of EV battery chargers on the power quality of distribution systems. *IEEE Trans. Power Deliv.* **2003**, *18*, 975–981. [CrossRef]
11. Blanco, A.M.; Stiegler, R.; Meyer, J. Power Quality Disturbances Caused by Modern Lighting Equipment (CFL and LED). In Proceedings of the 2013 IEEE Grenoble Conference, Grenoble, France, 16–20 June 2013; pp. 1–6. [CrossRef]
12. Peng, F.Z.; Akagi, H.; Nabae, A. A new approach to harmonic compensation in power systems-a combined system of shunt passive and series active filters. *IEEE Trans. Ind. Appl.* **1990**, *26*, 983–990. [CrossRef]
13. Sasaki, H.; Machida, T. A New Method to Eliminate AC Harmonic Currents by Magnetic Flux Compensation-Considerations on Basic Design. *IEEE Trans. Power Appar. Syst.* **1971**, *PAS-90*, 2009–2019. [CrossRef]
14. Gyugyi, L.; Strycula, E.C. Active AC Power Filters. In Proceedings of the IEEE IAS Annual Meeting, Chicago, IL, USA, 11–14 October 1976; p. 529.
15. Singh, B.; Al-Haddad, K.; Chandra, A. A review of active filters for power quality improvement. *IEEE Trans. Ind. Electron.* **1999**, *46*, 960–971. [CrossRef]
16. Akagi, H.; Kanazawa, Y.; Nabae, A. Generalized Theory of the Instantaneous Reactive Power in Three-Phase Circuits. In Proceedings of the 1983 International Power Electronics Conference, Tokyo, Japan, 27–31 March 1983; pp. 1375–1386.
17. Depenbrock, M. The FBD-method, a generally applicable tool for analyzing power relations. *IEEE Trans. Power Syst.* **1993**, *8*, 381–387. [CrossRef]
18. Czarnecki, L.S. Orthogonal decomposition of the currents in a 3-phase nonlinear asymmetrical circuit with a nonsinusoidal voltage source. *IEEE Trans. Instrum. Meas.* **1988**, *37*, 30–34. [CrossRef]
19. Bhattacharya, S.; Divan, D. Synchronous frame based controller implementation for a hybrid series active filter system. In Proceedings of the IAS '95. Conference Record of the 1995 IEEE Industry Applications Conference Thirtieth IAS Annual Meeting, Orlando, FL, USA, 8–12 October 1995; pp. 2531–2540. [CrossRef]

20. Hoon, Y.; Radzi, M.M.; Hassan, M.; Mailah, N. Control Algorithms of Shunt Active Power Filter for Harmonics Mitigation: A Review. *Energies* **2017**, *10*, 2038. [CrossRef]

21. Pinto, J.G.; Exposto, B.; Monteiro, V.; Monteiro, L.F.C.; Couto, C.; Afonso, J.L. Comparison of current-source and voltage-source Shunt Active Power Filters for harmonic compensation and reactive power control. In Proceedings of the IECON 2012—38th Annual Conference on IEEE Industrial Electronics Society, Montreal, QC, Canada, 25–28 October 2012; pp. 5161–5166. [CrossRef]

22. Lai, J.; Peng, F.Z. Multilevel converters-a new breed of power converters. In Proceedings of the 1995 Industry Applications Conference, Thirtieth IAS Annual Meeting, IAS '95, Orlando, FL, USA, 8–12 October 1995; pp. 2348–2356. [CrossRef]

23. Rodriguez, J.; Lai, J.; Peng, F.Z. Multilevel inverters: A survey of topologies, controls, and applications. *IEEE Trans. Ind. Electron.* **2002**, *49*, 724–738. [CrossRef]

24. Nabae, A.; Takahashi, I.; Akagi, H. A New Neutral-Point-Clamped PWM Inverter. *IEEE Trans. Ind. Appl.* **1981**, *IA-17*, 518–523. [CrossRef]

25. Hochgraf, C.; Lasseter, R.; Divan, D.; Lipo, T.A. Comparison of multilevel inverters for static VAr compensation. In Proceedings of the 1994 IEEE Industry Applications Society Annual Meeting, Denver, CO, USA, 2–6 October 1994; pp. 921–928. [CrossRef]

26. Teodorescu, R.; Blaabjerg, F.; Pedersen, J.K.; Cengelci, E.; Enjeti, P.N. Multilevel inverter by cascading industrial VSI. *IEEE Trans. Ind. Electron.* **2002**, *49*, 832–838. [CrossRef]

27. Hasan, M.; Abu-Siada, A.; Islam, S.; Muyeen, S. A Novel Concept for Three-Phase Cascaded Multilevel Inverter Topologies. *Energies* **2018**, *11*, 268. [CrossRef]

28. Reddy, K.B.M.; Pattnaik, S. Novel symmetric and asymmetric topology of multilevel inverter with reduced number of switches. In Proceedings of the 2017 IEEE International Conference on Industrial Technology (ICIT), Toronto, ON, Canada, 22–25 March 2017; pp. 165–170. [CrossRef]

29. Elias, M.F.M.; Rahim, N.A.; Ping, H.W.; Uddin, M.N. Asymmetrical transistor-clamped H-bridge cascaded multilevel inverter. In Proceedings of the 2012 IEEE Industry Applications Society Annual Meeting, Las Vegas, NV, USA, 7–11 October 2012; pp. 1–8. [CrossRef]

30. Dixon, J.; Moran, L. High-level multistep inverter optimization using a minimum number of power transistors. *IEEE Trans. Power Electron.* **2006**, *21*, 330–337. [CrossRef]

31. Haddad, M.; Rahmani, S.; Hamadi, A.; Al-Haddad, K. New Single Phase Multilevel reduced count devices to Perform Active Power Filter. In Proceedings of the SoutheastCon 2015, Fort Lauderdale, FL, USA, 9–12 April 2015; pp. 1–6. [CrossRef]

32. Martinez-Rodrigo, F.; Ramirez, D.; Rey-Boue, A.; de Pablo, S.; Lucas, L.H. Modular Multilevel Converters: Control and Applications. *Energies* **2017**, *10*, 1709. [CrossRef]

33. Moranchel, M.; Huerta, F.; Sanz, I.; Bueno, E.; Rodríguez, F. A Comparison of Modulation Techniques for Modular Multilevel Converters. *Energies* **2016**, *9*, 1091. [CrossRef]

34. Rolim, L.G.B.; da Costa, D.R.; Aredes, M. Analysis and Software Implementation of a Robust Synchronizing PLL Circuit Based on the pq Theory. *IEEE Trans. Ind. Electron.* **2006**, *53*, 1919–1926. [CrossRef]

35. Karimi-Ghartemani, M. Linear and Pseudolinear Enhanced Phased-Locked Loop (EPLL) Structures. *IEEE Trans. Ind. Electron.* **2014**, *61*, 1464–1474. [CrossRef]

36. Singh, B.N.; Khadkikar, V.; Chandra, A. Generalised single-phase p-q theory for active power filtering: Simulation and DSP-based experimental investigation. *IET Power Electron.* **2009**, *2*, 67–78. [CrossRef]

Article

Hybrid HVDC (H^{2}VDC) System Using Current and Voltage Source Converters

José Rafael Lebre [1], Paulo Max Maciel Portugal [2] and Edson Hirokazu Watanabe [1,*]

[1] Electrical Engineering Program, COPPE—Federal University of Rio de Janeiro, Athos da Silveira Ramos 149, 68504 Rio de Janeiro, Brazil; jr_lebre@coe.ufrj.br

[2] Furnas Centrais Elétricas, Departament of Operation Electrical Studies, Real Grandeza 219, Botafogo, 68504 Rio de Janeiro, Brazil; pmaxpo@furnas.com.br

* Correspondence: watanabe@coe.ufrj.br; Tel.: +55-21-3622-3477

Received: 27 April 2018; Accepted: 21 May 2018; Published: 23 May 2018

Abstract: This paper presents an analysis of a new high voltage DC (HVDC) transmission system, which is based on current and voltage source converters (CSC and VSC) in the same circuit. This proposed topology is composed of one CSC (rectifier) and one or more VSCs (inverters) connected through an overhead transmission line in a multiterminal configuration. The main purpose of this Hybrid HVDC (H^{2}VDC), as it was designed, is putting together the best benefits of both types of converters in the same circuit: no commutation failure and system's black start capability in the VSC side, high power converter capability and low cost at the rectifier side, etc. A monopole of the H^{2}VDC system with one CSC and two VSCs—here, the VSC is the Modular Multilevel Converter (MMC) considered with full-bridge submodules—in multiterminal configuration is studied. The study includes theoretical analyses, development of the CSC and VSCs control philosophies and simulations. The H^{2}VDC system's behavior is analyzed by computational simulations considering steady-state operation and short-circuit conditions at the AC and DC side. The obtained results and conclusions show a promising system for very high-power multiterminal HVDC transmission.

Keywords: Multiterminal HVDC; CSC; FBMMC; MMC; Hybrid HVDC; Full-bridge; power control; voltage control; DC short-circuit handling

---

1. Introduction

In the Brazilian electrical system, hydroelectric power plants are being built far (2000 to 3000 km) away from its main load centers. The long distances associated with the large amount of power that must be transmitted are making the HVDC transmission system more attractive in comparison with the conventional AC system. AC-to-DC and DC-to-AC conversion can be done by Current Source Converter (CSC) based on thyristor or by Voltage Source Converter (VSC) based, for instance, on IGBT (Insulated Gate Bipolar Transistor). So far, the majority of HVDC transmission systems have used the thyristor-based CSC technology and they are able to convert power up to 4 GW per pole [1]. The VSC-HVDC is relatively new and consequently has fewer projects in service if compared to the CSC-HVDC. Nowadays, VSC has its rating limited to about a quarter the power of a CSC [2]. The amount of research effort and new development in the application of this kind of converter is very high.

A hybrid DC transmission system based on CSC and VSC is proposed and analyzed in this paper. Figure 1 shows the hybrid system topology with one thyristor-based CSC operating as a rectifier connected through overhead transmission lines to VSCs operating as inverters.

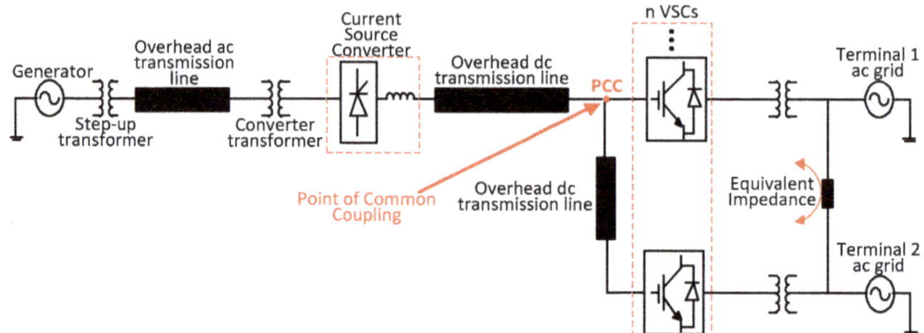

Figure 1. Proposed Hybrid HVDC System topology. PCC: point of common coupling.

The main purpose of using CSC at the rectifier side is that its state-of-the-art technology allows the conversion of higher power (in the range of a few GW) at relatively low cost if compared with the cost of VSC. At the inverter terminals, n VSCs can be used so the power can be delivered to different locations without problems of commutation failure, and the active and reactive power can be controlled independently by each VSC. It is also possible to interchange power between VSCs. The main objective of this paper is to analyze the steady and transient states of this new Hybrid HVDC (H^2VDC) system. This analysis is done by using digital simulation program.

Commutation failure is a problem for inverters based on thyristors, which may cause voltage sag in the AC network connected to the rectifier. This problem does not exist when VSC is used. Another important point about the proposed system is the independent control of active and reactive power in the VSCs within the four quadrants of the active and reactive power (PQ) plane.

The H^2VDC system shown in Figure 1 introduces some benefits and improvements to the operation of power systems, when compared with the conventional CSC- and pure VSC-HVDC. In Figure 1, in fact, the VSCs are the Modular Multilevel Converters (MMC) based on full-bridge submodules (FBMMC). Therefore, hereafter the VSC will be referred to as just MMC, or FBMMC or some cases HBMMC (half-bridge MMC). The benefits of the H^2VDC system are listed below.

- In the case of a collapsed receiving AC network, the H^2VDC can be totally or partially restored by the "black-start" capability of the MMC.
- Active and reactive power at the MMC can be controlled independently, limited only by its rating. The reactive power can be used to control the voltage at its AC terminal, and this characteristic reduces the reactive power compensation equipment. In addition, it gives more reliability to the AC receiving system. Naturally, when the MMC is operating with its maximum power, the AC voltage control cannot be done since the power factor is equal to unity.
- There is no need for minimum short-circuit ratio (SCR) at the AC receiving network, which may be even a passive network. Therefore, no equipment such as a synchronous generator or synchronous compensator is necessary to increase the SCR, which means that the footprint may be smaller if compared with conventional CSC-HVDC.
- DC short-circuit current can be controlled by using full-bridge (FB) submodules (SM) in the MMC.
- It is possible to transmit more power (4 GW) than in the case of a pure MMC-HVDC system at lower cost.

Since there is no commutation failure at the inverters of the H^2VDC system, the impact at the AC rectifier grid, because of a short circuit at the AC receiving system, is smaller when compared with a similar situation for the CSC-HVDC transmission system. This means that no additional equipment is necessary at the AC rectifier grid. Therefore, the global cost of the transmission system may be reduced also because of this fact.

The use of H^{2}VDC system may turn the power system operation more reliable, adding benefits to the electrical system including the reduction of the applied penalties in case of interruption in service, which means less cost to the energy company.

The configuration shown in Figure 1 was first published by the authors in [3,4]. However, all the previous papers about the H^{2}VDC system have considered the conventional thyristor-based CSC and the conventional two level VSCs. The present paper is, in fact, an upgrade from the previous papers since the conventional two level VSCs used before had been changed for full-bridge MMCs. This MMC has all the benefits of the conventional HBMMC added with the capability of interrupting the DC current in case of DC side short-circuit, almost no voltage harmonic content, high power capability, etc.

Several other papers discuss the viability of hybrid systems [5–9]. A hybrid multiterminal system with a CSC operating near to a power plant and MMCs operating at load centers is discussed in [5]. In that case, MMCs are composed by half-bridge submodules (HBSM), so, to protect the system against DC short-circuits, it was proposed the connection of high power diodes in series at the DC line. As HBMMC converters cannot be used to block the DC current inherently, as discussed for several SM topologies with DC fault handling [10,11], they have their control mode switched to Statcom, controlling the reactive power at their respective AC grids in consequence.

Some recent studies have discussed the use of full-bridge submodules (FBSM)-based MMC to handle DC faults by controlling the DC voltage reference and not just blocking the SM switches [12,13]. Considering this approach, the MMC converter is able to operate as a Statcom during DC short-circuits without blocking diodes or DC breakers. Also, by using FBMMC, the hybrid system can perform a power flow interchange among the converters as discussed in [13]. In the Brazilian electrical system, the newest hydroelectric power plants are being located specifically in the Amazon region, which is far from the load centers. Some of these load centers has a mix of power plants, which means that the power flow reversal capability (not possible if series power diodes are used) of HVDC systems is an interesting feature for the power system planning.

The Section 2 discuss the configuration of the H^{2}VDC system and its converter's control philosophy. Section 3 presents the simulation results of the study. Section 4 discuss a possible application of the proposed system in Brazil. Section 5 presents the conclusions. After the conclusions, there is a section with a glossary describing the names adopted for the variables in the control diagrams presented in the paper.

2. System Configuration and Control

2.1. System Configuration

Figure 2 shows the analyzed hybrid system topology with one 12 pulse thyristor-based CSC operating as rectifier connected through a DC overhead transmission line to two FBMMCs operating as inverters.

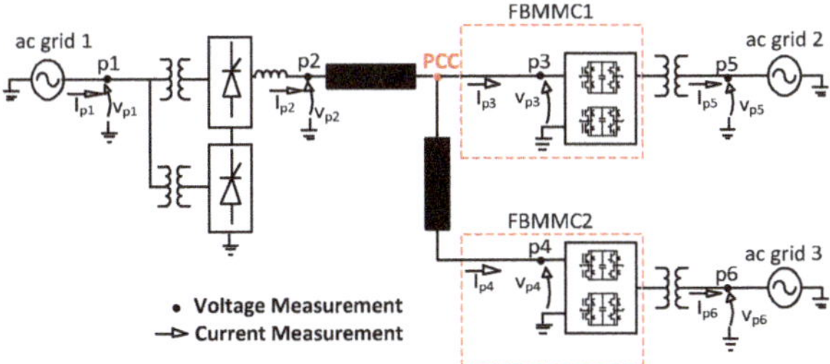

Figure 2. Analyzed Hybrid HVDC system topology.

The measurement points p1 to p6 in Figure 2 are used to guide the simulations. The connection among all the converters is done through an overhead DC transmission line that is represented by its distributed parameters. In this model, the resistance, inductance and capacitance parameters are given as per length (km). The point of common coupling (PCC) in Figure 2 is the point where the DC transmission line backbone is split to connect others MMCs trough DC transmission line branches.

The FBMMC configuration with N submodules in each arm is shown in Figure 3.

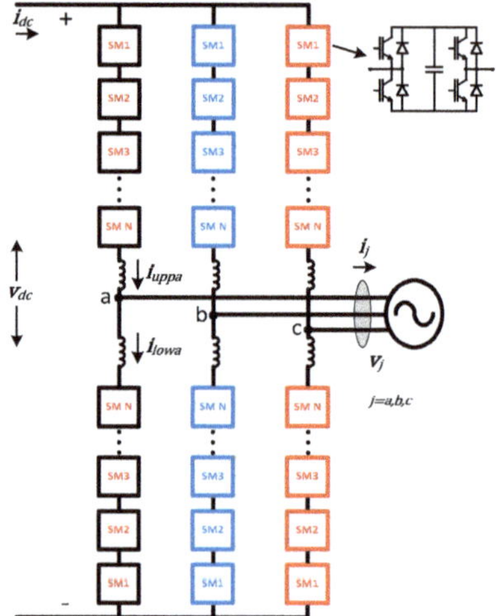

Figure 3. MMC topology with FBSM. SM: submodule.

Full-bridge MMC allows to control the DC short-circuit current and its application is suitable in case of a HVDC with overhead transmission line. The control of the DC short-circuit current is not possible if conventional two or three level VSC or HBMMC is used because they have uncontrollable DC short-circuit current paths.

2.2. Control System

2.2.1. Current Source Converter

The control of the 12-pulse current source converter, in this study, considers only DC current control mode with the same controller type as in [9]. In this mode, the DC current reference is the input of the controller and the firing angle value of the thyristor is its output.

The DC current reference value is defined as i^*_{dc}. In Figure 4, when the block DC fault detection control detects the overcurrent, the DC protection scheme is activated.

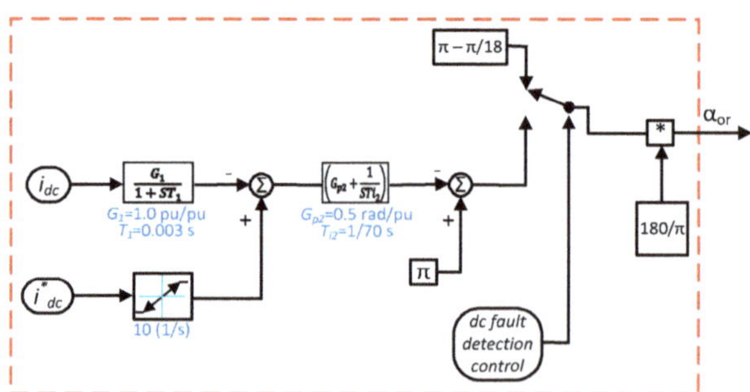

Figure 4. Rectifier DC current control block diagram.

2.2.2. The Voltage Source Converter (FBMMC)

The control philosophies applied to FBMMC1 and FBMMC2 are:

- FBMMC1 operates controlling DC voltage and reactive power;
- FBMMC2 operates controlling active and reactive power.

The control applied in the FBMMC1 and FBMMC2 was based on vector control theory with some special parts such as circulating current control and voltage balancing algorithm (VBA) before the MMCs receive the switching signals. Figure 5 shows the block diagram of the complete control sequence for both FBMMCs. In this figure, Outer Control means the control of the active and reactive power that flows into the converter; in the Inner Control, the NVL block stands for the nearest voltage level modulation control (NVL); and the VBA block is the SM capacitor voltage balancing block (VBA). More details on the control sequence applied to FBMMC1 and FBMMC2 can be seen in [13]. In both converters, active and reactive power are controlled. At FBMMC1 the real power is used to control the DC-link voltage. At FBMMC2 the real power is an order that this converter has to inject in the AC grid.

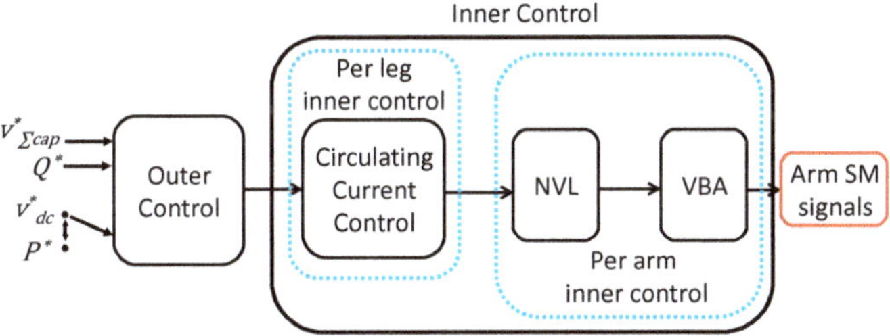

Figure 5. Control sequence of the FBMMC1 and FBMMC2. NVL: nearest voltage level; VBA: voltage balancing algoritm.

The reactive power control loop used in FBMMC1 and FBMMC2 is shown in Figure 6. The variable Q in Figure 6 is the reactive power calculated (measured) using dq reference frame.

FBMMC1 is set to control the DC voltage of the H^2VDC system, at the PCC, to give the voltage reference of the system. In steady-state condition, DC voltage at FBMMC2 and at CSC is given by Ohm's law and depends on the direction of the DC current flow. The block diagram for DC voltage control and for the capacitor's voltages control (CVC) used in FBMMC1 is shown in Figure 7 as in [12]. The gain K is used to set the sensibility of the CVC.

As it was said before, FBMMC2 is set to control its active power, which control block diagram is shown in Figure 8. The variable P in the sum block is the active power calculated using the dq reference frame. The CVC used in FBMMC2 is also presented in Figure 8 as in [13].

The control loops described in Figures 6–8 generate the current reference signals on d (i_d^*) and q (i_q^*) axes that are compared with the respective measured currents on d and q axes. These reference signals are the input of the current control loop that gives the voltage output on d and q axes (v_d^* and v_q^*), as shown in Figure 9. Figures 6–9 compose the outer control. The PI controllers are adopted here because they are widely used for HVDC applications.

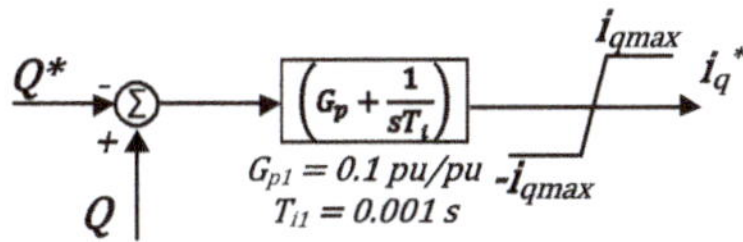

Figure 6. Reactive power control loop used in FBMMC1 and FBMMC2.

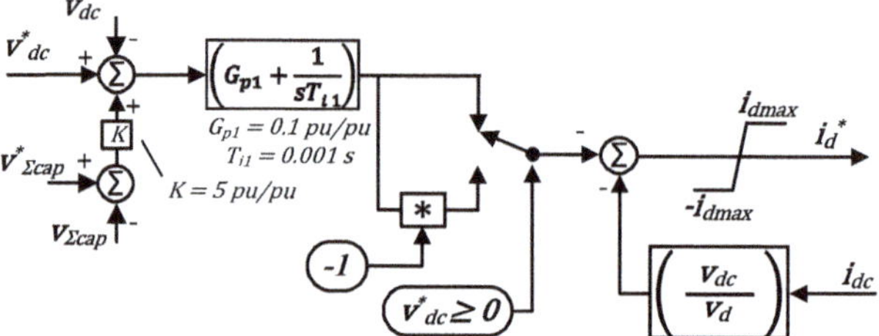

Figure 7. DC Voltage control loop for FBMMC1.

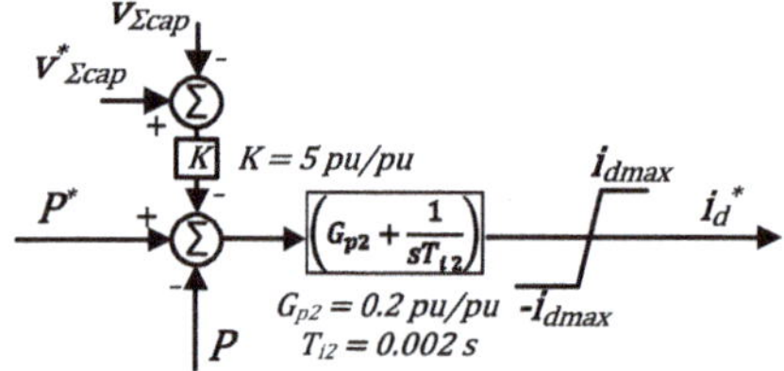

Figure 8. Active power control loop for FBMMC2.

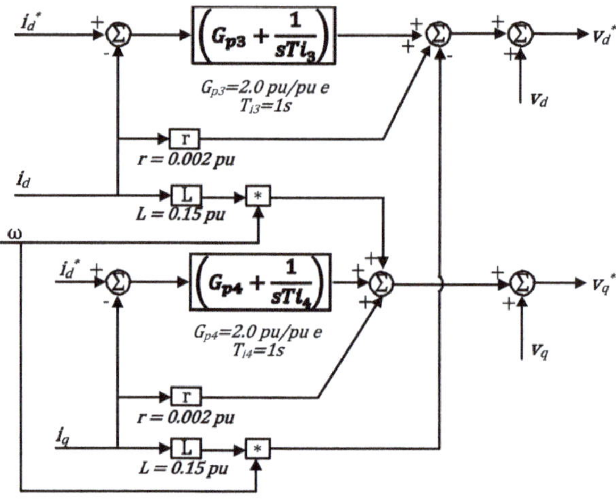

Figure 9. Inner current controller block diagram for FBMMC1 and FBMMC2.

Figure 10 shows the block diagram for the circulating current control. Figure 10a shows the leg common voltage v^*_{diff} calculation, which is composed basically by the control of the second order harmonic (2ω, where ω is the grid frequency) and for the zero-sequence control [14]. Figure 10b shows the voltage reference calculation for upper and lower arms. It is important to note that the DC voltage reference v^*_{dc} is the one ramped up during the start-up and re-start procedure of the FBMMCs and it is kept equal to zero during a DC fault.

As the FBMMC are composed by 20 SM per arm, there is no need for pulse width modulation (PWM)-based modulation techniques in order to have low harmonic voltage waveform, so a nearest voltage level (NVL) modulation technique was adopted in the simulation studies [15]. The time step for the simulation studies was 10 μs and the modulation verification was applied with intervals of 200 μs.

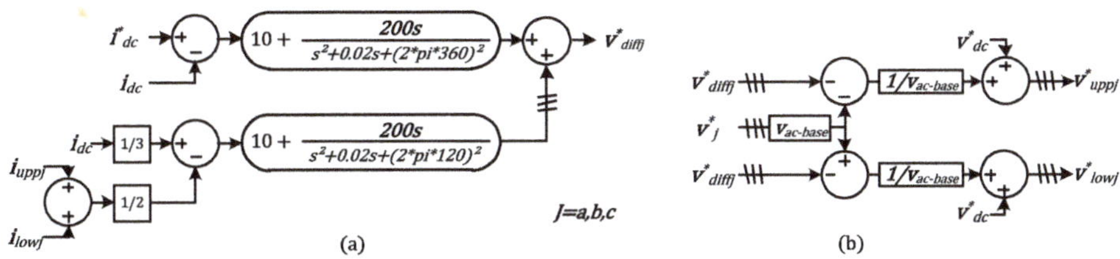

Figure 10. Circulating current control: (**a**) v^*_{diff} calculation; (**b**) upper and lower arms voltage reference calculation.

2.3. Master-Slave Control

HVDC multiterminal systems can be controlled through the so-called master-slave control philosophy. In this case, the master converter controls the system DC voltage and the other connected converters are slaves controlling active power, frequency or another variable. For the H^2VDC system discussed in this paper there are two main possibilities to implement the master-slave approach. The first one is to consider the CSC as the master, which means that it would control the system DC voltage and all FBMMC would control its active power. The second—the one adopted in this paper—is to consider the CSC (rectifier) controlling active power (which is conventional for CSC-HVDC) and one FBMMC as the master converter controlling the DC voltage and the other FBMMC controlling the active power. By using this strategy, one can note that the power to energize the DC line will come from the AC grid 2 in Figure 2. After the DC voltage reaches the rated value, the CSC is set to deliver power to the multiterminal system. After the previous action, the FBMMC2 is set to dispatch active power.

2.4. Dc Faults Protections and Management

The DC fault protection scheme adopted in this study considers overcurrent and undervoltage to detect the fault. The measured DC current value that triggers the DC fault scheme is set at 2 pu. The time delay to start the DC fault handling after the current crosses the overcurrent limit is 100 µs. After the fault detection, the CSC overcurrent strategy is to increase the firing angle up to its maximum value adopted in this study: 170 degrees.

There are two ways to interrupt the DC current during a DC fault using a FBMMC:

- Blocking all the IGBTs and forcing the DC current to flow through the SM capacitors; or
- By control actions, forcing the DC voltage or the DC current fall to zero.

In the first case, the usual strategy to interrupt DC short-circuit current using FBMMC is to block all switches so the current between the AC and DC sides are forced to flow through the series connection of all SM capacitors and the current is quickly blocked [10]. In this case, it is expected that the total current interruption takes approximately 5 ms after the detection of the DC short-circuit. However, considering this approach, the converter cannot control reactive power during the DC short-circuit. Therefore, in this paper it is proposed to use the second option, where the FBMMC DC voltage reference is set to zero and the DC current short-circuit goes to zero. This action of control is slower than the all switch blocking strategy (first option) and is expected that the short-circuit current interruption takes approximately 40~50 ms after DC short-circuit.

2.4.1. Dc Voltage Control Under Short-Circuit Condition

The DC fault handling without blocking all the switches comes up with a different issue: the control of DC voltage reference at all FBMMC. For conventional HBMMC, the DC voltage reference is generally kept at the rated value. However, in this case there is a need for surge arresters to avoid overvoltage transient when starting the system (or recovering after faults). For the FBMMC, the DC voltage may be softly controlled by regulating the desired DC voltage reference, which is the approach adopted to handle the DC current fault in this paper.

Figure 11 shows the influence of the DC fault detection control on the FBMMC overall control. After the fault detection, the FBMMC1 DC voltage reference is set to zero and the FBMMC2 DC current reference is set to zero in order to avoid the converters to feed the DC short-circuit. After the deionization time, the converter FBMMC1 is supposed to restore the voltage level at the DC line. This control action is done by ramping up the FBMMC1 DC voltage reference to the rated value. As the DC voltage starts to be restored to its rated value, the FBMMC2's DC current control regulates its own DC-side voltage by controlling its reference value v^*_{dc} and keeping its DC current controlled. The objective of this action is to avoid the FBMMC2 to become a short-circuit for the DC system or present an uncontrolled DC current.

It is important to discuss about v^*_{dc} in Figure 11. By using FBMMC it is possible to decouple the AC and DC sides of the converter as discussed in [12,13], so it is possible to control separately the active power from both sides. However, if the active power from one side is set independently from the other, the energy stored inside the converter's capacitors will fall or rise undesirably unless the capacitors are connected to an energy storage system such as batteries. The FBMMC1, as discussed in Section 2.3, is set to control the DC voltage. Therefore, the active power that flows at the FBMMC1's AC side is the amount of power left in the DC system, which makes FBMMC1 to work in slack mode at the DC system.

The concept of decoupling between AC and DC sides can be observed in the FBMMC2's control depicted in Figure 11. Usually, for a slave converter in an HVDC system, the AC active power reference P^* is controlled by the slave converter while the system DC voltage is kept controlled by the master converter. This is not the case for an FBMMC. The control of the FBMMC2 must take advantage of the decoupling between the converter AC and DC sides into account and set a reference for its DC voltage independently from the AC side active power. Once the DC voltage is controlled by the master converter (FBMMC1), FBMMC2 (and any other slave FBMMC connected to this multiterminal grid) must be set to control its DC-side current and then provide a DC voltage reference for the inner control as shown in Figure 11b. As the HVDC system's DC side current presents a constant mean value in normal operation, a PI controller has been adopted in this paper to set the DC voltage reference needed to regulate the DC-side current. One can note that the same reference is adopted for the active power in the outer control and for the DC-side current (with a gain K to convert the units) to generate the voltage reference v^*_{dc} for the FBMMC2. This strategy is supposed to guarantee power transfer between AC and DC sides without charging or discharging the SM capacitors. However, this approach still does not consider the losses inside the converter, which would provoke a slow decrease in the energy stored inside the capacitors. To avoid this problem, the CVC described in [13] is adopted for the whole operation inside the outer control, not only when $v^*_{dc} = 0$. For the FBMMC1, the CVC is only needed while $v^*_{dc} = 0$.

Once FBMMC2 is set to control the DC current, a DC short-circuit might not be sufficient to trigger the overcurrent protection. Therefore, this converter is also set to detect a DC fault when a sudden DC voltage dropdown occurs.

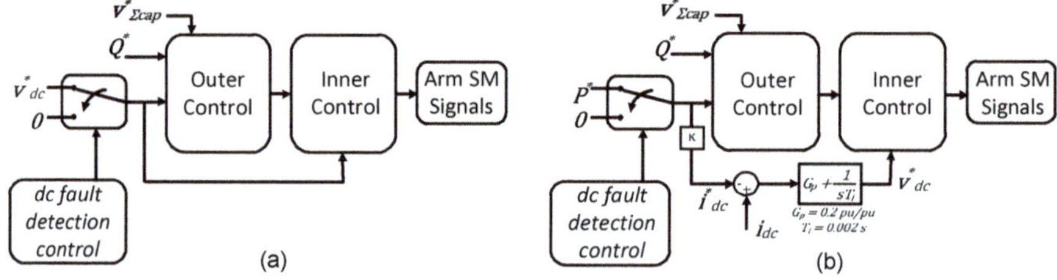

Figure 11. Overall FBMMC control considering the DC fault detection control: (**a**) DC voltage FBMMC controller—FBMMC1; (**b**) active power FBMMC controller—FBMMC2.

3. Simulations

The simulations developed here are used to study the performance of the H^2VDC system shown in Figure 2—with one CSC and 2 FBMMC—in normal and some emergency conditions. This section shows the DC fault handling capability of the proposed hybrid system. Besides, it shows the system behavior during a pole-to-ground fault. In the studied system the CSC and FBMMCs are rated at 1000 MVA and 500 kV DC voltage. The simulation considers that capacitors of both FBMMC1 and FBMMC2 are previously charged. Table 1 presents the main parameters. The SM voltage was chosen equal 25 kV and thus would need series switches, which is not normal for MMC. This choice decreases the number of SM and makes faster the simulation. The practical SM voltage in actual applications with single switches is in the order of 2 kV.

Table 1. Analyzed H^2VDC System parameters.

Parameter	Value
Rated DC voltage	500 kV
FBMMC Rated AC voltage	280 kV
Rated DC power	1000 MW
CSC AC system reactance	150 mH
FBMMC1 AC system reactance	42 mH
CSC transformers rated voltages	345/220/220 kV
FBMMC transformers rated voltages	280/280 kV
Transformers equivalent reactance	0.15 pu
Transformers equivalent resistance	0.001 pu
CSC smoothing reactance	500 mH
FBMMC smoothing reactance	50 mH
Number of SMs per arm	20
SM rated voltage	25 kV
SM capacitance	1 mF
Arm inductance	5 mH
FBMMC's inertia constant, H	37.5 ms
Line resistance per unit length	0.015 Ω/km
Line inductance per unit length	0.792 mH/km
Line inductance per unit length	14.4 nF/km
Line 1 length	1000 km
Line 2 length	200 km

3.1. Single Phase Short Circuit at the AC Grid 1

Figure 12 shows the H^2VDC system behavior during a single-phase short-circuit at phase a of the AC grid 2, as shown in Figure 2. It is considered a non-permanent AC short-circuit. The short-circuit is applied at t = 1 s and lasts for 100 ms. In this case, there is no special protection scheme developed for AC faults.

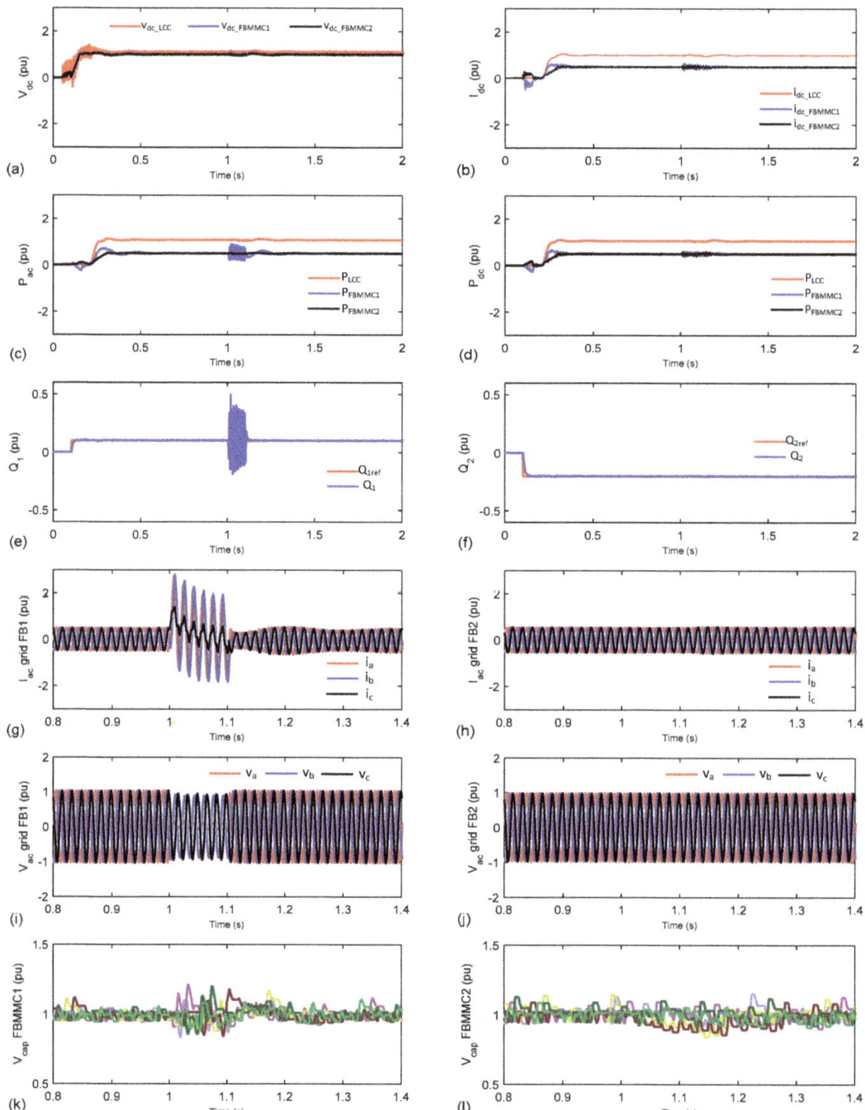

Figure 12. H^{2}VDC system behavior during a single-phase fault at t = 1 s for 0.1 s in phase a of the FBMMC1: (**a**) DC voltage; (**b**) DC current; (**c**) AC active power; (**d**) DC active power; (**e**) AC current at p5 (FBMMC1); (**f**) AC current at p6 (FBMMC2); (**g**) AC voltage at p5 (FBMMC1); (**h**) AC voltage at p6 (FBMMC2); (**i**) reactive power at the AC side of FBMMC1; (**j**) reactive power at the AC side of FBMMC2; (**k**) FBMMC1 capacitor voltages; (**l**) FBMMC2 capacitor voltages.

Based on the simulations in Figure 12, it is possible to conclude that the proposed H^{2}VDC system presents a stable dynamic performance during an AC short-circuit. After the AC fault extinction, the H^{2}VDC system returns to all its fault previous value. Figure 12a shows that the DC voltage is not affected significantly in this case. The transmitted DC power at the rectifier, as shown in Figure 12d, has just a little transient, which is a great advantage over pure CSC-HVDC system. Figure 12g–j show the AC currents and voltages with a zoom in from 0.8 s to 1.4 s. Figure 12k–l show the capacitors voltages on FBMMC 1 and 2 also with a zoom. The soft oscillatory behavior in both converters is due to the oscillation in the DC currents at the converter terminals during the DC short-circuit, as shown in Figure 12b.

3.2. Short Circuit At the DC Line

Figure 13 shows the H^{2}VDC system behavior during a short-circuit at the DC line backbone. In this case, it is considered a non-permanent DC short-circuit at the DC line, with duration of 100 ms. DC protection schemes described in detail in Section 2 are activated here.

The CSC converter is set to control the DC current. The FBMMC1 is set to control the DC voltage and FBMMC2 is set to control its active power flow. Both FBMMC keep controlling its reactive power. At the instant 0.05 s, CSC is turned on with its current reference set to zero. Then, at the instant 0.1 s, FBMMC1 DC voltage reference is ramped up to 1 pu at a rate of 20 pu/s, thus, it reaches 1 pu in 50 ms. The FBMMC2 was turned on at the instant 0.15 s. Also, at this time, the current reference at the CSC is ramped up from 0 to 1 pu at a rate of 10 pu/s, thus, it reaches 1 pu in 100 ms. The power reference for the FBMMC2 is ramped up at the instant 0.2 s with the rate of 5 pu/s. After these operating sequences, the H^{2}VDC system reaches its steady-state operation.

At t = 1 s, a DC short-circuit is applied at the DC line backbone. At this moment, the DC current grows up and the DC voltage goes down. Therefore, considering the measured overcurrent and undervoltage values, the DC protection of each converter get into operation: the firing angle of the CSC is set to its maximum value by its control action; the FBMMC1 DC voltage reference is set to zero; and the FBMMC2 active power reference (and i^*_{dc}) is set to zero. Then, after the set deionization time of 200 ms, the CSC is set to control zero active power, so, the CSC firing angle alfa goes around 90°. After 200 ms of the FBMMC1 fault detection, the FBMMC1 DC voltage reference is ramped up to 1 pu at a rate of 20 pu/s. After 250 ms of the CSC fault detection, the CSC power reference is ramped up at a rate of 10 pu/s. After 300 ms of the FBMMC2 fault detection, the FBMMC2 power reference is ramped up also with 5 pu/s.

Figure 13a shows that the DC voltage of the system is set to zero in order to eliminate the DC short-circuit current. In Figure 13a there is also a zoom that shows the DC voltages waveforms measured at all converters terminals in normal operation. In Figure 13b, it is possible to analyze that the DC short-circuit current is controlled by the control actions of all converters (CSC and FBMMCs) together. Figure 13c,d show the active power measured at both AC and DC sides of the converters. Figure 13e,f show that FBMMC1 and FBMMC2 operate as STATCOM during the DC short-circuit time, being it a great advantage over other VSCs. This is possible by the fact that the IGBTs are not blocked during the DC short-circuit. In fact, as discussed in Section 2.3, the FBMMC control adopted allows the independent control of its AC and DC voltages [12,13].

Figure 13g,h show FBMMC1 and FBMMC2 capacitors voltages. The initial short-circuit current causes the capacitors voltages to decrease, but, during the deionization time, the capacitor's voltage control is set to keep it in the nominal value.

Figure 13i shows the firing angle behavior of the CSC. When the CSC control detects an overcurrent more than the set value ($i_{dc} > 2$ pu), it elevates the firing angle to its maximum value (170 degrees) with the purpose of reducing the DC current. After the deionization time (200 ms), the CSC control sets the firing angle at 90 degrees to zero power during 50 ms to wait the system DC voltage be restored again to 1 pu. After the DC voltage restoration, the CSC firing angle is set to its nominal value again to transmit DC power normally.

In this case, the H^{2}VDC system returns to its normal operation approximately 400 ms after the DC short-circuit is applied. It is important to note that all the time spent in the system's restoration process can be set to be faster or slower depending on the system parameters and requirements.

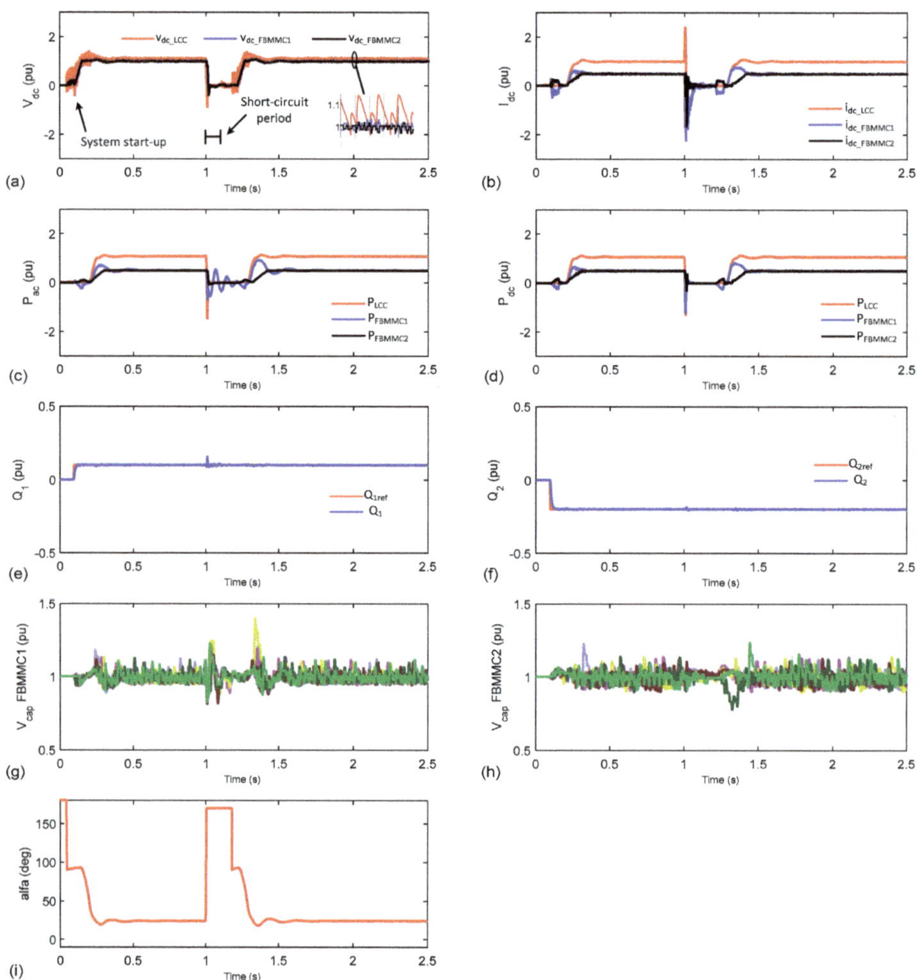

Figure 13. System behavior during a DC fault: (**a**) DC voltages; (**b**) DC currents; (**c**) AC active power; (**d**) DC active power; (**e**) reactive power at the AC grid connected to FBMMC1; (**f**) reactive power at the AC grid connected to FB2; (**g**) FBMMC1 capacitor voltages; (**h**) FBMMC2 capacitor voltages; (**i**) CSC firing angle α_{or}.

4. Possible Application

One possible practical application of the H^{2}VDC system is the power transmission from a large hydroelectric power plant located in the Amazon region to the main load centers in Brazil (Rio de Janeiro, São Paulo and Minas Gerais) in a multiterminal configuration (CSC and FBMMCs). Figure 14 shows a simplified map of the Brazilian electrical system with the H^{2}VDC system highlighted inside in black.

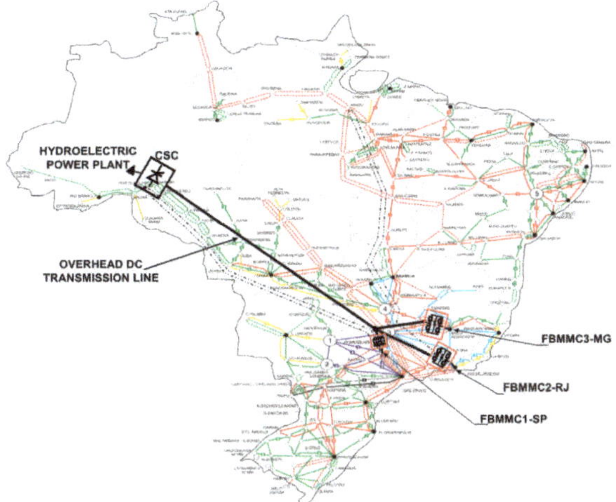

Figure 14. Hypothetical configuration of the H^{2}VDC system for transmitting power from the Amazon region to the main load centers of Brazil with three FBMMC terminals (Rio de Janeiro, São Paulo and Minas Gerais).

In Figure 14, the CSC is connected directly in a hydroelectric power plant to convert (AC to DC) all the generated power. This converted power is transmitted through an overhead DC transmission line directly to the 3 FBMMCs that are connected at the Brazilian main load centers. All the benefits of the FBMMCs shown before could be applied to the operation of each AC receiving system (hypothetically, Rio de Janeiro, São Paulo and Minas Gerais). As discussed in Section 2.4.1, the third FBMMC would be controlled with the same strategy used for FBMMC2 in the simulation studies. Considering strong AC systems and an operation as rectifier, the use of CSC is extremely recommended, more power can be transmitted, and the total cost will be reduced in comparison with a pure MMC system.

In steady-state condition is expected that the transmitted power is unidirectional, which means that the CSC operates as rectifier and the FBMMCs operate as inverters. In special conditions it could be possible that any FBMMC operates as rectifier interchanging power among the FBMMCs when an AC grid has power excess and the other AC grid has power deficit. All these characteristics can improve the operation of the power system.

5. Conclusions

The analyses of the H^{2}VDC system study and dynamic performance were shown in detail in the simulations. These simulations have considered operating conditions of single phase AC short-circuit at AC grid 1 and short-circuit at the DC line backbone. Considering these operating conditions, the H^{2}VDC system presented a stable dynamic performance returning to its normal operating conditions.

The AC single-phase short circuit simulation study has shown that the H^{2}VDC system is not affected by single-phase AC faults, so there is no need for further actions but wait until the fault condition to be extinct. The AC single-phase short-circuit was analyzed here because in many countries its response is considered to be a planning criterion.

This paper presented a discussion on how to handle DC faults in a hybrid multiterminal DC transmission system without the need of DC breakers, power diodes in series with the converter and surge arresters. In addition, this approach allows reactive power control at the AC grids connected to the FBMMC even during the DC faults, adding a great advantage to the system over blocking the IGBTs. The normal operation of the studied system was reached after approximately 400 ms the DC short-circuit happening.

The DC short-circuit analysis considered in this paper is a non-permanent fault, which means that the DC short-circuit is self-extinguished after the voltage is forced to zero. If the DC short-circuit at the overhead DC line persists after the recovering strategy discussed in this paper, some attempts to re-start should be implemented to confirm that the fault is permanent and then, turn off the converters in order to start the maintenance procedure.

The simulations have shown that the technology of FBMMC was successfully applied in the operation of the H^2VDC systems with overhead transmission line by the fact that the DC short-circuit current can be controlled.

The black start capability of the MMC is not analyzed in this paper. The used control does not allow the "black-start" and it would be necessary to develop this control to apply this function.

This paper proposes a DC voltage control strategy to restore the system normal operation after the DC faults without overcurrents or overvoltages and no need for critical communication between two stations. By controlling the DC voltage with the master FBMMC and the DC current with the slave converters, the proposed strategy assures that there will be no overcurrents among the FBMMC.

Based on the theoretical development and the simulation results shown here, the H^2VDC system may become a very promising HVDC multiterminal transmission system, which improves the operation of power systems, by making the system more reliable and safe.

Author Contributions: Conceptualization, J.R.L., P.M.M.P. and E.H.W.; Data curation, J.R.L.; Formal analysis, J.R.L. and P.M.M.P.; Funding acquisition, E.H.W.; Investigation, J.R.L., P.M.M.P. and E.H.W.; Methodology, J.R.L., P.M.M.P. and E.H.W.; Software, J.R.L.; Supervision, E.H.W.; Writing—original draft, J.R.L. and P.M.M.P.; Writing—Review & Editing, J.R.L., P.M.M.P. and E.H.W.

Acknowledgments: This work was partially supported by FAPERJ Grant CNE E02/2017 as well as by CNPq Grant No. 306243/2014-8 and No. 142147/2014-1.

Conflicts of Interest: The authors declare no conflicts of interest.

Glossary

v^*_{dc}, v_{dc}	DC voltage reference and measured, respectively.
i^*_{dc}, i_{dc}	DC current reference and measured, respectively.
v_j	abc phase-to-neutral voltages at the AC bar for $j = a, b, c$.
i_j	abc line currents at the AC bar for $j = a, b, c$.
i_{uppa}, i_{lowa}	Phase a upper and lower arm currents.
P^*, P	There-phase active power reference and measured, respectively.
Q^*, Q	There-phase reactive power reference and measured, respectively.
α_{or}	Alfa order for the CSC.
ω	Grid frequency.
v_d^*, v_d, i_d^*, i_d	d axis voltage and current references and measured, respectively.
v_q^*, v_q, i_q^*, i_q	q axis voltage and current references and measured, respectively.
G_p, T_i	Proportional gain and time constant for the PI controllers (empirically tuned).
K	Proportional gain to set the sensibility of the CVC error signal.
L, r	Reactance and resistance for decoupling the dq control.
i_{dmax}, i_{qmax}	d and q axis maximum current limit for the outer control.
i_{uppj}, i_{lowj}	Upper and lower arm currents for $j = a, b, c$.
$v^*_{\sum cap}$	Reference value for the sum of all capacitor's voltages in one FBMMC.
$v_{\sum cap}$	Sum of all capacitor's voltages measured in one FBMMC.
v^*_j	Voltages references from outer control output (v_d^*, v_q^*) for $j = a, b, c$.
v^*_{diffj}	Leg common voltage for $j = a, b, c$.
$v_{dc\text{-}base}$	DC base voltage.
$v_{ac\text{-}base}$	AC base voltage for the FBMMC.
v^*_{uppj}, v^*_{lowj}	Upper and lower arm reference voltages for the modulation control for $j = a, b, c$.
NVL	Nearest voltage level modulation control.
VBA	Voltage Balancing Algorithm.

References

1. C-EPRI. Available online: http://www.cepri.com.cn/aid/details_72_272.html (accessed on 10 March 2018).
2. Francos, P.L.; Verdugo, S.S.; Alvarez, H.F.; Guyomarch, S.; Loncle, J. INELFE Europe's first integrated onshore HVDC interconnection. In Proceedings of the 2012 IEEE Power and Energy Society General Meeting, San Diego, CA, USA, 22–26 July 2012; pp. 1–8. [CrossRef]
3. Portugal, P.M.M.; Watanabe, E.H.; Macedo, N.J.P. Hybrid HVDC system using current and voltage source converter. In Proceedings of the CIGRE CE-B4 Colloquium HVDC and Power Electronics to Boost Network Performance, Brasilia, Brazil, 2–3 October 2013.
4. Portugal, P.M.M.; Watanabe, E.H.; Macedo, N.J.P. Study and development of a hybrid HVDC System composed by current and voltage source converters. In Proceedings of the XIII Symposium of Specialists in Electric Operational and Expansion Planning, Foz do Iguaçu, Brazil, 18–21 May 2014.
5. Tang, G.; Xu, Z. A LCC and MMC hybrid HVDC topology with DC line fault clearance capability. *Int. J. Electr. Power Energy Syst.* **2014**, *62*, 419–428. [CrossRef]
6. Lee, Y.; Cui, S.; Kim, S.; Sul, S.K. Control of hybrid HVDC transmission system with LCC and FB-MMC. In Proceedings of the 2014 IEEE Energy Conversion Congress and Exposition (ECCE), Pittsburgh, PA, USA, 14–18 September 2014; pp. 475–482. [CrossRef]
7. Jung, J.J.; Cui, S.; Lee, J.H.; Sul, S.K. A New Topology of Multilevel VSC Converter for a Hybrid HVDC Transmission System. *IEEE Trans. Power Electron.* **2017**, *32*, 4199–4209. [CrossRef]
8. Xu, Z.; Wang, S.; Xiao, H. Hybrid high-voltage direct current topology with line commutated converter and modular multilevel converter in series connection suitable for bulk power overhead line transmission. *IET Power Electron.* **2016**, *9*, 2307–2317. [CrossRef]
9. Nguyen, M.H.; Saha, T.K.; Eghbal, M. Hybrid multi-terminal LCC HVDC with a VSC Converter: A case study of Simplified South East Australian system. In Proceedings of the IEEE Power and Energy Society General Meeting, San Diego, CA, USA, 22–26 July 2012. [CrossRef]
10. Marquardt, R. Modular Multilevel Converter: An universal concept for HVDC-Networks and extended DC-bus-applications. In Proceedings of the 2010 International Power Electronics Conference—ECCE Asia—(IPEC), Sapporo, Japan, 21–24 June 2010; pp. 502–507. [CrossRef]
11. Sharifabadi, K.; Harnefors, L.; Nee, H.P.; Norrga, S.; Teodorescu, R. *Design, Control and Application of Modular Multilevel Converters for HVDC Transmission Systems*; John Wiley & Sons: Hoboken, NJ, USA, 2016; ISBN 9781118851562.
12. Wenig, S.; Goertz, M.; Prieto, J.; Suriyah, M.; Leibfried, T. Effects of DC fault clearance methods on transients in a full-bridge monopolar MMC-HVDC link. In Proceedings of the 2016 IEEE PES Innovative Smart Grid Technologies Conference Europe, Melbourne, VIC, Australia, 28 November–1 December 2016; pp. 850–855. [CrossRef]
13. Lebre, J.R.; Watanabe, E.H. Fullbridge MMC control for hybrid HVDC systems. In Proceedings of the 2017 Brazilian Power Electronics Conference (COBEP), Juiz de Fora, Brazil, 19–22 November 2017; pp. 1–6. [CrossRef]
14. Moon, J.W.; Kim, C.S.; Park, J.W.; Kang, D.W.; Kim, J.M. Circulating current control in MMC under the unbalanced voltage. *IEEE Trans. Power Deliv.* **2013**, *28*, 1952–1959. [CrossRef]
15. Kouro, S.; Bernal, R.; Miranda, H.; Silva, C.A.; Rodríguez, J. High-performance torque and flux control for multilevel inverter fed induction motors. *IEEE Trans. Power Electron.* **2007**, *22*, 2116–2123. [CrossRef]

 *energies*

Article

New Hybrid Static VAR Compensator with Series Active Filter

Ayumu Tokiwa [1], Hiroaki Yamada [1], Toshihiko Tanaka [1,*], Makoto Watanabe [2], Masanao Shirai [2] and Yuji Teranishi [2]

[1]   Department of Electrical and Electronic Engineering, Yamaguchi University, 2-16-1 Tokiwadai, Ube, Yamaguchi 755-8611, Japan; w059vg@yamaguchi-u.ac.jp (A.T.); hiro-ymd@yamaguchi-u.ac.jp (H.Y.)

[2]   The Chugoku Electric Manufacturing Company, Incorporated, 4-4-32 Minamiku Osu, Hiroshima 732-8564, Japan; a001188@pnet.gr.energia.co.jp (M.W.); a200022@pnet.gr.energia.co.jp (M.S.); a001033@pnet.gr.energia.co.jp (Y.T.)

*   Correspondence: totanaka@yamaguchi-u.ac.jp; Tel.: +81-836-85-9400

Received: 7 August 2017; Accepted: 10 October 2017; Published: 16 October 2017

Abstract: This paper proposes a new hybrid static VAR compensator (SVC) with a series active filter (AF). The proposed hybrid SVC consists of a series AF and SVC. The series AF, which is connected in series to phase-leading capacitors in the SVC, performs for a resistor for source-side harmonic currents. A sinusoidal source current with a unity power factor is obtained with the series AF, although the thyristor-controlled reactor generates harmonic currents. A digital computer simulation was implemented to confirm the validity and high practicability of the proposed hybrid SVC using PSIM software. The simulation results demonstrate that sinusoidal source currents with a unity power factor are achieved with the proposed hybrid SVC.

Keywords: static var compensator; series active filter; thyristor-controlled reactor; phase-leading capacitor; hybrid static var compensator; static synchronous compensator; hybrid active filter

---

1. Introduction

Large-capacity electric arc furnaces and rolling mills cause rapid reactive-power fluctuations in distribution feeders. These rapid reactive-power fluctuations lead to reactive-power interferences such as flickers and voltage fluctuations in distribution feeders. Static VAR compensators (SVCs) with thyristor-controlled reactors (TCRs) and phase-leading capacitors (PLCs) are widely used to solve the reactive-power interferences in distribution feeders because of low costs [1]. However, TCRs generate harmonic currents on the source side [2]. Many topologies have been proposed to improve the compensation characteristics of the source-side harmonic currents for SVCs [3–6]. The passive filters, which consist of the 5th-, 7th-, and 11th-tuned filters, are combined with the TCR. A three-phase current-controlled voltage-source pulse-width-modulated (PWM) inverter is connected in series to the passive LC filters through matching transformers (MTs). The series-connected three-phase current-controlled voltage-source PWM inverter improves the compensation characteristics of the passive filters.

Hybrid active filters (AFs) have also been proposed by many researchers [7–11]. The previously proposed hybrid AF topologies are essentially a series connection of the LC tuned filters and a three-phase shunt AF. The rating of the shunt AF can be reduced because the fundamental reactive and designated-order harmonic currents are compensated for by the series-connected LC tuned filters. A hybrid SVC topology with the hybrid AF has been proposed [12,13]. The PLCs with TCRs control the fundamental reactive power on the source side. As the shunt AF compensates only for harmonic currents, the required rating of the shunt AF is low. In [14], a static synchronous compensator (STATCOM) is combined with TCRs. The STATCOM performs with PLCs compensating for harmonic currents on the source side. Thus, the required rating of the parallel-connected STATCOM is high. A combined system of shunt-passive and series AFs has also been proposed [15–17]. While the proposed topologies are practical and cost-effective, the fundamental reactive power on the source side cannot be controlled. Thus, a hybrid SVC topology consisting of TCRs and pure PLCs with a small-rated voltage-source PWM inverter has not been reported, as far as the authors know.

This paper proposes a new hybrid SVC topology comprising a small-rated series AF and SVC, which consists of TCRs and pure PLCs. The series AF is connected in series to the pure PLCs. In [17], Prof. H. Fujita, et al. proposed a combined system consisting of a shunt-passive filter and series AF for a current-source harmonic-producing load. The series AF performs for a resistor of K_C Ω for source-side harmonic currents. We note that there are two current-source harmonic-producing loads, namely, the TCR and the three-phase load, in the newly proposed hybrid SVC. Considering both the three-phase load and the TCR as a current-source harmonic-producing load, the previously proposed control strategy for the series AF in [17] is applicable to the newly proposed hybrid SVC. This is a simple and practical idea for the control strategy of the series AF in the newly proposed hybrid SVC. Thus, the series AF in the proposed hybrid SVC performs for a resistor of K_C Ω for source-side harmonic currents. The source-side harmonic currents are isolated by the series AF, while PLCs with TCRs compensate for the fundamental reactive currents on the source side. Sinusoidal source currents with a unity power factor are achieved by the newly proposed hybrid SVC. The basic principle of the proposed hybrid SVC is discussed in detail. The compensation characteristics of the harmonic currents are shown in detail, and these have been confirmed by digital computer simulation using PSIM software. The simulation results demonstrate that sinusoidal source currents with a unity power factor are obtained. From the simulation results, the required capacity of the series AF is 2.8% as compared to the rating of the three-phase load. The addition of the small-rated series AF isolates the source-side harmonics perfectly. Only the newly proposed topology can be applied to the SVCs that are under commercial operations, although the required rating of the series AF is slightly higher than that of the series AF in [3–6]. This demonstrates that the proposed hybrid SVC is useful and cost-effective for practical distribution feeders.

2. Previously Proposed Topologies of the Hybrid Static VAR Compensator

Figure 1 shows a power circuit diagram of the previously proposed hybrid SVC topology [3–6]. The previously proposed hybrid SVC topology consists of TCRs and 5th-, 7th-, and 11th-tuned filters with series-connected AFs. In [3], the added series AF performs for a resistor for the source-side currents. The purpose of the series AF is to suppress the anti-resonance between the LC tuned filters and source impedance.

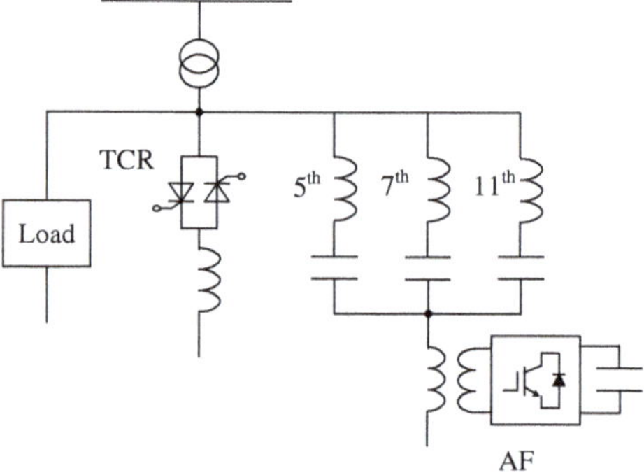

Figure 1. Power circuit diagram of the previously proposed hybrid static VAR compensator (SVC) with a series active filter (AF) [3–6].

The source-side harmonic current compensations with the series AF under normal operation were not discussed or demonstrated. In [4–6], the added series AFs perform as current sources for harmonic currents, which flow into the LC tuned filters. This injection improves the compensation characteristics of the LC tuned filters. The required rating of the series AF was 1.4% as compared to the rating of the three-phase load. The previously proposed hybrid SVC topology in Figure 1, however, cannot be applied to the SVCs that are under commercial operations, because PLCs should be replaced by the LC tuned filters.

Figure 2 shows a power circuit diagram of the previously proposed hybrid SVC topology [12,13]. The PLCs with TCRs control the fundamental reactive power on the source side. The inductors are connected in series to the PLCs with TCRs. These allow the fifth-tuned LC filters to sink the fifth-harmonics generated by the load. The shunt AFs inject the harmonic currents. Thus, the shunt AFs improve the compensation characteristics of the fifth-tuned filters. The required rating of the shunt AFs was 4.0% as compared to the rating of the three-phase load. The previously proposed hybrid SVC topology is not applicable to the SVCs that are under commercial operations, because the conventional PLCs with TCRs also should be replaced with those shown in Figure 2.

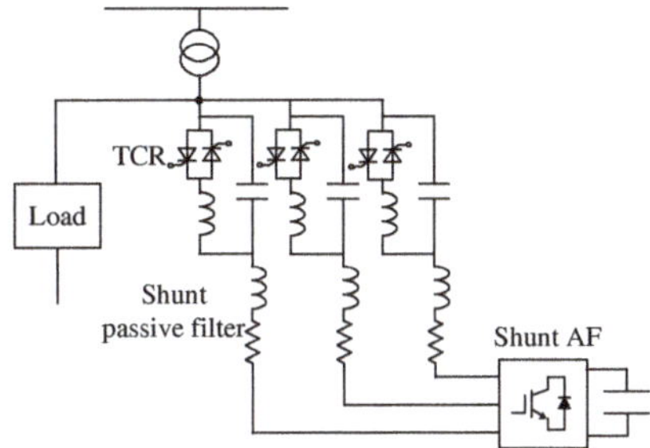

Figure 2. Power circuit diagram of the previously proposed hybrid static VAR compensator (SVC) with thyristor-controlled reactor (TCR) and phase-leading capacitor (PLC) with series AF [12,13].

Figure 3 shows a power circuit diagram of the previously proposed hybrid SVC with the STATCOM [14]. The STATCOM performs with PLCs compensating for harmonic currents on the source side. Thus, the required rating of the parallel-connected STATCOM is much higher than those of the added AFs in Figures 1 and 2. In [14], the required rating of the STATCOM is 80% as compared to the rating of the three-phase load. Using a large capacity, the STATCOM results in a high-cost hybrid SVC.

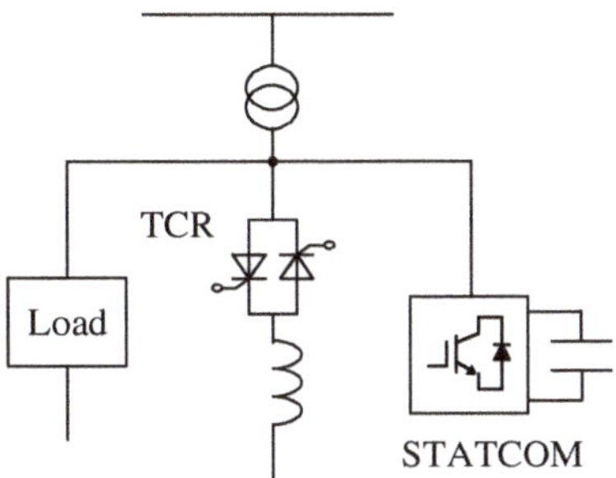

Figure 3. Power circuit diagram of the previously proposed hybrid static VAR compensator (SVC) with a static synchronous compensator (STATCOM) [14].

Table 1 shows summaries of the previously proposed hybrid SVC. As described in the Introduction, a hybrid SVC topology with a small-rated voltage-source PWM inverter, which can be applied to the conventional SVC consisting of TCRs and pure PLCs, has not been reported, as far as the authors know.

Table 1. Previously proposed hybrid static VAR compensator (SVC) topologies and functions of added active filter (AF).

	Topology	Functions of Added AF	Required Rating of Added AF for Three-Phase Load Rating
Figure 1	Thyristor-controlled reactor (TCR) and passive LC filter with series AF	Improvement of harmonic voltage compensation characteristics of passive LC filter	1.4%
Figure 2	TCR and phase-leading capacitor (PLC) with series AF	Harmonic current compensation	4.0%
Figure 3	TCR and parallel-connected static synchronous compensator (STATCOM)	Fundamental reactive-power control and harmonic current compensation	80%

3. Newly Proposed Hybrid Static VAR Compensator

Figure 4 shows a circuit diagram of the proposed hybrid SVC. Table 2 shows circuit constants for Figure 4, which are used in the following simulation results. The proposed hybrid SVC comprises a series AF and SVC, which consists of the Δ-connected TCR and Δ-connected pure PLCs. The series AF consists of a three-phase voltage-source PWM inverter with insulated-gate bipolar transistors (IGBTs). The series AF is connected in series to the three-phase PLCs through MTs, where the turns ratio is 1:2. The small-rated LC filter (inducer L_f and capacitor C_f) suppresses switching ripples that are generated by the PWM inverter, which performs for the series AF. The purpose of this paper is to demonstrate the compensation performance of the reactive and harmonic currents for the proposed hybrid SVC. Thus, ideal models for IGBTs, inductors, and capacitors are used. Table 3 shows data for the MTs used in Figure 4. Leakage inductors, winding resistors, and magnetizing admittance are considered in the MT model of Figure 4. A three-phase load generates the fundamental reactive currents, and fifth- and seventh-order harmonic currents. A three-phase current source is used to demonstrate the three-phase load. The authors, who are with the Chugoku Electric Manufacturing Company, have had extensive experience with developing commercial SVCs and selling SVCs to customers. Generally, the capacities of TCRs and PLCs are decided by considering the true load conditions of the customers. Thus, the capacities of TCRs and PLCs were decided with the published paper [18].

The rating of the three-phase load is 176 MVA. The rating of the three-phase source voltage is 33 kVrms, 60 Hz, while the rating of the inductor of the TCRs is 80 MVA and that of the PLCs is 140 MVA. Thus, the per-phase base rated current and impedance are 3.1 kArms and 6.2 Ω, which are both 1 pu. The pure PLCs with TCRs compensate for the fundamental reactive power on the source side. The series AF performs for a resistor of K_C Ω. The source-side harmonic currents are isolated by the series AF, while PLCs with TCRs compensate for the fundamental reactive currents on the source side. Sinusoidal source currents with a unity power factor are achieved by the newly proposed hybrid SVC.

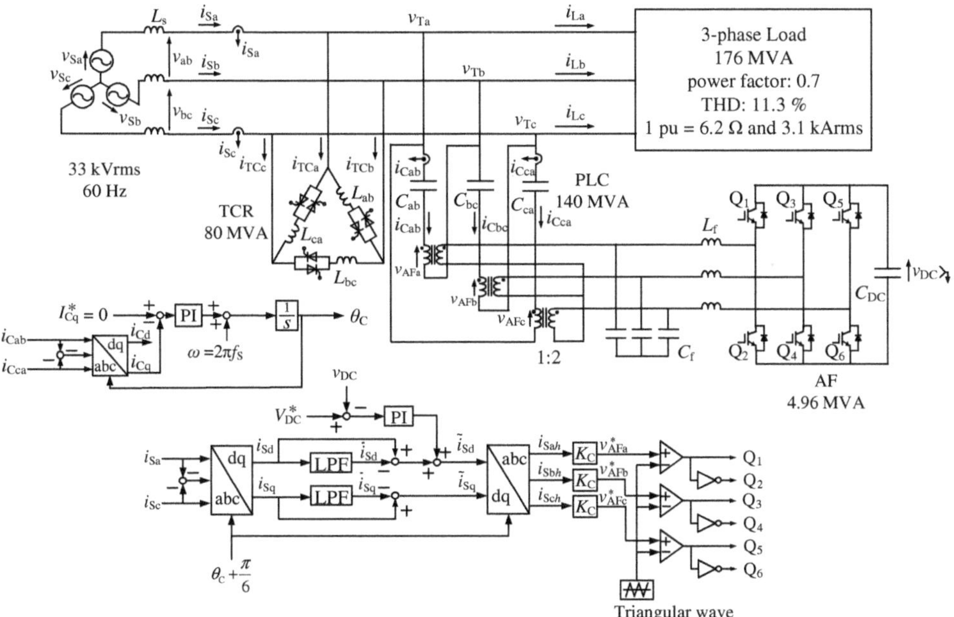

Figure 4. Circuit diagram of the proposed hybrid static VAR compensator (SVC) with a series active filter (AF).

Table 2. Circuit Constants for Figure 4.

Item	Symbol	Value
Source inductor	L_S	1.64 mH
Filter inductor	L_f	0.066 mH
Filter capacitor	C_f	96 µF
Inductors of SVC	L_{ab}, L_{bc}, L_{ca}	108 mH
PLCs	C_{ab}, C_{bc}, C_{ca}	114 µF
DC capacitor	C_{DC}	4700 µF
DC-capacitor voltage	V_{DC}^*	10 kVdc
Switching frequency	f_S	12 kHz
Cut-off frequency of second-order LPF	f_C	179 Hz
Damping factor of second-order LPF	ζ	0.7
Proportional gain for constant DC-capacitor voltage controller	K_P	0.6
Integral gain for constant DC-capacitor voltage controller	T_I	0.01
Proportional gain for PLL	K_P	5
Integral gain for PLL	T_I	0.01
Control gain for series AF	K_C	10, 20, 30 Ω

Table 3. Data for the matching transformers (MTs) used in Figure 4.

Primary-side voltage	1.19 kVrms
Primary-side rated current	1.42 kArms
Turn ratio	1:2
No load current	56.8 Arms (4.0%)
Iron loss	17 kW
Copper loss	34 kW
Impedance voltage	35.6 Vrms (3.0%)

Figure 5 shows per-phase equivalent circuits for Figure 4. Figure 5a shows a per-phase equivalent circuit for Figure 4, where $\dot{V}_S$ is the source voltage, $\dot{I}_S$ is the source current, $\dot{I}_L$ is the load current, $\dot{I}_{TC}$ is the TCR current, $\dot{I}_C$ is the PLC current, $\dot{V}_{Ch}$ is the output value of the series AF, $\dot{Z}_S$ is the source-side impedance, and $\dot{Z}_C$ is the impedance of a per-phase PLC. The three-phase load generates the fundamental reactive currents and harmonic currents. TCRs with PLCs compensate for the fundamental reactive currents. However, TCRs also generate harmonic currents. Thus, there are two current-source harmonic-producing loads, namely, the TCR and three-phase load, in Figure 5a. Harmonic currents generated by both the three-phase load and TCRs flow into PLCs and the source voltage $\dot{V}_S$. Thus, the three-phase load and TCRs are considered by a current-source harmonic-producing load. Figure 5b shows a per-phase equivalent circuit for Figure 5a, where $\dot{I}_{LTh}$ is the sum of $\dot{I}_{TCh}$ generated by the TCR and $\dot{I}_{Lh}$ generated by the three-phase load. Professor H. Fujita et al. proposed a combined system of a shunt-passive filter and series AF for the current-source harmonic-producing load, which is a large-capacity thyristor rectifier, with a novel control method of the series AF [17]. The control method proposed in [17] is applicable for Figure 4 because two current-source harmonic-producing loads, namely, the TCR and three-phase load, can be considered as a current-source harmonic-producing load, as shown in Figure 5b. The series AF performs for a resistor of K_C Ω for the harmonic source currents. The output voltage $\dot{V}_{Ch}$ of the series AF in Figure 5b is given by

$$\dot{V}_{Ch} \;=\; K_C \cdot \dot{I}_{Sh} \tag{1}$$

When no harmonics are included in the source voltage $\dot{V}_S$, a per-phase base equivalent circuit for load-side harmonic currents is shown in Figure 5c. In Figure 5c, the source-side harmonic current $\dot{I}_{Sh}$ is given by

$$\dot{I}_{Sh} \;=\; \frac{\dot{Z}_C}{K_C + \dot{Z}_S + \dot{Z}_C} \cdot \dot{I}_{LTh} \tag{2}$$

If $K_C \gg (\dot{Z}_S + \dot{Z}_C)$ in Equation (2),

$$\dot{I}_{Sh} \;=\; 0 \tag{3}$$

Thus, the sinusoidal source currents i_{Sa}, i_{Sb}, and i_{Sc} are obtained with the series AF connected to the pure PLCs. The transfer function $G(s)$ of the low-pass filter (LPF) in Figure 4 is considered in Equation (2). The transfer function $G(s)$ of the LPF is expressed as

$$G(s) \;=\; \frac{\omega_C^2}{s^2 + 2\xi\omega_C s + \omega_C^2} \tag{4}$$

where $\omega_C = 2\pi f_C$. The cut-off frequency f_C is 179 Hz, and the damping factor ζ is 0.7. With the transfer function $G(s)$, the source-side harmonic currents $\dot{I}_{Sh}$ are rewritten as

$$\dot{I}_{Sh} = \frac{\dot{Z}_C}{K_C \cdot (1 - G(s)) + \dot{Z}_S + \dot{Z}_C} \cdot \dot{I}_{LTh} \tag{5}$$

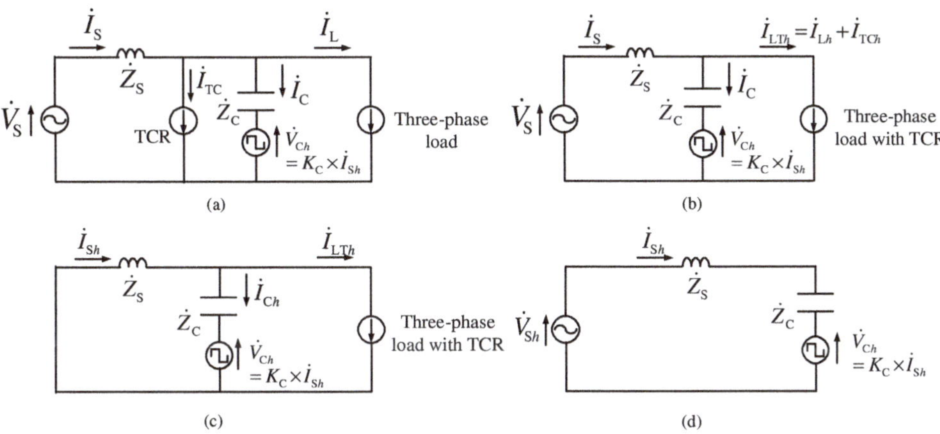

Figure 5. Per-phase equivalent circuits for Figure 4 with the control gain K_C Ω. (a) Per-phase equivalent circuit. (b) Per-phase equivalent circuit for current-source harmonic-producing load. (c) Equivalent circuit for $\dot{I}_{LTh}$. (d) Equivalent circuit for $\dot{V}_{Sh}$.

Figure 6a shows gain plots for Figure 5c, where the source-side impedance $\dot{Z}_S = 0.1$ pu. In Figure 6, f_0 is the source-voltage frequency, which is 60 Hz. As described before, the per-phase rated impedance is 6.2 Ω in Figure 4. To satisfy that $K_C \gg (\dot{Z}_S + \dot{Z}_C)$ in Equation (2), the control gain K_C is varied from 10 to 30 Ω. The line with K_C=0 Ω shows the compensation characteristics for the source-side harmonic currents when the series AF is not connected. From the second- to fifth-order components, the gains are positive. This means that the harmonic currents $\dot{I}_{LTh}$ generated by the TCRs and three-phase load are magnified on the source side. Increasing the control gain K_C suppresses the harmonic current magnifications on the source side. In particular, when K_C is 30 Ω, there is no positive area for the fundamental and harmonic components. Thus, the series AF with the control gain K_C of 30 Ω perfectly isolates the harmonic currents generated by both the TCRs and the three-phase load on the source side. The sinusoidal source currents with a unity power factor are obtained with the proposed hybrid SVC.

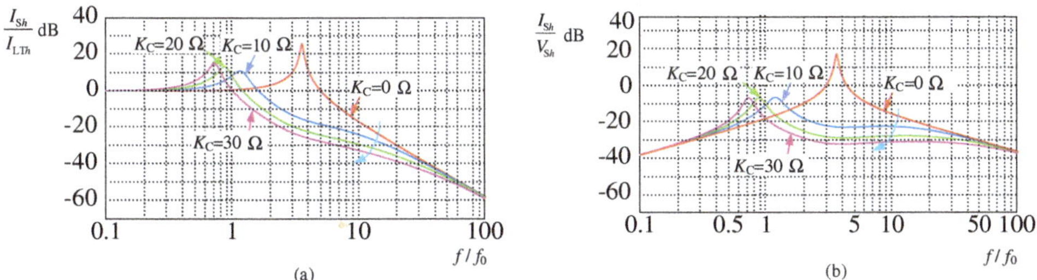

Figure 6. Gain plots for Figure 5. (a) Gain plots for Figure 5c. (b) Gain plots for Figure 5d.

Figure 5d shows a per-phase base equivalent circuit for the source-voltage harmonics $\dot{V}_{Sh}$. In Figure 5d, $\dot{I}_{Sh}$ is expressed by

$$\dot{I}_{Sh} = \frac{\dot{V}_{Sh}}{K_C \cdot (1 - G(s)) + \dot{Z}_S + \dot{Z}_C} \tag{6}$$

Figure 6b shows gain plots for Figure 5d. The line with $K_C = 0\ \Omega$ shows the compensation characteristics for the source-side harmonic voltages $\dot{V}_{Sh}$ when the series AF is not connected. From the third- to fourth-order components, the source-side harmonic current $\dot{I}_{Sh}$ flows into the PLC. The lines with $K_C = 10, 20, 30\ \Omega$ show the compensation characteristics for the source-side harmonic voltages $\dot{V}_{Sh}$ with the series AF. The source-side harmonic currents $\dot{I}_{Sh}$ are isolated by the series AF in the proposed hybrid SVC. Therefore, sinusoidal source currents with a unity power factor are obtained with the proposed hybrid SVC. We note that no fundamental source voltage appears across the series AF, and this results in a significant reduction in the required rating of the series AF.

Here, the control strategy of the series AF in the time domain, which was proposed in [17], is briefly introduced. A three-phase phase-locked loop (PLL) is used to detect the electric angle θ_C of a-phase PLC current [19]. Here, only the PLC currents are detected to generate the electric angle θ_C. The terminal voltages v_{Ta}, v_{Tb}, and v_{Tc} are not detected. Thus, any sensors for the external voltages detections are not needed in the proposed hybrid SVC shown in Figure 4. The firing angle of the TCR is also controlled using the electric angle θ_C.

The source currents i_{Sa}, i_{Sb}, and i_{Sc} are expressed as

$$
\begin{aligned}
i_{Sa} &= \sqrt{2}I_{SF}\cos(\omega_S t - \phi_F) + \sqrt{2}\sum_{h=2}^{\infty} I_{Sh}\cos(h\omega_S t - \phi_h) \\
&= i_{SaF} + i_{Sah} \\
i_{Sb} &= \sqrt{2}I_{SF}\cos\left(\omega_S t - \frac{2}{3}\pi - \phi_F\right) \\
&\quad + \sqrt{2}\sum_{h=2}^{\infty} I_{Sh}\cos\left(h\omega_S t - \frac{2}{3}h\pi - \phi_h\right) \\
&= i_{SbF} + i_{Sbh} \\
i_{Sc} &= \sqrt{2}I_{SF}\cos\left(\omega_S t + \frac{2}{3}\pi - \phi_F\right) \\
&\quad + \sqrt{2}\sum_{h=2}^{\infty} I_{Sh}\cos\left(h\omega_S t + \frac{2}{3}h\pi - \phi_h\right) \\
&= i_{ScF} + i_{Sch}
\end{aligned}
\tag{7}
$$

Three-phase source currents, i_{Sa}, i_{Sb}, and i_{Sc} are detected, and then the detected source currents are transformed into d-q coordinates. Generally, to transform i_{Sa}, i_{Sb}, and i_{Sc} into d-q coordinates, the electric angle θ_S of a-phase terminal voltage v_{Ta} is required. In Figure 4, $\theta_C + \frac{\pi}{6}$ equals θ_S. Using the detected electric angle θ_C, i_{Sd} and i_{Sq} can be calculated; i_{Sd} and i_{Sq} are given by

$$
\begin{aligned}
i_{Sd} &= \sqrt{3}I_{SF}\cos\left(\frac{2}{3}\pi + \phi_F\right) \\
&\quad + \frac{2}{\sqrt{3}}\Big\{\cos\left(\omega_S t + \frac{2}{3}\pi\right)\cdot\sum_{h=2}^{\infty} I_{Sh}\cos(h\omega_S t - \phi_h) \\
&\quad + \cos\omega_S t \cdot \sum_{h=2}^{\infty} I_{Sh}\cos\left(h\omega_S t - \frac{2}{3}h\pi - \phi_h\right) \\
&\quad + \cos\left(\omega_S t + \frac{4}{3}\pi\right)\cdot\sum_{h=2}^{\infty} I_{Sh}\cos\left(h\omega_S t + \frac{2}{3}h\pi - \phi_h\right)\Big\} \\
&= \bar{i}_{Sd} + \tilde{i}_{Sd}
\end{aligned}
$$

$$
\begin{aligned}
i_{Sq} =\ & -\sqrt{3}I_{SF}\sin\!\left(\frac{2}{3}\pi + \phi_F\right) \\[4pt]
& - \frac{2}{\sqrt{3}}\Big\{ \sin\!\left(\omega_S t + \frac{2}{3}h\pi\right) \cdot \sum_{h=2}^{\infty} I_{Sh}\cos(h\omega_S t - \phi_h) \\[4pt]
& \quad + \sin \omega_S t \cdot \sum_{h=2}^{\infty} I_{Sh}\cos\!\left(h\omega_S t - \frac{2}{3}h\pi - \phi_h\right) \\[4pt]
& \quad + \sin\!\left(\omega_S t + \frac{4}{3}\pi\right) \cdot \sum_{h=2}^{\infty} I_{Sh}\cos\!\left(h\omega_S t + \frac{2}{3}h\pi - \phi_h\right) \Big\} \\[4pt]
=\ & \bar{i}_{Sq} + \tilde{i}_{Sq}
\end{aligned}
\tag{8}
$$

The DC components $\bar{i}_{Sd}$ and $\bar{i}_{Sq}$ in *d-q* coordinates originate from the fundamental components i_{SaF}, i_{SbF}, and i_{ScF} of the source currents i_{Sa}, i_{Sb}, and i_{Sc}, respectively, in a-b-c coordinates. The AC components $\tilde{i}_{Sd}$ and $\tilde{i}_{Sq}$ in *d-q* coordinates originate from the harmonic currents i_{Sah}, i_{Sbh}, and i_{Sch} of the source currents. These AC components are extracted by high-pass filters (HPFs) with a second-order LPF. The extracted AC components are then retransformed into a-b-c coordinates. Retransforming $\tilde{i}_{Sd}$ and $\tilde{i}_{Sq}$ into a-b-c coordinates gives the source-side harmonic currents i_{Sah}, i_{Sbh}, and i_{Sch}. With the extracted source-side harmonic currents i_{Sah}, i_{Sbh}, and i_{Sch}, the reference values v^*_{AFa}, v^*_{AFb}, and v^*_{AFc} for the series AF are given by

$$
\begin{aligned}
v^*_{AFa} &= K_C \cdot i_{Sah} \\
v^*_{AFb} &= K_C \cdot i_{Sbh} \\
v^*_{AFc} &= K_C \cdot i_{Sch}
\end{aligned}
\tag{9}
$$

Therefore, the series AF performs for a resistor of K_C Ω for source-side harmonic currents i_{Sah}, i_{Sbh}, and i_{Sch}. A sine-triangle intercept technique is used to generate the gate signals for the three-phase voltage-source PWM inverter. The switching frequency f_S of the three-phase voltage-source PWM inverter is 12 kHz.

4. Simulation Results

The validity and high practicability of the proposed hybrid SVC were confirmed by digital computer simulation using PSIM software. PSIM is an electronic circuit simulation software package, designed specifically for use in power electronics and motor drive simulations [20]. PSIM is developed and released by Powersim [21].

The root-mean-square (RMS) value of the fundamental components of the load currents is 3.1 kArms. The fifth-order components of 334 Arms and seventh-order components of 100 Arms are also included in the load currents i_{La}, i_{Lb}, and i_{Lc}. Thus, the total harmonic distortion (THD) value of i_{La}, i_{Lb}, and i_{Lc} is 11.3%, respectively. The power factor is 0.7. Circuit constants for Figure 4 shown in Table 2 are used in the following simulation results. The small-rated LC filter (inducer L_f and capacitor C_f) suppresses switching ripples generated by the PWM inverter, which performs for the series AF. As shown in Figure 4, the constant DC-capacitor voltage control block is added in the control circuit for the series AF. The DC-capacitor voltage v_{DC} is controlled using the *d*-axis component in *d-q* coordinates of the PLC currents i_{Cab}, i_{Cbc}, and i_{Cca}.

Figures 7–9 show simulation results for Figure 4 before/after the series AF, which is a three-phase voltage-source PWM inverter, was started. We note that the DC-capacitor voltage v_{DC} should be 10 kVdc, which is the reference value V^*_{DC} of 10 kVdc, before the series AF is started. The constant DC-capacitor voltage control had already started before the series AF was started, as shown in Figures 7–9. Thus, the series AF was smoothly started without any transient phenomena of the small-rated LC filter (L_f and C_f).

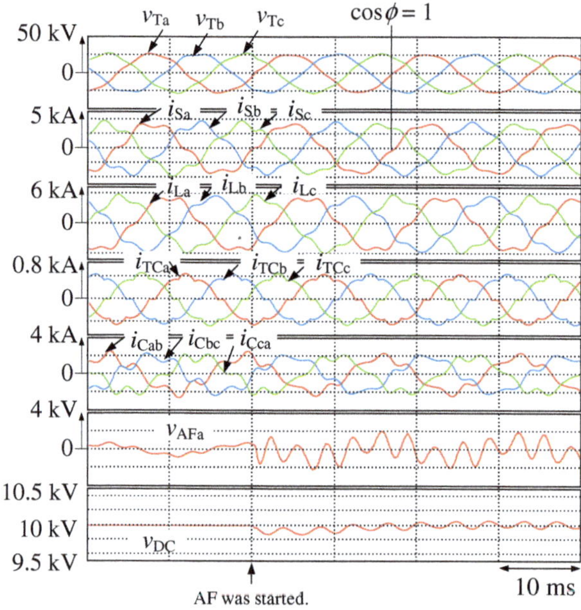

Figure 7. Simulation waveforms for Figure 4 before/after the series active filter (AF) with K_C of 10 Ω was started.

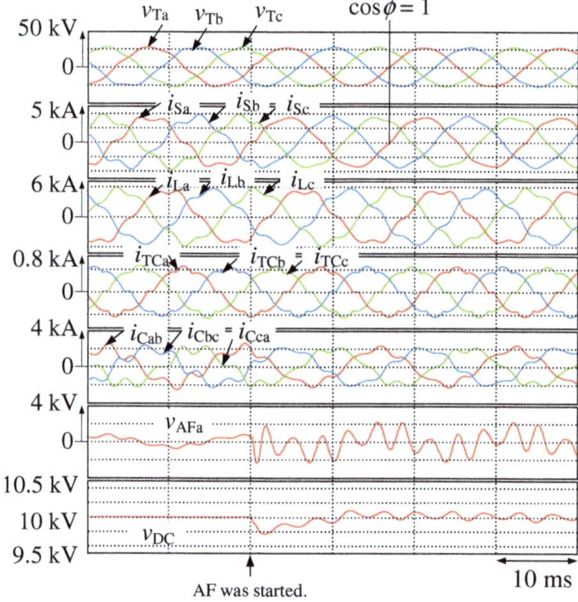

Figure 8. Simulation waveforms for Figure 4 before/after the series active filter (AF) with K_C of 20 Ω was started.

The receiving-end voltage waveforms are v_{Ta}, v_{Tb}, and v_{Tc}; i_{Sa}, i_{Sb}, and i_{Sc} are the source-current waveforms; i_{La}, i_{Lb}, and i_{Lc} are the load current waveforms; i_{TCa}, i_{TCb}, and i_{TCc} are the TCR current waveforms; i_{Cab}, i_{Cbc}, and i_{Cca} are the PLC current waveforms; v_{AFa} is an a-phase output-voltage waveform of the series AF; v_{DC} is the DC-capacitor voltage waveform.

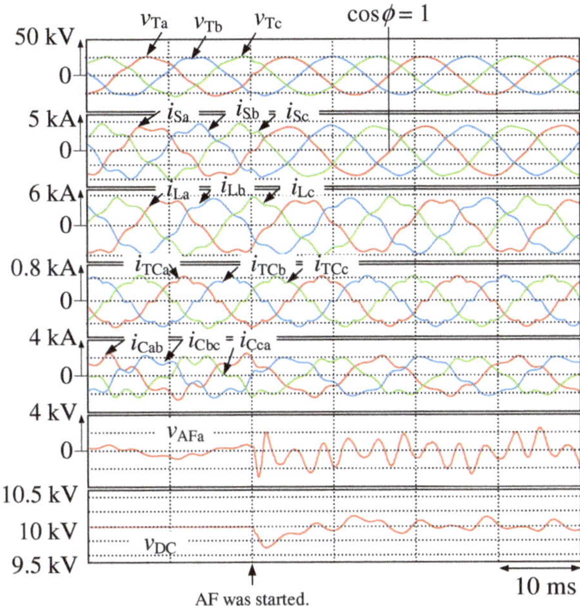

Figure 9. Simulation waveforms for Figure 4 before/after the series active filter (AF) with K_C of 30 Ω was started.

In Figures 7–9, the source currents i_{Sa}, i_{Sb}, and i_{Sc} are heavily distorted because the three-phase load and TCRs generate harmonic currents and inject these to the source side. The THD value of i_{Sa}, i_{Sb}, and i_{Sc} is 14.2%, respectively. In Figure 7, the source currents i_{Sa}, i_{Sb}, and i_{Sc}, however, were distorted after the series AF was started. The THD values of i_{Sa}, i_{Sb}, and i_{Sc} are 8.6%, 8.6%, and 8.7%, respectively. With the control gain K_C of 10 Ω, as shown in Figure 6, the insufficient compensation performance for harmonic currents was achieved. In Figure 8, the source currents i_{Sa}, i_{Sb}, and i_{Sc} were distorted after the series AF was started. The THD values of i_{Sa}, i_{Sb}, and i_{Sc} are 4.7%, 4.8%, and 4.8%, respectively. With the control gain K_C of 20 Ω, as shown in Figure 6, insufficient compensation performance for harmonic currents was achieved. In Figure 9, the source currents i_{Sa}, i_{Sb}, and i_{Sc} were sinusoidal with a unity power factor after the series AF was started. The THD values of i_{Sa}, i_{Sb}, and i_{Sc} are 3.0%, 3.1%, and 3.2%, respectively. As shown in Figure 6, the sufficient compensation performance for harmonic currents was achieved with the control gain K_C of 30 Ω. The DC-capacitor voltage v_{DC} was well controlled to its reference value $V_{DC}^* = 10$ kVdc. The ripple of the DC-capacitor voltage was 3.0% in the transient state and $\pm 1.0\%$ in the steady state.

Figure 10 shows spectra of a-phase source current i_{Sa}. When the series AF was not started, when $K_C = 0$ Ω, the RMS value of the fifth-order harmonics was 336 Arms and the seventh-order harmonics was 36.8 Arms. A higher control gain K_C achieves sufficient compensation characteristics for source-side harmonic currents.

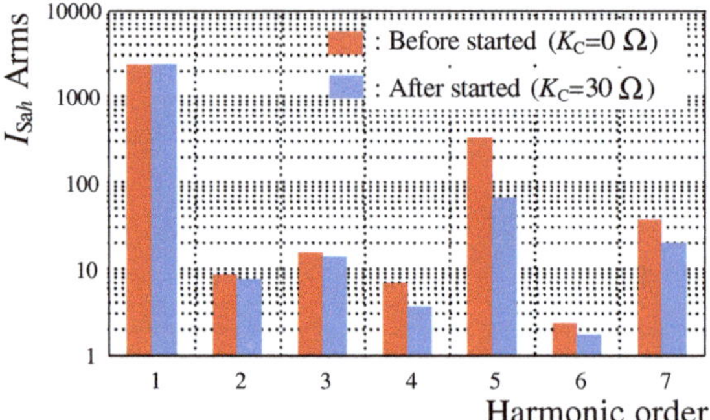

Figure 10. Spectra of a-phase source current i_{Sa}.

Figure 11 shows the simulation results for Figure 4 with the three-phase load variation from 0.6 to 1.0 pu, where the control gain K_C was 30 Ω. The per-phase base rated current was 3.1 kArms. Thus, each phase current of the three-phase load was varied from 1.86 kArms, which is 0.6 pu, to 3.1 kArms, which is 1 pu. Before and after the load variations, the source currents i_{Sa}, i_{Sb}, and i_{Sc} were sinusoidal with a unity power factor. The DC-capacitor voltage v_{DC} was well controlled to its reference value $V_{DC}^* = 10$ kVdc. The ripple of the DC-capacitor voltage was 7.0% in the transient state and $\pm$1.0% in the steady state.

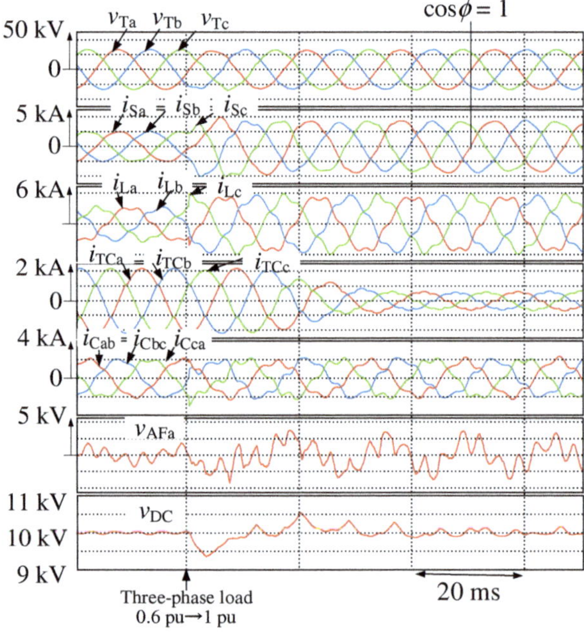

Figure 11. Simulation waveforms for Figure 4 with the three-phase load variation from 0.6 to 1 pu, when the control gain K_C was 30 Ω.

In the simulation results of Figures 7–9 and 11, no harmonic voltages in the three-phase source voltages were considered. However, harmonics are included in the true high-voltage three-phase distribution feeders [22]. As shown in Figure 6b, harmonic currents flow into the PLCs if the source voltages are distorted. Thus, it is desirable that the series AF is always under operation despite a few switching losses are caused by the series AF, which consists of IGBTs.

The required rating of the three-phase voltage-source PWM inverter, which performs for the series AF, is now calculated. The required rating S_{AF} is expressed as

$$S_{AF} = V_{AFa} \cdot I_{Cab} + V_{AFb} \cdot I_{Cbc} + V_{AFc} \cdot I_{Cca} \tag{10}$$

where V_{AFa}, V_{AFb}, V_{AFc}, I_{Cab}, I_{Cbc}, and I_{Cca} are all RMS values. From the simulation results shown in Figure 9, S_{AF} = 4.96 MVA. This is 2.8%, as compared to the rating of the three-phase load. The addition of the small-rated three-phase and low-cost voltage-source PWM inverter significantly improves the power quality on the source side.

5. Comparisons between the Newly Proposed Topology and Previously Proposed Topologies

Table 1 has summarized the previously proposed hybrid SVC topologies. It is desirable that the rating of the added power converter, which performs for the AF or STATCOM, is low from the viewpoint of the cost. Although a topology as shown in Figure 3 achieves excellent compensation characteristics of fundamental reactive and harmonic currents with a more rapid response than those of the other topologies of Figures 1 and 2, the large-capacity STATCOM with a high price is required. In the previously proposed topology shown in Figure 1, the required rating of the added AF is only 1.4% as compared to that of the three-phase load. However, LC tuned filters are required rather than pure PLCs. This means that the previously proposed topology shown in Figure 1 cannot be applied to the SVCs that are under commercial operations. The topology of Figure 2 also cannot be applied to the SVCs that are under commercial operations. Only the newly proposed topology can be applied to the SVCs that are under commercial operations, although the required rating of the series AF is slightly higher than that of the series AF in Figure 1. As described in Section 3, a three-phase PLL is used to detect the electric angle θ_C of a-phase PLC current. Only the PLC currents are detected, and any sensors for the external voltages and currents are not needed in the proposed hybrid SVC shown in Figure 4. This also demonstrates that the newly proposed topology can be applied to the SVCs that are under commercial operations. The authors thus conclude that the proposed hybrid SVC is useful and cost-effective for practical distribution feeders.

6. Conclusions

This paper has proposed a new hybrid SVC topology comprising a series AF and SVC, which consists of TCRs and pure PLCs. The series AF is connected in series to the pure PLCs. The series AF performs for a resistor for source-side harmonic currents. A sinusoidal source current with a unity power factor is obtained. Any sensors for the external voltages and currents are not needed in the proposed hybrid SVC. The basic principle of the proposed hybrid SVC has been discussed in detail. The compensation characteristics of the harmonic currents have been shown and were confirmed by digital computer simulation using PSIM software. Simulation results have demonstrated that the sinusoidal source currents with a unity power factor are obtained. From the simulation results, the required-capacity of the series AF is 2.8% as compared to the rating of the three-phase load. Only the newly proposed topology can be applied to the SVCs that are under commercial operations, although the required rating of the series AF is slightly higher than that of the series AF in [3–6]. It is thus concluded that the proposed hybrid SVC is useful and cost-effective for practical distribution feeders.

Author Contributions: Ayumu Tokiwa significantly contributed to demonstrate the basic principle of new hybrid static var compensator with series active filter and contributed to the design of the proposed circuit topology and the implementation of the digital computer simulation. Toshihiko Tanaka proposed the circuit topology and helped with the writing of this paper. Hiroaki Yamada, Makoto Watanabe, Masanao Shirai, and Yuji Teranishi were responsible for guidance and key suggestions.

Conflicts of Interest: The authors declare no conflict of interest.

References

1. Chen, J.H.; Lee, W.J.; Chen, M.S. Using a static var compensator to balance a distribution system. *IEEE Trans. Ind. Appl.* **1999**, *35*, 298–304.
2. Xu, W.; Marti, J.R.; Dommel, H.W. Harmonic analysis of systems with static compensators. *IEEE Trans. Power Syst.* **2009**, *6*, 183–190.
3. Zhou, L.; Qi, S.; Huang, L.; Zhou, Z.; Zhu, Y. Design and simulation of a control system with a comprehensive power quality in power system. In Proceedings of the Electrical Machines and Systems (ICEMS), Seoul, Korea, 8–11 October 2007; pp. 376–381.
4. Zhang, X.; Wang, Y.; Lei, W.; Yang, J.; Tang, X.; Si, W.; Hou, J. A comprehensive power quality controller for substations in power system. In Proceedings of the Applied Power Electronics Conference and Exposition, Dallas, TX, USA, 19–23 March 2006; pp. 1758–1762.
5. Luo, A.; Shuai, Z.; Zhu, W.; Shen, Z.J. Combined system for harmonic suppression and reactive power compensation. *IEEE Trans. Ind. Electron.* **2009**, *56*, 418–428.
6. Luo, A.; Peng, S.; Wu, C.; Wu, J.; Shuai, Z. Power electronic hybrid system for load balancing compensation and frequency-selective harmonic suppression. *IEEE Trans. Ind. Electron.* **2012**, *59*, 723–732.
7. Jou, H.-L.; Wu, J.-C.; Wu, K.-D. Parallel operation of passive power filter and hybrid power filter for harmonic suppression generation. *IEE Proc. Transm. Distrib.* **2001**, *148*, 8–14.
8. Senini, S.; Wolfs, P.J. Hybrid active filter for harmonically unbalanced three phase three wire railway traction loads. *IEEE Trans. Power Electron.* **2000**, *15*, 702–710.
9. Varschavsky, A.; Dixon, J.; Rotella, M.; Moran, L. Cascaded nine-level inverter for hybrid-series active power filter, using industrial controller. *IEEE Trans. Ind. Electron.* **2010**, *57*, 2761–2767.
10. Luo, A.; Tang, C.; Shuani, Z.; Zhao, W.; Rong, F.; Zhou, K. A novel three-phase hybrid active power filter with a series resonance circuit tuned at the fundamental frequency. *IEEE Trans. Ind. Electron.* **2009**, *56*, 2431–2440.
11. Corasaniti, V.F.; Barbieri, M.B.; Arnera, P.L.; Valla, M.I. Hybrid power filter to enhance power quality in a medium-voltage distribution network. *IEEE Trans. Ind. Electron.* **2009**, *56*, 2885–2893.
12. Rahmani, S.; Hamadi, A.; Al-Haddad, K.; Dessaint, L.A. A combination of shunt hybrid power filter and thyristor-controlled reactor for power quality. *IEEE Trans. Ind. Electron.* **2014**, *61*, 2152–2164.
13. Yang, J.; Wang, Y.; Duan, Y.; Fu, Z.; Wang, Z. A three-phase comprehensive reactive power and harmonics compensator based on a comparatively small rating APF. In Proceedings of the Applied Power Electronics Conference and Exposition, Anaheim, CA, USA, 22–26 February 2004.
14. Al-Mubarak, A.H.; Thorvaldsson, B.; Halonen, M.; Al-Kadhem, M.Z. Hybrid and classic SVC technology for improved efficiency and reliability in Saudi transmission grid. In Proceedings of the T&D Conference and Exposition, Chicago, IL, USA, 14–17 April 2014.
15. Peng, F.Z.; Akagi, H.; Nabae, A. A new approach to harmonic compensation in power systems-A combined system of shunt passive and series active filters. *IEEE Trans. Ind. Appl.* **1990**, *26*, 983–990.
16. Peng, F.Z.; Akagi, H.; Nabae, A. Compensation characteristics of the combined system of shunt passive and series active filters. *IEEE Trans. Ind. Appl.* **1993**, *29*, 144–152.
17. Fujita, H.; Akagi, H. A practical approach to harmonic compensation in power systems-series connection of passive and active filters. *IEEE Trans. on Ind. Appl.* **1991**, *27*, 1020–1025.
18. Konishi, S.; Baba, K.; Daiguji, M. Static var compensator. *Fuji Electr. J.* **2001**, *74*, 289–295. (In Japanese)
19. Arruda, L.N.; Silva, S.M.; Filho, B.J.C. PLL structure for utility connected systems. In Proceedings of the Industry Applications Conference, Chicago, IL, USA, 30 September–4 October 2001; pp. 2655–2660.
20. Jin, H. Computer simulation of power electronic circuits and systems using PSIM. In Proceedings of the International Conference on Power Electronics (ICPE), Seoul, Korea, 10–14 October 1995; pp. 79–84.
21. PSIM Accelerates Your Pace of Innovation. Available online: https://powersimtech.com (accessed on 16 October 2017).
22. *IEEE 519-2014, Recommended Practices and Requirements for Harmonic Control in Electrical Power System*; IEEE: New York, NY, USA, 2014.

Article

Reactive Power and Current Harmonic Control Using a Dual Hybrid Power Filter for Unbalanced Non-Linear Loads

Leonardo Rodrigues Limongi, Fabricio Bradaschia, Calebe Hermann de Oliveira Lima and Marcelo Cabral Cavalcanti *

Department of Electrical Engineering, Federal University of Pernambuco, Recife 50740-550, Brazil; leonardo.limongi@ufpe.br (L.R.L.); fabricio.bradaschia@ufpe.br (F.B.); calebe_03@hotmail.com (C.H.d.O.L.)
* Correspondence: marcelo.ccavalcanti@ufpe.br; Tel.: +55-81-99215-8083

Received: 25 April 2018; Accepted: 28 May 2018; Published: 30 May 2018

Abstract: An important power quality issue is related to current harmonic components demanded by non-linear loads. A solution to mitigate this issue is to use hybrid power filters (HPFs), that apply low power active filters with passive filters. Some dual-converter topologies have been shown to be attractive due to a better compensation performance compared with single filters, where the HPFs give a reactive power support (an extra feature) together with harmonic compensation. On the other hand, the drawback of dual converters is the high number of active switches. Besides that, due to the high number of unbalanced non-linear loads connected to the electrical grid, triplen harmonics can appear. However, traditional HPFs do not compensate triplen harmonics, which usually have considerable values. Therefore, in this paper, a dual HPF based on the nine-switch inverter (DHPF-NSI) is proposed to compensate current harmonics and to provide reactive power support. The NSI presents a reduced number of switches when compared with classical dual topologies. The compensation of the third harmonic caused by unbalanced nonlinear loads was also inserted in the control system. Experimental results are presented for the DHPF-NSI in order to demonstrate the reactive power and harmonic compensation performances.

Keywords: hybrid power filter; power quality; reactive power

---

1. Introduction

Disturbances that compromise the power quality may be classified as phenomena that can affect the voltage and/or the current of the grid, such as voltage sags, swells, spikes, interruptions, harmonic distortion, and fluctuations [1]. One of the most important issues related to power quality is the excessive current harmonics in the grid caused by local non-linear loads. The current harmonics could damage the conductors at high temperatures and overload the electrical system. Another problem that impairs the power quality is the voltage drop in the electrical grid caused by the excess of reactive power demanded by local inductive loads. Most loads have inductive characteristics such as motors, transformers, and industrial furnaces that require reactive power to operate. However, the reactive power does not produce work and the consequences of excess reactive power in the grid include a loss of electric energy as heat, voltage drops, underutilization of installed capacity, new loads being preventing from being connected without a further expansion of the utility grid, and many others.

The solutions to these power quality issues are generally classified according to the electrical supply voltage level, i.e., at transmission and distribution levels. The Flexible Alternating Current Transmission System (FACTS) consists in power conditioning devices usually employed to compensate reactive power at transmission level, guaranteeing the stability of the electrical grid [2]. On the other hand, Custom Power Devices (CPD) are based on the use of static controllers at distribution systems for electric utilities to supply value-added power with the reliability and power quality requested by customers [3–5]. Among the available CPDs, the passive filters, shunt Active Power Filters (APFs), Hybrid Power Filters (HPFs), and Unified Power Quality Conditioners (UPQCs) are commonly used solutions for current harmonic compensation, having the reactive power support as a possible extra feature.

The passive filters, composed of inductors and capacitors, are commonly used for current harmonic compensation and power factor correction [6]. Those passive filters are tuned in the frequency of the harmonic to be compensated, creating a low impedance path that absorbs the currents demanded by the non-linear load. However, this solution is not viable for several reasons:

- In order to mitigate all the current harmonic components generated by unbalanced non-linear loads, a great number of passive filters must be used for each harmonic to be compensated.
- Some problems related to the peak resonance of the filter with the electrical grid, a phenomenon known as harmonic-amplifying, can occur [7–9]. This phenomenon happens because the passive filters drain the current harmonics not only from the local non-linear loads but also from other non-linear loads connected to the grid.
- Depending on the power level, passive filters are bulkier and heavier and present inferior performance when compared with active solutions, such as active power filters and hybrid power filters.

The advancement of controlled semiconductor devices enable the development of shunt active power filters, which are considered a superior solution for current harmonics mitigation and for power factor correction in low and medium power when compared with passive filters [10–16]. The shunt active filters have an adjustable compensation performance for both current harmonic and reactive power, being able to change the family of harmonic components to be compensated depending on the load connected locally. The conventional topology consists of a three-phase voltage source inverter connected to the Point of Common Coupling (PCC) through an inductive filter. The shunt active filter supplies the current harmonics required by the non-linear load, while the grid only supplies the fundamental current. Nevertheless, those filters have power switches that increase the cost, making them more expensive than passive filters. Besides that, the shunt active filter needs a control system to activate the switches by performing current and voltage measurements, increasing its complexity.

Hybrid power filters have been proposed as a low-cost alternative to the shunt active filters [7,8,17–22]. HPFs are composed of the combination of one or more passive filters with an active filter of reduced power. In the HPFs, the passive filter is tuned to provide a specific current harmonic component consumed by the load. The active filter aims to compensate the harmonic components near the resonant frequency of the passive filter and to perform the reactive power compensation, in addition to avoiding the harmonic-amplifying phenomenon caused by the passive filter. Another advantage of HPFs is that the power switches have lower installed power when compared to conventional active filters.

In the literature, one of the first hybrid filters has been presented by Peng, Akagi, and Nabae [17], in which a combination of a series active filter and a shunt passive filter has been performed. This configuration provided a good compensation of harmonics. Fujita and Akagi [7] proposed a hybrid filter using a series active filter together with a passive filter. In this configuration, the active filter can dampen the resonant peak between the grid and the passive filter. As the passive filter assists in the filtering of current harmonics, the power of the active filter is reduced when compared to the configuration using only the active filter. In this case, the switches have to withstand less voltage than that in the shunt active filters because the passive filter capacitors divide the line voltage with the active parts of the converter. Therefore, the configuration proposed in [7] is a topology with low design cost due to power reduction of the active parts. Srianthumrong and Akagi [9] used an HPF with a reduced amount of passive elements. This configuration is composed of a voltage source inverter and only one passive LC filter tuned to the seventh harmonic component. This topology has low cost and volume, due to the reduction of some passive elements, when compared to the previous configurations. Besides the current harmonics compensation, HPFs have been also given a reactive power support, correcting the power factor of the load as an extra feature [18,23,24].

Other important active filtering solution comes from the idea of using dual converters configurations [19,20,25–27]. The main advantage of using such configurations comes from their enhanced compensation capabilities. An HPF configuration using two converters, known as back-to-back converter, has been developed in [26]. The back-to-back configuration consists of two voltage source inverters sharing the same dc-link and it has two three-phase output sets, one being responsible for current harmonics compensation and the other responsible for the reactive power control. The purpose of this HPF topology is to improve the harmonic and reactive power compensations. For the power factor correction, the HPF has a great advantage in relation to the capacitor banks, since it is capable of fine-tuning the reactive consumption [19].

The nine-switch inverter (NSI) has been proposed by Liu et al. [28]. This converter uses three fewer switches and has the same number of output terminals in comparison to the back-to-back configuration. With the reduction in the number of switches, this topology has a low design cost. Limongi et al. [27] presented a dual HPF configuration based on the NSI, as can be seen in Figure 1. The NSI can be divided into two parts. The first part is the top unit that is composed of the upper and intermediate switches, with the terminals A, B, and C. The second part is the bottom unit that is composed of the lower and intermediates switches, with the terminals R, S, and T, as shown in Figure 1. The purpose of this configuration is to connect the two sets of three-phase outputs with two passive LC filters for the current harmonics compensation. Originally, the top unit was responsible for mitigating the 5th and 7th harmonic components and controlling the dc-link voltage, while the bottom unit was responsible for compensating the 11th and 13th harmonics. However, this configuration is not able to compensate the triplen current harmonics (3rd, 9th, etc.) caused by unbalanced non-linear loads, and its control system has not been designed to compensate reactive power.

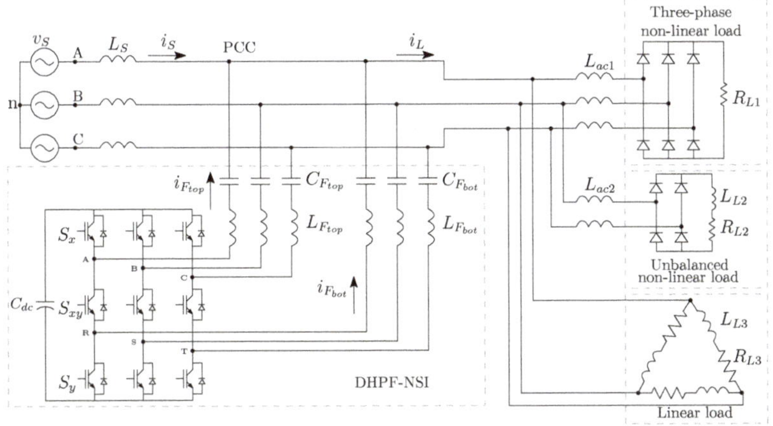

Figure 1. Dual HPF based on the NSI (DHPF-NSI) connected to a three-phase electrical system.

In this paper, two extra features are proposed to the dual HPF based on the NSI (named here DHPF-NSI, shown in Figure 1): the compensation of triplen current harmonics and the reactive power support. The reason for using a dual HPF is because dual filters have a better compensation capacity than single filters, necessary to compensate triplen current harmonics, and HPFs have lower installed power when compared with APFs. The NSI is the chosen topology, since it presents a lower number of switches than back-to-back converters, implying a lower cost. As mentioned before, unbalanced non-linear loads connected to the electrical grid consume typical triplen harmonics currents. For this reason, this work inserts a current compensation for the 3rd harmonic component. In the proposed control system, the top unit is responsible for mitigating the 3rd and 5th harmonic components, while the bottom unit is responsible for mitigating the 7th, 9th, and 11th harmonic components. Both units are also used for compensating the reactive power consumed by the loads and to control the dc-link voltage. In order to control the reactive power, the mathematical model of DHPF-NSI and the controller design are also developed. The mathematical modeling and the controller design are the main original contributions of this paper. In order to validate the capability of the DHPF-NSI to compensate current triplen harmonics and to provide reactive power support, experimental results are obtained in a prototype built in laboratory.

2. Dual Hybrid Power Filter

The dual HPF based on the NSI [27] connected to a three-phase electrical system is shown in Figure 1. There are three different loads connected to the system: (1) a linear load to absorb reactive power, imposing a low power factor; (2) a three-phase non-linear load that produces mainly 5th, 7th, 11th, and 13th harmonics and that causes distortion in the grid currents; (3) a single-phase non-linear load that produces mainly 3rd harmonics and that becomes dominant with great contribution to the total harmonic distortion. The harmonics produced by the non-linear loads make the grid current harmonics levels above the recommended limits [29]. Therefore, the dual HPF is connected at the PCC to provide harmonics currents and to compensate reactive power. Each HPF unit has output terminals connected in series with an LC passive filter, where the LC filters for each unit are tuned in different frequencies.

2.1. Nine-Switch Inverter Analysis

The NSI proposed by Liu et al. [28] is an alternative to the back-to-back inverter, having the advantage of a reduced number of switches. The upper switches are called S_A, S_B, and S_C, and the lower switches are called S_R, S_S, and S_T. The intermediate switches are shared between the two units (top and bottom) of the NSI. Table 1 shows the possible switching states for leg AR and the respective output voltages [27,28].

Table 1. Possible switching states and the respective output voltages for leg AR of the NSI.

NSI Switching States	S_A	S_{AR}	S_R	v_{Ao}	v_{Ro}
1	1	1	0	$\frac{v_{dc}}{2}$	$\frac{v_{dc}}{2}$
2	0	1	1	$\frac{v_{dc}}{2}$	$\frac{-v_{dc}}{2}$
3	1	0	1	$\frac{-v_{dc}}{2}$	$\frac{-v_{dc}}{2}$

From the possible combinations shown in Table 1, the top and bottom switches duty cycles are given by [27]:

$$D_x = \frac{3}{4} + \frac{\hat{V}_x \sin\left(\omega_x t + \varphi_x\right)}{v_{dc}} \tag{1}$$

$$\bar{D}_y = \frac{1}{4} + \frac{\hat{V}_y \sin\left(\omega_y t + \varphi_y\right)}{v_{dc}}. \tag{2}$$

The gate signals of the switches S_x and S_y are defined by Equations (1) and (2), respectively, and the gate signals of the switches S_{xy} are determined by the logic operation XOR between the gate signals of S_x and S_y.

2.2. Passive Filter Analysis

The peak voltages of the capacitors of the filters are evaluated from two analyses: the maximum dc voltage and the maximum fundamental ac voltage, across the filter. The analysis of the dc equivalent circuit is made in a steady state, the same as performed in [27]. The inductors L_{top} and L_{bot} and the grid voltages $v_{S_{ABC}}$ can be considered as short circuits. Therefore, the dc voltage components in each capacitor are given by

$$v_{C_{top}} = -\left(\frac{C_{bot}}{C_{top} + C_{bot}}\right)\frac{v_{dc}}{2} \tag{3}$$

$$v_{C_{bot}} = \left(\frac{C_{top}}{C_{top} + C_{bot}}\right)\frac{v_{dc}}{2}. \tag{4}$$

The equivalent circuit for the fundamental frequency is shown in Figure 2. In the fundamental frequency analysis, the inductors of the passive filter and grid are disregarded, so their reactances in the fundamental frequency are negligible in comparison to the capacitive reactance of the passive filter. The resistance value is low so that the voltage drop is also negligible. The capacitors ac voltages $v_{AC_{top}}$ and $v_{AC_{bot}}$ should be considered as the difference between the grid voltage and the output voltages synthesized by the NSI, $v_{F_{top}}$ and $v_{F_{bot}}$), as follows:

$$v_{AC_{top}} = \hat{V}_S - v_{F_{top}} \tag{5}$$

where $v_{F_{top}}$ has the amplitude range given by

$$-\frac{v_{dc}}{4} \leq v_{F_{top}} \leq \frac{v_{dc}}{4}. \tag{6}$$

The analysis of the bottom capacitor voltage is similar to the top capacitor voltage.

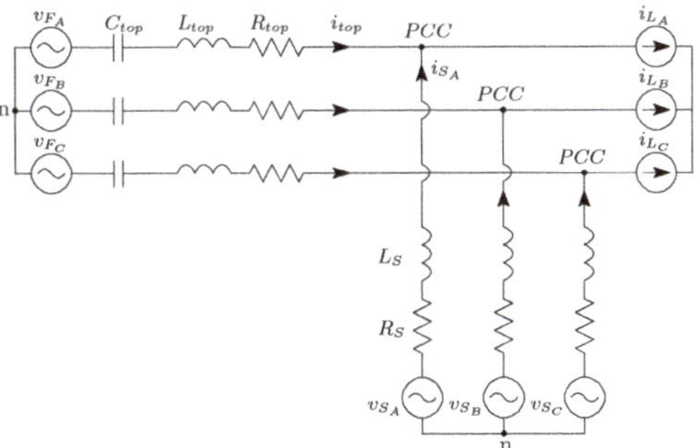

Figure 2. Equivalent circuit for the fundamental ac voltage.

Another important requirement in the design of the passive filter is the reactive power supplied by the dual HPF based on the NSI. The reactive power supplied to the grid depends directly on the capacitance values and indirectly on the dc-link voltage. Therefore, the three-phase reactive power that the dual HPF can provide to the grid is given by

$$Q_{3\phi} = Q_{3\phi_{top}} + Q_{3\phi_{bot}} = 3\omega(C_{top} + C_{bot})V_S^{rms}(V_S^{rms} - v_F^{rms}) \tag{7}$$

where ω is the fundamental angular frequency of the electrical grid.

3. Dual Hybrid Power Filter Mathematical Model for Reactive Power Control

The strategy of the reactive power control is based on the instantaneous reactive power theory proposed by Akagi et al. [30]. The reactive power can be written in the rotating dq reference frame according to

$$Q = v_{PCC_q}\bar{i}_{Sd} - v_{PCC_d}\bar{i}_{Sq} \tag{8}$$

where v_{PCC_d} and v_{PCC_q} are the dq-axis PCC voltages, i_{Sd} and i_{Sq} are the dq-axis grid currents. The signal of Equation (8) is changed in relation to that proposed in [30] for better coupling between the instantaneous and conventional power theories. Thus, the positive reactive power is related to inductive loads and negative reactive power is related to capacitive loads. In this paper, the voltage v_{PCC_d} is in phase with the grid voltage, i.e., $v_{PCC_q} = 0$. Thus, it is observed that the reactive power control can be performed by controlling i_{Sq}.

The HPF modeling is developed from the equivalent circuit for reactive power control shown in Figure 2:

$$\vec{i}_{F_{top ABC}} = \vec{i}_{L_{ABC}} - \vec{i}_{S_{ABC}} \tag{9}$$

$$\begin{aligned} \vec{v}_{F_{ABC}} = &\vec{v}_{S_{ABC}} + R_{top}\vec{i}_{top ABC} + L_{top}\frac{d\vec{i}_{top ABC}}{dt} \\ &+ \frac{1}{C_{top}} \int \vec{i}_{F_{top ABC}} dt - R_S\vec{i}_{S_{ABC}} - L_S\frac{d\vec{i}_{S_{ABC}}}{dt}. \end{aligned} \tag{10}$$

Using Equation (9) in Equation (10) yields

$$\vec{v}_{F_{ABC}} = \vec{v}_{S_{ABC}} + P_L(t) + P_S(t). \tag{11}$$

The terms $P_L(t)$ and $P_S(t)$ are written as follows:

$$P_L(t) = R_{top}\vec{i}_{L_{ABC}} + L_{top}\frac{d\vec{i}_{L_{ABC}}}{dt} + \frac{1}{C_{top}} \int \vec{i}_{L_{ABC}} dt \tag{12}$$

$$P_S(t) = -R_{eq}\vec{i}_{S_{ABC}} - L_{eq}\frac{d\vec{i}_{S_{ABC}}}{dt} - \frac{1}{C_{top}} \int \vec{i}_{S_{ABC}} dt \tag{13}$$

where $R_{eq} = R_S + R_{top}$ and $L_{eq} = L_S + L_{top}$.

To simplify the abc reference frame to the $\alpha\beta$ reference frame transformations, all terms in Equation (11) are derived so that the terms with integrals are eliminated. Since Clarke's inverse transformation matrix is composed of constant elements in relation to time [31], the derivative of this matrix is null. Therefore, Equation (11) is rewritten as

$$\frac{d\vec{v}_{F_{\alpha\beta}}}{dt} = \frac{d\vec{v}_{S_{\alpha\beta}}}{dt} + P_L^{\alpha\beta}(t) + P_S^{\alpha\beta}(t) \tag{14}$$

where the terms $P_L^{\alpha\beta}(t)$ and $P_S^{\alpha\beta}(t)$ are given by

$$P_L^{\alpha\beta}(t) = T_{\alpha\beta}\frac{dP_L(t)}{dt} = R_{top}\frac{d\vec{i}_{L_{\alpha\beta}}}{dt} + L_{top}\frac{d^2\vec{i}_{L_{\alpha\beta}}}{dt^2} + \frac{1}{C_{top}}\vec{i}_{L_{\alpha\beta}} \tag{15}$$

$$P_S^{\alpha\beta}(t) = T_{\alpha\beta}\frac{dP_S(t)}{dt} = -R_{eq}\frac{d\vec{i}_{S_{\alpha\beta}}}{dt} - L_{eq}\frac{d^2\vec{i}_{S_{\alpha\beta}}}{dt^2} - \frac{1}{C_{top}}\vec{i}_{S_{\alpha\beta}}. \tag{16}$$

As the control is performed in the dq rotating reference frame, the dq voltages and currents are obtained from Park's transformation [32]:

$$\vec{f}_{dq}(t) = e^{-j\theta}\vec{f}_{\alpha\beta}(t) \tag{17}$$

where $\vec{f}(t)$ can assume any three-phase voltage or current as a function of time.

Using Equation (17) in Equation (14):

$$\frac{de^{j\theta}\vec{v}_{F_{dq}}}{dt} = \frac{de^{j\theta}\vec{v}_{S_{dq}}}{dt} + P_L^{dq}(t) + P_S^{dq}(t) \tag{18}$$

where $P_L^{dq}(t)$ and $P_S^{dq}(t)$ are given by

$$P_L^{dq}(t) = R_{top}\frac{de^{j\theta}\vec{i}_{L_{dq}}}{dt} + L_{top}\frac{d^2 e^{j\theta}\vec{i}_{L_{dq}}}{dt^2} + \frac{1}{C_{top}}e^{j\theta}\vec{i}_{L_{dq}} \tag{19}$$

$$P_S^{dq}(t) = -R_{eq}\frac{de^{j\theta}\vec{i}_{S_{dq}}}{dt} - L_{eq}\frac{d^2 e^{j\theta}\vec{i}_{S_{dq}}}{dt^2} - \frac{1}{C_{top}}e^{j\theta}\vec{i}_{S_{dq}}. \tag{20}$$

Using the derivatives $\frac{de^{j\theta}\vec{f}(t)}{dt}$ and $\frac{d^2 e^{j\theta}\vec{f}(t)}{dt^2}$ in Equation (18) and eliminating the term $e^{j\theta}$ from this equation, the terms of the voltages and currents in the frequency domain (s) are

$$j\omega\vec{V}_{F_{dq}} + s\vec{V}_{F_{dq}} = j\omega\vec{V}_{S_{dq}} + s\vec{V}_{S_{dq}} + P_L^{dq}(s) + P_S^{dq}(s) \tag{21}$$

where $P_L^{dq}(s)$ and $P_S^{dq}(s)$ are given by

$$P_L^{dq}(s) = R_{top}\left(j\omega\vec{I}_{L_{dq}} + s\vec{I}_{L_{dq}}\right) + \frac{1}{C_{top}}\vec{I}_{L_{dq}} + L_{top}\left(-\omega^2\vec{I}_{L_{dq}} + j2\omega s\vec{I}_{L_{dq}} + s^2\vec{I}_{L_{dq}}\right) \tag{22}$$

$$P_S^{dq}(s) = -R_{eq}\left(j\omega\vec{I}_{S_{dq}} + s\vec{I}_{S_{dq}}\right) - \frac{1}{C_{top}}\vec{I}_{S_{dq}} - L_{eq}\left(-\omega^2\vec{I}_{S_{dq}} + j2\omega s\vec{I}_{S_{dq}} + s^2\vec{i}_{S_{dq}}\right). \tag{23}$$

The model obtained in Equation (21) is compared to the complete system (Figure 1) through simulations performed in MATLAB/Simulink. For the simulations, all non-linear loads in the system have been disconnected making the load current (I_L) and $P_L^{dq}(s)$ null. The grid voltage V_S and inverter output voltage V_F are in phase with the d-axis, guaranteeing that the q-axis voltage components are null. In this paper, V_{Sd} is a not null constant and V_{Fd} is constant. However, the reference value of V_{Fd} can change for the reactive power control.

The currents of the model obtained in Equation (21) and the real system currents during the transient period are seen in Figure 3. This transient period occurs due to a step in V_{Fd}. The currents I_{Sd} and I_{Sq} obtained from the mathematical model and the real system are similar and both stabilize and converge to the same reference. Therefore, it can be stated that the mathematical model is consistent with the real system dynamics.

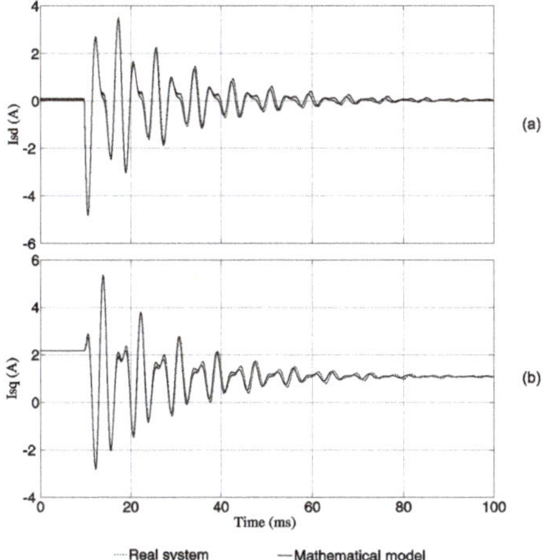

Figure 3. Dynamics of the model obtained in Equation (21) and the real system: (**a**) Current I_{Sd}; (**b**) Current I_{Sq}.

In Figure 3, it is observed that the current I_{Sq} changes with the step in V_{Fd}. Therefore, it is possible to control the current I_{Sq} through the voltage V_{Fd}. Consequently, the reactive power control can be realized by controlling the current I_{Sq}. The transfer function of the plant obtained in Equation (21) is given by

$$G(s) = \frac{I_{Sq}}{V_{Fd}} = \frac{\dfrac{-\omega}{L_{eq}}}{s^2 + \dfrac{R_{eq}}{L_{eq}}s + \dfrac{\dfrac{1}{C_{top}} - L_{eq}\omega^2}{L_{eq}}} = \frac{K}{R(s)}. \tag{24}$$

4. Hybrid Power Filter Control System

The control system for the HPF based on the NSI to compensate the harmonic components of currents and the reactive power consumed by the loads is shown in Figure 4. This control is divided in two subsystems: The first subsystem is responsible for the fundamental components control (dc-link voltage control fixed in a reference value and reactive power control). The second subsystem is responsible for the harmonic components control [9].

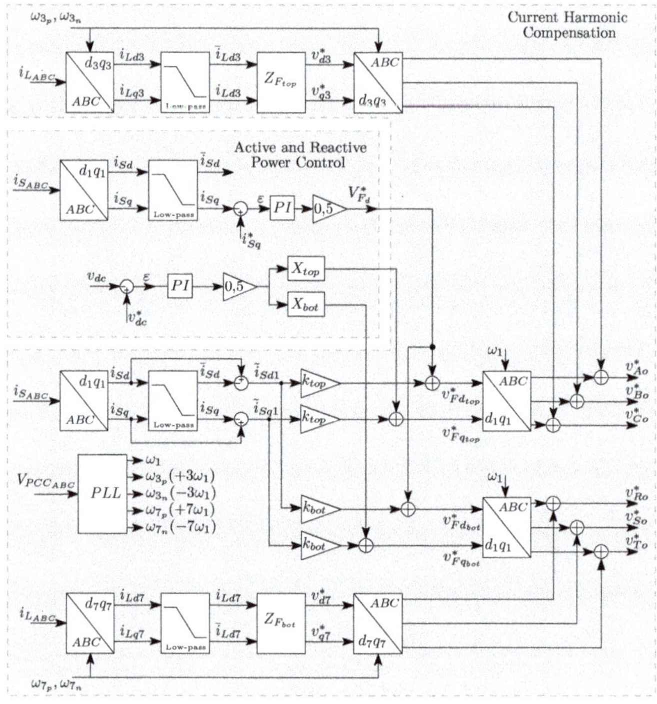

Figure 4. Control diagram of the DHPF-NSI with fundamental and harmonic components control.

4.1. Active and Reactive Power Control

4.1.1. Reactive Power Control

A closed loop analysis is performed by utilizing a Proportional-Integral (PI) controller with unitary feedback, and the transfer function of the controller is given by

$$C(s) = \frac{k_{pQ}s + k_{iQ}}{s} = \frac{Z(s)}{L(s)} \tag{25}$$

where k_{pQ} and k_{iQ} are the proportional and integral gains of the PI controller, respectively. Thus, the closed-loop transfer function can be obtained as follows:

$$\frac{I_{S_q}}{I_{S_q}^*} = \frac{C(s)G(s)}{1 + C(s)G(s)} = \frac{KZ(s)}{L(s)R(s) + KZ(s)} = \frac{B(s)}{Q(s)} \tag{26}$$

where $I_{S_q}^*$ is the reference value of I_{S_q} in the frequency domain. k_{pQ} and k_{iQ} gains is designed using the polynomial roots Q(s) of the closed-loop transfer function expressed by Equation (26), which is given by

$$Q(s) = s^3 + \frac{R_{eq}}{L_{eq}}s^2 + \frac{\frac{1}{C_{top}} - L_{eq}\omega^2 - \omega k_{pQ}}{L_{eq}}s - \frac{\omega k_{iQ}}{L_{eq}}. \tag{27}$$

The polynomial in Equation (27) has three roots due to the two poles of the plant in Equation (24) and one pole of the controller in Equation (25). The criterion for the determination of the gains is the pole placement design [33], the same as performed in [34]. Therefore, the required polynomial A(s) has two complex conjugated poles (with aT and ω_d being the real and imaginary parts, respectively), two complex poles, and one real pole. The polynomial A(s) required by the closed-loop system is given by

$$A(s) = (s + (aT + j\omega_d))(s + (aT - j\omega_d))(s + a) \tag{28}$$

where T is the relation between the position of the real part of the complex conjugated poles and the real pole.

Using the polynomial coefficients in Equations (27) and (28), the following relations are obtained:

$$T = \frac{1}{2}\left(\frac{R_{eq}}{a.Leq} - 1\right) \tag{29}$$

$$k_{pQ} = \frac{1}{\omega}\left[\frac{1}{C_{top}} - Leq\left(w^2 + a^2T^2 + \omega_d^2 + 2a^2T\right)\right] \tag{30}$$

$$k_{iQ} = -\frac{Leq}{\omega}(a^3T^2 + a.\omega_d^2). \tag{31}$$

The variable T is defined from the relation obtained in Equation (29). The gains k_{pQ} and k_{iQ} are calculated based on Equations (30) and (31).

The position of the real pole is defined from the stabilization time of the plant at 2%, which is given by

$$t_{2\%} = \frac{4}{a}. \tag{32}$$

4.1.2. Dc-Link Voltage Control

The top unit is responsible for controlling the dc-link voltage in a desirable value. The control strategy is the same as that used in [27], where the dc-link voltage control is based on the assumption that the power on the dc side is the same as the active power on the ac side of the converter. Therefore, if the dc-link voltage remains constant, the average value of the current that passes through the dc capacitor must be null. The resulting model for dc-link voltage control is based on an ideal capacitor and the closed-loop transfer function is given by

$$G_{CL}(s) = \frac{V_{dc}}{V_{dc}^*} = \frac{k_{cv}}{C}\frac{k_p s + k_i}{s^2 + \frac{k_p k_{cv}}{C}s + \frac{k_i k_{cv}}{C}}. \tag{33}$$

The controller is a conventional PI to cancel the error between the desired reference voltage V_{dc}^* and the measured dc-link voltage V_{dc}.

4.2. Current Harmonic Compensation

4.2.1. Feedback Control

In this control, the grid harmonic currents components ($\tilde{i}_{Sd}$ and $\tilde{i}_{Sq}$) are the current references of the proportional feedback control, as proposed in [9]. Thus, each extracted current harmonic is multiplied by a proportional gain (K), so that the reference voltages can be written as follows:

$$v^*_{Fdq} = K\tilde{i}_{Sdq} \tag{34}$$

where v^*_{Fdq} are the reference voltages produced by the feedback control.

4.2.2. Top and Bottom Feedforward Controls

The feedforward action is inserted into the control system for the improvement in harmonic current compensation. In the top unit, the feedforward control compensates the third harmonic of the load currents. The feedforward action of this unit has the function of creating a low impedance path for the third harmonic, preventing this harmonic from circulating through the grid. The reference voltages of the feedforward control to compensate the third harmonic are obtained as proposed in [9]:

$$\vec{v}^*_{dq3} = R_{F_{top}} + j\omega_3 L_{F_{top}} - \frac{1}{\omega_3 C_{F_{top}}}\vec{i}_{L_{dq3}} \tag{35}$$

where $X_{top} = \omega_3 L_{F_{top}} - \frac{1}{\omega_3 C_{F_{top}}}$.

In the bottom unit, the feedforward control compensates the seventh harmonic of the load currents. The feedforward action of this unit has the function of creating a low impedance path for the seventh harmonic, preventing this harmonic from circulating through the grid. The reference voltages of the feedforward control to compensate the seventh harmonic are obtained as

$$\vec{v}^*_{dq7} = R_{F_{bot}} + j\omega_7 L_{F_{bot}} - \frac{1}{\omega_7 C_{F_{bot}}}\vec{i}_{L_{dq7}} \tag{36}$$

where $X_{bot} = \omega_7 L_{F_{bot}} - \frac{1}{\omega_7 C_{F_{bot}}}$.

The reference voltages of the reactive power, dc-link voltage, feedback, and feedforward (third harmonic) controls are added for composing the reference voltages of the top unit. The reference voltages of the reactive power and feedforward (seventh harmonic) controls are added for composing the reference voltages of the bottom unit.

5. Experimental Prototype

A DHPF-NSI prototype was built using an insulated-gate bipolar transistor (IGBT) with a switching frequency of 20 kHz [35,36]. The general diagram of the experimental setup is shown in Figure 5. The hardware platform used to control the Dual Hybrid Power Filter inverter is a dSPACE development modular system based on a DS1005 processor board [37] and several boards for each special hardware task, i.e., a DS5101 board for Pulse-Width Modulation (PWM) generation [38], a DS2004 board for Analog/Digital (A/D) conversion [39], and a DS4002 board for Digital Input/Output (I/O) [40]. All boards are hosted in a dSpace PX10 expansion box that uses the DS817 board for bidirectional communication with a Personal Computer (PC) through optical fibers. The signals measured from the system are the grid currents $i_{S_{ABC}}$, load currents $i_{L_{ABC}}$, filter currents $i_{F_{ABC}}$, and line-to-line voltages $v_{S_{ABC}}$. A general view of the experimental test bench is shown in Figure 6. The prototype parameters can be seen in Table 2. The top unit LC filter is tuned to 300 Hz to mitigate the 5th harmonic and to contribute to low order current harmonics compensation. The bottom unit

LC filter is tuned to 660 Hz to mitigate the 11th harmonic and to contribute to high-order current harmonics compensation.

Table 2. Prototype parameters.

Parameter	Symbol	Value
Line-to-line grid voltage (rms)	$\hat{V}_S$	220 V
Grid frequency	f_R	60 Hz
Switching frequency	f_S	20 kHz
Dc-link voltage	v_{dc}	260 V
Dc-link capacitor	C_{dc}	4700 µF
Top filter capacitor (5th harmonic)	$C_{F_{top}}$	46 µF
Top filter inductor (5th harmonic)	$L_{F_{top}}$	6.12 mH
Top filter Resistor (5th harmonic)	$R_{F_{top}}$	620 mΩ
Bottom filter capacitor (11th harmonic)	$C_{F_{bot}}$	46 µF
Bottom filter inductor (11th harmonic)	$L_{F_{bot}}$	1.26 mH
Bottom filter resistor (11th harmonic)	$R_{F_{bot}}$	300 mΩ
Nonlinear three-phase load input inductor	L_{ac_1}	5 mH
Nonlinear three-phase load dc-link resistor	R_{L1}	31 Ω
Unbalance load input inductor	L_{ac_2}	7 mH
Unbalance load inductor	L_{L2}	128 mH
Unbalance load dc-link resistor	R_{L2}	60 Ω
Linear three-phase load inductor	L_{L3}	128 mH
Linear three-phase load dc-link resistor	R_{L3}	24 Ω

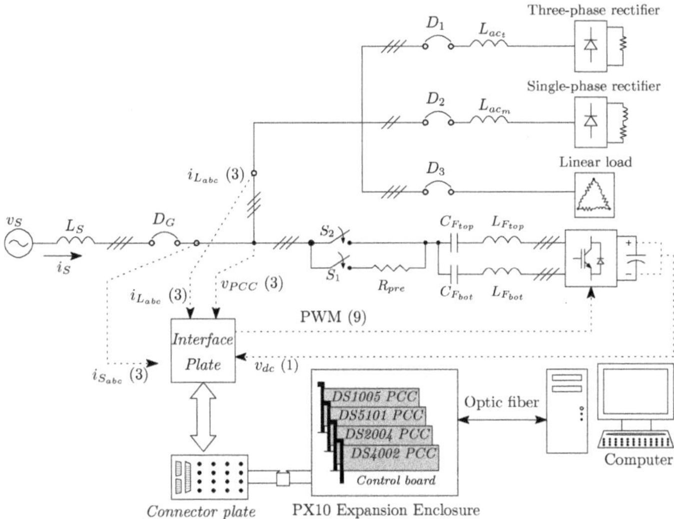

Figure 5. Single-line diagram of the experiment prototype showing the interface between the DHPF-NSI and dSPACE.

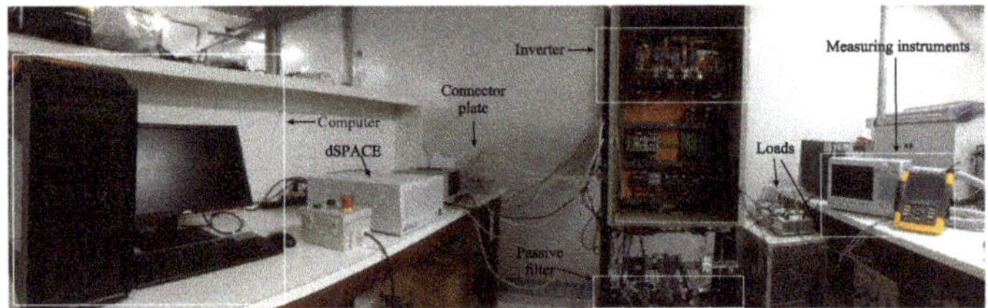

Figure 6. General view of the experimental test bench.

5.1. Reactive Power Compensation Performance

Based on the parameters of the system presented in Table 2, the gains of the reactive power controller were calculated. The value of ω_d defined for the design is 1820 rad/s, which is the same as in Equation (24). For controller design, the stabilization time ($t_{2\%}$) is equal to 150 ms. Based on Equation (32), the position of the real pole is obtained as 26.67 rad/s. The value of the variable T, calculated in Equation (29), is 1.64 s. Therefore, by substituting the values of the system parameters and the values of a, T, and ω_d in Equations (30) and (31), the values of the controller gains were obtained. The controller gains are given in Table 3.

Table 3. Gains of the reactive power controller for the prototype parameters.

Parameter	Symbol	Value
Real pole position	a	26.67 rad/s
Variable T	T	1.64 s
Position of the imaginary part of the complex pole	ω_d	1820 rad/s
Proportional gain of reactive power control	k_{pQ}	$-0.075\ \Omega$
Integral gain of reactive power control	k_{iQ}	$-1475.2\ \Omega\cdot\mathrm{s}^{-1}$

In order to verify the stability of the reactive power control system, the frequency response of the plant, $G(s)$ in Equation (24), the controller, $C(s)$ in Equation (25), and the open-loop transfer function, $C(s)G(s)$, are shown in Figure 7, using the data in Tables 2 and 3. As can be seen, the designed controller has an infinite gain for dc (0 Hz) components, ensuring zero steady-state error in the reactive power control, and imposes a phase margin (PM) around 90° for the open-loop transfer function, which denotes a stable system. The cut-off frequency, 4.2 Hz at 0 dB, is sufficient to produce a step response with a settling time around 147 ms, very close to 150 ms, which is the value used for the controller design.

In order to validate the reactive control design, a step in $i_{S_q}^*$ was applied in the experimental prototype, producing the responses shown in Figure 8. Initially, the $i_{S_q}^*$ value was set to 3 A. At 25 ms, the value of $i_{S_q}^*$ was changed to 6 A, taking around 150 ms to i_{S_q} to reach the reference value, as can be seen in Figure 8a. The corresponding reactive power generated by the DHPF-NSI is shown in Figure 8b. It should be noted that the DHPF-NSI can compensate the reactive power consumed by any type of load in the range of 500–1050 var.

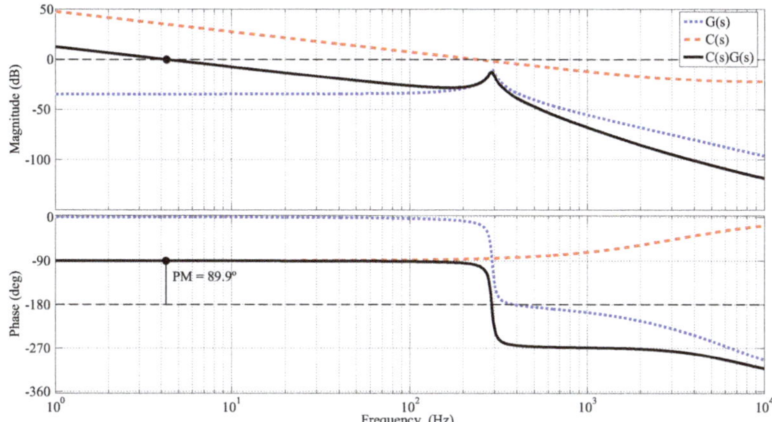

Figure 7. Frequency response of the plant, $G(s)$, the controller, $C(s)$, and the open-loop transfer function, $C(s)G(s)$, for the reactive power control.

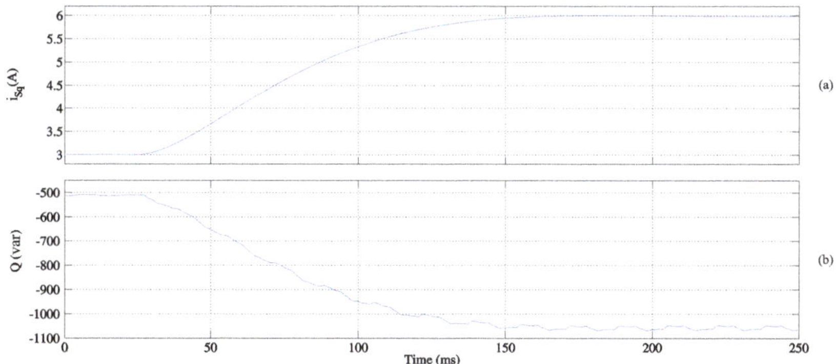

Figure 8. Step responses for a change from 3 A to 6 A in $i^*_{S_q}$ at 25 ms: (**a**) current i_{S_q} injected by the DHPF-NSI; (**b**) reactive power generated by the DHPF-NSI.

5.2. Current Harmonic Compensation Performance

The results in this section have been obtained using YOKOGAWA DL850 ScopeCorder [41] with a High-Speed 100 M/s, 12-Bit Isolation Module 720211 [42] and a power quality analyzer Fluke 434 series II [43] to measure the total harmonic distortion and the percentage value of each current harmonic. The IEEE 519-1992 standard [29] recommends that any load connected to general distribution system (up to 69 kV, which is the majority of industries) could consume currents with the harmonic limits defined by Table 4, which is a reproduction of Table 10.3 of [29]. The worst-case scenario would be the first line of Table 4, where each odd current harmonic up to the 9th order has a maximum distortion limit of 4%. From the 11th to the 15th order, the maximum distortion limit falls to 2%, and the THD (Total Harmonic Distortion, with the load at nominal power condition) is limited to 5%. All values are percentages of the fundamental component of the load current at nominal power condition.

With the DHPF-NSI is disabled, the currents waveforms and the THDs for the three-phase grid currents are shown in Figures 9 and 10. The load currents present high harmonic values that are well above the limits recommended in Table 4 [29]. Based on Table 4 and Figure 9, it is possible to build the current harmonic profile of the grid currents with the DHPF-NSI disabled, which is shown in Figure 11.

Table 4. Current Distortion Limits for General Distribution Systems (120 V Through 69 kV) [29].

Maximum Harmonic Current Distortion in Percent of I_L						
Individual Harmonic Order (Odd Harmonics)						
I_{sc}/I_L	<11	$11 \leq h < 17$	$17 \leq h < 23$	$23 \leq h < 35$	$35 \leq h$	TDD
<20 *	4.0	2.0	1.5	0.6	0.3	5.0
$\geq$20 and <50	7.0	3.5	2.5	1.0	0.5	8.0
$\geq$50 and <100	10.0	4.5	4.0	1.5	0.7	12.0
$\geq$100 and <1000	12.0	5.5	5.0	2.0	1.0	15.0
>1000	15.0	7.0	6.0	2.5	1.4	20.0

Even harmonics are limited to 25% of the odd harmonic limits above. Current distortions that result in a dc offset, e.g., half-wave converters, are not allowed. * All power generation equipment is limited to these values of current distortion, regardless of actual I_{sc}/I_L, where I_{sc} = maximum short-circuit current at PCC, and I_L = maximum demand load current (fundamental frequency component) at PCC.

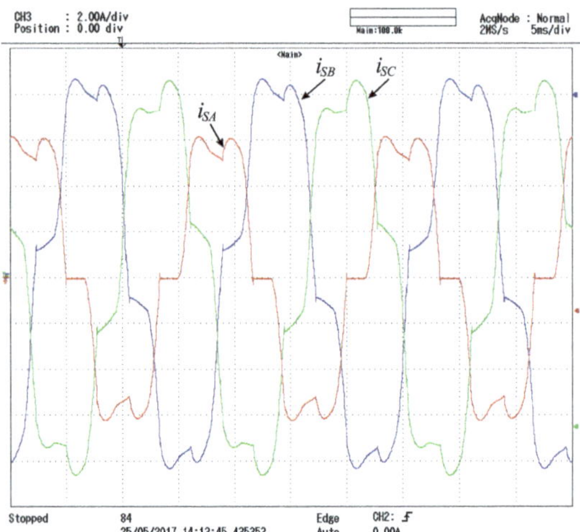

Figure 9. Three-phase grid current harmonic components with the DHPF-NSI disabled.

Figure 10. Three-phase grid current waveforms with the DHPF-NSI disabled.

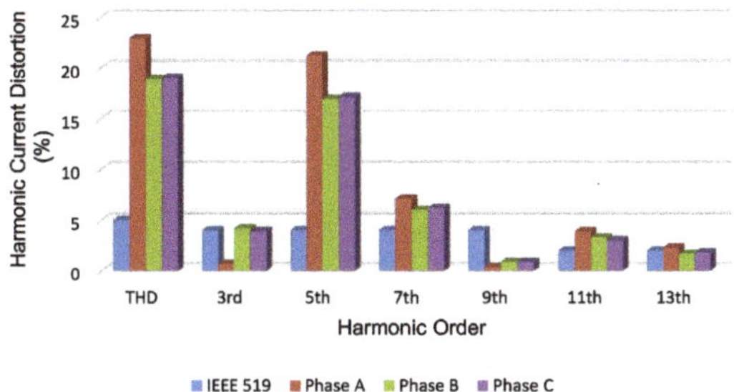

Figure 11. Three-phase grid current harmonic profile with the DHPF-NSI disabled.

As can be seen, not only are the 3rd, 5th, 7th, 11th, and 13th harmonic components of the grid currents above the IEEE 519-1992, but also the THDs of all three phases are as well. On the other hand, when the DHPF-NSI is enabled, the THDs and the waveforms of the grid currents are considerably improved, as shown in Figures 12 and 13. For example, the THDs are reduced from 18.9–22.9% (Figure 9) to less than 5% (Figure 12) in all phases, respecting the THD limit recommended in Table 4 [29]. Based on Table 4 and Figure 12, it is possible to build the current harmonic profile of the grid currents with the DHPF-NSI enabled, which is shown in Figure 14. It is possible to see that the influences from the 5th to the 13th harmonic components are strongly reduced, below the limits of the IEEE 519-1992. For instance, the 5th harmonic component has a reduction from 16.9–21.2% (Figure 9) to less than 4% (Figure 12) in all phases, proving the effectiveness of the proposed DHPF-NSI. The current waveform results (grid, load, and filters) for Phase A are shown in Figure 15 in order to illustrate the individual performance of each hybrid filter in the load current harmonic compensation.

Figure 12. Three-phase grid current harmonic components with the DHPF-NSI enabled.

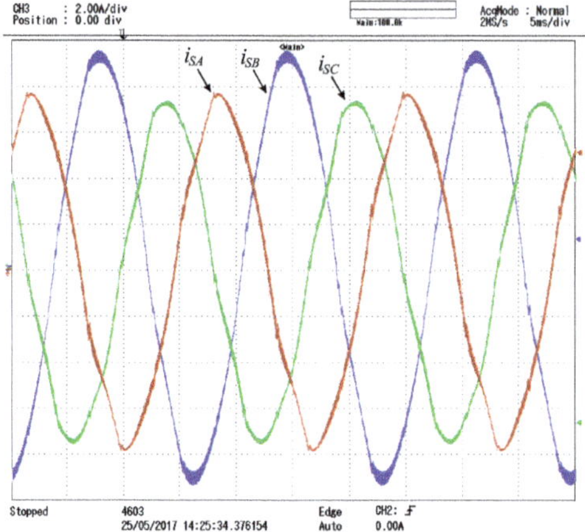

Figure 13. Three-phase grid current waveforms with the DHPF-NSI enabled.

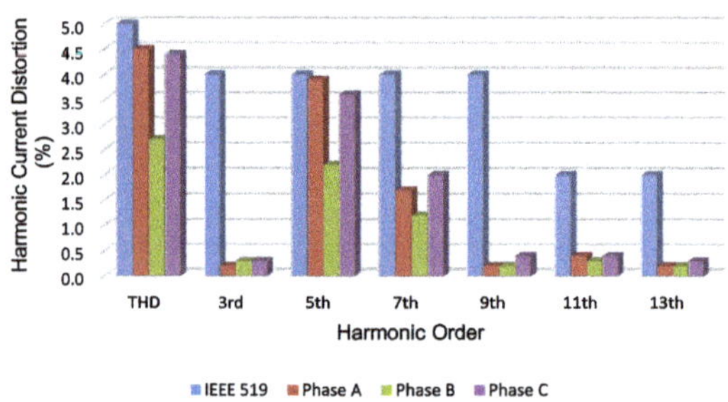

Figure 14. Three-phase grid current harmonic profile with the DHPF-NSI enabled.

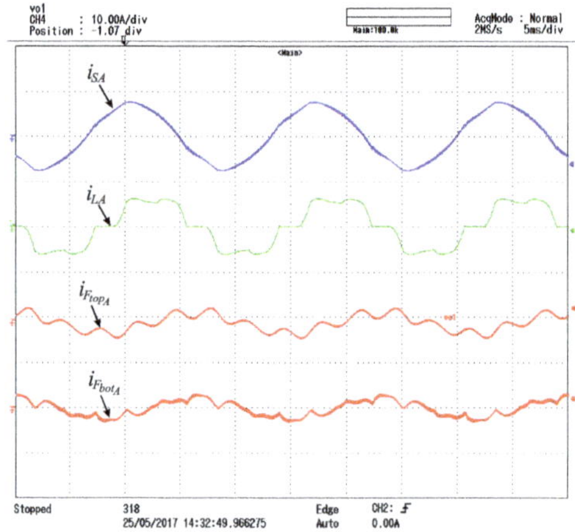

Figure 15. Current waveforms for Phase A. From top to bottom: grid (i_{SA}), load (i_{LA}), top unit ($i_{F_{topA}}$) and bottom unit ($i_{F_{botA}}$).

6. Conclusions

In this paper, two extra features have been proposed for the dual HPF based on the NSI: the compensation of triplen current harmonics and the reactive power support. Dual HPF are used because dual filters have a better compensation capacity than single filters, and this is needed to compensate triplen current harmonics. HPFs have a lower installed power when compared with APFs. The limits of reactive power that the DHPF-NSI can compensate as well as its mathematical model and the controller design for reactive power compensation have also been demonstrated. Experiments testing for the overall response of the proposed system, among other parameters, have been carried out, proving its feasibility in terms of reactive power support and harmonic compensation.

Author Contributions: Conceptualization, L.R.L.; Formal analysis, F.B. and C.H.d.O.L.; Funding acquisition, M.C.C.; Investigation, L.R.L.; Supervision, F.B.; Validation, C.H.d.O.L. and M.C.C.; Writing – original draft, C.H.d.O.L. and M.C.C.; Writing – review & editing, L.R.L. and F.B.

Acknowledgments: This work was partially supported by CAPES, CNPq Grant No. 305901/2015-0, No. 306106/2015-9 and No. 309789/2017-6 and FACEPE Grant No. APQ-0777-3.04/14, No. APQ-0896-3.04/14 and No. IBPG-1262-3.04/16.

Conflicts of Interest: The authors declare no conflict of interest.

References

1. Jewell, W. Electrical power systems quality. *IEEE Power Energy Mag.* **2003**, *99*, 63–64. [CrossRef]
2. Hingorani, N.G.; Gyugyi, L. *Understanding FACTS Concepts and Technology of Flexible AC Transmission Systems*; IEEE Press: Piscataway, NJ, USA, 2000; pp. 1–429.
3. Conseil International des Grands Reseaux Electriques (CIGRE). *Custom Power-State of the Art*; CIGRE WG14.31; CIGRE: Paris, France, 2002; pp. 1–104.
4. Institute of Electrical and Electronics Engineers (IEEE). *IEEE Guide for Application of Power Electronics for Power Quality Improvement on Distribution Systems Rated 1 kV Through 38 kV*; IEEE Standard 1409-2012; IEEE: New York, NY, USA, 2012; pp. 1–90.
5. Hossain, E.; Tür, M.R.; Padmanaban, S.; Ay, S.; Khan, I. Analysis and Mitigation of Power Quality Issues in Distributed Generation Systems Using Custom Power Devices. *IEEE Access* **2018**, *6*, 16816–16833. [CrossRef]
6. Rivas, D.; Moran, L.; Dixon, J.; Espinoza, J. Improving passive filter compensation performance with active techniques. *IEEE Trans. Ind. Electron.* **2003**, *50*, 161–170. [CrossRef]
7. Fujita, H.; Akagi, H. A practical approach to harmonic compensation in power systems - series connection of passive and active filters. *IEEE Trans. Ind. Appl.* **1991**, *27*, 1020–1025. [CrossRef]
8. Chen, L.; Jouanne, A. A comparison and assessment of hybrid filter topologies and control algorithms. In Proceedings of the IEEE 32nd Annual Power Electronics Specialists Conference, Vancouver, BC, Canada, 17–21 June 2001; pp. 565–570.
9. Srianthumrong, S.; Akagi, H. A medium-voltage transformerless AC/DC power conversion system consisting of a diode rectifier and a shunt hybrid filter. *IEEE Trans. Ind. Appl.* **2003**, *39*, 874–882. [CrossRef]
10. Limongi, L.R.; Roiu, D.; Bojoi, R.; Tenconi, A. Analysis of active power filters operating with unbalanced loads. In Proceedings of the IEEE Energy Conversion Congress and Exposition, San Jose, CA, USA, 20–24 September 2009; pp. 584–591.
11. Ribeiro, R.L.A.; Azevedo, C.C.; Sousa, R.M. A robust adaptive control strategy of active power filters for power-factor correction, harmonic compensation, and balancing of nonlinear loads. *IEEE Trans. Power Electron.* **2012**, *27*, 718–730. [CrossRef]
12. Furtado, P.C.S.; Rodrigues, M.C.B.P.; Braga, H.A.C.; Barbosa, P.G. Two-phase, three-wire shunt active power filter using the single-phase p-q theory. In Proceedings of the Brazilian Power Electronics Conference, Gramado, Brazil, 27–31 October 2013; pp. 1245–1250.
13. Rahman, N.F.A.; Radzi, M.A.M.; Soh, A.C.; Mariun, N.; Rahim, N.A. Adaptive hybrid fuzzy-proportional plus crisp-integral current control algorithm for shunt active power filter operation. *Energies* **2016**, *9*, 737. [CrossRef]

14. Monroy-Morales, J.L.; Campos-Gaona, D.; Hernández-Ángeles, M.; Peña-Alzola, R.; Guardado-Zavala, J.L. An active power filter based on a three-level inverter and 3D-SVPWM for selective harmonic and reactive compensation. *Energies* **2017**, *10*, 297. [CrossRef]

15. Tan, K.-H.; Lin, F.-J.; Chen, J.-H. A three-phase four-leg inverter-based active power filter for unbalanced current compensation using a Petri probabilistic fuzzy neural network. *Energies* **2017**, *10*, 2005. [CrossRef]

16. Hoon, Y.; Mohd Radzi, M.A.; Hassan, M.K.; Mailah, N.F. Control algorithms of shunt active power filter for harmonics mitigation: A review. *Energies* **2017**, *10*, 2038. [CrossRef]

17. Peng, F.Z.; Akagi, H.; Nabae, A. A new approach to harmonic compensation in power systems—A combined system of shunt passive and series active filters. *IEEE Trans. Ind. Appl.* **1990**, *26*, 983–990. [CrossRef]

18. Lam, C.S.; Wong, M.C. A novel b-shaped l-type transformerless hybrid active power filter in three-phase four-wire systems. In Proceedings of the 38th North American Power Symposium, Carbondale, IL, USA, 17–19 September 2006; pp. 235–241.

19. Bhattacharya, A., Chakraborty, C.; Bhattacharya, S. Parallel-connected shunt hybrid active power filters operating at different switching frequencies for improved performance. *IEEE Trans. Ind. Electron.* **2012**, *59*, 4007–4019. [CrossRef]

20. Limongi, L.R.; Silva Filho, L.R.; Genu, L.G.B.; Bradaschia, F.; Cavalcanti, M.C. Transformerless hybrid power filter based on a six-switch two-leg inverter for improved harmonic compensation performance. *IEEE Trans. Ind. Electron.* **2015**, *62*, 40–51. [CrossRef]

21. Luo, Z.; Su, M.; Yang, J.; Sun, Y.; Hou, X.; Guerrero, J.M. A repetitive control scheme aimed at compensating the 6k + 1 harmonics for a three-phase hybrid active filter. *Energies* **2016**, *9*, 787. [CrossRef]

22. Tokiwa, A.; Yamada, H.; Tanaka, T.; Watanabe, M.; Shirai, M.; Teranishi, Y. New hybrid static VAR compensator with series active filter. *Energies* **2017**, *10*, 1617. [CrossRef]

23. Corasaniti, V.F.; Barbieri, M.B.; Arnera, P.L. Hybrid active filter for reactive and harmonics compensation in a distribution network. *IEEE Trans. Ind. Electron.* **2009**, *56*, 670–677. [CrossRef]

24. Salmeron, P.; Litran, S.P. A control strategy for hybrid power filter to compensate four-wires three-phase systems. *IEEE Trans. Power Electron.* **2010**, *25*, 1923–1931. [CrossRef]

25. Asiminoaei, B.L.; Lascu, C; Blaabjerg, F; Boldea, I. Performance improvement of shunt active power filter with dual parallel topology. *IEEE Trans. Power Electron.* **2007**, *22*, 247–259. [CrossRef]

26. Kim, G.-T.; Lipo, T.A. VSI-PWM rectifier/inverter system with a reduced switch count. *IEEE Trans. Ind. Appl.* **1996**, *32*, 1331–1337.

27. Limongi, L.; Bradaschia, F.; Azevedo, G.M.S.; Genu, L.G.B.; Silva Filho, L.R. Dual hybrid power filter based on a nine-switch inverter. *Electr. Power Syst. Res.* **2014**, *117*, 154–162. [CrossRef]

28. Liu, C.; Wu, B.; Zargari, N.R.; Xu, D.; Wang, J. A novel three-phase three-leg AC/AC converter using nine igbts. *IEEE Trans. Power Electron.* **2009**, *24*, 1151–1160.

29. Institute of Electrical and Electronics Engineers (IEEE). *IEEE Recommended Practices and Requirements for Harmonic Control in Electrical Power Systems*; IEEE Standard 519-1992; IEEE: New York, NY, USA, 1993; pp. 1–112.

30. Akagi, H.; Kanazawa, Y.; Nabae, A. Instantaneous Reactive Power Compensators Comprising Switching Devices without Energy Storage Components. *IEEE Trans. Ind. Appl.* **1984**, *IA-20*, 625–630. [CrossRef]

31. Duesterhoeft, W.C.; Schulz, M.W.; Clarke, E. Determination of Instantaneous Currents and Voltages by Means of Alpha, Beta, and Zero Components. *Trans. Am. Inst. Electr. Eng.* **1951**, *70*, 1248–1255. [CrossRef]

32. Park, R.H. Two-reaction theory of synchronous machines generalized method of analysis-part I. *Trans. Am. Inst. Electr. Eng.* **1929**, *48*, 716–727. [CrossRef]

33. Ioannou, P.A.; Sun, J.A. *Robust Adaptive Control*; PTR Prentice-Hall: Upper Saddle River, NJ, USA, 1996; pp. 803–813, ISBN 9780134391007.

34. Ribeiro, R.L.A.; Rocha, T.O.A.; Sousa, R.M.; Santos, E.C.; Lima, A.M.N. A Robust DC-Link Voltage Control Strategy to Enhance the Performance of Shunt Active Power Filters Without Harmonic Detection Schemes. *IEEE Trans. Ind. Electron.* **2015**, *62*, 803–813. [CrossRef]

35. Semikron IGBT Module SKM50GB12T4. Available online: https://www.semikron.com/dl/service-support/downloads/download/semikron-datasheet-skm50gb12t4-22892000/ (accessed on 25 May 2018).

36. Semikron Gate Drive SKPC 22/2. Available online: http://astronix.biz/datasheets/SKPC22.PDF (accessed on 25 May 2018).

37. Modular dSPACE System DS1005 Processor Board. Available online: https://www.dspace.com/shared/data/bkm/catalog2014_en/files/assets/basic-html/page332.html (accessed on 25 May 2018).
38. dSPACE DS5101 Board PWM Generation. Available online: https://www.dspace.com/shared/data/pdf/2018/dSPACE_DS5101_Catalog2018.pdf (accessed on 25 May 2018).
39. dSPACE DS2004 Board for A/D Conversion. Available online: https://www.dspace.com/shared/data/pdf/2018/dSPACE_DS2004_Catalog2018.pdf (accessed on 25 May 2018).
40. dSPACE DS4002 Board for Digital I/O. Available online: https://www.dspace.com/shared/data/pdf/2018/dSPACE_DS4002_Catalog2018.pdf (accessed on 25 May 2018).
41. YOKOGAWA DL850 ScopeCorder. Available online: https://cdn.tmi.yokogawa.com/BUDL850E-00EN.pdf (accessed on 25 May 2018).
42. High-Speed 100 M/s, 12-Bit Isolation Module 720211. Available online: https://cdn.tmi.yokogawa.com/BUDL850E-01EN.pdf (accessed on 25 May 2018).
43. Fluke 434-II Power Quality Analyzer. Available online: http://media.fluke.com/documents/F430-II_umeng0100.pdf (accessed on 25 May 2018).

Article

Power Quality Improvement in a Cascaded Multilevel Inverter Interfaced Grid Connected System Using a Modified Inductive–Capacitive–Inductive Filter with Reduced Power Loss and Improved Harmonic Attenuation

Meenakshi Jayaraman and Sreedevi VT *

School of Electrical Engineering, Vellore Institute of Technology University (VIT University), Chennai 600127, India; jayaraman.meenakshi@gmail.com
* Correspondence: sreedevi.vt@vit.ac.in; Tel.: +91-44-3993-1310

Received: 5 October 2017; Accepted: 6 November 2017; Published: 10 November 2017

Abstract: Recently, multilevel inverters are more researched due to the advantages they offer over conventional voltage source inverters in grid connected applications. Passive filters are connected at the output of these inverters to produce sinusoidal waveforms with reduced harmonics and to satisfy grid interconnection standard requirements. This work proposes a new passive filter topology for a pulse width modulated five-level cascaded inverter interfaced grid connected system. The proposed passive filter inserts an additional resistance-capacitance branch in parallel to the filter capacitor of the traditional inductive–capacitive–inductive filter in addition to a resistance in series with it to reduce damping power loss. It can attenuate the switching frequency harmonic current components much better than the traditional filter while maintaining the same overall inductance, reduced capacitance and resistance values. The basic parameter design procedure and an approach to discover the parameters of the proposed filter is introduced. Further, a novel methodology using Particle Swarm Optimization (PSO) is recommended to guarantee minimum damping loss while ensuring reduced peak during resonance. In addition, PSO algorithm is newly employed in this work to maximize harmonic attenuation in and around the switching frequency on the premise of allowable values of filter inductance and capacitance. A comparative discussion considering traditional passive filters and the proposed filter is presented and evaluated through experiments conducted on a 110 V, 1 kW five-level grid connected inverter. The modulation algorithm for the multilevel inverter is implemented using a SPARTAN 6-XC6SLX25 Field Programmable Gate Array (FPGA) processor. The analysis shows that the proposed filter not only provides decreased damping power loss but also is capable of providing considerable harmonic ripple reduction in the high frequency band, improved output waveforms and lesser Total Harmonic Distortion (THD) with improved power quality for the multilevel inverter based grid connected system.

Keywords: Pulse Width Modulation (PWM); Field Programmable Gate Array (FPGA); Total Harmonic Distortion (THD); harmonics

---

1. Introduction

Voltage source inverters are a key component in most PV systems installed for grid connected and standalone applications [1–4]. These inverters use sinusoidal Pulse Width Modulation (PWM) strategy which involves high speed switching of semiconductor devices to produce an AC output which generates high frequency noise, harmonics and cause high switching losses. A passive filter, which is cost effective, is connected between the inverter and the grid to attenuate the PWM carrier and sideband harmonics to meet IEEE standards. Earlier, first order passive L type filters were used on the AC side of the PWM inverters for attenuation of these switching harmonics. However, L filter has a restriction in low switching frequency applications due to its inevitable large size [5,6]. A higher order LC or LCL filter provides better harmonic suppression at lower switching frequencies with a reduction of overall filter size. The LCL filter instead of L filter is more attractive because it can provide high frequency harmonic attenuation with the same inductance value [7–9]. Conversely, these higher order filters trigger a resonance between the inverter and the grid which needs to be damped either actively [10,11] or passively [12–24]. Active damping methods being flexible, involve a well-designed control algorithm to dampen the resonance but are limited by high cost, control complexity and the use of additional sensors. Passive damping schemes are economical, less complex and carry an increased reliability.

Various filter topologies and damping techniques for passive filters have been proposed in the literature for voltage source inverters. Different damping schemes for a passive filter used in grid connected inverters is discussed [9,12,13]. However, the procedure to design the filter parameters and the damping component is not provided. A structured procedural analysis and comparison of active and passive damping methods for LCL filters employed in a grid connected voltage source converter is explained in [14]. However, the analysis described in the paper is specific to low power applications operating at high switching frequencies. Passive LCL filters with a resistance connected across the filter inductor on the grid side have been more popular for three phase grid connected inverter systems working at a particular resonant and switching frequency [15,16]. In [15], the resistance value is chosen approximately from bode plots obtained for a set of resistance values selected using trial-and-error method, whereas plots are presented for the loss obtained in the inverter and filter system and compared with the loss obtained using active damping technique in [16]. Neither of them provide information on the selection of optimal damping components [15,16]. In [8], the value of the damping resistance in a LCL filter used in a three phase rectifier system is assumed to be one-third of the LCL filter capacitor impedance at resonant frequency. Although the performance of the filter is proven to be efficient, the selection procedure may not be suitable for all conditions. Another design approach where a virtual resistor is assumed to be on the output side of the LC filter side to damp the switching oscillations in a PWM inverter is proposed in [17]. On the other hand, an extra control algorithm that simulates the role of a real resistor makes the system complex. To overcome this, a method that uses a resistor to help dissipate the energy stored in the resonant circuit of a LC filter used in an IGBT inverter is proposed for motor drive applications [18]. However, the above filter is designed with a resonant frequency above the switching frequency, since the purpose is to limit merely dv/dt and not to suppress the switching harmonics. Another design method where a resistor is connected in series with the filter inductor to damp harmonic resonance at the DC link of a rectifier-inverter system is proposed in [19]. A detailed analysis of a split-capacitor with resistive (SC-R) damping for a LCL filter used in a three-phase grid connected inverter that includes power loss in the damping resistance in the analysis is done in [20]. A design-oriented analysis for the selection of the split-capacitor resistive (SC-R) damping circuit parameters is presented in [6]. Since the capacitor in the damping branch is split from the filtering capacitor in [6,20], the high-frequency harmonic attenuation of the passively damped LCL filter is reduced, compared to that with an undamped LCL filter. Another simple design criterion to find the optimized damping resistor values used in a LCL and a higher order LLCL filter is proposed in [21]. In the above work, an additional RC branch in connected across the trap filter to perform passive damping. However, LLCL filter provides a lesser harmonic attenuation performance in the high frequency band [22]. To overcome this, a new high-order filter, named the LTCL filter, is proposed in [23] with multiple LC traps inserted in parallel with the capacitor branch. However, multiple traps tuned to different frequencies require judiciously selected components to attain desired harmonic attenuation. Another novel passive filter, namely, L(LCL)^2, is employed for a grid inverter in [24], where a design bench mark is utilized to obtain the filter component values and the inductance present in the filter design is optimized using Genetic algorithm. Although optimization techniques such as

Particle Swarm Optimization (PSO) and Genetic Algorithms have been utilized to obtain filtering component values in LCL filters employed in rectifiers, and power filters [24–26], these algorithms work based on allowable components value ranges for choosing the reactive elements in the filter. Poor design of output filters leads to reduced attenuation at the switching frequency, larger filter size, higher cost and more losses. Hence, it becomes a challenging task to design the filter components with reduced size, good damping and improved attenuation.

Recently, multilevel inverters are more researched due to the advantages they offer over conventional voltage source inverters in grid connected applications [27–29]. Among the different multilevel inverter topologies [30], Cascaded H-Bridge (CHB) inverters [31] are advantageous owing to their modularity, simplicity and they require fewer components to achieve the same number of voltage levels. Although there are various modulation techniques [30] available for multilevel inverters, modulation based on multi carrier signal arrangements with traditional sinusoidal PWM technique such as Phase Disposition PWM (PDPWM), Phase Opposition Disposition PWM (PODPWM) and Alternative Phase Opposition Disposition PWM (APODPWM) [32,33] are quite popular and simple to implement, which again necessitates a filter at the inverter output to limit the harmonics caused by PWM. Several methods based on the basic modulation techniques have been used for minimization of harmonics in multilevel inverters [34–37]. However, all these methods involve a well-designed modulation scheme and involve a complicated power circuit design to obtain better quality. Instead, passive filters are economical and simple to improve power quality. Passive filter designs considering LC trap/LCR filter and LCL filters for a neutral point clamped three level inverter is presented in literature [38,39]. The analysis and performance of a LCL filter for a cascaded inverter operating as a DSTATCOM is studied in [40]. However, the analysis is restricted to low frequency operation in [38] and [40]. Most of the literature has focused on filter designs for conventional voltage source inverters and filter designs considering multilevel inverters is limited. To address the above challenges, this work proposes a new L-(CR)^2-L passive filter structure for a pulse width modulated inverter used in grid connected applications. The design consists of an additional resistance–capacitance branch across the filter capacitor of the traditional LCL filter apart from an additional resistance in series with the filter capacitor. The proposed filter can attenuate the switching frequency harmonics much better than the traditional LCL filter, while maintaining the same overall inductance value and reduced capacitance value of the traditional LCL filter. It can reduce the damping power loss with lesser value of damping resistance and provides considerable damping power loss reduction compared to conventional topologies. Further, this work includes the basic parameter design procedure with expressions to arrive at individual filter component values of the proposed filter in addition to a novel technique using PSO to minimize damping power loss, while ensuring reduced resonant peaking. Further, the same algorithm is also employed to ensure maximize harmonic attenuation in and around the switching frequency based on permissible filter inductance and capacitance values. A performance comparison of the proposed filter against two other passive filters is presented. Results obtained from experiments conducted on a 1 kW, 110 V, 50 Hz five-level inverter set up based on a SPARTAN 6-XC6SLX25 Field Programmable Gate Array (FPGA) processor are included to evaluate the performance of the proposed filter. This work considers a five-level CHB inverter for the filter implementation. The structure of a five-level CHB inverter interfaced with the grid through a traditional LCL filter is presented in Figure 1. In Figure 1, V_{s1} and V_{s2} represents the DC voltage applied to the upper and lower H-bridges of the five-level inverter, S_1 to S_8 are the inverter switches, L_{inv} and L_{grid} represents the inverter side and grid side inductance of the LCL filter respectively and C represents the LCL filter capacitance. PDPWM modulation technique [33], which is well discussed in the literature, is used to generate gate signals for the five-level inverter.

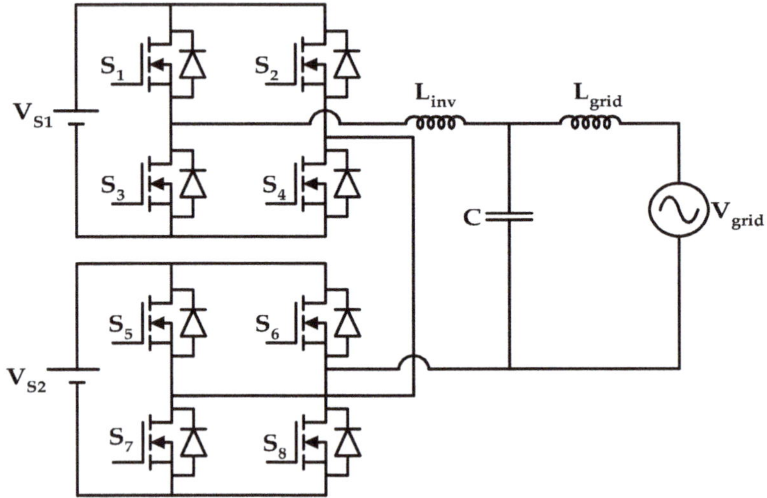

Figure 1. Five-level Cascaded H-Bridge (CHB) inverter with LCL filter connected to grid.

The rest of the paper is organized as follows: Section 2 provides the complete analysis and design of the proposed passive filter structure. Section 3 focuses on the filter parameter design and verification. Simulation and experimental results are provided in Section 4. The work is concluded in Section 5.

2. Proposed Filter Design

The proposed filter is presented in Figure 2. In Figure 2, L_{inv} represents the inverter side inductance, L_{grid} is the grid side inductance and C_r is the filter capacitance. R_r is the series resistance to damp resonance. R_h-C_h represents the additional R-C branch connected across the R_r-C_r branch mainly to bypass harmonic components and to reduce damping power loss. i_{out} represents the grid side current.

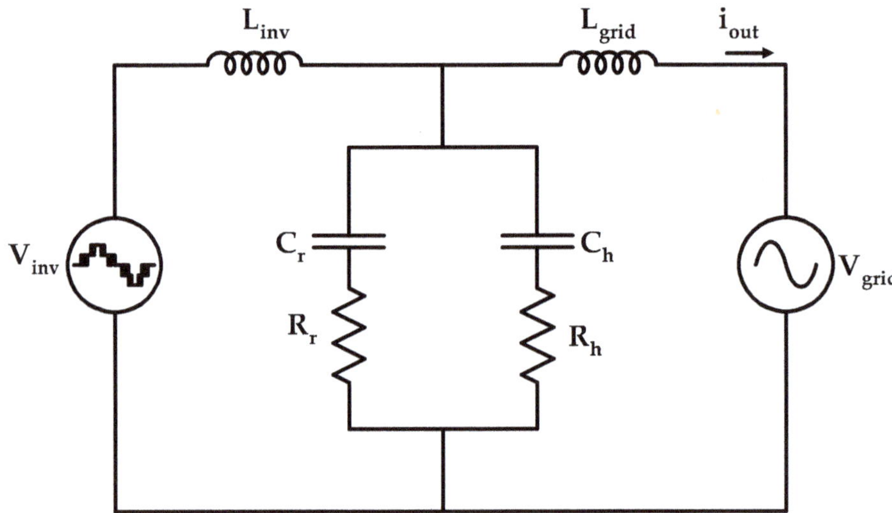

Figure 2. Proposed filter structure for the five-level inverter.

The transfer function of the proposed filter structure is derived as in Equation (1).

$$\frac{i_{out}(s)}{V_{inv}(s)} = \frac{R_r C_r R_h C_h s^2 + (R_r C_r + R_h C_h)s + 1}{\begin{aligned}&L_{inv}L_{grid}C_r C_h(R_r + R_h)s^4 + (C_r L_{inv}L_{grid} + C_h L_{inv}L_{grid}\\ &+ C_r C_h L_{inv}R_r R_h + C_r C_h L_{grid}R_r R_h)s^3\\ &+ (C_r R_r L_{inv} + C_h R_h L_{inv} + C_r R_r L_{grid} + C_h R_h L_{grid})s^2 + (L_{inv} + L_{grid})s\end{aligned}} \tag{1}$$

2.1. Selection of Overall Inductor and Capacitor Values for the Proposed Filter

Larger values of capacitance for the proposed configuration provide better higher order harmonic attenuation. However, it results in higher reactive power and increased demand of current from the inverter side inductance, resulting in the decrease of efficiency of the overall filter system. In addition, smaller capacitor size requires larger inductance to meet the harmonic mitigation requirement. Thus, the selection of capacitor value for the proposed filter is a tradeoff between the reactive power and the selection of inductor values. The overall capacitance value ($C_t = C_r + C_h$) for the proposed filter configuration is chosen by calculating the reactive power absorbed by the filter at rated conditions as given by Equation (2).

$$C_t = \frac{Q}{2\pi f_{out}(V_{grid})^2} \tag{2}$$

where Q is the reactive power absorbed by the system, which is usually chosen to be less than 5% of the rated power [21,23]; V_{grid} is the rated system voltage; and f_{out} is the output frequency of the system.

A number of methods have been used in the literature to select inductor values in a LCL filter [6,9,12,20]. This work considers some of those aspects to design suitable inductor values for the proposed structure. The inverter side inductance is usually designed by calculating the current ripple. Selection of small values of ripple current decreases the switching losses. However, it results in large values of inductors. Thus, the selection of the ripple current is a trade-off between the size of the inductors and the switching losses. The inverter output voltage and inductor current waveforms during one switching cycle for a carrier based modulation is shown in Figure 3.

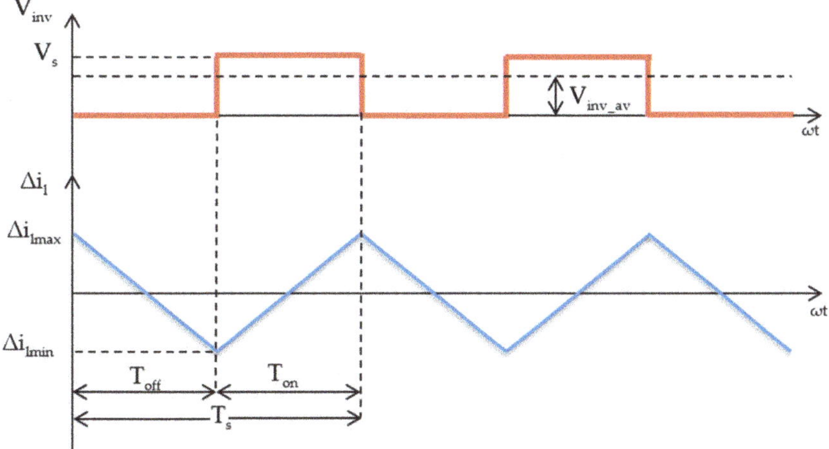

Figure 3. Inverter output voltage and inductor current waveforms during one switching cycle.

The inverter switching frequency is usually greater than the system output frequency. Hence, the average value of the inverter output voltage can be treated constant during the switching period T_s. In Figure 3, V_{inv_av} represents the average inverter output voltage, while T_{on} and T_{off} represent the ON and OFF periods. The turn ON and OFF times are related as $T_{off}/T_{on} = 1 - \alpha$ and $T_{on}/T_s = \alpha$, where α represents the duty cycle. The peak-to-peak value of the inductor current during PWM switching is obtained using Equation (3) [24].

$$\Delta i_{l(peak\_peak)} = \frac{V_s - V_{inv\_av}}{L_{inv}} \alpha T_s \tag{3}$$

where L_{inv} represents the inverter side inductance of the passive filter and V_s ($V_{s1} = V_{s2} = V_s$) is the voltage applied to one of the H-Bridges of the inverter. As the grid voltage is usually assumed to be purely sinusoidal, the fundamental component of the grid current can be assumed to be zero. Hence, the fundamental component of voltage appearing across the inductor is zero. Hence, the inverter average output voltage and grid voltage are given as $v_{inv\_av} = \alpha V_s$ and $V_{out}(\omega t) = m V_s \sin(\omega t)$ where m is the modulation index. Using the above expressions, Equation (3) is rewritten as Equation (4).

$$\Delta i_{l(peak\_peak)} = \frac{V_s T_s}{4 L_{inv}} (1 - m^2 \sin^2(\omega t)) \tag{4}$$

Equation (4) provides the peak-to-peak value of the inductor ripple current. Thus, inverter side inductance value L_{inv} of the proposed filter is using Equation (5).

$$L_{inv} = \frac{1}{4} \frac{2 V_s}{f_{switch} \times h} \tag{5}$$

where V_s is the input voltage to one H-Bridge of the inverter, "h" is the amount of ripple current which is generally lower than 40% of the peak value of the rated system current [21,22] and f_{switch} is the switching frequency of the inverter.

The value of grid side inductance, L_{grid} is chosen such that $L_{grid} = a\, L_{inv}$, where "a" is the inductance ratio factor. With the factor "a" less than unity ($L_{grid} < L_{inv}$), it is possible to reduce the overrating of the switches and improve stability. This design approach is used for grid side inductor design in [12,23]. With the factor chosen as unity, the inverter and grid side inductances are made equal to maximize the attenuation capability of the LCL filter [21,22]. However, this results in an increased cost and volume of the filter elements. Choosing $a > 1$ results in a smaller inverter side inductance [14] when compared to the grid side. This design may introduce significant current ripples on the inverter output leading to serious overheating and losses on the inductor [5]. It also increases the ratings of the switches raising the cost and switching losses. Considering another criterion, the overall inductance value of the proposed filter is to be kept less than 0.1 per unit to limit the voltage drop across the inductor [20].

2.2. Selection of Resonant Frequency

A good selection of the resonant frequency is another important design criterion considered for the design of the proposed filter. The resonant frequency is usually chosen based on the fundamental frequency and switching frequency of the inverter. It is chosen such that it is in the range $10 f_{out} \leq f_{res} \leq (f_{switch}/2)$ to ensure that resonant and switching harmonic currents flow through the respective branches of the filter circuit. The resonant frequency of the proposed filter is given by Equation (6).

$$f_{res} = \frac{1}{2\pi} \sqrt{\frac{L_{inv} + L_{grid}}{L_{inv} L_{grid}(C_h + C_r)}} \tag{6}$$

2.3. Selection of Individual Filter Components

The R_h-C_h branch in the proposed filter is designed to provide a path for the current harmonics at and around the switching frequency, $\omega = \omega_{switch}$, as given by Equation (7).

$$Z_{R_h C_h switch} = R_h + \frac{1}{\omega_{switch} C_h} < Z_{R_r C_r switch} = R_r + \frac{1}{\omega_{switch} C_r} \tag{7}$$

where $Z_{R_h C_h switch}$ is the impedance offered by the R_h-C_h branch at the switching frequency and $Z_{R_r C_r switch}$ is the impedance of the R_r-C_r branch at the switching frequency. The R_r-C_r branch is designed to provide a path for the current around the resonant frequency, $\omega = \omega_{res}$ as given by Equation (8).

$$Z_{R_r C_r res} = R_r + \frac{1}{\omega_{res} C_r} < Z_{R_h C_h res} = R_h + \frac{1}{\omega_{res} C_h} \tag{8}$$

where $Z_{R_r C_r res}$ and $Z_{R_h C_h res}$ are the impedances offered by the R_r-C_r and R_h-C_h branches at resonant frequency. A careful design is required for the proper choice of R_r-C_r and R_h-C_h branch values. Equal capacitor values are considered for both the branches in [20]. Further, the overall capacitance of the filter circuit should also meet the reactive power absorption of lesser than 5% as given in Equation (2). Choice of equal capacitor values for the proposed structure may not offer the required impedances at the resonant and switching frequencies. Hence, as per Equations (7) and (8), a new factor called capacitance ratio is introduced, which is given by C_r/C_h with C_h less than the C_r branch value. From the knowledge of the switching frequency and the resonant frequency, which is usually near to half of the switching frequency value, the design of C_h and C_r is carried out. The values of R_r and R_h are required to compute the impedance offered by the two parallel branches which influences the harmonic and resonant current flow path and subsequently the losses.

Again, the selection of R_r and R_h requires careful steps to follow Equations (7) and (8). The impedance of the R_r-C_r and R_h-C_h circuit branch is calculated and given by Equation (9).

$$Z_{parallel}(s) = \frac{(1 + R_r C_r s)(1 + R_h C_h s)}{R_r C_r C_h s^2 + R_h C_r C_h s^2 + C_h s + C_r s} \tag{9}$$

The resistance R_r can vary between zero and infinity ohms. Therefore, the impedance is limited within the range given by Equation (10).

$$\frac{1 + R_h C_h s}{R_h C_r C_h s^2 + s(C_r + C_h)} \leq Z_{parallel}(s) \leq \frac{1 + R_h C_h s}{C_h s} \tag{10}$$

For a minimum value of R_h, accordingly, for a good damping effect, the resonant frequency should fall in the range calculated and given by Equation (11).

$$\frac{1}{R_r(C_r + C_h)} \leq \omega_{res} \leq \frac{1}{R_r C_h} \tag{11}$$

Equation (11) is rewritten as Equation (12) which is useful to fix the range of R_r.

$$\frac{1}{\omega_{res}(C_h + C_r)} \leq R_r \leq \frac{1}{\omega_{res} C_h} \tag{12}$$

where $\omega_{res} = 2\pi f_{res}$ and f_{res} is given by Equation (6). For a minimum value of R_h, as R_r varies between zero ohms and infinity ohms, the resonant frequency varies between a minimum value and maximum value as given by Equation (13).

$$\sqrt{\frac{L_{inv} + L_{grid}}{L_{inv} L_{grid}(C_h + C_r)}} \leq \omega_{res} \leq \sqrt{\frac{L_{inv} + L_{grid}}{L_{grid} L_{inv} C_h}} \tag{13}$$

From Equations (12) and (13), the value of R_r can be fixed. For lower values of resonant frequency, higher damping resistance value is obtained and vice versa.

The resistance R_h can vary between zero ohms and infinity ohms. Therefore, the impedance of the parallel filter branch is limited within values given by Equations (14) and (15).

$$Z_{imp}(s) = \frac{1 + R_r C_r s}{R_r C_r C_h s^2 + s(C_r + C_h)} \tag{14}$$

$$Z_{imp}(s) = \frac{1 + R_r C_r s}{C_r s} \tag{15}$$

The aim of $R_h C_h$ branch is to provide path for high frequency harmonics. Further, the value of R_h should be lesser and simultaneously minimize damping loss with reduced R_r. Thus, for minimum R_h, based on impedance offered, the switching frequency is limited within the range given using Equation (16).

$$\frac{1}{R_r C_r} \leq \omega_{switch} \leq \frac{C_r + C_h}{R_r C_r C_h} \tag{16}$$

Introducing resistance ratio factor as

$$\beta = \frac{R_h}{R_r} \tag{17}$$

Combining Equations (16) and (17), the selection of R_h falls within the range as given by Equation (18).

$$\frac{\beta}{\omega_{switch} C_r} \leq R_h \leq \frac{\beta(C_r + C_h)}{\omega_{switch} C_r C_h} \tag{18}$$

where the value of resistance ratio factor is less than unity to minimize damping loss. The next section discusses the effect of resistance ratio factor on damping power loss.

Since the aim of R_h-C_h branch is to pass on the maximum switching ripple components through it, the value of R_h is chosen to be smaller than R_r to provide the necessary lesser impedance at the switching frequency as $C_h < C_r$. Now, with two parallel branches, R_r-C_r and R_h-C_h, the current through R_r and R_h varies based on their impedance. In the proposed damping scheme, the switching harmonic currents largely pass through the R_h-C_h branch and the resonant current flows predominantly through the R_r-C_r branch. The R_h-C_h branch, with a lesser resistance and capacitance value provides an alternate non dissipative path (with a smaller resistance which also improves harmonic attenuation, the loss in the resistance is almost negligible) for the high frequency currents thereby reducing the harmonic damping losses. The R_r-C_r branch, with a higher resistance and a slightly high capacitance value than the R_h-C_h branch (In this branch, the resistance value required becomes small when compared to traditional R_d-C_d damping) provides an alternate path for the resonance currents thereby reducing the losses in the damping circuitry.

2.4. Damping Power Loss

The power dissipated in the R_r-C_r and R_h-C_h circuit is calculated separately at the fundamental and switching frequencies.

(i) Fundamental frequency: The power loss at fundamental frequency is given in Equation (19).

$$P_{fundamental} = \omega_{out}^2 V_{grid}^2 \left(\frac{C_r^2 R_r}{(1 + \omega_{out}^2 C_r^2 R_r^2)} + \frac{C_h^2 R_h}{(1 + \omega_{out}^2 C_h^2 R_h^2)} \right) \tag{19}$$

where ω_{out} is the fundamental frequency in rad/s. Further, with two parallel branches C_r and C_h, the resulting current through R_r is halved, thereby reducing the damping loss at fundamental frequency.

(ii) Switching frequency: The power loss at switching frequency is given by Equation (20).

$$P_{switching} = Re(V_{AM}Y_{damp}I^*{}_{damp})$$ (20)

where I_{damp} is the current through the damping branch comprising of parallel combination of R_r, C_r and R_h, C_h and is given by Equation (21).

$$I_{damp} = V_{damp}/Z_{damp}$$ (21)

where Z_{damp} is calculated from Equation (9) at switching frequency.

The voltage across the damping branch V_{damp} is given by Equation (22).

$$V_{damp} = V_{AM}Y_{dampM}$$ (22)

where V_{AM} gives the harmonic amplitudes of the five-level inverter output voltage given by [33] and Y_{dampM} (at $V_{grid} = 0$) is given by Equation (23)

$$Y_{dampM} = \frac{V_{damp}}{V_{out}}$$ (23)

and given as

$$Y_{damp} = \frac{Z_{damp} + Z_2}{Z_1Z_2 + Z_{damp}Z_1 + +Z_{damp}Z_2}$$ (24)

where Z_1 and Z_2 are the impedances offered by L_{inv} and L_{grid}, respectively.

(iii) Total Power Loss

The total power loss in the damping circuit is given the sum of the fundamental frequency power loss and the switching frequency power loss as Equation (25).

$$P_{total} = P_{fundamental} + P_{switching}$$ (25)

2.5. Optimization to Maximize Harmonic Attenuation using PSO

Frequent trial and error in a design procedure increases the need to optimize parameters. PSO is a stochastic optimization algorithm based on deterministic points of natural selection that simulate the natural process of group communication of a swarm of animals when they flock or hunt with a velocity. If one member finds a desired path, the rest of the swarm will follow it. The PSO algorithm imitates the behavior of such animals by particles with an initial velocity and position in space. With a random population (called swarm), each particle (member of swarm) flies through the search space and remembers the best position [25]. The members communicate among themselves these positions and adjust their own position and velocity in a dynamic fashion. This way the particles fly towards an optimum position. In passive filter designs, it is challenging to achieve maximum harmonic attenuation while selecting the filter components judiciously within their allowable limits. The selection of filter components affects the design performance index such as THD, size, harmonic attenuation ratio, etc. To maximize attenuation around the switching frequency and its multiples, PSO algorithm is employed in this work. The target of optimization is to obtain maximum value of harmonic attenuation considering the values of the filter inductance and capacitance. The objective function to obtain maximum harmonic attenuation for the proposed filter is expressed as Equation (26).

$$Obj = \max|G(j\omega)|_{\omega=\omega_{switch}}$$ (26)

where $G(j\omega)$ is given by is given by

$$G(j\omega)\,at(\omega = \omega_{switch}) = \frac{1}{(L_{inv}L_{grid}C_r + L_{inv}L_{grid}C_h)(j\omega)^3 + (L_{inv} + L_{grid})(j\omega)}$$

Subject to

(i) $L_{min} < L_{inv} < L_{max}$, where L_{min} and L_{max} are the minimum and maximum allowable inverter side inductance values, respectively.

(ii) $C_{min} < C_t < C_{max}$, where C_{min} and C_{max} are the minimum and maximum allowable capacitance values, respectively.

The relation between inverter side and grid side inductances and maximum/minimum value of inductance and capacitance is presented in Section 2.1.

(iii)   Harmonics greater than 35 should be less than <0.3% of the rated fundamental current as per IEEE 519-1992 standards. The current harmonic attenuation is computed by considering that, at high frequencies, the converter is a harmonic generator, while the grid is considered as short circuit source. The magnitude of switching ripple current to inverter voltage ripple at the switching frequency is evaluated using Equation (27).

$$\left.\frac{i_{out}(j\omega)}{V_{inv}(j\omega)}\right|_{\omega-\omega_{switch}} = \frac{C_rC_h(j\omega)^2 + (C_r + C_h)(j\omega) + 1}{\begin{aligned}&L_{inv}L_{grid}C_rC_h(j\omega)^4 \\ &+ (C_rL_{inv}L_{grid} + C_hL_{inv}L_{grid} + C_rC_hL_{inv} + C_rC_hL_{grid})(j\omega)^3 \\ &+ (C_rL_{inv} + C_hL_{inv} + C_rL_{grid} + C_hL_{grid})(j\omega)^2 \\ &+ (L_{inv} + L_{grid})(j\omega)\end{aligned}} \tag{27}$$

The magnitude of the switching ripple current at the switching frequency is guided by the recommendations of IEEE 519-1992 (harmonic magnitude to be lesser than 0.3% of fundamental). With inverter side inductance values of 1 mH, 2 mH and 3 mH computed as discussed in Section 2.1, the maximum magnitude of harmonics greater than 35 (since harmonics are mostly around switching frequency, the magnitude is evaluated at 10 kHz) is found to be 0.37%, 0.16% and 0.28% of rated fundamental current, respectively. Hence, the inverter inductance is chosen as 2 mH. Compared to step-by-step method, this value has 1% error.

2.6. Optimization to Minimize Damping Power Loss Using PSO

Reactive components used in passive filters inherently produce switching oscillations that needs to be damped. These oscillations appear as a peak amplitude at the resonant frequency. The damping resistance used to suppress these oscillations produces power loss and weakens the high frequency attenuation ability. To efficiently use the damping circuit in the proposed design, this work considers the damping power loss, damping resistance value and the damping of resonant oscillations. In the proposed work, care is taken that the damping resistance does not weaken the harmonic attenuation and does not contribute to damping power loss. The damped filter should show more attenuation at the resonant frequency as well as in the high frequency band. This work uses PSO algorithm to minimize power loss on the premise of allowable damping resistance value and minimum peaking in the resonance. An objective function is defined that minimizes the power loss. In the following section, the steps towards an optimal design to minimize damping power loss is described:

Particle: The particle of particle swarm optimization is denoted by a set of two-dimensional vector as $X = \left[P_{total}, R_d, |G(j\omega)|_{\omega=\omega_{res}}\right]$.

Objective Function: This paper uses Equation (28) as the objective function that minimizes the damping power loss.

$$Obj = \min(P_{total}) \tag{28}$$

where P_{total} is defined from Equation (25) that includes fundamental and switching frequency power loss given by Equations (23) and (24).

Subject to the Constraints:

(a) The harmonics produced due to PWM is mainly located at the switching frequency and its multiples. The frequency response characteristics of the proposed filter shows that the filter produces attenuation with a slope of 60 dB/decade after the resonant frequency. Therefore, if the switching frequency is located far from the resonant frequency, the greater the attenuation at the switching frequency would be. The amplitude function of the grid current to the inverter voltage at the switching frequency is given using Equation (29).

$$G(j\omega)_{\omega=\omega_{res}} = \frac{R_r C_r R_h C_h (j\omega)^2 + (R_r C_r + R_h C_h)(j\omega) + 1}{\begin{aligned}&L_{inv} L_{grid} C_r C_h (R_r + R_h)(j\omega)^4 + (C_r L_{inv} L_{grid} + C_h L_{inv} L_{grid}\\ &+ C_r C_h L_{inv} R_r R_h + C_r C_h L_{grid} R_r R_h)(j\omega)^3\\ &+ (C_r R_r L_{inv} + C_h R_h L_{inv} + C_r R_r L_{grid} + C_h R_h L_{grid})(j\omega)^2\\ &+ (L_{inv} + L_{grid})(j\omega)\end{aligned}} \tag{29}$$

(b) The shunt damping resistance is limited as $R_{dmin} \leq R_d \leq R_{dmax}$, where $R_d = R_r + R_h$ and R_{dmin} and R_{dmax} are the minimum and maximum allowable values of the shunt damping resistance (sum of R_r and R_h) value, respectively, as discussed in Section 2.2.

The search process begins with the initialization of the particles, number of populations, generations, best values of the objective function in the local search and global search [28]. The defined objective function is calculated for various values of resistance and is stored as an array. If the constraints are satisfied, a comparison is established between the obtained objective function value and the best value that is searched by the particles and is stored as P_{best}. A comparison is also made with the global best value that is searched by all the particles and stored as G_{best}. The search is repeated until all the populations and generations are exhausted. The PSO algorithm renews the position and velocity to search the whole space by $V = \lambda.V + \alpha 1 * rand() * (P_{best} - Present) + \alpha 2 * rand() * (G_{best} - Present)$ where "V" is the velocity of the particle, Present is the present position of the particle, P_{best} and G_{best} represents the partial optimal solution and global optimal solution. λ is weighted factor ranging from 0.1 to 0.9. $\alpha 1$ and $\alpha 2$ are the learning factors, commonly taken as a value equal to two [25].

3. Parameter Design, Verification and Discussion

With above design and constraints considered, the proposed filter configuration is constructed for a five-level inverter. Table 1 depicts the system parameters. Table 2 shows the designed filter component values. To assess the performance of the proposed filter, two filters namely LCL filter with resistance damping [21,22] (Filter I), and R_d-C_d damped LCL filter [20,21] (Filter II) are compared with the proposed filter (Filter III). The overall inductance and capacitance values of all filters (for Filter III, capacitance becomes lesser as per design) are maintained the same. The damping circuit parameters of Filter I and Filter II are based on [21] and are listed in Table 2.

Table 1. System Design Parameters.

Parameters	Specifications
Output Voltage	110 V
Output Power	1 kW
Input voltage	78 V to each H-Bridge
Switching frequency	10 kHz

Table 2. Filter Design Parameters.

Parameter	Filter I	Filter II	Filter III
L_{inv}	1.5 mH	1.5 mH	2 mH
L_{grid}	1.5 mH	1.5 mH	1 mH
C_r	8 µF	4 µF	4 µF
R_r	20 Ω	20 Ω	9 Ω
C_h	-	4 µF	2 µF
R_h	-	-	1 Ω

The frequency response characteristics obtained for the three filters based on the above design parameters is depicted in Figure 4. It is clear that the proposed filter provides greater attenuating effect on the harmonics when compared to the R_d-C_d damped LCL filter. The proposed filter exhibits high harmonic attenuation capability in the high frequency band while maintaining the same overall inductance and capacitance values compared to Filter I and Filter II.

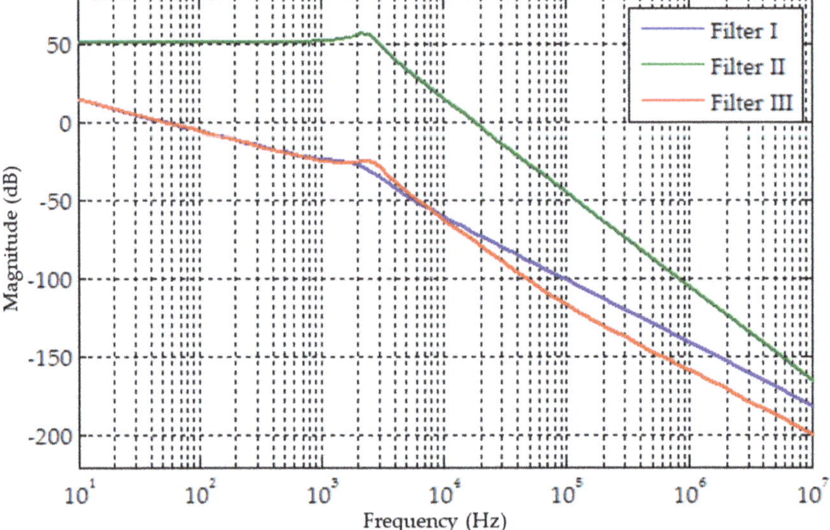

Figure 4. Frequency response characteristics with the three filters: Filter I, resistance damped LCL filter; Filter II, R_d-C_d damped LCL filter; and Filter III, proposed filter.

From Equation (25), it is clear that changes in inverter switching frequency changes the damping power loss. If switching frequency varies, the parameters of the filters also vary. High switching frequency operation necessitates lesser values of filter components. Figure 5 presents the damping loss achieved with all the three filters for different values of switching frequency. It is seen that Filter III provides reduced damping loss compared to Filter I and Filter II at the switching frequencies considered because the impedance of the R_h-C_h branch is less at the switching frequency, providing a bypass path for high frequency harmonic currents as per the proposed design.

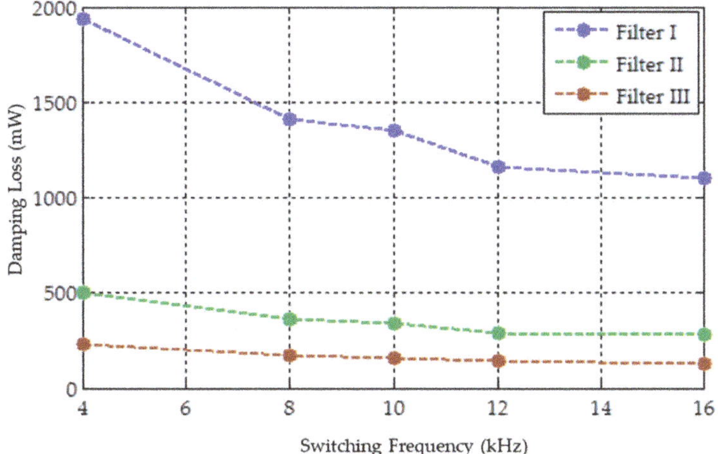

Figure 5. Damping loss with switching frequency variation.

Figure 6 presents the damping loss achieved with all three filters for different values of modulation index. With the proposed filter, lower damping power loss is obtained compared to the other two filters at all of the modulation indices considered, as the majority of the high frequency harmonic currents circulate through the R_h-C_h branch, bypassing the R_r branch.

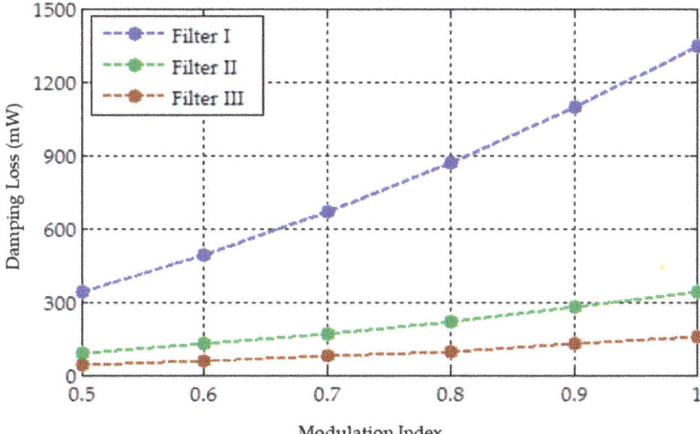

Figure 6. Damping loss for different modulation index values.

Figure 7 presents the damping loss variation achieved for variation in the values of the resistance ratio by keeping R_r constant. It is seen in Figure 7 that, at higher values of β, the damping power loss is reduced for any value of R_r. Again, for a particular β, as the value of R_r increases, the damping losses increase. This is expected, as increase in the resistance value increases the damping power loss. Thus, to reduce the damping power loss, lower values of R_r is chosen and, at the same time, higher value of resistance loss ratio is chosen. Choosing a higher resistance ratio factor helps in choosing lower values of R_r in the proposed design. Hence, the resistance parameter selection for both the parallel branch circuits of the proposed filter plays a vital role in reducing the overall damping losses. The curves validate the design value selection of the resistances used in the proposed filter design.

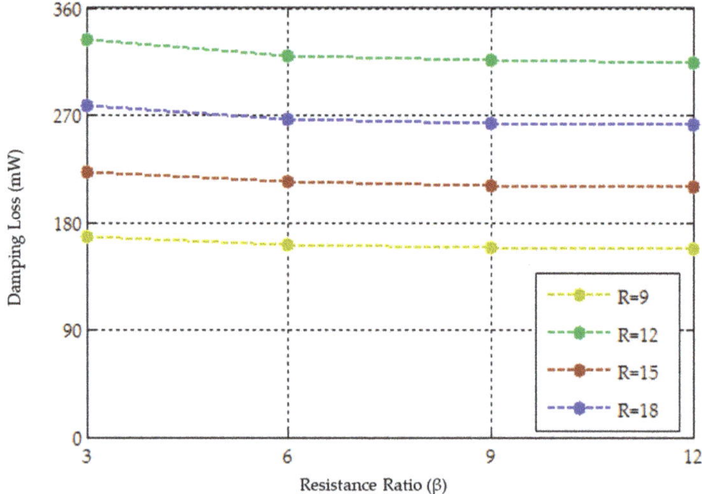

Figure 7. Damping loss variation with resistance ratio.

Figures 8 and 9 present the variation of damping loss and harmonics (maximum harmonic magnitude with respect to fundamental) achieved for different values of C_r keeping capacitance ratio constant. It is seen in Figure 8 that, as the value of C_r increases, the damping losses increase at a capacitance ratio value, whereas the percentage magnitude of the maximum harmonic with respect to the fundamental decreases. Again, for the same C_r, as the capacitance ratio increases, the damping power losses increase with a decrease in the harmonic magnitude percentage. Small capacitance value contributes to significant ripple on the output of the inverter. There is a minimum capacitance value on these curves which attributes to lower damping loss and lesser percentage of harmonic magnitude with respect to the fundamental. Thus, the capacitance parameter selection for both the parallel branch circuits of the proposed filter plays a vital role in reducing the overall damping losses. The curves validate the design value selection of the capacitance used in the proposed filter design.

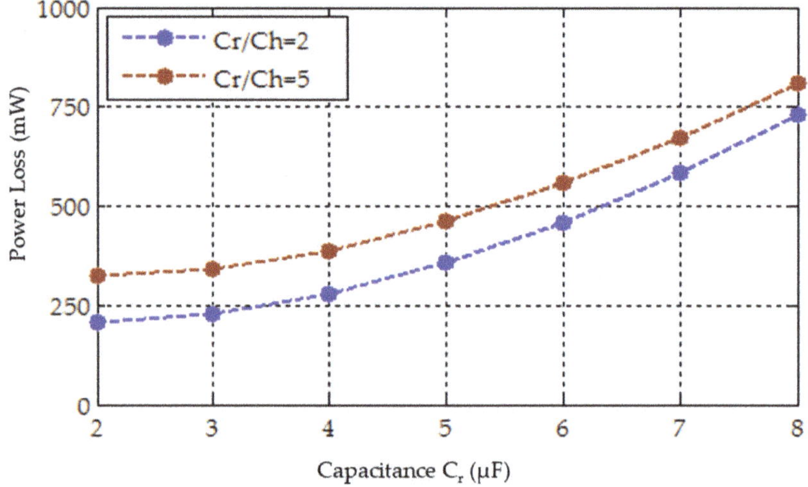

Figure 8. Damping loss variation with capacitance ratio.

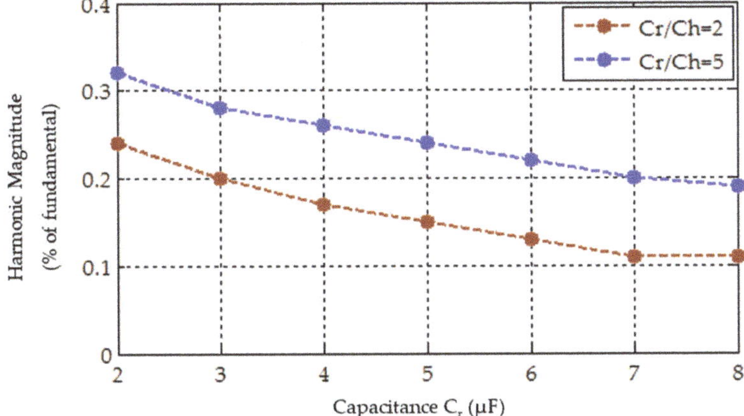

Figure 9. Harmonic variation with capacitance ratio.

Figure 10 shows the frequency response plot obtained with the proposed filter for two varying values of R_r (depicted as R1 and R2 in the plot for convenience) with the resistance ratio remaining constant. It can be seen that, at constant resistance ratio, for smaller values of R_r (represented as R1 in the plot), the attenuation in the high frequency band is high.

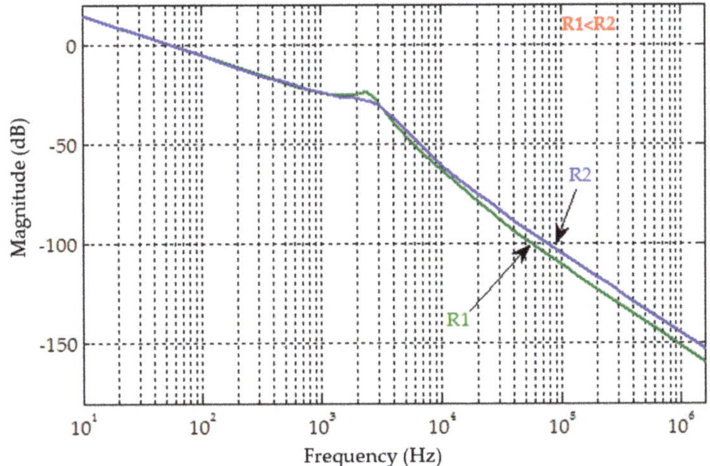

Figure 10. Frequency response characteristics for varying R_r with resistance ratio constant.

Figure 11 shows the frequency response plot of the proposed filter for two different values of resistance ratio (depicted as G1 and G2 in the plot for convenience) for constant values of R_r. It can be seen that, at higher values of resistance ratio (pictured as G1 in the plot), the attenuation in the high frequency band is high.

Thus, to reduce the damping power loss and simultaneously achieve high harmonic attenuation, lower values of R_r is chosen and, at the same time, higher value of resistance loss ratio is chosen. The frequency response plots also validate the design value selection of the resistances used in the proposed filter design.

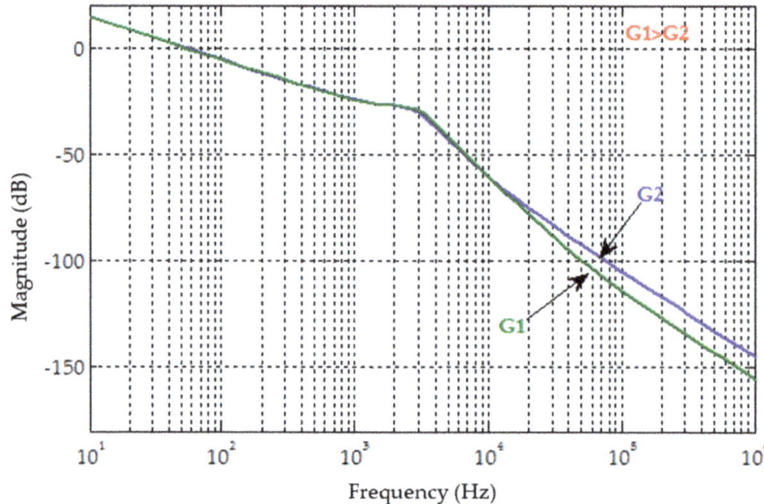

Figure 11. Frequency response characteristics for varying resistance ratio with R_r constant.

Table 3 shows the variation of harmonics around the switching frequency and twice the switching frequency for the three filters. It is evident that, with the proposed filter, the dominating current harmonics can meet the recommendations of IEEE 519-1992. The other two filters also follow the recommendations, except for Filter I where the percentage near the switching frequency is slightly higher. The proposed filter provides a much lesser harmonic magnitude percentage when compared with the conventional LCL filter with simple resistance damping. Thus, the proposed filtering scheme with a carefully designed filter parameter values is an attempt to reduce the damping power loss and at the same time provide a higher harmonic attenuation rate in the high frequency band of around switching frequency and its multiples.

Table 3. Variation of harmonics around switching frequency.

	Max Harmonics Around Switching Frequency	Max Harmonics Around Twice Switching Frequency
Filter I	0.45	0.32
Filter II	0.29	0.17
Filter III	0.16	0.15

Table 4 presents the comparison of the proposed filter configuration with Filter I and Filter II based on number of components, attenuation and damping loss.

Table 4. Comparison of different filter configurations.

Filter/Parameter	Filter I	Filter II	Filter III
Number of inductors	2	2	2
Number of capacitors	1	2	2
Number of resistors	1	1	2
Total volume of inductance	100 cm^3	100 cm^3	84 cm^3
Total volume of capacitance	36 cm^3	32 cm^3	26 cm^3
Harmonic Attenuation in high frequency band	60 dB/decade	40 dB/decade	100 dB/decade
Damping Loss	2.45 W	1.08 W	0.16 W

4. Experimental Results

With reference to the design procedure outlined in Section 2, a laboratory setup of a 1 kW five-level CHB inverter based on FPGA processor is developed as per specifications in Tables 1 and 2. The grid is simulated using a programmable ac source (Chroma 61511). The filter parameter selection is based on the aforementioned design described in Section 2. PDPWM technique is employed to generate switching pulses for the inverter. The hardware implementation of PDPWM using FPGA offers computational simplicity, fast prototyping, simple hardware and software design. An image of the experiment set up is shown in Figure 12.

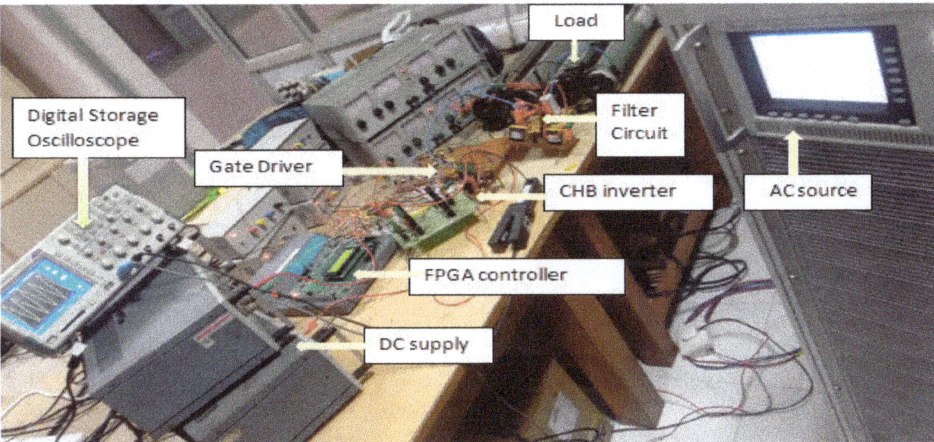

Figure 12. Picture of the experimental set-up.

The experimental output voltage/current waveforms, which are captured before connecting the proposed passive filter to the inverter output, are shown in Figure 13a, and the corresponding spectrum of the inverter output current is shown in Figure 13b. As expected, the waveforms in Figure 13a show a five-level inverter output. From the harmonic spectrum, it is evident that the uppermost harmonics of the five-level inverter output current are around the switching frequency of 10 kHz. The spectrum presents harmonic components not only at the switching frequency and respective multiples (20 kHz, 30 kHz, 40 kHz, etc.) but also in the neighborhood of these frequencies in the form of sidebands. The THD content on the output current waveform is measured as 26.90% before connecting the proposed passive filter.

The proposed filter structure is designed to attenuate the higher order harmonics from the inverter output. Experiments are carried out to compare the performance of the proposed filter with two other passive filter configurations. The experimental output waveforms and the current harmonic spectra obtained by connecting the passive filters are shown in Figures 14–16. Filter I is the resistive damped LCL filter strategy shown in Figure 14. Filter II is the R_d-C_d damped LCL filter configuration shown in Figure 15. Filter III is the proposed filter strategy shown in Figure 16. In Figures 13b, 14b, 15b and 16b, the significant harmonics of the output currents are highlighted. It can be inferred from the spectra of Figures 13b, 14b, 15b and 16b that, before connecting the passive filter, the inverter output has the maximum harmonic amplitude lying at the switching frequency of 10 kHz. With the implementation of passive filter, the current harmonic components are reduced by a factor of nearly 6 dB/Hz from Filter II to Filter III and by a factor of almost 12 dB/Hz from Filter I to Filter III at the switching frequency of 10 kHz. With the proposed filter, the percentage of harmonics in the high frequency band (in and around multiples of switching frequency) is reduced compared to the spectra of Filter I and Filter II, showing that the proposed filter has better harmonic attenuating effect in the high frequency band.

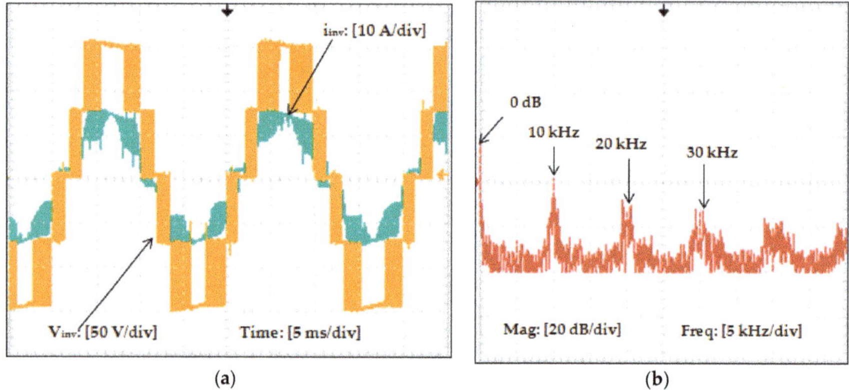

Figure 13. Experimental waveforms of the inverter output before connecting passive filter: (**a**) output voltage and current waveforms; and (**b**) output current harmonic spectrum.

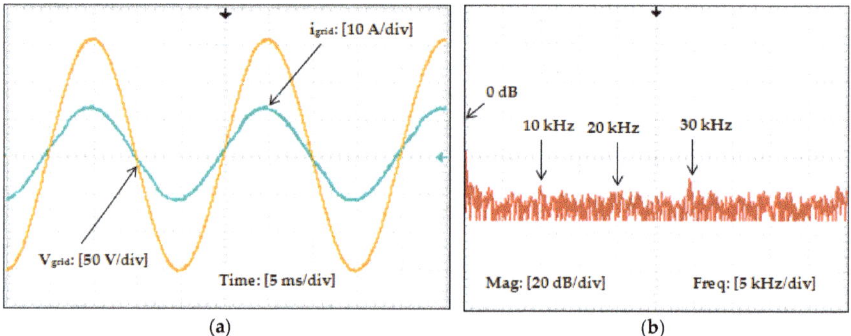

Figure 14. Experimental waveforms with Filter I: (**a**) output voltage and current waveforms; and (**b**) output current harmonic spectrum.

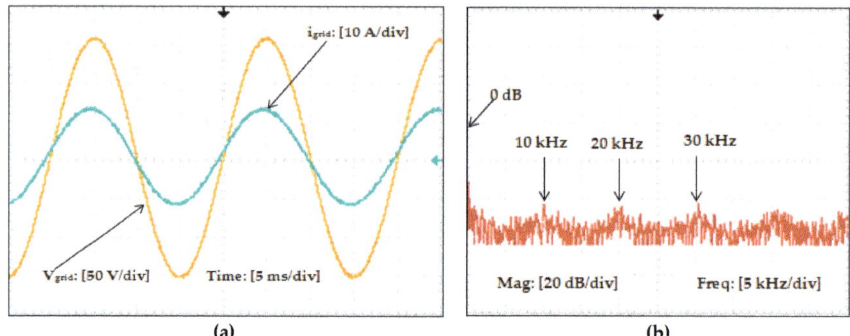

Figure 15. Experimental waveforms with Filter II: (**a**) output voltage and current waveforms; and (**b**) output current harmonic spectrum.

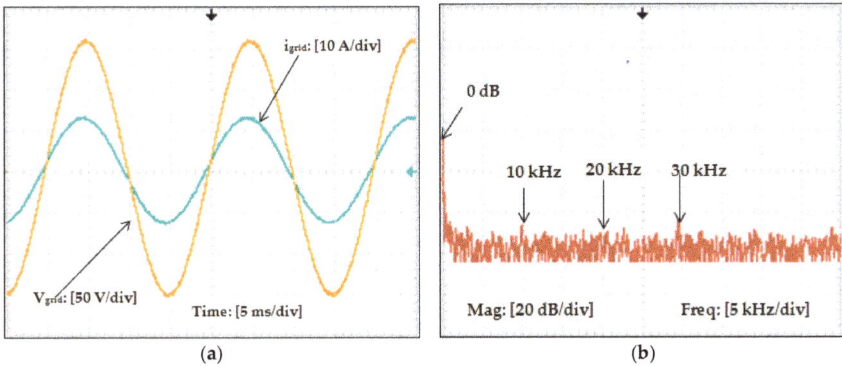

Figure 16. Experimental waveforms with Filter III: (**a**) output voltage and current waveforms; and (**b**) output current harmonic spectrum.

From the spectra of Figures 14b, 15b and 16b, the amplitudes of the dominant harmonic current (greater than or equal to 35th) are listed in Table 5 for all three filters. It can be seen that the dominating harmonic current meets the IEEE recommendations in Filter III. The measured THD for the three filters is depicted in Table 5, which shows that the inverter output current with Filter III has the lowest THD of 0.82%.

Table 5. Percent Total Harmonic Distortion (THD)and dominant harmonic's magnitude in different cases.

	Filter I	Filter II	Filter III
THD (%)	0.98	0.99	0.82
Dominant Harmonic's Magnitude (%)	0.47	0.29	0.18

With Filter I, the amplitude of the harmonics at the switching frequency of 10 kHz is measured as 0.47% of fundamental; with Filter II, the harmonic amplitude at the switching frequency is measured to be 0.29% of fundamental; and, with Filter III, the harmonic amplitude at the switching frequency is measured to be 0.18% of fundamental. The amplitude of the harmonics (percent of the fundamental current) at the switching frequency, and twice, thrice, four times and five times the switching frequency with all the three filters are listed in Table 6.

Observing the output waveforms and the harmonic spectra in Figure 16, it can be seen that the designed filter has good attenuation effects on the harmonics. The percentage magnitude of harmonics around the switching frequency (10 kHz) is greatly reduced. The output waveforms have become sinusoidal and the THD content on the inverter output current is measured as 0.82% after connecting the proposed filter.

Table 6. Maximum harmonic magnitude as percentage of fundamental with the three filters.

Filter	Maximum Harmonics around f_{switch}	Maximum Harmonics around $2f_{switch}$	Maximum Harmonics around $3f_{switch}$	Maximum Harmonics around $4f_{switch}$	Maximum Harmonics around $5f_{switch}$
Filter I	0.47%	0.32%	0.28%	0.18%	0.07%
Filter II	0.29%	0.18%	0.08%	0.06%	0.04%
Filter III	0.18%	0.16%	0.02%	0.01%	0.01%

Figure 17a–c shows the measured damping branch currents obtained with all three filters when the output voltage is 110 V. The RMS values of the damping branch currents (i_{dam}) are measured as

353 mA, 237 mA and 138 mA with Filter I, Filter II and Filter III, respectively. Thus, the loss in the damping circuit is computed as 2.49 W, 1.12 W and 0.17 W for Filter I, Filter II and Filter III, respectively, which is in very close agreement with the theoretical analysis.

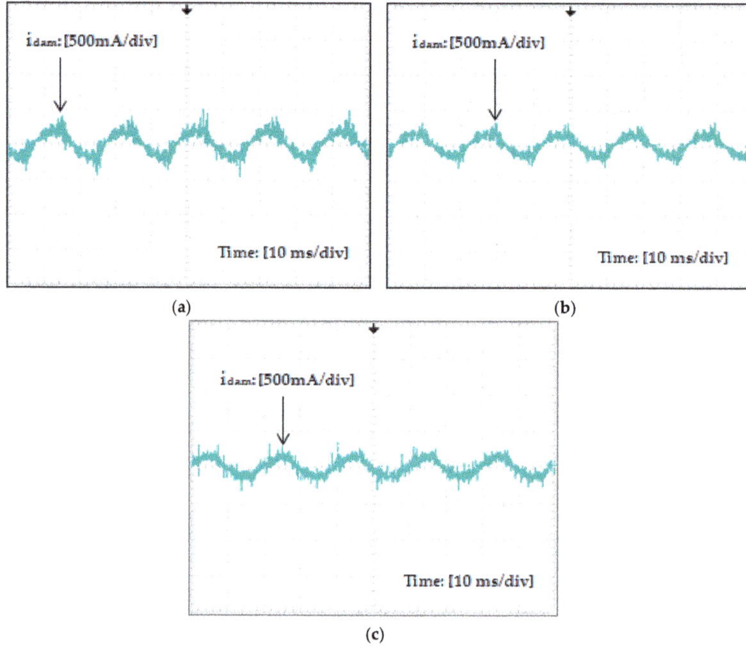

Figure 17. Experimental damping branch current waveforms with: (**a**) Filter I; (**b**) Filter II; and (**c**) Filter III.

The percentage damping loss reduction achieved with the proposed filter structure with respect to Filter I and Filter II is shown in Figure 18. It is clear in Figure 18 that Filter III achieves 93% damping power loss reduction with respect to Filter I. With respect to Filter II, 85% damping power loss reduction is achieved, which is in good agreement with the theoretical analysis.

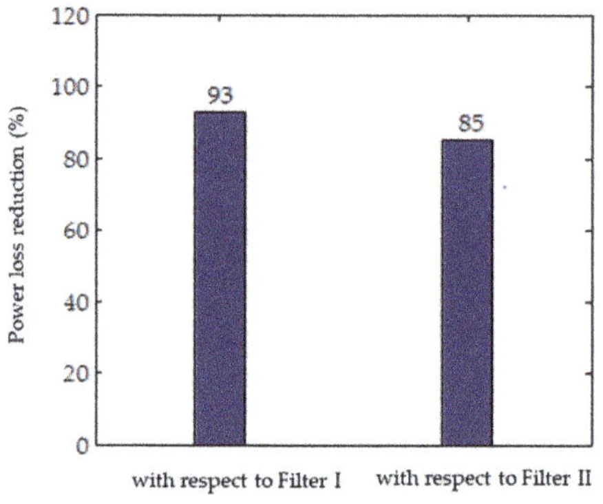

Figure 18. Damping loss reduction in Filter III.

5. Conclusions

In this paper, a novel filter structure with improved power quality and low damping power loss is proposed for a five-level CHB inverter used in grid connected applications. An additional resistance–capacitance branch in parallel to the filter capacitor of the traditional LCL filter in addition to a resistance in series with the capacitor reduces the damping power loss and provides improved harmonic attenuation. A detailed parameter design procedure of the filter components is introduced providing expressions to obtain individual filter component values. PSO algorithm is employed to minimize damping power loss on the premise of lower resonant peak and increased harmonic attenuation. Similarly, PSO algorithm maximizes the harmonic attenuation in and around the switching frequency on the premise of allowable inductance and capacitance values. A 1 kW, 110 V, 50 Hz grid connected five-level CHB inverter with the proposed filter structure is implemented on an experimental set up to verify the validity of the filter configuration through experimental results. SPARTAN 6-XC6SLX25 FPGA processor is used to implement the modulation algorithm for the five-level inverter. The results show that the proposed filter structure is beneficial in providing a significant amount of harmonic attenuation in the high frequency band compared to alternative filter topologies, while maintaining the same overall inductance, reduced capacitance and resistance values thereby reducing the overall filter size. The THD content on the output current waveforms with the designed filter is 0.82% and the high frequency spectrum is well within the IEEE requirements. An approximate damping power loss reduction of nearly 89% is achieved with the proposed filter structure compared to the conventional topologies for the multilevel inverter based grid connected system.

Author Contributions: Meenakshi Jayaraman and Sreedevi VT conceived and designed the work. Meenakshi Jayaraman conducted the experiments and analysed the data. Both authors contributed to article writing.

Conflicts of Interest: The authors declare no conflict of interest.

References

1. Rockhill, A.A.; Liserre, M.; Teodorescu, R.; Rodriguez, P. Grid-filter design for a multimegawatt medium-voltage voltage-source inverter. *IEEE Trans. Ind. Electron.* **2011**, *58*, 1205–1217. [CrossRef]
2. Gabe, I.J.; Montagner, V.F.; Pinheiro, H. Design and implementation of a robust current controller for VSI connected to the grid through an LCL filter. *IEEE Trans. Power Electron.* **2009**, *24*, 1444–1452. [CrossRef]
3. Turkay, B.E.; Telli, A.Y. Economic analysis of standalone and grid connected hybrid energy systems. *Renew. Energy* **2011**, *36*, 1931–1943. [CrossRef]
4. Lai, N.B.; Kim, K.H. An improved current control strategy for a grid-connected inverter under distorted grid conditions. *Energies* **2016**, *9*, 190. [CrossRef]
5. Tang, Y.; Loh, P.C.; Wang, P.; Choo, F.H.; Gao, F. Exploring inherent damping characteristic of LCL-filters for three-phase grid-connected voltage source inverters. *IEEE Trans. Power Electron.* **2012**, *27*, 1433–1443. [CrossRef]
6. Balasubramanian, A.K.; John, V. Analysis and design of split-capacitor resistive-inductive passive damping for LCL filters in grid-connected inverters. *IET Power Electron.* **2013**, *6*, 1822–1832. [CrossRef]
7. Yu, Y.; Li, H.; Li, Z.; Zhao, Z. Modeling and analysis of resonance in LCL-type grid-connected inverters under different control schemes. *Energies* **2017**, *10*, 104. [CrossRef]
8. Liserre, M.; Blaabjerg, F.; Hansen, S. Design and control of an LCL-filter-based three-phase active rectifier. *IEEE Trans. Ind. Appl.* **2005**, *41*, 1281–1291. [CrossRef]
9. Cha, H.; Vu, T.K. Comparative analysis of low-pass output filter for single-phase grid-connected photovoltaic inverter. In Proceedings of the Twenty-Fifth Annual IEEE Applied Power Electronics Conference and Exposition (APEC), Palm Springs, CA, USA, 21–25 February 2010; pp. 1659–1665.
10. Zhang, N.; Tang, H.; Yao, C. Analysis of active damping of LCL filter used in single-phase PV system in discrete domain. *Int. J. Electron.* **2015**, *102*, 1022–1043. [CrossRef]
11. Zhang, N.; Tang, H.; Yao, C. A systematic method for designing a PR controller and active damping of the LCL Filter for single-phase grid-connected PV inverters. *Energies* **2014**, *7*, 3934–3954. [CrossRef]

12. Wang, T.C.Y.; Ye, Z.; Sinha, G.; Yuan, X. Output filter design for a grid-interconnected three-phase inverter. In Proceedings of the IEEE 34th Annual Power Electronics Specialists Conference, Acapulco, Mexico, 15–19 June 2003; Volume 2, pp. 779–784.

13. Ahmed, K.H.; Finney, S.J.; Williams, B.W. Passive filter design for three-phase inverter interfacing in distributed generation. In Proceedings of the IEEE 5th International Conference-Workshop on Compatibility in Power Electronics, Gdansk, Poland, 29 May–1 June 2007; pp. 1–9.

14. Beres, R.; Wang, X.; Blaabjerg, F.; Bak, C.L.; Liserrie, M. A review of passive filters for grid-connected voltage source converters. In Proceedings of the Twenty-Ninth Annual IEEE Applied Power Electronics Conference and Exposition (APEC), Fort Worth, TX, USA, 16–20 March 2014; pp. 2208–2215.

15. Petterson, S.; Salo, M.; Tuusa, H. Applying an LCL-filter to a four-wire active power filter. In Proceedings of the 37th IEEE Power Electronics Specialists Conference, Jeju, Korea, 18–22 June 2006; pp. 1–7.

16. Routimo, M.; Tuusa, H. LCL type supply filter for active power filter—Comparison of an active and a passive method for resonance damping. In Proceedings of the IEEE Power Electronics Specialists Conference, Orlando, FL, USA, 17–21 June 2007; pp. 2939–2945.

17. Dahono, P.A.; Bahar, Y.R.; Sato, Y.; Kataoka, T. Damping of transient oscillations on the output LC filter of PWM inverters by using a virtual resistor. In Proceedings of the 4th IEEE International Conference on Power Electronics and Drive Systems, Denpasar, Indonesia, 25–25 October 2001; Volume 1, pp. 403–407.

18. Habetler, T.G.; Naik, R.; Nondahl, T.A. Design and implementation of an inverter output LC Filter used for dv/dt reduction. *IEEE Trans. Power Electron.* **2002**, *17*, 327–331. [CrossRef]

19. Tanaka, T.; Fujikawa, S.; Funabiki, S. A new method of damping harmonic resonance at the DC link in large-capacity rectifier-inverter systems using a novel regenerating scheme. *IEEE Trans. Ind. Appl.* **2002**, *38*, 1131–1138. [CrossRef]

20. Channegowda, P.; John, V. Filter optimization for grid interactive voltage source inverters. *IEEE Trans. Ind. Electron.* **2010**, *57*, 4106–4114. [CrossRef]

21. Wu, W.; He, Y.; Tang, T.; Blaabjerg, F. A new design method for the passive damped LCL and LLCL filter-based single-phase grid-tied inverter. *IEEE Trans. Ind. Electron.* **2013**, *60*, 4339–4350. [CrossRef]

22. Wu, W.; He, Y.; Blaabjerg, F. An LLCL power filter for single-phase grid-tied inverter. *IEEE Trans. Power Electron.* **2012**, *27*, 782–789. [CrossRef]

23. Xu, J.; Yang, J.; Ye, J.; Zhang, Z.; Shen, A. An LTCL filter for three-phase grid-connected converters. *IEEE Trans. Power Electron.* **2014**, *29*, 4322–4338. [CrossRef]

24. Anzalchi, A.; Moghaddami, M.; Moghaddasi, A.; Sarwat, A.I.; Rathore, A.K. A new topology of higher order power filter for single-phase grid-tied voltage-source inverters. *IEEE Trans. Ind. Electron.* **2016**, *63*, 7511–7522. [CrossRef]

25. Singh, S.; Singh, B. Optimized passive filter design using modified particle swarm optimization algorithm for a 12-pulse converter-fed LCI-synchronous motor drive. *IEEE Trans. Ind. Appl.* **2014**, *50*, 2681–2689. [CrossRef]

26. Liserre, M.; Dell'Aquila, D.; Blaabjerg, F. Genetic algorithm-based design of the active damping for an LCL-filter three-phase active rectifier. *IEEE Trans. Power Electron.* **2004**, *19*, 76–86. [CrossRef]

27. Coppola, M.; Napoli, F.D.; Guerriero, P.; Iannuzzi, D.; Daliento, S.; Pizzo, A.D. An FPGA-based advanced control strategy of a grid tied PV CHB inverter. *IEEE Trans. Power Electron.* **2016**, *31*, 806–816. [CrossRef]

28. Alepuz, S.; Busquets-Monge, S.; Bordonau, J.; Gago, J.; Gonzalez, D.; Balcells, J. Interfacing renewable energy sources to the utility grid using a three-level inverter. *IEEE Trans. Ind. Electron.* **2006**, *53*, 1504–1511. [CrossRef]

29. Chavarria, J.; Biel, D.; Guinjoan, F.; Meza, C.; Negroni, J.J. Energy-balance control of PV cascaded multilevel grid-connected inverters under level-shifted and phase-shifted PWMs. *IEEE Trans. Ind. Electron.* **2013**, *60*, 98–111. [CrossRef]

30. Rodriguez, J.; Lai, J.S.; Peng, F.Z. Multilevel inverters: A survey of topologies, controls, and applications. *IEEE Trans. Ind. Electron.* **2002**, *49*, 724–738. [CrossRef]

31. Malinowski, M.; Gopakumar, K.; Rodriguez, J.; Perez, M.A. A survey on cascaded multilevel inverters. *IEEE Trans. Ind. Electron.* **2010**, *57*, 2197–2206. [CrossRef]

32. McGrath, B.P.; Holmes, D.G. Multicarrier PWM strategies for multilevel inverters. *IEEE Trans. Ind. Electron.* **2002**, *49*, 858–867. [CrossRef]

33. Holmes, D.G.; Lipo, T.A. *Pulse width Modulation for Power Converters: Principles and Practice*, 1st ed.; Wiley-IEEE Press: Hoboken, NJ, USA, 2003; pp. 99–104.

34. Odeh, C.I.; Nnadi, D.B.N. Single-phase 9-level hybridised cascaded multilevel inverter. *IET Power Electron.* **2013**, *6*, 468–477. [CrossRef]

35. Rahim, N.A.; Selvaraj, J.; Krismadinata, C. Five-level inverter with dual reference modulation technique for grid-connected PV system. *Renew. Energy* **2010**, *35*, 712–720. [CrossRef]

36. Palanivel, P.; Dash, S.S. Analysis of THD and output voltage performance for cascaded multilevel inverter using carrier pulse width modulation techniques. *IET Power Electron.* **2011**, *4*, 951–958. [CrossRef]

37. Gupta, K.K.; Ranjan, A.; Bhatnagar, P.; Sahu, L.K.; Jain, S. Multilevel inverter topologies with reduced device count: A review. *IEEE Trans. Power Electron.* **2016**, *31*, 135–151. [CrossRef]

38. Kim, S.H.; Kim, Y.H. Design and analysis of an LC trap/LCR output filter for a single-phase NPC three-level inverter. *Int. J. Electron.* **2008**, *95*, 1279–1292. [CrossRef]

39. Jiao, Y.; Lee, F.C. LCL filter design and inductor current ripple analysis for a three-level NPC grid interface converter. *IEEE Trans. Power Electron.* **2015**, *30*, 4659–4668. [CrossRef]

40. Pandey, R.; Tripathi, R.N.; Hanamoto, T. Comprehensive analysis of LCL filter interfaced cascaded H-bridge multilevel inverter-based DSTATCOM. *Energies* **2017**, *10*, 346. [CrossRef]

Article

Space Vector Modulation for an Indirect Matrix Converter with Improved Input Power Factor

Nguyen Dinh Tuyen * and Phan Quoc Dzung

Faculty of Electrical and Electronics Engineering, Ho Chi Minh City University of Technology,
VNU-HCM Ho Chi Minh City 700000, Vietnam; pqdung@hcmut.edu.vn
* Correspondence: ndtuyen@hcmut.edu.vn; Tel.: +84-919-142-110

Academic Editor: José Gabriel Oliveira Pinto
Received: 16 February 2017; Accepted: 19 April 2017; Published: 25 April 2017

Abstract: Pulse width modulation strategies have been developed for indirect matrix converters (IMCs) in order to improve their performance. In indirect matrix converters, the LC input filter is used to remove input current harmonics and electromagnetic interference problems. Unfortunately, due to the existence of the input filter, the input power factor is diminished, especially during operation at low voltage outputs. In this paper, a new space vector modulation (SVM) is proposed to compensate for the input power factor of the indirect matrix converter. Both computer simulation and experimental studies through hardware implementation were performed to verify the effectiveness of the proposed modulation strategy.

Keywords: indirect matrix converter (IMC); input filter; input power factor; matrix converter (MC); space vector modulation (SVM); power quality

---

1. Introduction

Matrix converters (MCs) provide a number of advantages, including sinusoidal input and output currents, regeneration capability, and compact size with good power-to-weight ratio. The development of MCs began in the early 1980s when Alensia and Venturini introduced the basic principles of operation [1]. Afterwards, the MCs were applied to the adjustable motor speed drive, for renewable energy applications, power supplies, and many others application [2,3].

MCs are classified into direct matrix converters (DMCs) and indirect matrix converters (IMCs) [4]. Both converters are able to generate input/output waveforms with the same performance and the same voltage transfer ratio capability. The IMC is now preferable to the DMC because it has many extra advantages over the DMC, such as easy implementation, more secure computation, the possibility power switch number reduction, and the possibility of constructing AC-AC converters with multiple three-phase outputs and multiphase output voltages [5–7].

The modulation strategies for IMC have recently been discussed, and many researchers have developed various methods for IMC modulation. In [8], space vector modulation (SVM) methods were proposed for common mode voltage reduction. In order to reduce the harmonic level in the load side, a new modulation method based on the virtual DC-link voltage was applied [9]. However, the analysis of the input power factor of the IMC has not been presented in the literature.

Figure 1 shows the main circuit of the IMC. The IMC system consists of an input LC filter, a rectifier stage, and an inverter stage. In Figure 1, the LC input filter, which consists of three inductors and three capacitors, acts as an interface between the power supply and converter, so that it smoothens the input currents, and removes the electromagnetic interference (EMI) problems. However, it diminishes the power factor at the input power supply, and the input power factor cannot be unified anymore. Particularly, in the case of operation at low voltage output, the input power factor is greatly worsened.

It is well known that the poor input power factor induces negative effects on AC power source utilization, as well as increasing the magnitude of the electric current more than is necessary.

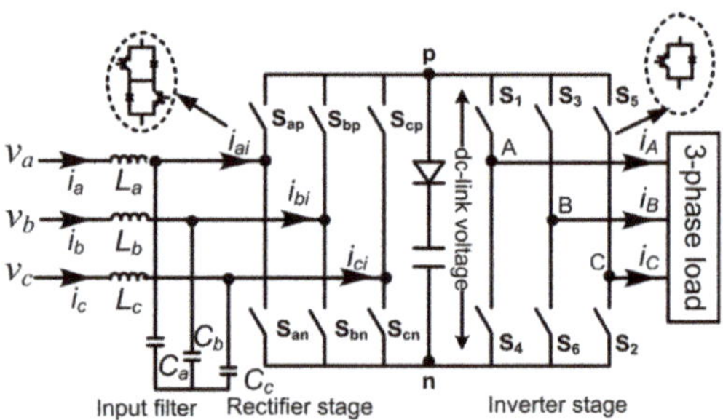

Figure 1. The indirect matrix converter (IMC) topology.

In case of the conventional space vector modulation method for the IMC, the input sectors and the duty cycles of the modulated switches in the rectifier stage are determined by the input phase voltage magnitude and its phase angle [5–9]. Although there is much work required to ensure good performance on the output voltage and input current by the IMC with the conventional SVM, papers investigating the input power factor problem are rare.

In this paper, we propose a new SVM method for the IMC, which can easily compensate for the input power factor without IMC performance degradation. The compensation algorithm is based on the calculation of the optimal compensation angle, which depends on input filter parameters and the demanded output load.

This paper is organized as follows: The IMC operation and the conventional SVM method are presented in Section 2. The effects of the input filter in the input side and the proposed SVM method are described in Section 3. The simulation and experimental results are provided in Sections 4 and 5, respectively. Some conclusions are given in the last section.

2. Operational Principles of IMC and Conventional SVM Method

2.1. Operational Principles of IMC

The power circuit of the IMC feeding a three-phase inductive load is shown in Figure 1. The voltages and currents at the power supply side are denoted by (v_a, v_b, v_c) and (i_a, i_b, i_c), respectively, while those of the load side are denoted by (v_A, v_B, v_C) and (i_A, i_B, i_C). The IMC consists of two stages, i.e., rectifier stage and inverter stage, as shown in Figure 1. The purpose of the rectifier stage is to generate a sinusoidal input current, as well as the positive DC-link voltage. The rectifier stage successively connects the positive input voltage to the positive pole (p), and the negative voltage to the negative pole (n) of DC-link bus. Based on DC-link voltage, which is determined by segment of two positive line-to-line input voltages, the inverter stage is modulated to generate the three-phase output voltage with the desired magnitude and variable frequency.

2.2. Conventional SVM Method

In the conventional SVM approach for the IMC, the operation of the rectifier stage depends only on the phase angle and instantaneous value of the input voltage. The balanced three–phase input voltage is given in Equation (1).

$$v_a = V_{in} \cos(\omega_{in} t)$$
$$v_b = V_{in} \cos(\omega_{in} t - 2\pi/3)$$
$$v_c = V_{in} \cos(\omega_{in} t - 4\pi/3)$$

(1)

where V_{in} is the input voltage magnitude and ω_{in} is the input angular frequency.

From the input voltage, the input sectors are defined as shown in Figure 2. The six input sectors shown in Figure 2 can be classified into two cases: In the first case, one input voltage is positive and two input voltages are negative (sectors 1, 3, 5). In the second case, two input voltages are positive and one input voltage is negative (sectors 2, 4, 6).

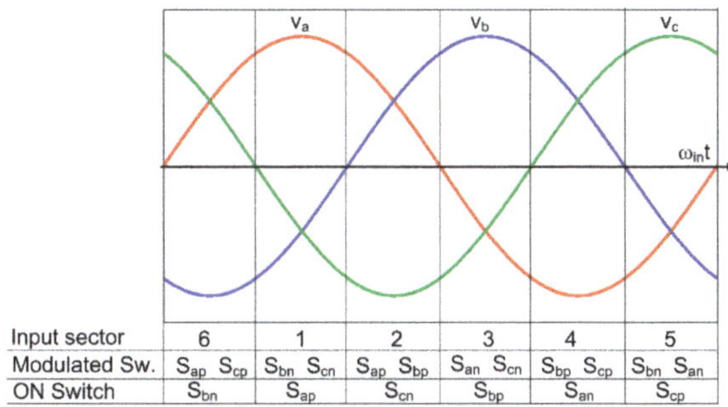

Input sector	6	1	2	3	4	5
Modulated Sw.	S_{ap} S_{cp}	S_{bn} S_{cn}	S_{ap} S_{bp}	S_{an} S_{cn}	S_{bp} S_{cp}	S_{bn} S_{an}
ON Switch	S_{bn}	S_{ap}	S_{cn}	S_{bp}	S_{an}	S_{cp}

Figure 2. The definition of input sectors.

To explain easily how to find the duty cycle of all switches in the rectifier stage, for example, we assume that the rectifier stage operates in sector 1 without missing the generality of the analysis. In this sector, the instantaneous supply voltage v_a is positive, while v_b and v_c are negative. Under this condition, the switch S_{ap} is always on while S_{bn} and S_{cn} are modulated. Based on the analysis in [5–9], the duty cycle of two switches S_{bn} and S_{cn} are given as:

$$d_{ab} = -\frac{v_b}{v_a} \tag{2}$$

$$d_{ac} = -\frac{v_c}{v_a} \tag{3}$$

During one sampling period, the DC-link voltage is modulated with two line-to-line input voltages; while S_{bn} is turned on, the DC-link voltage equals to v_{ab}, and while S_{cn} is turned on, the DC-link voltage equals to v_{ac}.

The average DC-link voltage is obtained as follows:

$$V_{dc} = d_{ab}(v_a - v_b) + d_{ac}(v_a - v_c) = \frac{3}{2}\frac{V_{in}^2}{v_a} \tag{4}$$

From Equation (4), the minimum value of the average DC-link voltage is

$$V_{dc(\min)} = \frac{3}{2}V_{in} \tag{5}$$

We can find the switching states, the corresponding DC-link voltage, and its average value for any other input sector, by utilizing the same approach, and the results are summarized in Table 1. Once the switching state of the rectifier stage is determined, the traditional space vector pulsewidth modulation (SVPWM) can be applied to control the inverter stage. For calculating the duty cycles of the active and zero vectors in the inverter stage, it is necessary to refer to the local average DC-link voltage value. The eight space vectors with the six active vectors ($V_1 \sim V_6$) and the two zero vectors (V_0, V_7) are used in the SVPWM method.

Table 1. The switching states and their DC-link voltages according to the input sector.

Input Sector	Conducting Switch	Modulated Switches	Dc-Link Voltage	Average Value (V_{dc})
1	S_{ap}	S_{bn}, S_{cn}	v_{ab}, v_{ac}	$3V_{in}^2/2v_a$
2	S_{cn}	S_{ap}, S_{bp}	v_{ac}, v_{bc}	$-3V_{in}^2/2v_c$
3	S_{bp}	S_{an}, S_{cn}	v_{ba}, v_{bc}	$3V_{in}^2/2v_b$
4	S_{an}	S_{bp}, S_{cp}	v_{ba}, v_{ca}	$-3V_{in}^2/2v_a$
5	S_{cp}	S_{bn}, S_{an}	v_{cb}, v_{ca}	$3V_{in}^2/2v_c$
6	S_{bn}	S_{ap}, S_{cp}	v_{ab}, v_{cb}	$-3V_{in}^2/2v_b$

For a reference output voltage vector, $\vec{v}_{out}$, with the magnitude, V_{out}, and the phase angle, θ_{out}, in sector 1 shown in Figure 3, the duty cycles of two active vectors V_1, V_2 and two zero vectors V_0, V_7 are given as follows:

$$d_1 = \sqrt{3}\frac{V_{out}}{V_{dc}}\sin(\pi/3 - \theta_{out})$$

$$d_2 = \sqrt{3}\frac{V_{out}}{V_{dc}}\sin(\theta_{out}) \tag{6}$$

$$d_0 = d_7 = 0.5(1 - d_1 - d_2)$$

where d_0, d_1, d_2 and d_7 are duty cycles of V_0, V_1, V_2 and V_7, respectively.

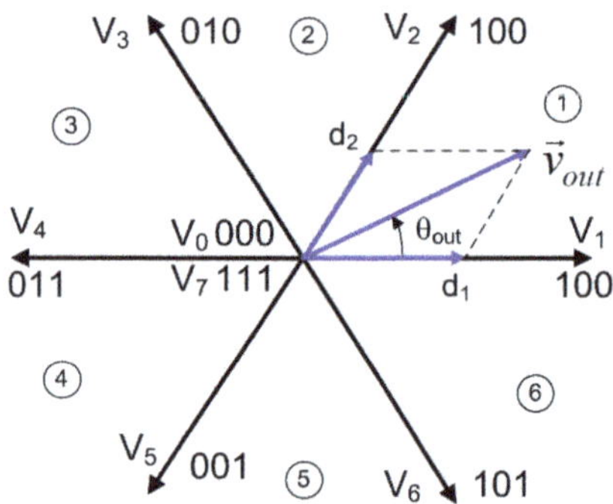

Figure 3. The space vector diagram of inverter stage.

The voltage transfer ratio of the IMC, *m*, is defined as follows:

$$m = \frac{V_{out}}{V_{in}} \tag{7}$$

According to Equations (4)–(7), the voltage transfer ratio should be smaller than 0.866 in order to maintain all duty cycles positive.

To obtain the balanced input current and output voltage, the switching patterns for the rectifier and the inverter stage should be combined effectively. As mentioned before, the DC-link voltage has two values, v_{ab} and v_{ac}, during one sampling period with the duty cycles d_{ab} and d_{ac}, respectively. Therefore, the switching states at the inverter stage are divided into two groups, as shown in Figure 4. The duty cycles of two active and two zero vectors in each group are calculated as follows:

$$d_{1(ab)} = d_1 d_{ab};\ d_{2(ab)} = d_2 d_{ab};\ d_{0(ab)} = d_{7(ab)} = d_7 d_{ab} \tag{8}$$

$$d_{1(ac)} = d_1 d_{ac};\ d_{2(ac)} = d_2 d_{ac};\ d_{0(ac)} = d_{7(ac)} = d_7 d_{ac} \tag{9}$$

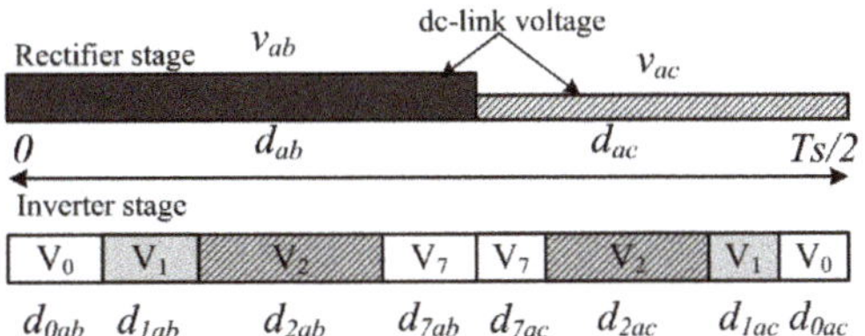

Figure 4. The switching pattern of IMC.

3. SVM to Compensate Input Power Factor

3.1. Input Filter Analysis

Figure 5 shows the equivalent circuit of the input filter per phase: v_a and i_a are the voltage and current of the power supply, v_{ai}, and i_{ai} are the rectifier input voltage and current after the filter. The equivalent single-phase model in Figure 5 can be used to derive the displacement angle between the source line current i_a and the input voltage v_a [10].

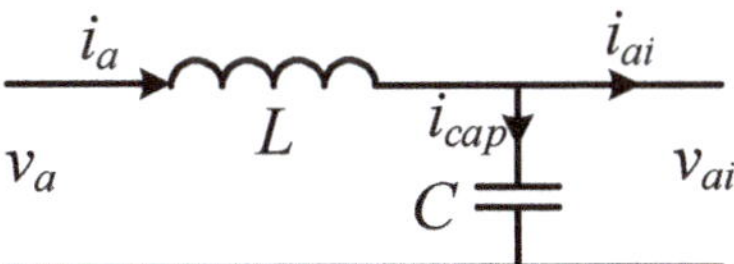

Figure 5. The equivalent circuit of the input filter for phase a.

From Figure 5, it follows:

$$v_a = v_{ai} + L(di_a/dt)$$
$$i_a = i_{ai} + i_{cap} \tag{10}$$
$$i_{cap} = C(dv_{ai}/dt)$$

Therefore, the input current and voltage can be rewritten as:

$$
\begin{aligned}
v_a &= V_{in}e^{j0}\\
i_a &= I_{in} - w_{in}^2 LCI_{in} - j\omega CV_{in}\\
&= \sqrt{[(1 - w_{in}^2 LC)I_{in}]^2 + (w_{in}CV_{in})^2}\,e^{-j\mathrm{atan}[w_{in}CV_{in}/(1-w_{in}^2 LC)I_{in}]}
\end{aligned}
\tag{11}
$$

Hence, the input filter causes the phase angle difference between the voltage and current of the power source, which is given as:

$$\delta = \arctan\left[\frac{w_{in}CV_{in}}{(1 - w_{in}^2 LC)}\frac{1}{I_{in}}\right] \tag{12}$$

where V_{in}, I_{in} are the input voltage and the current amplitude of the power supply, respectively. L and C are the inductance and capacitance of the input filter.

In this paper, we consider a new SVM method to compensate the displacement angle between the input current and the voltage in Equation (12).

3.2. The proposed SVM Method

For convenience, the input space vectors are defined as follows:

$$\overrightarrow{i}_{in} = \frac{2}{3}\left(i_a + i_b e^{j2\pi/3} + i_c e^{j4\pi/3}\right) = I_{in} e^{j\alpha_{in}} \tag{13}$$

$$\overrightarrow{v}_{in} = \frac{2}{3}\left(v_a + v_b e^{j2\pi/3} + v_c e^{j4\pi/3}\right) = V_{in} e^{j\beta_{in}} \tag{14}$$

where $\overrightarrow{i}_{in}$ and $\overrightarrow{v}_{in}$ are the input current and the input voltage space vectors, respectively.

Table 2 shows the entire possible switching states and the corresponding space vectors for input current. From Table 2, the input current space vectors are composed of six active current vectors with fixed directions and three zero vectors, where I_{dc} refers to the DC-link current. Figure 6 shows the space vector diagram of the rectifier stage. Each active current vector represents the switching condition between the input phase voltage and the DC-link bus.

Table 2. The input current space vectors according to the switching state.

Switching State						Input Current $\overrightarrow{i}_{in}$	
S_{ap}	S_{bp}	S_{cp}	S_{an}	S_{bn}	S_{cn}	I_{in}	α_{in}
1	0	0	0	1	0	$2/\sqrt{3}I_{dc}$	$-\pi/6$
1	0	0	0	0	1	$2/\sqrt{3}I_{dc}$	$\pi/6$
0	1	0	0	0	1	$2/\sqrt{3}I_{dc}$	$\pi/2$
0	1	0	1	0	0	$2/\sqrt{3}I_{dc}$	$5\pi/6$
0	0	1	1	0	0	$2/\sqrt{3}I_{dc}$	$7\pi/6$
0	0	1	0	1	0	$2/\sqrt{3}I_{dc}$	$3\pi/2$
1	0	0	1	0	0	0	x
0	1	0	0	1	0	0	x
0	0	1	0	0	1	0	x

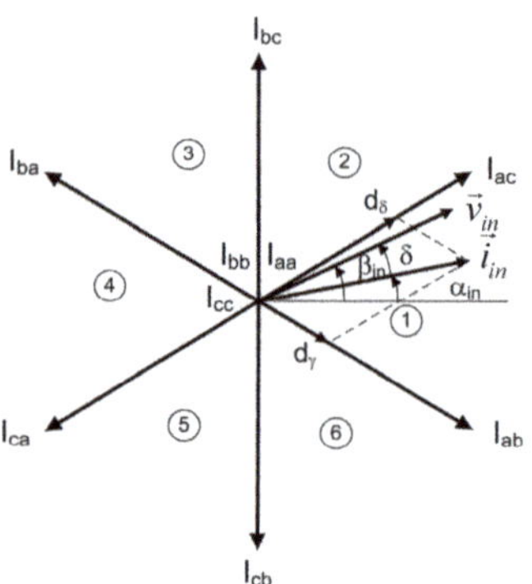

Figure 6. The space vector diagram of rectifier stage.

In order to compensate for the displacement angle caused by the input filer, the input current vector has to be lagged behind the input voltage vector with the angle δ as shown in Figure 6. If we assume that the input current space vector, I_{in}, is located in sector 1, I_{in} is synthesized by two neighbor active vectors I_{ab} and I_{ac}. The duty cycles d_δ and d_γ for two active vectors are given by:

$$d_\delta = m_{rec} \sin(\beta_{in} - \delta + \pi/6) \tag{15}$$

$$d_\gamma = m_{rec} \sin(\pi/6 - (\beta_{in} - \delta)) \tag{16}$$

where m_{rec} is the modulation index of the rectifier stage.

In the modulation of the rectifier stage, the zero current vectors are not used to obtain the maximum DC-link voltage.

Therefore, in order to complete the sampling period, the duty cycles of two adjacent active vectors are adjusted as follows:

$$d_{ac} = \frac{d_\delta}{d_\delta + d_\gamma} = \frac{\cos(\beta_{in} - 4\pi/3 - \delta)}{\cos(\beta_{in} - \delta)} \tag{17}$$

$$d_{ab} = \frac{d_\gamma}{d_\delta + d_\gamma} = \frac{\cos(\beta_{in} - 2\pi/3 - \delta)}{\cos(\beta_{in} - \delta)} \tag{18}$$

And, the average DC-link voltage becomes:

$$V_{dc} = d_{ab}(v_a - v_b) + d_{ac}(v_a - v_c) = \frac{3}{2} \frac{V_{in}}{\cos(\beta_{in} - \delta)} \cos \delta \tag{19}$$

The calculation of duty cycles for the active and zero vectors is similar to that of the conventional method, and they are given in Equation (20).

$$d_1 = \sqrt{3} \tfrac{V_{out}}{V_{dc}} \sin(\pi/3 - \theta_{out})$$
$$d_2 = \sqrt{3} \tfrac{V_{out}}{V_{dc}} \sin(\theta_{out}) \tag{20}$$
$$d_0 = d_7 = 0.5(1 - d_1 - d_2)$$

Similar to the conventional method, the voltage space vectors in the inverter stage are arranged in a double-side switching sequence V_0–V_1–V_2–V_7–V_7–V_2–V_1–V_0, but with unequal halves.

The duty cycles of the active and the zero vectors in the inverter stage are given in Equation (21) during which the first active current vector I_{ab} is applied to the rectifier stage.

$$d_{1(ab)} = \sqrt{3} \tfrac{V_{out}}{V_{dc}} \sin(\pi/3 - \theta_{out}) \tfrac{\cos(\beta_{in} - 2\pi/3 - \delta)}{\cos(\beta_{in} - \delta)}$$
$$d_{2(ab)} = \sqrt{3} \tfrac{V_{out}}{V_{dc}} \sin(\theta_{out}) \tfrac{\cos(\beta_{in} - 2\pi/3 - \delta)}{\cos(\beta_{in} - \delta)} \tag{21}$$
$$d_{0(ab)} = d_{7(ab)} = \tfrac{1}{2}\left(d_{ab} - d_{1(ab)} - d_{2(ab)}\right)$$

In contrary, when the second active current vector I_{ac} is applied to the rectifier stage, the duty cycles are obtained as Equation (22).

$$d_{1(ac)} = \sqrt{3} \tfrac{V_{out}}{V_{dc}} \sin(\pi/3 - \theta_{out}) \tfrac{\cos(\beta_{in} - 4\pi/3 - \delta)}{\cos(\beta_{in} - \delta)}$$
$$d_{2(ac)} = \sqrt{3} \tfrac{V_{out}}{V_{dc}} \sin(\theta_{out}) \tfrac{\cos(\beta_{in} - 4\pi/3 - \delta)}{\cos(\beta_{in} - \delta)} \tag{22}$$
$$d_{0(ac)} = d_{7(ac)} = \tfrac{1}{2}\left(d_{ac} - d_{1(ac)} - d_{2(ac)}\right)$$

Figure 7 shows the block diagram of the proposed SVM method. First, after detecting the three-phase input voltages/currents using voltage/current sensors, the displacement angle δ is calculated based on Equation (12). Then, the duty cycles of the active vectors in the rectifier stage and average dc-link voltage are determined based on Equations (18) and (19). In the inverter stage, the duty cycles of active and zero vectors are calculated based on the desired output voltage, desired compensation angle δ_{com}, and the duty cycles of the rectifier stages. Finally, the gating signals are generated by using the switching pattern.

From Equation (1) and Figure 6, the line-to-line input voltages v_{ab} and v_{ac} are obtained as follows:

$$v_{ab} = \sqrt{3}V_{in}\cos(\beta_{in} + \pi/6) = \sqrt{3}V_{in}\cos(\alpha_{in} + \delta + \pi/6) \tag{23}$$

$$v_{ac} = \sqrt{3}V_{in}\cos(\beta_{in} - \pi/6) = \sqrt{3}V_{in}\cos(\alpha_{in} + \delta - \pi/6) \tag{24}$$

From Figure 6, the phase angle of input current vector, α_{in}, in sector 1 is given as:

$$-\frac{\pi}{6} \leq \alpha_{in} \leq \frac{\pi}{6} \tag{25}$$

From Equation (23)–(25), in order to keep v_{ab} and v_{ac} positive, the compensation angle should satisfy the Equation (26).

$$\delta_{com} \leq \frac{\pi}{6} \tag{26}$$

Therefore, the compensation angle is chosen as following:

$$\begin{cases} \delta_{com} = \delta & \text{if } \delta \leq \pi/6 \\ \delta_{com} = \pi/6 & \text{if } \delta > \pi/6 \end{cases} \tag{27}$$

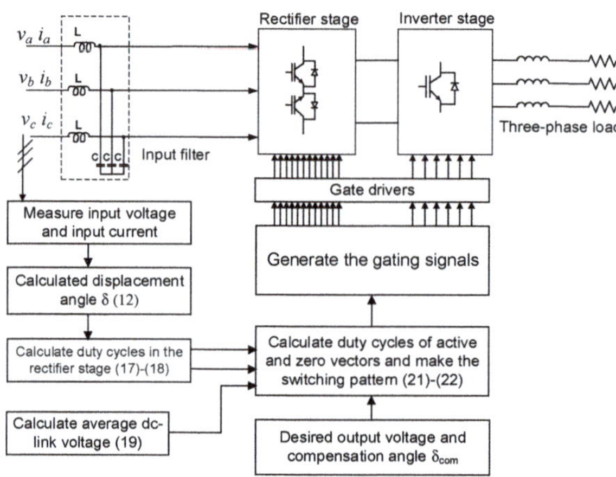

Figure 7. The block diagram of the proposed SVM method.

4. Simulation Results

In order to verify the performance and the effectiveness of the proposed method, simulations were carried out using Psim 9.0 software (Powersim Inc., Rockville, MD, USA) with three-phase RL load. The parameters of the IMC system were as follows:

- Three-phase load: $R = 12\ \Omega$, $L = 10$ mH.
- Input filter inductance: $L = 1$ mH, input filter capacitance: $C = 25\ \mu$F.
- Switching frequency: 10 kHz ($T_s = 100\ \mu$s).
- Three-phase power supply voltage for the IMC set to 100 V (line-to-neutral voltage), frequency: 60 Hz.

Figure 8a,b shows the input and the output side waveforms with using the conventional method when the output voltage transfer ratio, m is 0.6, and the out frequency, f_{out} is 50 Hz. In Figure 8a, the input current of the IMC (i_{ai}) contained many switching harmonics, and its fundamental component was in phase with input voltage (v_a), while the input current (i_a) of the power supply led the input voltage by $\delta = \pi/9$. Therefore, the power factor (*PF*) in the input side was not at unity; the achievable power factor was 0.94.

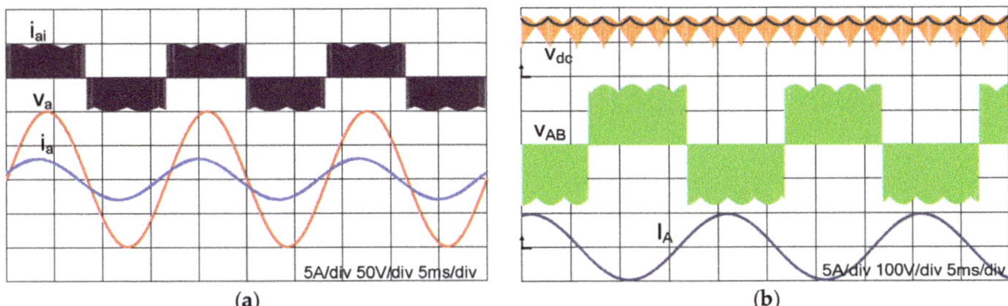

Figure 8. (**a**) Input waveforms; and (**b**) DC-link voltage and output waveforms with the conventional method at $m = 0.6$ and $f_{out} = 50$ Hz.

Figure 9a,b shows the input and the output side waveforms with the proposed modulation method under the same output voltage condition, with the conventional method in Figure 8. From Equation (27), the compensation angle was selected to be $\delta_{com} = \pi/9$. The input current of power supply was in phase with the input voltage, meaning that the input power factor became at unity, so that the input current magnitude became smaller under the same load condition. Contrarily, the input current of the IMC lagged behind the input voltage of power supply by $\pi/9$. Also, the output voltage/current waveforms in Figure 9b had good performance, like those in Figure 8b, although the DC-link waveform was different from that of the conventional method, due to the different modulation strategy.

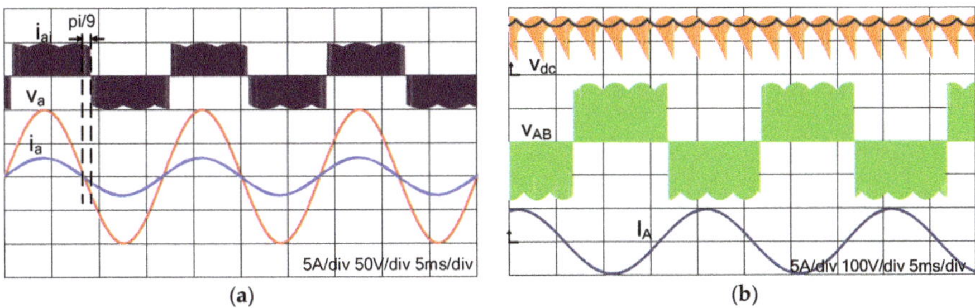

Figure 9. (**a**) Input waveforms; and (**b**) DC-link voltage and output waveforms with the proposed method at $m = 0.6$ and $f_{out} = 50$ Hz.

In order to investigate the performance of the low output voltage operation, the output voltage references were changed to $m = 0.35$ and $f_{out} = 50$ Hz. Figures 10 and 11 show the currents and voltages at the input and output side with the conventional and the proposed methods, respectively. As mentioned before, in Figure 10a, the displacement angle between the input and output current became larger at the low output voltage $\delta = \pi/4$; the input power factor was 0.71. When the proposed method was applied with the maximum compensating angle ($\delta_{com} = \pi/6$) from Figure 11, the input power factor improved to 0.91. Figure 11a shows that the IMC input current i_{ai} lagged behind the input voltage v_a with the desired compensated angle $\pi/6$. From Figure 11b, the proposed method provided the same output voltage performance as that in Figure 10b.

From the simulation, it was clear that the proposed method is effective in improving the input power factor without deteriorating the output performance in both cases of low and high output voltage transfer ratio.

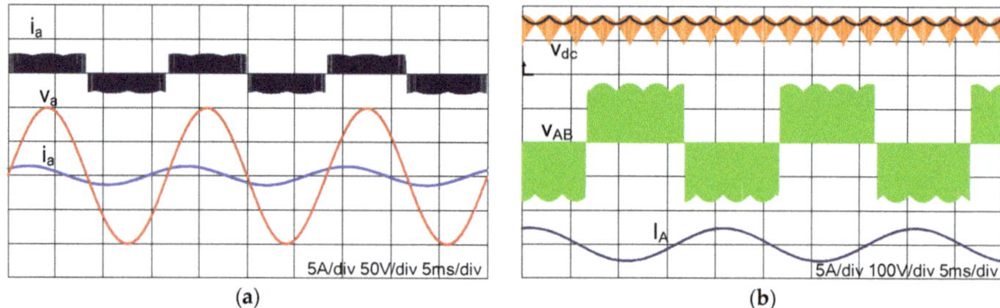

Figure 10. (**a**) Input waveforms; and (**b**) DC-link voltage and output waveforms with the conventional method at $m = 0.35$ and $f_{out} = 50$ Hz.

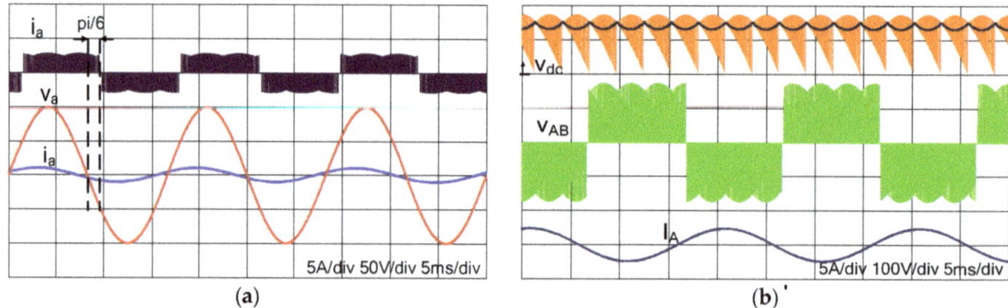

Figure 11. (**a**) Input waveforms; and (**b**) DC-link voltage and output waveforms with the proposed method at $m = 0.35$ and $f_{out} = 50$ Hz.

5. Experimental Results

In order to validate the effectiveness of the proposed SVM method, a laboratory prototype was set up experimentally. Figure 12 shows the experimental setup for the IMC topology. The controller board was developed with high performance DSP (TMS320F28335, Texas Instruments, Dallas, TX, USA) and CPLD (EPM128S, Altera, San Jose, CA, USA). The power circuit of the rectifier stage was constructed by six insulated-gate bipolar transistor (IGBT) modules (SK 60GM123, Semikron, Nürnberg, Germany) and the inverter stage was composed of three dual-IGBTs (FMG2G100US60, Fairchild Semiconductor, Sunnyvale, CA, USA).

Figures 13 and 14 show the input and the output waveforms with the conventional method and the proposed method, respectively, under the same load condition: $m = 0.6$ and $f_{out} = 50$ Hz. We saw that the input power factor improved from 0.91 to unity, with the compensation angle at $\pi/9$.

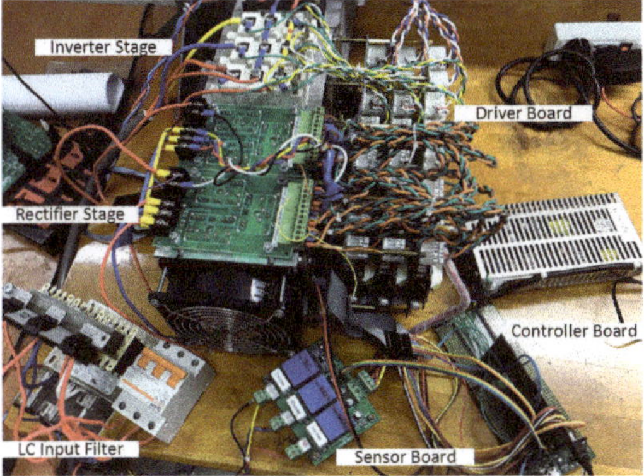

Figure 12. Experimental setup for the IMC topology.

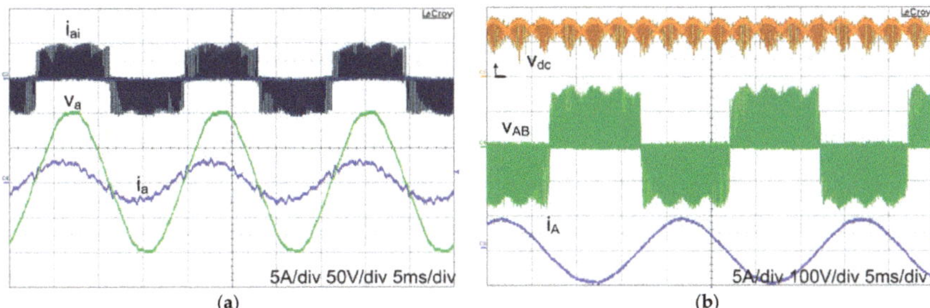

Figure 13. (**a**) Input waveforms; and (**b**) DC-link voltage and output waveforms with the conventional method at $m = 0.6$ and $f_{out} = 50$ Hz ($PF = 0.94$).

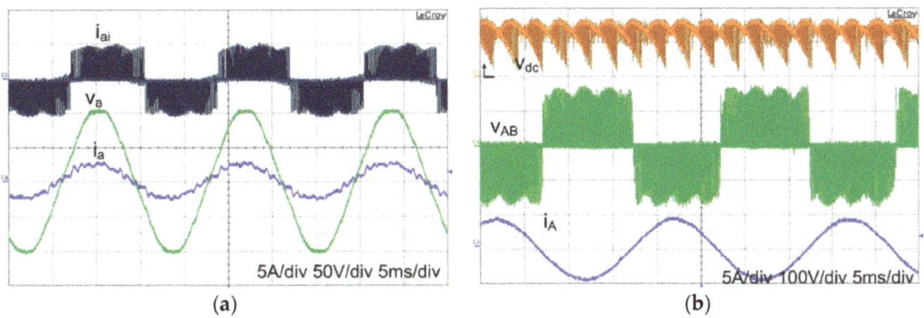

Figure 14. (**a**) Input waveforms; and (**b**) DC-link voltage and output waveforms with the proposed method at $m = 0.6$ and $f_{out} = 50$ Hz ($PF = 1.0$).

Figures 15 and 16 show the experimental performance of the low voltage operation at voltage transfer ratio $m = 0.35$ and output frequency $f_{out} = 50$Hz, with the conventional method and the proposed method, respectively. In Figure 15a, the maximum compensation angle $\pi/6$ was applied, so that the power factor improved from 0.71, which was obtained from the conventional method in Figure 15a, to 0.91. Comparing Figure 15a with Figure 16a, the input current magnitude became smaller under the proposed method, due to the improved input power factor.

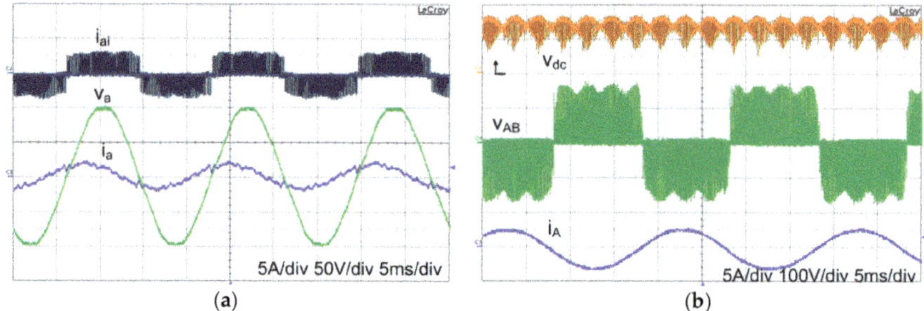

(a)   (b)

Figure 15. (**a**) Input waveforms; and (**b**) DC-link voltage and output waveforms with the conventional method at $m = 0.35$ and $f_{out} = 50$ Hz (PF = 0.71).

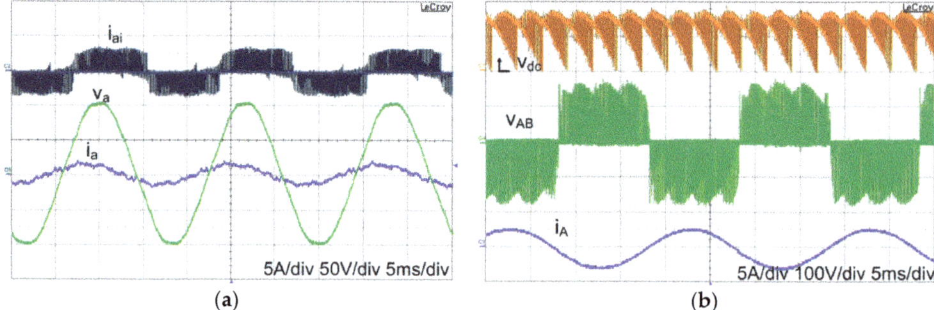

(a)   (b)

Figure 16. (**a**) Input waveforms; and (**b**) DC-link voltage and output waveforms with the proposed method at $m = 0.35$ and $f_{out} = 50$ Hz ($PF = 0.91$).

As a result, the experimental results matched the simulated results very well, and the effectiveness of the proposed method was also verified by experiment as well as simulation.

6. Conclusions

This paper presents a new SVM method for the IMC to compensate for the input power factor, which is implemented by using the input voltage and the input current. Analysis, design and implementation of the proposed SVM for the three-phase IMC with the improvement of the input power factor, were presented. And also, the allowable of compensating angle according to the displacement angle caused by LC input filter is analyzed. By using the proposed SVM method, the input power factor was compensated significantly regardless of the existence of the input filter, without any performance degradation on the input/output waveforms. The unity input power factor was obtained with the proposed SVM method. Thanks to the improved power factor, the magnitude of the input current became smaller compared to the conventional method. The performance of the proposed SVM was verified with both simulation and experiment.

Acknowledgments: This research is funded by Vietnam National Foundation for Science and Technology Development (NAFOSTED) under grant number 103.99-2015.102.

Author Contributions: Phan Quoc Dzung developed the idea of this paper and performed the simulation. Nguyen Dinh Tuyen implemented the main research, performed the experiment. All authors contributed to the writing of the manuscript, revised and approved the final manuscript.

Conflicts of Interest: The authors declare no conflict of interest.

1. Alesina, A.; Venturini, M.G.B. Solid-state power conversion: A Fourier analysis approach to generalized transformer synthesis. *IEEE Trans. Circuits Syst.* **1981**, *28*, 319–330. [CrossRef]
2. Arevalo, S.L.; Zanchetta, P.; Wheeler, P.W. Control of a matrix converter-based AC power supply for aircrafts under unbalanced conditions. In Proceedings of the IECON 2007, Taipei, Taiwan, 5–8 November 2007; pp. 1823–1828.
3. Sebtahmadi, S.S.; Pirasteh, H.; Kaboli, S.H.A.; Radan, A.; Mekhilef, S. A 12-sector space vector switching scheme for performance improvement of matrix-converter-based DTC of IM drive. *IEEE Trans. Power Electron.* **2015**, *30*, 3804–3817. [CrossRef]
4. Jussila, M.; Eskola, M.; Tuusa, H. Analysis of non-idealities in direct and indirect matrix converters. In Proceedings of the EPE–ECCE Europe 2005, Dresden, Germany, 11–14 September 2005; pp. 1–10.
5. Kolar, J.W.; Schafmeister, F.; Round, S.D.; Ertl, H. Novel three-phase AC-AC sparse matrix converters. *IEEE Trans. Power Electron.* **2007**, *22*, 1649–1661. [CrossRef]
6. Nguyen, T.D.; Lee, H.H. Development of a three-to-five-phase indirect matrix converter with carrier-based PWM based on space-vector modulation analysis. *IEEE Trans. Ind. Electron.* **2016**, *63*, 13–24. [CrossRef]
7. Nguyen, T.D.; Lee, H.H. Dual three-phase indirect matrix converter with carrier-based PWM method. *IEEE Trans. Power Electron.* **2014**, *29*, 569–581. [CrossRef]
8. Nguyen, T.D.; Lee, H.H. Modulation strategies to reduce common-mode voltage for indirect matrix converters. *IEEE Trans. Ind. Electron.* **2012**, *59*, 129–140. [CrossRef]
9. Kim, S.; Yoon, Y.D.; Sul, S.K. Pulsewidth modulation method of matrix converter for reducing output current ripple. *IEEE Trans. Power Electron.* **2010**, *25*, 2620–2629. [CrossRef]
10. Nguyen, H.M.; Lee, H.H.; Chun, T.W. Input power factor compensation algorithms using a new direct-SVM method for matrix converter. *IEEE Trans. Ind. Electron.* **2011**, *58*, 232–243. [CrossRef]

Article

Power Quality Event Detection Using a Fast Extreme Learning Machine

Ferhat Ucar [1,*]**, Omer F. Alcin** [2]**, Besir Dandil** [3] **and Fikret Ata** [2]

[1]   Department of Electrical and Electronics Engineering, Technology Faculty, Firat University, 23119 Elazig, Turkey

[2]   Department of Electrical and Electronics Engineering, Faculty of Engineering and Architecture, Bingol University, 12000 Bingol, Turkey; ofalcin@bingol.edu.tr (O.F.A.); fata@bingol.edu.tr (F.A.)

[3]   Department of Mechatronics Engineering, Technology Faculty, Firat University, 23119 Elazig, Turkey; bdandil@firat.edu.tr

*   Correspondence: fucar@firat.edu.tr; Tel.: +90-424-237-0000

Received: 9 November 2017; Accepted: 3 January 2018; Published: 7 January 2018

Abstract: Monitoring Power Quality Events (PQE) is a crucial task for sustainable and resilient smart grid. This paper proposes a fast and accurate algorithm for monitoring PQEs from a pattern recognition perspective. The proposed method consists of two stages: feature extraction (FE) and decision-making. In the first phase, this paper focuses on utilizing a histogram based method that can detect the majority of PQE classes while combining it with a Discrete Wavelet Transform (DWT) based technique that uses a multi-resolution analysis to boost its performance. In the decision stage, Extreme Learning Machine (ELM) classifies the PQE dataset, resulting in high detection performance. A real-world like PQE database is used for a thorough test performance analysis. Results of the study show that the proposed intelligent pattern recognition system makes the classification task accurately. For validation and comparison purposes, a classic neural network based classifier is applied.

Keywords: event detection; power quality; histogram; machine learning; wavelet transform

---

1. Introduction

The Smart grid is a complex network that needs an advanced monitoring system to assure its reliability, security, and sustainability. The increase in stream of data from the smart devices makes more amount of knowledge to be processed by network operators. Considering the industrial Internet of Things (IoT) architecture, intelligent devices use embedded processing and communication capabilities that produce exceptionally large amounts of data, rising the necessity of fast processing algorithms. Furthermore, today's power system experts face with a new paradigm. The government is not the only provider of energy, there are also non–public companies providing grid demand. Common purpose is effective energy consumption. Thus, power providers bring service quality, resilience, sustainability and reliability into the forefront. Blackouts and power quality issues inherently create significant financial loses because modern industrial area and electrical energy are tightly coupled [1–4].

In the research field of power quality monitoring, Power Quality Event (PQE) classification has an important position. In monitoring centers, measured PQ signals are collected and transformed to knowledge for managing the whole grid with the help of intelligent systems. Researchers investigate event classification in terms of feature space and decision space [5–12]. Feature space includes extracting distinctive features of signal and in a decision space the classifier performs discrimination. Construction of feature set relies on different signal processing methods [5,13]. In literature, there has been many studies based on transform and model–based methods [14–19]. In addition to data–driven methods, such models using micro–synchrophasor measurement data [20] are also proposed. Conventionally, Fast Fourier Transform (FFT) and Root Mean Square (RMS) variation tracking methods exist and have a long-term usage in feature extracting [21,22]. In addition, FFT and RMS methods have no ability in signal analysing of time and frequency domains [23]. Short time FFT (ST–FFT), which is proposed to upgrade the FFT method with a time domain analysis, has a fixed window width when analyzing a raw signal. Kalman Filter, Hilbert–Huang Transform, and S–Transform are enumerable among the most used methods [3,5,24,25]. In power systems, Wavelet Transform (WT) was first used in the year 1996 with its Multi–Resolution Analysis (MRA) structure [26,27]. WT is a time–frequency analysis method that uses a variable window width to gain a robust frequency tracking [28,29]. The histogram is a representation, which briefs distribution of a numerical array by counting the same values of data within specified intervals. Sturges Rule defines the choice of those intervals in data range [30]. This article uses a histogram method as a crucial part of the feature set and proposes the method as its contribution. Histogram and commonly used feature extraction method WT are integrated to obtain an effective feature set. The histogram method is easy to implement and it has a less computational time. This study proposes a fast algorithm for feature extraction that is the most important phase of PQE classification.

Developing technology of computer hardware systems brings powerful components, which have a high process capability. Following this, intelligent systems are able to implement complex artificial methods. Conventional Artificial Neural Network (ANN) structure, Support Vector Machine (SVM) classifier, and Fuzzy and Expert system based classifiers are commonly used decision makers in the literature [5,13,23,25,31–33]. Today, Machine Learning (ML) based classifiers are challenging topics for researchers. Furthermore, one method that has presented top performance is Extreme Learning Machine (ELM). ELM is a learning algorithm that covers the Single Layer Feed Forward Neural Network (SLFN) structure and it has an adequate performance without any necessity of iterative process [34]. Since it was first proposed, ELM has been applied to classification and regression models in the various field of research as computer vision, biomedical signal processing and so on [35–42].

In this article, a novel feature extracting method is highlighted, which is combined with Discrete Wavelet Transform (DWT) Entropy details. Decision-making is held by ELM with a high performance value. The histogram method retrieves distinctive features from the raw PQE data and has never been used before in PQE classification. With an ELM based classifier, the proposed pattern recognition system compiles PQE classification process with an acceptable performance improving. Using ELM and the histogram, this study expresses its novelty among other studies in the literature. The processed database has been simulated via an elaborate software. Simulation model generates the more frequent voltage disturbances such as sag, swell, interruption, harmonics, and flickers. In literature, there has been so many studies using transform based methods in feature extraction. The study in Ref. [8] uses Discrete Gabor Transform (DGT) with a type-2 fuzzy based SVM classifier. They experiment two different level of noise conditions using a synthetic dataset. In our study, we use a non-transform based easy to implement method using an extremely fast ELM classifier, our proposed system outperforms the DGT with SVM method. (Please see Section 6).

The proposed system utilizes a single phase event classification that is compatible with a multiple usage in three–phase systems. DWT–Entropy and Histogram methods generate a distinctive feature set from raw synthetic data. We designed the dataset using a comprehensive model in MATLAB (R2015a, MathWorks, MA, US) [43]. In our study, we may list our contributions as: (1) we propose a non-transform based feature extraction method that uses a histogram with an effective computational cost. Using a conventional DWT based method has improved the overall performance; (2) in decision-making, we use a machine learning based non-iterative ELM classifier. In comparison with classical algorithms like ANN, ELM solves a single linear equation to reach the solution; and (3) an intelligent classifier system uses a detailed dataset that we designed through a PQE generator toolbox elaborately. For the next stages, we have planned to prepare it as a virtual toolbox for Power Systems lectures. In Figure 1, we present the general block sheme of the proposed intelligent event classification system. The three main steps—database construction, feature extraction and decision steps—are demonstrated with the included methods.

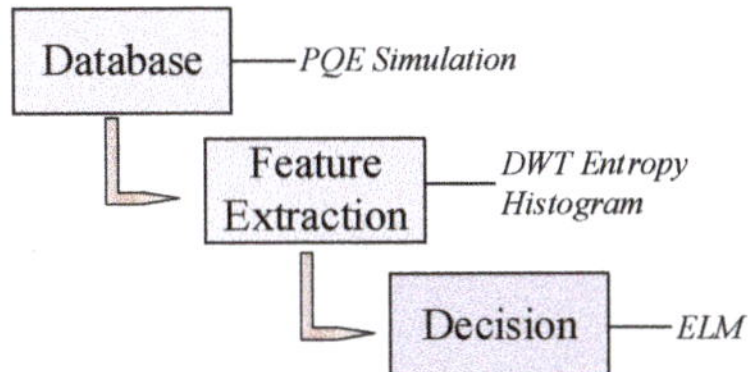

Figure 1. General block scheme of the proposed intelligent event classification system.

Following this chapter, the rest of this article is structured in this case: Sections 2–4 describe the methodology of FE and decision-making under the topics of DWT and the Entropy method, histogram, and ELM structure, respectively. Section 5 describes the PQE dataset and designated PQE generator; Section 6 emphasizes analyses and results of the proposed pattern recognition system. The last chapter is a brief conclusion for the study.

2. Feature Extraction: Wavelet–Entropy

WT operates a resilient time-frequency analysis and reveals the implicit partitions that signal includes. While FFT only performs frequency analysis, ST–FFT fills the gap and performs its analysis in the time-frequency domain. ST–FFT utilizes a fixed width window when tracking the signal. WT overcomes this issue by means of a scalable window width. Thus, analysis continues with extended window width to probe low-frequency divisions of signal and with reduced window width to probe high-frequency divisions. In power system signal processing, WT is a useful tool because it can clearly detect beginning and ending points of events [44]. WT governs a scalable wavelet model when healing the constant resolution affair and gives a flexible time–frequency analysis at different resolution levels [22,26,44–46]. The discrete form of WT is expressed as:

$$DWT(a,b) = 2^{-a/2} \int y(t)\psi(2^{-a}t - b)dt,\tag{1}$$

where a is the scaling parameter of frequency, and b is a time offset. $y(t)$ represents the processed signal while $\psi(t)$ is the wavelet function. DWT method uses MRA, dividing the signal into lower frequency levels. Theoretically, levels of frequency sub–bands are unlimited whereas in practice sampling frequency restricts the levels of MRA [44]. In this study, 8-level decomposition is used in DWT MRA and the wavelet function is "Daubechies 4" (*db4*) based on former works in literature [5]. Figure 2 shows DWT–MRA analysis in graphical representation for chosen sample events. As it can be seen in depth, details in d_1–d_4 range indicate start and end moments of PQE, clearly. In Figure 2, "*s*" is the raw signal.

Raw signals should be subject to a size reduction process before serving as classifier input. In this study, the entropy method is applied to detail vectors of DWT (for detail vectors, see Figure 2). In terms of statistical explanation, entropy states the "disorder" in a signal. The usage of signal processing field, Shannon is one of the first proponents of the entropy approach [47]. Entropy computation is an optimum way to measure the disorder in a non-stationary signal. Commonly used entropy calculations in signal processing are Shannon, Threshold, Norm, Sure method and Logarithmical Energy [46,48]. Shannon Entropy is preferred in this study, which is described as:

$$E(y) = \sum_i y_i^2 \cdot \log_2(y_i^2),\tag{2}$$

where y_i is the element of signal number i. Entropy computation generates eight features based on DWT coefficient vectors for each PQE.

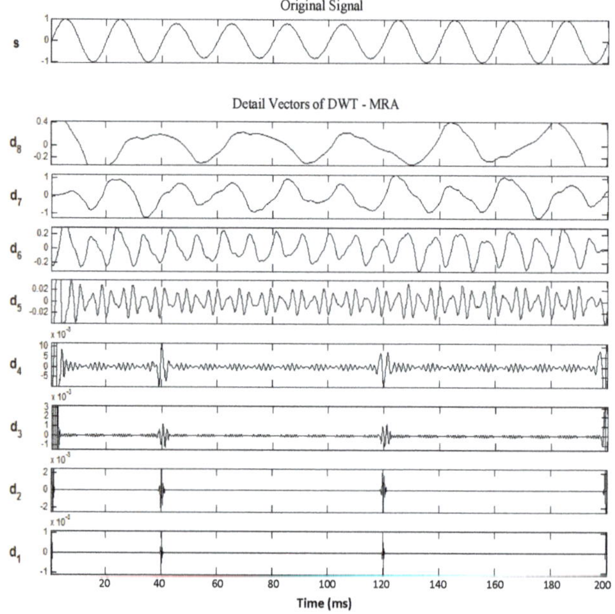

Figure 2. Discrete wavelet transform–multi resolution analysis of power quality event data in graphical view: 8-level decomposition details of a sample voltage sag event.

Figure 3 illustrates a graphical representation of DWT details' entropy of four selected sample events in the dataset. It can be clearly seen that DWT–Entropy features characterize PQE data effectively. In this study, DWT is preferred for performance boosting of the histogram method.

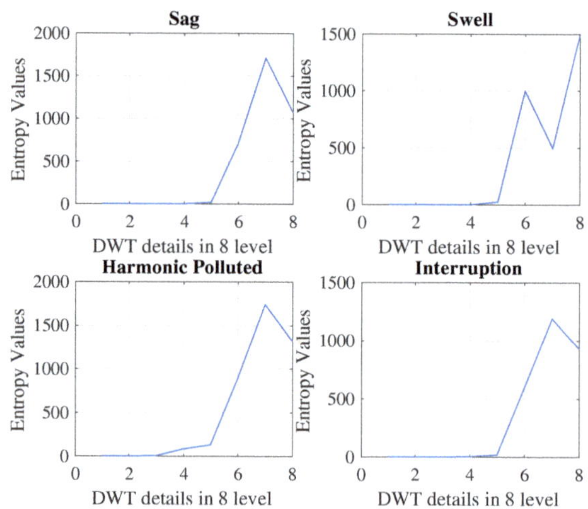

Figure 3. DWT–Entropy features of 8-level decomposition for sample events.

3. Feature Extraction:Histogram

Using histogram, a graphical distribution is achieved that indicates the counts of samples in specific intervals throughout complete data array [30,49,50]. In our PQE dataset, histogram features characterize nearly whole events individually. Figure 4 illustrates the general histogram bars of sample events.

Figure 4 shows a unique distribution scene for each event, making feature extraction more distinctive.

In this study, counting points of each interval, so called "bins", are specified by Sturges' Rule, which is defined inclusively in [30]. With this designation according to Equation (3), the histogram feature set has consisted of 14 elements for each PQE , where C is the interval number and k is the samples of each signal here is 10.001. Figure 5 shows us histogram features of sample events:

$$C = 1 + 3.322 log(k).\tag{3}$$

As one can see in Figure 5, histogram features have the ability to emphasize nearly all PQE individually. Figure 5 includes the counts of signal magnitudes according to chosen 14 bins.

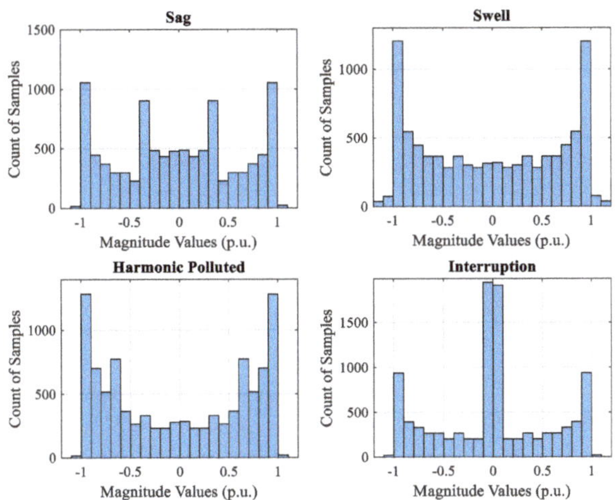

Figure 4. General histograms for sample events.

Algorithm 1 briefs the whole process of feature extracting. Algorithm 1 runs for every sample of the dataset, which is a number of 600.

Algorithm 1 Applied FE method using DWT–Entropy and histogram

---

Input: Loading PQE Dataset
Output: Total feature set to be classified

    Feature Extraction :
1: **for** $i = 1$ **to** 600 **do**

2:    Calculate the DWT details of PQE signals using (1),
3:    Calculate the Entropy values of DWT details with (2)
4:    Form the feature set $[E_1 \dots E_8]$,
5:    Calculate the Histogram counts of PQE signals
6:    Form the feature set $[H_1 \dots H_{14}]$
7: **end for**
8: Compose the total feature vector $[E_1 \dots E_8, H_1 \dots H_{14}]$

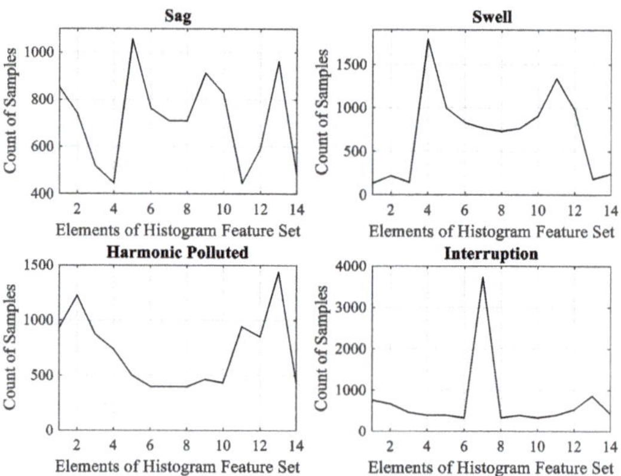

Figure 5. Histogram features of sample events.

4. Decision-Making: Extreme Learning Machine

ELM was firstly proposed by Huang et al. [34] and is a learning structure applied to SLFNs. In the ELM algorithm, weights and biases of the input layer are arbitrary while only the outputs are calculated [51]. The fact that the first layer is assigned arbitrarily has been stated, and the learning time for ELM is extremely short. Additionally, the ELM structure has accurate generalization ability compared to Feed Forward ANN (FF–ANN) based conventional learning algorithm [38,39]. Figure 6 briefs a basic SLFN frame. Inputs and outputs of the classifier are shown as x_i and y_i.

The basic SLFN frame, which contains M total of hidden nodes and operates with $g(x)$ activation function, can be described in mathematical form as:

$$\sum_{i=1}^{M} \beta_i g(w_i \cdot x_j + b_i) = o_j, \quad j = 1 \ldots N, \tag{4}$$

where w is the input weights of the layer, and β is weights of the output layer. b_i is bias values of the input layer. o defines the expected output of ELM. $(w_i \cdot x_j)$ operand is the inner product of w_i and x_j so-called weighted inputs. Given the structure of SLFN can establish the *"zero error"* theoretically, i.e., o value is equal to y output vector. Thus, Equation (4) can be reformulated as:

$$\sum_{i=1}^{M} \beta_i g(w_i \cdot x_j + b_i) = y_j, \quad j = 1 \ldots N. \tag{5}$$

Equation (5) exhibits that there are suitable output weights able to form measured outputs or real outputs of SLFN. If a facilitation is implemented as in (6), Equation (5) can be reformed as in (7):

$$g(w_i \cdot x_j + b_i) = H_{ij}, \tag{6}$$

$$Y = H \cdot \beta. \tag{7}$$

Equation (7) refers to a linear equation whose solution takes us to output values of ELM. In usual learning frames, there is a need for iterative processes to obtain expected outputs, but ELM solves only a linear equation to execute the similar process at one time without any iteration. Equation (8) describes the solution for getting β value from (7):

$$\beta = H^{\diamond} \cdot Y, \tag{8}$$

where $H^{\diamond}$ is operated via the "Moore–Penrose inverse" so-called generalized inverse of H matrix [22,52].

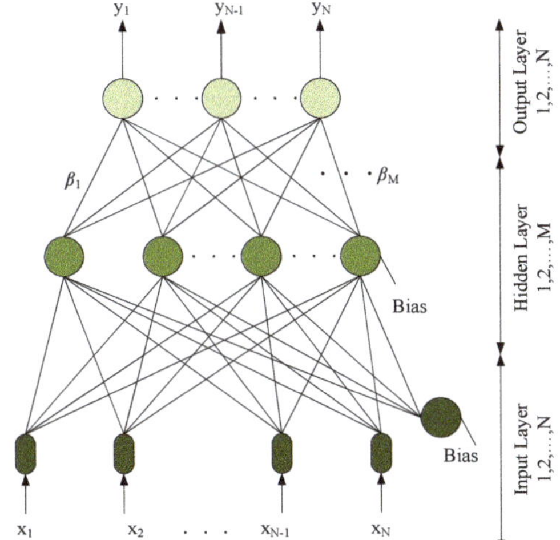

Figure 6. Basic single layer feed-forward neural network frame.

Algorithm 2 briefs the ELM learning. In decision-making, we use the last feature vector with a length of 22 that includes eight features $(E_1, \ldots, E_8)$ from DWT–Entropy and 14 features $(H_1, \ldots, H_{14})$ from the histogram method. Process loop runs for every sample of the dataset.

Algorithm 2 ELM Method

Input: Training set, $[t = 1, 2, ..., T]$,
Output: Output weights of ELM structure: Calculation of β from $\mathbf{Y} = \mathbf{H}\beta$.

   Initialisation :
1: Defining input weights and biases randomly.
2: **for** $t = 1, 2, ..., T$. **do**

3:    Compute $\mathbf{H}$ matrix using (4) and (6),
4:    Compute the output weights from (8),
5: **end for**

   Test :
6: Predict an unlabeled test input
7: Decide the type of PQE

---

5. Power Quality Event Data Description

The PQE simulation model presented in this paper has three steps: generating events using mathematical equations, normalization, and building last datasets to be processed. All three steps of the model have been designed in MATLAB [43]. Built simulation model generates five categories of voltage events such as sag, swell, interruption, harmonic polluted voltage, and flicker. In addition, a pure sinusoidal voltage is generated in order to depict normal operating conditions. PQE generator is operated at 10 kHz sampling frequency. Sampling frequency value can be thought of as measurement devices' operating frequency. The built model composes the dataset using mathematical models of events [8,25]. The frequency of the grid model is considered as 50 Hz; thus, a data array includes 200 samples in a period duration and measured time is set to 1-s. This operation time gives $1 \times 10{,}001$ length of raw data vector. The complete dataset includes six different classes for a total of 600 events, each with a length of 10,001 samples. At the end of the feature extraction process, the dataset is subject to a size reduction and as a result the feature vector has a length of 22 before processing in the classifier stage. In a 1-s period of an operation window, data rows have three sections as pre–event, event and post–event. Occurring durations are different from each other for every single event data. The built simulation model sets different durations of events in every data row. This makes every PQE unique in each class. Table 1 briefs the dataset in terms of Event Class (EC) types.

Table 1. Mathematical model based power quality event types and numbers.

PQE Type	Class	Number of Signals
Normal Sine	EC1	100
Sag	EC2	100
Swell	EC3	100
Harmonics	EC4	100
Interruption	EC5	100
Flicker	EC6	100
Total		600

In order to resemble a real-field dataset, noise distortion is considered at values of Signal-to-Noise Ratio (SNR) 10 dB, 20 dB, and 30 dB. Noise addition makes dataset closer to real-site signals so that classifier performance is forced to various difficulty levels. Figure 7 illustrates three type of exemplary PQE signals in the dataset.

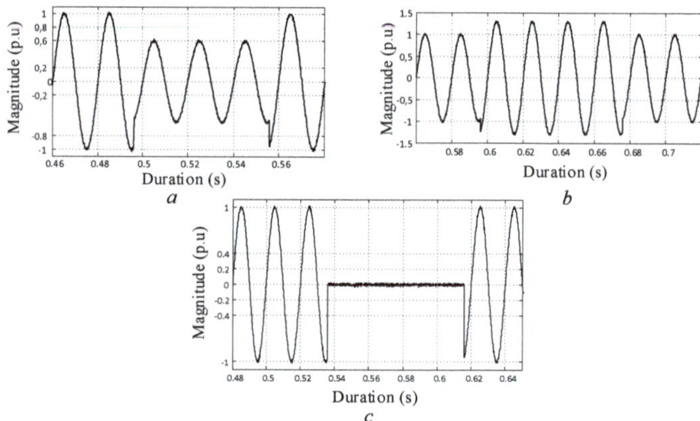

Figure 7. Three samples of PQE data in graphical view (Signal-to-noise ratio 30 dB) (**a**) sag; (**b**) swell; (**c**) interruption.

6. Results and Discussion

After feature extraction process using DWT–Entropy and histogram methods, a set of distinctive features is obtained to be classified in a decision-making phase. Whole feature set matrix includes $E_1, \ldots, E_8$ set as DWT–Entropy features, and $H_1, \ldots, H_{14}$ set as the histogram features. The feature matrix has a row number of 600, the same as samples of the dataset. In this study, a feature vector that consists of 22 elements is used for classification.

The feature set has a pre-processing period containing normalization and a cross-validation procedure. A 10-fold cross validation algorithm is used to get a better test performance and to force classifiers to more complicated test periods. Because of using cross-validation, accuracy values are given for 10-fold on average. FF–ANN and ELM have the same form as the SLFN structure. Because of this reality, we compare the proposed method just with the classic FF–ANN topology. In our experiments, number of hidden neurons are 225 and 20 for ELM and FF–ANN. Both classifiers use tangent sigmoid activation function in the hidden layer. Given parameters are acquired empirically as optimum values of several experiments. All the simulation is held by a work station hardware including a dual-processor with a clock of 2.1 GHz and 32 GB of RAM value. Results for SNR 10 dB, SNR 20 dB, and SNR30 dB are given using those classifier parameters above.

In Table 2, results for SNR 30 dB, 20 dB, and 10 dB noise conditions are presented collectively for both classifiers. It is said that ELM has a robust structure for different noise states. In Table 2, results for SNR 30 dB conditions show that ELM has a superior performance with 100% accuracy and classifies all the classes correctly. In addition, ELM is good at a time cost in both training and test phases. FF–ANN has an adequate performance, but it is nearly three times slower in the training period. This time costs when processing the big data concept. In Table 2, we can see the accuracy performance of SNR 20 dB as 99.66%. In Ref. [8], the performance value that belongs to the same condition is 96.22%. Using an easy to implement method with less features and less computational time, our proposed system outperforms the DGT with type-2 fuzzy based SVM. Because our classifier demonstrates its robustness in Table 2 with different noise levels, we set the SNR 30 dB value as the benchmark level. Other result tables present the results with the benchmark noise level as SNR 30 dB.

Table 2. The result table for signal-to-noise ratio 30 dB, 20 dB, and 10 dB conditions.

Classifier Type	Average Accuracy (%)	Time (s)	
		Training	Test
Results for 30 dB Conditions			
ELM	100	0.3266	0.0250
FF–ANN	99.0	0.9007	0.0111
Results for 20 dB Conditions			
ELM	99.66	0.3141	0.0000
FF–ANN	98.00	0.8883	0.0262
Results for 10 dB Conditions			
ELM	99.83	0.2906	0.0000
FF–ANN	99.17	0.9641	0.0138

When making a comparison between two similar classifiers, it is always considered whether the test is running under equal circumstances. The results carried out using the optimal classifier parameters are above. In this part of the experiments, two more different operation conditions are provided for classifiers: (1) when ELM has a number of 20 hidden layer neurons just the same as FF–ANN; and (2) when FF–ANN has a number of 225 hidden neurons just the same as ELM. This allows both classifiers for evaluating with the same situations. Table 3 presents the results of those two conditions.

The most important argument of this evaluation approach is about time cost of FF–ANN. In Table 3, when FF–ANN has 225 neurons, training time is nearly four hundred times more. In comparison with FF–ANN, when ELM has 20 neurons, it generates nearly accuracy of 98% with a fast training time. Dealing with a large-scale dataset, training time is a crucial value of interest.

Table 3. The result table for SNR 30 dB and both ELM & FF–ANN are under equal terms of neuron numbers.

	20 Neurons			225 Neurons		
Classifier	Average Accuracy (%)	Training Time (s)	Testing Time (s)	Average Accuracy (%)	Training Time (s)	Testing Time (s)
ELM	97.50	0.0578	0.0000	100	0.3266	0.0250
FF–ANN	99.00	0.9007	0.0111	96.00	129.353	0.0325

The results above are obtained via using of the full feature set. Now, the ongoing analysis has a feature searching starting with DWT–Entropy features. Figure 8 shows $E_1, \ldots, E_8$ feature sub-set for all the samples of the dataset.

One can see from Figure 8 that E_1 to E_5 features are less distinctive comparing to E_6 to E_8 features. Magnitudes of E_1–E_5 are low values and show a little change only for $EC4$ class, but E_6–E_8 features differ from each other for all the classes of the dataset. Table 4 shows the ELM results of classification using only DWT–Entropy for SNR 30 dB with different combinations.

Table 4. ELM results for DWT–Entropy features (SNR 30 dB).

Feature Set	Average Accuracy (%)	Time (s)	
		Training	Test
E_6–E_8	72.70	0.3609	0.0156
E_1–E_5	63.80	0.3094	0.0000
E_1–E_8	96.20	0.3563	0.0000

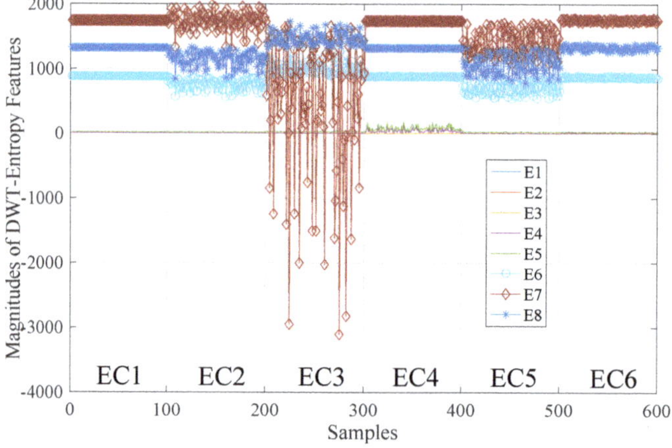

Figure 8. Variation of DWT–Entropy feature sub-set according to samples, Classes: *EC1–EC6*.

The meaning of Table 4 differs from graphical projection. Figure 8 tells us that E_1–E_5 features are less distinctive but classification results refute that estimation. Using just E_6–E_8 features gives 72.7% of average accuracy. When E_1–E_5 features are added and using the whole DWT–Entropy sub-set, average accuracy rises to 96.2% value and the result for using only E_1–E_5 sub-set has an average accuracy of 63.8%.

General distribution of the histogram features is given in Figure 9. It can be clearly seen that all histogram features (H_1–H_{14}) are distinctive for nearly all classes. The features of $EC2$ are less distinctive among all feature sub-sets, but if we zoom in its distribution, it is seen that they differ from each other.

Now, the next results are obtained using just the histogram features, H_1–H_{14}. Table 5 lists the ELM classification according to H_1–H_{14} feature sub-set with SNR 30 dB. The histogram feature sub-set with an average accuracy of 98.7% is adequate for the proposed PQE classification system.

Table 5. ELM result table for histogram features (SNR 30 dB).

Feature Set	Average Accuracy (%)	Time (s)	
		Training	**Test**
H_1–H_{14}	98.70	0.3734	0.0266

A general comparison is given in Table 6 according to processing time and average accuracy values; specs of total feature sets are also listed. As it can be seen in Table 6, extracting features from the histogram method is 15 times faster than DWT–Entropy. Average accuracy values are close to each other so the important point is time cost, most particularly dealing with big data. For the whole feature set, the proposed system for PQE classification reaches the perfect classification. However, the proposed intelligent recognition system can be used just preferring the histogram method based feature set, which is the novelty of this paper. Its accuracy reaches an adequate performance value and the DWT method supports this for more.

Table 6. Feature extraction method comparison.

Method	Feature Set Length	Processing Time (s)	Average Accuracy of ELM (%)
DWT–Entropy	8	7.66	96.2
Histogram	14	0.51	98.7
Total feature set	22	8.2	100

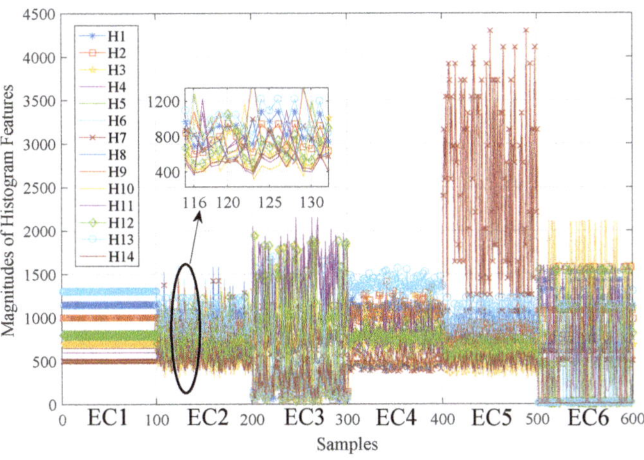

Figure 9. Variation of histogram feature subset according to samples, Classes: *EC1–EC6*.

7. Conclusions

In this paper, a machine learning based ELM classifier coupled with DWT–Entropy and histogram methods for classification of PQE signals is proposed. With its MRA nature, the DWT method is used to establish a time-frequency analysis suitable for event signals. Feature extraction from DWT is based on entropy calculation. This study proposes the histogram, which is a novel feature extracting method used in power system signal processing with a machine learning based classifier. Histogram features characterize PQE data in an accurate manner. Feature extraction methods provide a size reduction in the raw dataset before classification process. In addition, the proposed feature extraction method of histogram comes to the fore with being fast in time, and its easy implementation for embedded systems. Designing a power quality event monitoring device using just histogram features is one of the future aims of this study. The conventional DWT method is used for performance improvement of the histogram. Given results prove the increase in performance.

The results from the proposed pattern recognition system prove that it carries an efficient classification process with six categories of PQE. In addition, it can be said that the proposed system has a robust structure for different noisy conditions. Future works will include building an embedded system for the histogram method. Thus, it would be able to run the algorithm on a field device.

Acknowledgments: This work was supported by the Firat University Scientific Research Projects Unit (FUBAP) funding under the Ph.D. Thesis grant program with a project number of TEKF.16.18. This paper is also a part of the Ph.D. thesis of candidate F. Ucar in Firat University, EEE Department, Elazig, Turkey.

Author Contributions: All the authors have contributions to this study in every step of the organisation process. Ferhat Ucar, Omer F. Alcin, and Besir Dandil conceived and designed the experiments; Ferhat Ucar and Omer F. Alcin performed the experiments; Besir Dandil and Fikret Ata analysed the data and contributed to the methodology and experiments; Ferhat Ucar, Omer F. Alcin, Besir Dandil, and Fikret Ata wrote the paper.

Conflicts of Interest: The authors declare no conflict of interest.

References

1. Stimmel, C.L. *Big Data Analytics Strategies for the Smart Grid*; Auerbach Publications: Boston, MA, USA, 2015; Volume 53, p. 160.
2. Keyhani, A.; Marwali, M. (Eds.); *Smart Power Grids 2011*; Springer: Berlin/Heidelberg, Germany, 2011.
3. Ribeiro, P.F.; Duque, C.A.; Ribeiro, P.M.; Cerqueira, A.S. *Power Systems Signal Processing for Smart Grids*; Wiley: Hoboken, NJ, USA, 2013; pp. 1–442.
4. Arghandeh, R. Micro-Synchrophasors for Power Distribution Monitoring, a Technology Review. *arXiv* **2016**, 1–18, arXiv:1605.02813.
5. Mahela, O.P.; Shaik, A.G.; Gupta, N. A critical review of detection and classification of power quality events. *Renew. Sustain. Energy Rev.* **2015**, *41*, 495–505.
6. Khokhar, S.; Mohd Zin, A.A.B.; Mokhtar, A.S.B.; Pesaran, M. A comprehensive overview on signal processing and artificial intelligence techniques applications in classification of power quality disturbances. *Renew. Sustain. Energy Rev.* **2015**, *51*, 1650–1663.
7. Zhou, Y.; Arghandeh, R.; Spanos, C.J. Partial Knowledge Data-driven Event Detection for Power Distribution Networks. *IEEE Trans. Smart Grid* **2017**, *PP*, 1.
8. Naderian, S.; Salemnia, A. Method for classification of PQ events based on discrete Gabor transform with FIR window and T2FK-based SVM and its experimental verification. *IET Gener. Transm. Distrib.* **2017**, *11*, 133–141.
9. Li, Y.; Song, X.; Meng, X. Application of signal processing and analysis in detecting single line-to-ground (SLG) fault location in high-impedance grounded distribution network. *IET Gener. Transm. Distrib.* **2016**, *10*, 382–389.
10. Mitra, R.; Goswami, A.K.; Tiwari, P.K. Voltage sag assessment using type-2 fuzzy system considering uncertainties in distribution system. *IET Gener. Transm. Distrib.* **2017**, *11*, 1409–1419.
11. Nasiri, S.; Seifi, H. Robust probabilistic optimal voltage sag monitoring in presence of uncertainties. *IET Gener. Transm. Distrib.* **2016**, *10*, 4240–4248.
12. Duda, R.O.; Hart, P.E.; Stork, D.G. *Pattern Classification*; Ricoh Innovations, Inc.: Menlo Park, CA, USA, 2001; p. 680.
13. Bollen, M.; Gu, I. *Signal Processing of Power Quality Disturbances*; Wiley: Hoboken, NJ, USA, 2006; p. 861.
14. Agüera-Pérez, A.; Carlos Palomares-Salas, J.; de la Rosa, J.J.G.; María Sierra-Fernández, J.; Ayora-Sedeño, D.; Moreno-Muñoz, A. Characterization of electrical sags and swells using higher-order statistical estimators. *Measurement* **2011**, *44*, 1453–1460.
15. Hajian, M.; Akbari Foroud, A. A new hybrid pattern recognition scheme for automatic discrimination of power quality disturbances. *Measurement* **2014**, *51*, 265–280.
16. Barros, J.; de Apraiz, M.; Diego, R.I. A virtual measurement instrument for electrical power quality analysis using wavelets. *Measurement* **2009**, *42*, 298–307.

17. Moravej, Z.; Abdoos, A.A.; Pazoki, M. Detection and Classification of Power Quality Disturbances Using Wavelet Transform and Support Vector Machines. *Electr. Power Compon. Syst.* **2009**, *38*, 182–196.

18. Şengüler, T.; Şeker, S. Continuous wavelet transform for ferroresonance detection in power systems. *Electr. Eng.* **2016**, *99*, 595–600.

19. Petrović, P.; Damljanović, N. New procedure for harmonics estimation based on Hilbert transformation. *Electr. Eng.* **2016**, *99*, 313–323.

20. Zhou, Y.; Arghandeh, R.; Konstantakopoulos, I.; Abdullah, S.; Von Meier, A.; Spanos, C.J. Abnormal event detection with high resolution micro-PMU data. In Proceedings of the 19th Power Systems Computation Conference (PSCC), Genoa, Italy, 20–24 June 2016.

21. Styvaktakis, E.; Bollen, M.; Gu, I. Automatic classification of power system events using RMS voltage measurements. In Proceedings of the IEEE Power Engineering Society Summer Meeting, Chicago, IL, USA, 21–25 July 2002; Volume 2, pp. 824–829.

22. Uçar, F.; Alçin, Ö.F.; Dandil, B.; Ata, F. Machine learning based power quality event classification using wavelet—Entropy and basic statistical features. In Proceedings of the 21st International Conference on Methods and Models in Automation and Robotics (MMAR), Miedzyzdroje, Poland, 29 August–1 September 2016; pp. 414–419.

23. Ekici, S. Classification of power system disturbances using support vector machines. *Expert Syst. Appl.* **2009**, *36*, 9859–9868.

24. Reaz, M.B.I.; Choong, F.; Sulaiman, M.S.; Mohd-Yasin, F. Prototyping of wavelet transform, artificial neural network and fuzzy logic for power quality disturbance classifier. *Electr. Power Compon. Syst.* **2007**, *35*, 1–17.

25. Uyar, M.; Yildirim, S.; Gencoglu, M.T. An effective wavelet-based feature extraction method for classification of power quality disturbance signals. *Electr. Power Syst. Res.* **2008**, *78*, 1747–1755.

26. Gaouda, A.; Salama, M. Power quality detection and classification using wavelet-multiresolution signal decomposition. *Power Deliv. IEEE* **1999**, *14*, 1469–1476.

27. Santoso, S.; Powers, E.; Grady, W.; Hofmann, P. Power quality assessment via wavelet transform analysis. *IEEE Trans. Power Deliv.* **1996**, *11*, 924–930.

28. Barros, J.; Diego, R.I.; de Apráiz, M. Applications of wavelets in electric power quality: Voltage events. *Electr. Power Syst. Res.* **2012**, *88*, 130–136.

29. Erişti, H.; Demir, Y. Automatic classification of power quality events and disturbances using wavelet transform and support vector machines. *IET Gener. Transm. Distrib.* **2012**, *6*, 968.

30. Sturges, H.A. The Choice of a Class Interval. *J. Am. Stat. Assoc.* **1926**, *21*, 65–66.

31. Yalcin, T.; Ozdemir, M. Pattern recognition method for identifying smart grid power quality disturbance. In Proceedings of the International Conference on Harmonics and Quality of Power (ICHQP), Belo Horizonte, Brazil, 16–19 October 2016; pp. 903–907.

32. Meher, S.K.; Pradhan, A.K. Fuzzy classifiers for power quality events analysis. *Electr. Power Syst. Res.* **2010**, *80*, 71–76.

33. Styvaktakis, E.; Bollen, M.; Gu, I. Expert system for classification and analysis of power system events. *IEEE Trans. Power Deliv.* **2002**, *17*, 423–428.

34. Huang, G.B.; Zhu, Q.Y.; Siew, C.K. Extreme learning machine: Theory and applications. *Neurocomputing* **2006**, *70*, 489–501.

35. Lu, J.; Huang, J.; Lu, F. Sensor Fault Diagnosis for Aero Engine Based on Online Sequential Extreme Learning Machine with Memory Principle. *Energies* **2017**, *10*, 39.

36. Lopez-Ramirez, M.; Ledesma-Carrillo, L.; Cabal-Yepez, E.; Rodriguez-Donate, C.; Miranda-Vidales, H.; Garcia-Perez, A. EMD-based feature extraction for power quality disturbance classification using moments. *Energies* **2016**, *9*, 565.

37. Alcin, O.F.; Siuly, S.; Bajaj, V.; Guo, Y.; Sengur, A.; Zhang, Y. Multi-category EEG signal classification developing time-frequency texture features based Fisher Vector encoding method. *Neurocomputing* **2016**, *218*, 251–258.

38. Huang, G.; Huang, G.B.; Song, S.; You, K. Trends in extreme learning machines: A review. *Neural Netw.* **2015**, *61*, 32–48.

39. Zhang, L.; He, Z.; Liu, Y. Deep object recognition across domains based on adaptive extreme learning machine. *Neurocomputing* **2017**, *239*, 194–203.

40. Zhang, L.; Zhang, D. Evolutionary cost-sensitive extreme learning machine. *IEEE Trans. Neural Netw. Learn. Syst.* **2017**, *28*, 3045–3060.

41. Zhang, L.; Zhang, D. Robust visual knowledge transfer via extreme learning machine-based domain adaptation. *IEEE Trans. Image Process.* **2016**, *25*, 4959–4973.

42. Zhang, L.; Zhang, D. Domain adaptation extreme learning machines for drift compensation in E-nose systems. *IEEE Trans. Instrum. Meas.* **2015**, *64*, 1790–1801.

43. *MATLAB*, R2015a [Computer Software]; MathWorks: Natick, MA, USA, 2015.

44. He, Z. *Wavelet Analysis and Transient Signal Processing Applications for Power Systems*; John Wiley & Sons: Singapore, 2016.

45. Thirumala, K.; Shantanu; Jain, T.; Umarikar, A.C. Visualizing time-varying power quality indices using generalized empirical wavelet transform. *Electr. Power Syst. Res.* **2017**, *143*, 99–109.

46. Chen, J.; Dou, Y.; Li, Y.; Li, J. Application of Shannon Wavelet Entropy and Shannon Wavelet Packet Entropy in Analysis of Power System Transient Signals. *Entropy* **2016**, *18*, 437.

47. Vapnik, V.N. *Statistical Learning Theory*; John Wiley & Sons: Hoboken, NJ, USA, 1998; p. 760.

48. Volkenstein, M.V. *Entropy and Information*; Springer: Berlin, Germany, 2009.

49. Pearson, K. Contributions to the Mathematical Theory of Evolution. II. Skew Variation in Homogeneous Material. *Philos. Trans. R. Soc. A* **1895**, *186*, 343–414.

50. Scott, D.W. Sturges' rule. *Wiley Interdiscip. Rev. Comput. Stat.* **2009**, *1*, 303–306.

51. Alcin, O.F.; Sengur, A.; Ghofrani, S.; Ince, M.C. GA-SELM: Greedy algorithms for sparse extreme learning machine. *Measurement* **2014**, *55*, 126–132.

52. Penrose, R. A generalized inverse for matrices. *Math. Proc. Camb. Philos. Soc.* **1955**, *51*, 406–413.

Article

An Efficient Phase-Locked Loop for Distorted Three-Phase Systems

Yijia Cao [1], Jiaqi Yu [1], Yong Xu [1], Yong Li [1,*] and Jingrong Yu [2]

[1] College of Electrical and Information Engineering, Hunan University, Changsha 410082, China;
    yijiacao@hnu.edu.cn (Y.C.); yujq629@hnu.edu.cn (J.Y.); xuyong@hnu.edu.cn (Y.X.)
[2] School of Information Science and Engineering, Central South University, Changsha 410083, China;
    Jingrong@csu.edu.cn
* Correspondence: yongli@hnu.edu.cn

Academic Editor: José Gabriel Oliveira Pinto
Received: 6 December 2016; Accepted: 20 February 2017; Published: 27 February 2017

Abstract: This paper proposed an efficient phase-locked loop (PLL) that features zero steady-state error of phase and frequency under voltage sag, phase jump, harmonics, DC offsets and step-and ramp-changed frequency. The PLL includes the sliding Goertzel discrete Fourier transform (SGDFT) filter-based fundamental positive sequence component separator (FPSCS), the synchronous-reference-frame PLL (SRF-PLL) and the secondary control path (SCP). In order to obtain an accurate fundamental positive sequence component, SGDFT filter is introduced as it features better filtering ability at the frequencies that are integer times of fundamental frequency. Meanwhile, the second order Lagrange-interpolation method is employed to approximate the actual sampling number including both integer and fractional parts as grid frequency may deviate from the rated value. Moreover, an improved SCP with single-step comparison filtering algorithm is employed as it updates reference angular frequency according to the FPSC, which promises a zero steady-state error of phase and improves the frequency tracking speed. In this paper, the mathematical model of the proposed PLL is constructed, its stability is analyzed. Also, design procedure of the control parameters is presented. The effectiveness of the proposed PLL is confirmed by experimental results and comparison with advanced pre-filtering PLLs.

Keywords: distorted grid conditions; SGDFT; Lagrange-interpolation method; frequency adaption; SCP

1. Introduction

The information about instantaneous grid voltage phase and frequency are usually obtained via phase lock loop (PLL), which is of vital importance to maintain synchronization and stable operation for grid-connected power electronic devices [1–5]. Recently, the presence of DC offsets and harmonics in grid voltages caused by measurement devices, nonlinear loads and grid faults throws down a new challenge to the synchronization technique [6].

Synchronous-reference-frame PLL (SRF-PLL) is probably the most popular synchronization technique under ideal grid condition [7,8]. However, the disturbance rejection capability of SRF-PLL is poor for unbalanced voltages, harmonics, step-and ramp-changed frequency and DC offsets. Under unbalanced condition, the fundamental negative sequence component imposes the second harmonic ripple on variables in dq axis. Under polluted condition, the nth order harmonic components of input voltages become $(n-1)$th harmonics (if it is a positive sequence harmonic) or $(n+1)$th harmonics (if it is a negative sequence harmonic) in dq axis [9]. Especially, DC offsets become fundamental component in dq axis. Under step-and ramp-changed frequency, there is an error between the reference angular frequency and the actual one. As a result, the SRF-PLL cannot track phase and frequency precisely.

To solve this problem, numerous advanced PLLs have been intensively studied [10–27]. The improved methods generally fall into two classes. One representative method is the in-loop filtering technique which adds various specific filters in the phase control loop, such as adaptive notch filter-based PLL [10], moving average filter-based PLL (MAF-PLL) [11], Type-1 PLL [12], *dq*-frame delayed signal cancellation operator based-PLL [13] and the variable sampling period filter based PLL (VSPF-PLL) [14]. These PLLs show satisfactory performances but they are not applicable to the situation where precise fundamental positive sequence component (FPSC) is needed.

The other method is the pre-filtering technique which employs various filters to extract FPSC from the non-ideal voltages [15–27]. References [15,16] present a multiple complex-coefficient filter-based (MCCF) synchronization technique with no need of the symmetrical component method and rotating frame transformations. Reference [17] proposes a generalized second-order and third-order complex-vector filter based on reference [15] for better dynamic performance and higher harmonic attenuation, but DC offsets of input signal are not considered. Moreover, as CCF gives relatively limited gains for each order harmonics especially for DC offsets, it cannot maintain tracking precision of grid phase under harmonics and DC offsets. As a representative pre-filtering SRF-PLL, second-order generalized integrator-based (SOGI) PLL is presented in [16] and improved in references [19,20]. Dual SOGI in [18] is the building block of the quadrature signal generator (QSG) and offers harmonic blocking capability to the system. References [19,20] make SOGI-PLL frequency adaptive by adding a harmonic decoupling network and an angular frequency feed forward loop, respectively. Consequently, the SOGI-based PLLs exhibit a relatively precise and frequency-adaptive response under unbalanced condition, but they also cannot track the phase precisely under harmonics and DC offsets due to similar filtering characteristics with CCF based PLL. The moving average filter-based (MAF) pre-filtering PLL has precise accuracy under unbalanced and heavily polluted conditions when the filtering window width is integer times of the input AC signal's period [21]. The filtering window width in [22,23] is set to one time of the input AC signal's period in *dq* axis, which can eliminate DC offsets of input signal. But this will give one period delay and increase the response time. References [24] proposes a generalized delay signal cancelation-based (GDSC) pre-filtering PLL, which can eliminate the negative-sequence component and any given harmonics under unbalanced voltages, harmonics and DC offsets. References [25–27] focus on the frequency-adaptive scheme of GDSC-PLL. They all track frequency precisely and give an acceptable response time when input frequency varies. However, it bears burdensome digital computation time, as it needs 4 to 5 cascaded DSC modules to suppress all-field harmonics. Furthermore, the aforementioned pre-filtering PLLs give steady-state error on estimated phase when input signal's frequency varies due to the fixed reference angular frequency. Thus, some improved PLLs adopt a secondary control path (SCP) for better tracking accuracy [28,29]. But the reference angular frequency is calculated directly from the input signals without pre-filter, which would make the reference angular frequency inaccuracy and consequently mislead the detected frequency under variable frequency.

With the aim of further improvement in tracking accuracy and disturbance rejection capability under unbalanced voltages, harmonics, step-and ramp-changed frequency and DC offsets, an improved PLL based on sliding Goertzel discrete Fourier transform (SGDFT) pre-filter is proposed. The FPSC is extracted by SGDFT pre-filter as it features unit gain on specified frequency and negative infinite gain on other frequencies; Second order Lagrange-interpolation method is used to approximate the actual sample number as grid frequency may deviate from the normal value; In order to obtain an accurate reference angular frequency, the extracted FPSC is adopted as the input of SCP rather than the one without being pre-filtered. Meanwhile, single-step comparison filtering algorithm is proposed according to the SCP characteristics, which gives lower delay and higher accuracy on the calculation of reference angular frequency than traditional low pass filter (LPF). By means of these schemes, zero steady-state error and rapid transient response in phase and frequency are guaranteed under distorted conditions.

The rest of this paper is organized as follows: Section 2 analyzes the SGDFT filter characteristics. Section 3 proposes the new PLL structure and its mathematical model. Section 4 describes the design and implementation method. The comparative experimental results of five PLLs obtained from the prototype comprised of signal generator and digital signal processor (DSP) TMS320F28335 control board are given in Section 5. Finally, the conclusions are drawn in Section 6.

2. SGDFT Overview

The quadrature signal generator (QSG) is widely used in pre-filtering PLL. However, it has limited performance under polluted condition, since the separated FPSCs still contain harmonics. The SGDFT filter is derived from the standard DFT equation and commonly used to compute DFT spectra [30]. Compared with QSG, it has better filtering ability at the frequencies that are integer times of fundamental frequency. The transfer function $H_{\mathrm{SGDFT}}(z)$ and the structure of SGDFT filter are shown in Equation (1) and Figure 1, respectively.

$$H_{\mathrm{SGDFT}}(z) = \frac{(1 - e^{-j2k\pi/N}z^{-1})(1 - z^{-N})}{1 - 2\cos(2k\pi/N)z^{-1} + z^{-2}} \tag{1}$$

where k is frequency domain index and N is the sampling number.

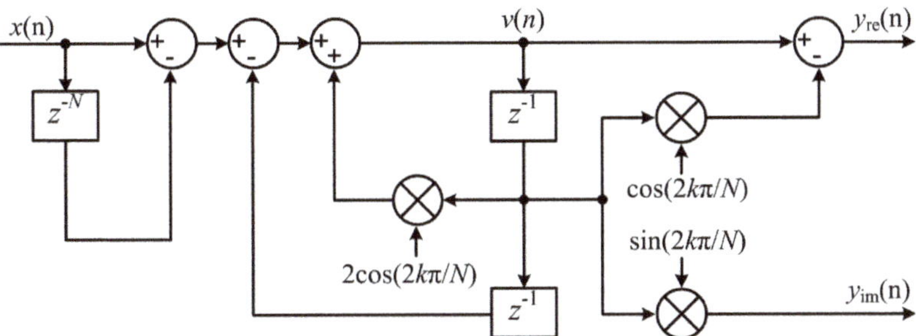

Figure 1. Block diagram description of SGDFT.

Figure 2 shows the frequency response and z-domain zero/pole of SGDFT. As shown in Figure 2a, $H_{\mathrm{re}}(z)$ and $H_{\mathrm{im}}(z)$ are the real and imaginary part of $H_{\mathrm{SGDFT}}(z)$, while $G_{\mathrm{d}}(s)$ and $G_{\mathrm{q}}(s)$ represent the transfer function of QSG in d and q axis, respectively. The magnitude of the four transfer functions are all 0 dB at the fundamental frequency $f_0 = 50$ Hz. It is shown that the frequency response of SGDFT is same with QSG except at the frequencies that are integer times of f_0. Meanwhile, SGDFT has much smaller gain than QSG at the integer multiple of f_0, which means that SGDFT has better filtering ability than QSG under polluted condition. In Figure 2b, there are 256 zeros (blue circle) of the transfer function equally spaced around the unit circle on z-domain. Moreover, it has two conjugate poles (red cross) cancelling zeros at $z = e^{\pm j2\pi/256}$. Thus, one can conclude that SGDFT can calculate the fundamental coefficient of the input signal precisely. Therefore, SGDFT is introduced to separate the FPSC.

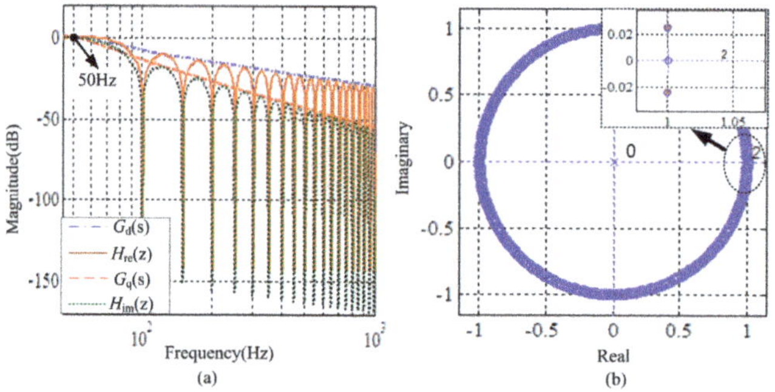

Figure 2. Frequency response and z-domain zero/pole of SGDFT when $k = 1$, $N = 256$. (**a**) Real and imaginary frequency response of SGDFT; (**b**) z-domain zero/pole.

3. The Proposed PLL

The block diagram of the proposed PLL is shown in Figure 3. This PLL structure consists of three main parts: FPSCS, SRF-PLL and SCP. The FPSCS module uses SGDFT filter and the symmetrical component method to separate the FPSC precisely even under distorted conditions. Considering that the practical grid frequency may deviate from the rated value, Lagrange-interpolation method is adopted to approximate the actual sampling number N_r, in order to make the SGDFT filter effective under variable frequency. The SCP module updates reference angular frequency, which improves its tracking performance.

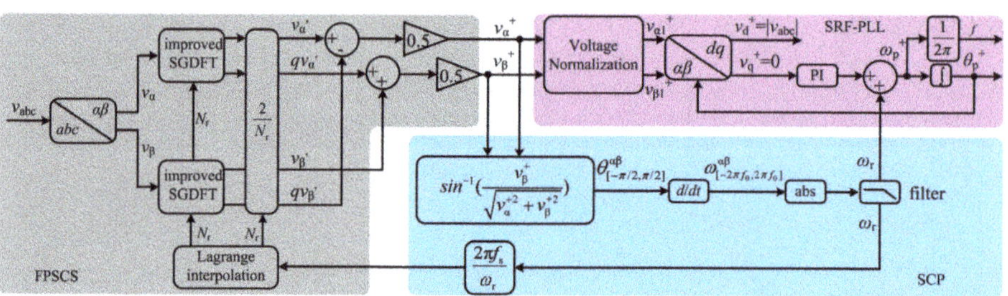

Figure 3. Block diagram of the proposed PLL.

3.1. SGDFT-Based FPSCS

The SGDFT structure is shown in Figure 1. v_α and v_β are input signals in two-phase stationary frame. v'_α and qv'_α are the filtered direct and quadrature parameters of v_α. v'_β and qv'_β are the filtered direct and quadrature parameters of v_β. v^+_α and v^+_β are FPSC in two-phase stationary frame, $v^+_{\alpha 1}$ and $v^+_{\beta 1}$ are normalized FPSC. The FPSC v^+_α and v^+_β in $\alpha\beta$ axis are obtained by:

$$\begin{bmatrix} v^+_\alpha \\ v^+_\beta \end{bmatrix} = \frac{1}{2} \begin{bmatrix} v'_\alpha - qv'_\beta \\ qv'_\alpha + v'_\beta \end{bmatrix} \tag{2}$$

In practical power system, the grid frequency f_0 is a time-varying parameter. The problem arises when f_0 cannot be divisible by the sampling frequency f_s, i.e., the order N_r would not be a integer and can be described as $N_r = N_a + D$, where $N_a = \text{floor}\,(N_r)$ is the integer and $D = N_r - N_a$ $(0 \le D < 1)$ is the fractional part. Thus, the delay is given as: $z^{-N_r} = z^{-N_a} z^{-D}$. The Lagrange-interpolation method is introduced because it is an effective way to approximate the fractional delay for FIR filter design [31]. The fraction delay z^{-D} can be approximated by:

$$z^{-D} \approx \sum_{k=0}^{n} \mathrm{H}(k)z^{-k} \qquad k = 0,1,\cdots,n \tag{3}$$

The coefficients H(k) are calculated as:

$$\mathrm{H}(k) = \prod_{\substack{i=0 \\ i\neq k}}^{n} \frac{D-i}{k-i} \qquad k = 0,1,\cdots,n \tag{4}$$

Specifically, the case $n = 1$ corresponds to the linear interpolation and the two coefficients are H(0) = 1 − D and H(1) = D. Figure 4 shows the frequency responses of Lagrange-interpolating fraction delay z^{-D} at different D values.

In Figure 4, the fractional D are 0.3 (red line), 0.5 (green line) and 0.75 (blue line) for validation. It shows that the magnitude of fraction delay with $n = 2$ (dashed line) is closer to the case with $D = 0$ (the real value), meanwhile the phase has smaller change at different D values than $n = 1$ (solid line). The magnitude of fraction delay with the order $n = 3$ is close to the case with $D = 0$ while it exceeds the unit amplitude, which may introduce stability problem. Moreover, the fraction delay with the order $n = 3$ consumes more addition and multiplication operations than $n = 2$. Considering these two aspects, $n = 2$ is chosen to be the Lagrange interpolation order. Hence, the corresponding fractional delay is:

$$z^{-N_r} = \mathrm{H}(0)z^{-N_a} + \mathrm{H}(1)z^{-N_a-1} + \mathrm{H}(2)z^{-N_a-2} \tag{5}$$

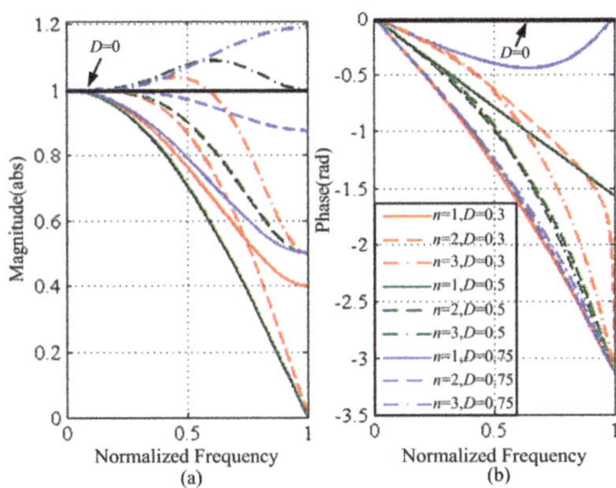

Figure 4. Frequency responses of Lagrange-interpolating fraction delay. (**a**) Magnitude responses; (**b**) Phase responses.

Therefore, the structure of SGDFT can be improved, as shown in Figure 5.

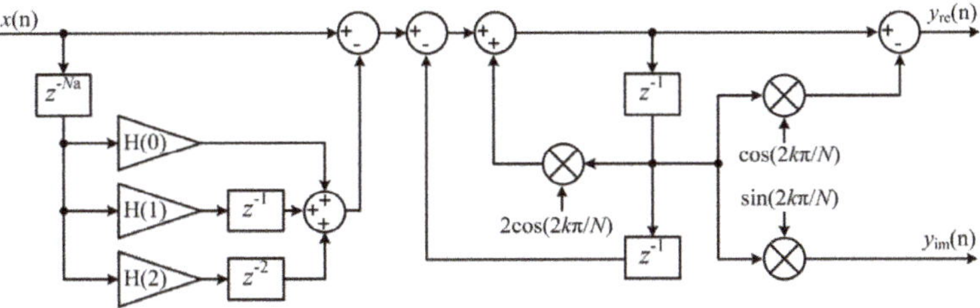

Figure 5. The improved structure of SGDFT.

3.2. Description of SCP

The secondary control path is introduced to improve the transient response rate. Moreover, it updates the reference angular frequency to decrease phase error under variable frequency.

In reference [28], the phase $\theta \in [-\pi/2, \pi/2]$ can be deduced as $\theta = \tan^{-1}(v_\beta^+/v_\alpha^+)$. However, *arctangent* function gives rapid change at $\pm n\pi/2$, which results in differential errors in digital implementation. Therefore, this paper uses *arcsin* function to calculate the phase. Besides, an *abs* function is adopted to regulate the negative angular frequency. In order to eliminate the digital differential error in the regulated angular frequency, the single-step comparison filter is realized as follows:

$$\omega_r(k_n) = \{\omega(k_n) \geq \omega(k_n - 1)\}?\{\omega_r(k_n) = \omega(k_n)\}:\{\omega_r(k_n) = \omega(k_n - 1)\} \tag{6}$$

where k_n is the current cycle counter. Both the differential part d/dt and Equation (6) need one control period delay, thus its transfer function can be described as $G_f(s) = 1/(2T_s s + 1)$. As analyzed in Section 2, the model of SGDFT is equivalent to QSG. Thus, the pre-filtering stage small signal model in the proposed PLL can be used as $G_o(s) = \omega_o/(s + \omega_o)$, where ω_o equals to $0.707\omega_r$. Therefore, the secondary control path transfer function is $\omega_r(s) = G_o(s) \times G_f(s) \times \omega(s)$.

Moreover, the introduction of SCP changes the type and pole-zero location of tracking loop [29]. Thus, the voltage normalization part is employed to remove this adverse effect. It can be achieved as:

$$\begin{cases} v_{\alpha 1}^+ = \dfrac{v_\alpha^+}{\sqrt{v_\alpha^{+2}+v_\beta^{+2}}} \\[3mm] v_{\beta 1}^+ = \dfrac{v_\beta^+}{\sqrt{v_\alpha^{+2}+v_\beta^{+2}}} \end{cases} \text{or} \quad \begin{cases} v_{\alpha 1}^+ = \dfrac{v_\alpha^+}{0.5(\sqrt{v_\alpha^{\prime 2}+qv_\alpha^{\prime 2}}+\sqrt{v_\beta^{\prime 2}+qv_\beta^{\prime 2}})} \\[3mm] v_{\beta 1}^+ = \dfrac{v_\beta^+}{0.5(\sqrt{v_\alpha^{\prime 2}+qv_\alpha^{\prime 2}}+\sqrt{v_\beta^{\prime 2}+qv_\beta^{\prime 2}})} \end{cases} \tag{7}$$

3.3. Proposed PLL Model

According to the aforementioned deduction, the small-signal model and equivalent model of the proposed PLL are shown in Figure 6.

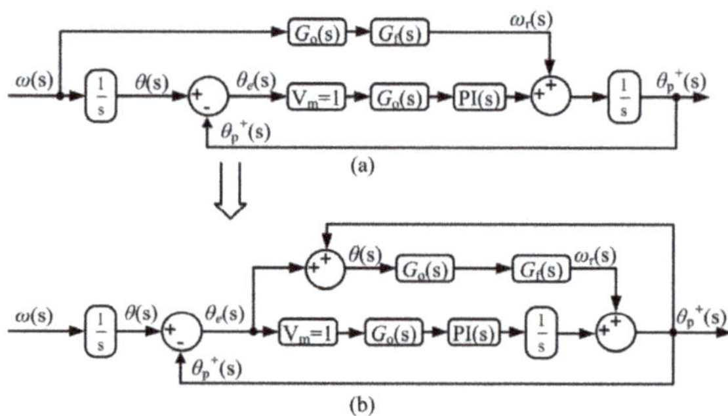

Figure 6. Small-signal model of the proposed PLL. (**a**) Original model; (**b**) Equivalent model.

It can be deduced from Figure 6b as:

$$\theta_p^+(s) = \underbrace{V_m G_o(s)G_{PI}(s)\frac{1}{s}\theta_e(s)}_{G_a(s)} + G_o(s)G_f(s)\{\theta_e(s) + \theta_p^+(s)\} \tag{8}$$

Then the open-loop transfer function can be obtained as:

$$G_{ol}(s) = \frac{\theta_p^+(s)}{\theta_e(s)} = \frac{G_o(s)G_f(s) + G_a(s)}{1 - G_o(s)G_f(s)} \tag{9}$$

The term $G_o(s) \times G_f(s)$ can be replaced by $G_e(s) = 1/(T_e s + 1)$ because these two terms are small inertial elements, where T_e is the equivalent delay that equals to $2T_s + 1/\omega_o$. Therefore, the complete open-loop and closed-loop transfer functions are:

$$G_{ol}(s) = \frac{\theta_p^+(s)}{\theta_e(s)} = \frac{s^3 + s^2\omega_o(1 + k_p T_e) + s\omega_o(k_p + k_i T_e) + \omega_o k_i}{T_e s^3(s + \omega_o)} \tag{10}$$

$$G_{cl}(s) = \frac{\theta_p^+(s)}{\theta(s)} = \frac{s^3 + s^2\omega_o(1 + k_p T_e) + s\omega_o(k_p + k_i T_e) + \omega_o k_i}{s^4 T_e + s^3(1 + \omega_o T_e) + s^2\omega_o(1 + k_p T_e) + s\omega_o(k_p + k_i T_e) + \omega_o k_i} \tag{11}$$

4. Systematic Design Approach

The aim of this section consists of four aspects: parameter design guidelines for the proposed PLL, system stability analysis, study of bandwidth and dynamic responses, and discrete implementation method.

4.1. Parameters Design

It can be seen from Equation (10) that the open-loop transfer function is a four-order expression. Thus, zero-pole cancellation which is convenient for parameters design is adopted to simplify the system. Suppose that numerator polynomial has three real zeros, and one of them equals to ω_o. Thus, the open-loop transfer function would be:

$$G_{ol}(s) = \frac{(s + \omega_{z1})(s + \omega_{z2})(s + \omega_o)}{T_e s^3(s + \omega_o)} \tag{12}$$

It has been proved that the coincident zeros (i.e., $\omega_{z1} = \omega_{z2}$) can provide a higher stability margin than the spread ones [32]. Thus, combining Equations (10), (12) and $\omega_{z1} = \omega_{z2} = \omega_z$, Equation (10) can be rewritten as:

$$G_{ol}(s) = \frac{(s + \omega_z)^2}{T_e s^3} = \frac{s^2 + \omega_o k_p T_e s + k_i}{T_e s^3} \tag{13}$$

where k_p and k_i are the proportional and integral parameters of PI regulator shown in Figure 6. From Equation (13), the phase margin (PM) and the crossover frequency ω_c of the proposed PLL can be determined as:

$$\begin{aligned} \mathrm{PM} &= -90° + 2\tan^{-1}(h) \\ \omega_c &= \frac{1}{T_e \sin^2\left(\tan^{-1}(h)\right)} \end{aligned} \tag{14}$$

where $\omega_c = h\omega_z$. Figure 7 illustrates that PM is a function of h.

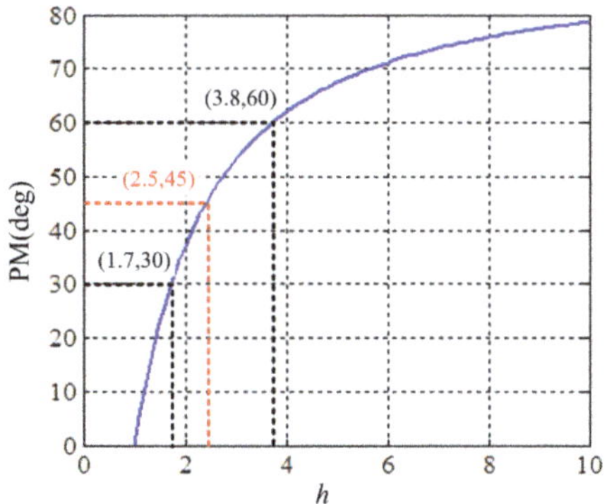

Figure 7. PM is a function of h.

The PM within the range of 30°~60° is the recommended range for stable operation [32]. Generally, PM = 45° is selected, which corresponds to $h = 2.5$, as shown in Figure 7. When $h = 2.5$, it can be derived that $\omega_c = 246.8$ rad/s, $\omega_z = 98.7$ rad/s, $k_p = 2\omega_z/(\omega_o \times T_e) = 189.2$ and $k_i = \omega_z{}^2 = 9746$ according to Equations (13) and (14).

4.2. System Stability

It is known that the introduction of SCP aggravates the stability problem as the voltage amplitude V_m changes the zeros of the open-loop and closed-loop transfer function. Thus, the voltage normalization is utilized to remove this effect. From Equations (10) and (11), it can be seen that V_m is no longer an influence factor. According to the analysis in Section 4.1, the open-loop bode plots are shown in Figure 8.

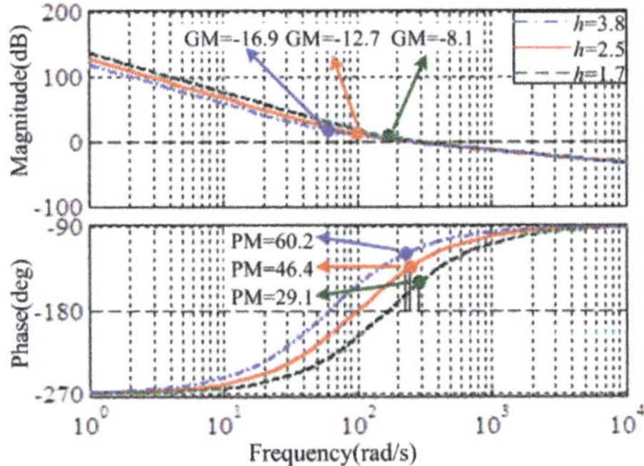

Figure 8. Open-loop bode plots of the proposed PLL.

It can be observed that the gain margin (GM) is between 8.1 and 16.9 dB within the PM range of 29.1°~60.2°, which are coincide with the aforementioned calculation. Therefore, one can conclude that the design guideline for the proposed PLL gives satisfactory PM and GM.

4.3. Bandwidth and Dynamic Response Evaluation

In order to analyze the bandwidth of the proposed PLL, the closed-loop bode plots of SRF-PLL, DSOGI-PLL and the proposed PLL are drawn. From Figure 9, it can be seen that the proposed PLL obtains wider bandwidth than the other two PLLs.

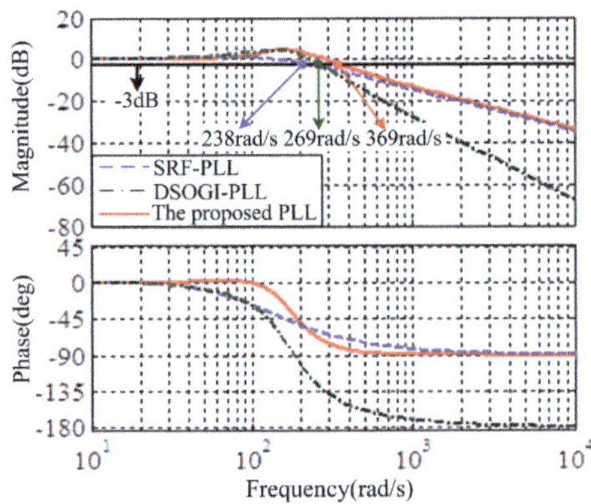

Figure 9. Closed-loop Bode plots of three kinds of PLL for $h = 2.5$.

The dynamic responses of the proposed PLL are evaluated by the unit step and ramp response. The corresponding transient responses are shown in Figure 10. It can be seen that the settling time are about 2 fundamental periods in these two conditions. Also, the steady-state value of unit step response equals to 1, and ramp response tracks the input ramp function $1/s^2$ precisely. These results validate that the proposed PLL obtains good dynamic performance.

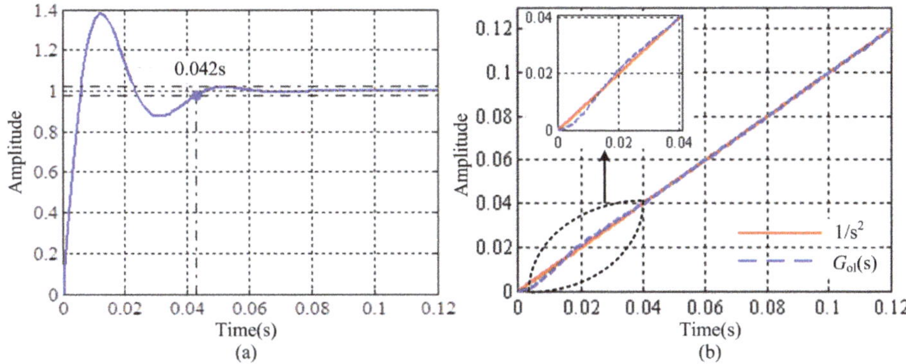

Figure 10. Transient response of closed-loop transfer function for $h = 2.5$. (**a**) The unit step response; (**b**) The unit impulse response.

4.4. Discrete Implementation of the Proposed PLL

Performance of the proposed PLL highly depends on digital discretization approach. The Tustin with pre-warping method ($s = \frac{\omega_1}{\tan(0.5\omega_1 T_s)} \frac{z-1}{z+1}$) gives better accuracy and frequency characteristics than the forward Euler and the backward Euler methods [33]. It is worth noting that SRF-PLL is regulated in q-axis. Therefore, $\omega_1 = 0$ and Tustin with pre-warping method has the same effect as Tustin (trapezoidal) method ($s = \frac{2}{T_s} \frac{z-1}{z+1}$). The discretization implementations of the proposed PLL are shown in Table 1. It is worth noting that n means the current period and y means α or β.

Table 1. The digital implementation.

	Discretization Expression	Difference Equation Implementations
SGDFT	$N_r(z) = \frac{2\pi f_s}{\omega_r(z)z^{-1}}$ $y(z) = \frac{(1-\cos(2k\pi/N_r)z^{-1})(1-z^{-N_r})}{1-2\cos(2k\pi/N_r)z^{-1}+z^{-2}}x(z)$ $qy(z) = \frac{\sin(2k\pi/N_r)z^{-1}(1-z^{-N_r})}{1-2\cos(2k\pi/N_r)z^{-1}+z^{-2}}x(z)$	$N_r(n) = \frac{2\pi f_s}{\omega_r(n-1)}$ $v(n) = 2\cos(2k\pi/N_r)v(n-1) - v(n-2)$ $\quad + x(n) - x(n-N_r)$ $y(n) = v(n) - \cos(2k\pi/N_r)v(n-1)$ $qy(n) = \sin(2k\pi/N_r)v(n-1)$
SRF-PLL	$v_{\alpha 1}^+(z) = v_\alpha^+(z)/\sqrt{v_\alpha^{+2}(z) + v_\beta^{+2}(z)}$ $v_{\beta 1}^+(z) = v_\beta^+(z)/\sqrt{v_\alpha^{+2}(z) + v_\beta^{+2}(z)}$ $v_q^+(z) = -\sin(\theta_p^+(z)z^{-1})v_{\alpha 1}^+(z)$ $\quad + \cos(\theta_p^+(z)z^{-1})v_{\beta 1}^+(z)$ $\omega_p^+(z) = v_q^+(z)[k_p + k_i T_s(1+z^{-1})/(2-2z^{-1})] + \omega_r(z)$ $\theta_p^+(z) = T_s(1+z^{-1})/(2-2z^{-1})\omega_p^+(z)$	$v_{\alpha 1}^+(n) = v_\alpha^+(n)/\sqrt{v_\alpha^{+2}(n) + v_\beta^{+2}(n)}$ $v_{\beta 1}^+(n) = v_\beta^+(n)/\sqrt{v_\alpha^{+2}(n) + v_\beta^{+2}(n)}$ $v_q^+(n) = -\sin(\theta_p^+(n-1))v_{\alpha 1}^+(n) + \cos(\theta_p^+(n-1))v_{\beta 1}^+(n)$ $\omega_p^+(n) = \omega_p^+(n-1) + \omega_r(n) - \omega_r(n-1)$ $\quad + v_q^+(n)(k_p + k_i T_s/2) - v_q^+(n-1)(k_p - k_i T_s/2)$ $\theta_p^+(n) = \theta_p^+(n-1) + T_s/2\left[\omega_p^+(n) + \omega_p^+(n-1)\right]$
SCP	$\theta_{\alpha\beta}^+(z) = v_\alpha^+(z)/\sqrt{v_\alpha^{+2}(z) + v_\beta^{+2}(z)}$ $\omega_{\alpha\beta}^+(z) = (2-2z^{-1})/(T_s + T_s z^{-1})\theta_{\alpha\beta}^+(z)$ $\omega_r(z) = \left|\omega_{\alpha\beta}^+(z)\right| LPF(z)$	$\theta_{\alpha\beta}^+(n) = v_\alpha^+(n)/\sqrt{v_\alpha^{+2}(n) + v_\beta^{+2}(n)}$ $\omega_{\alpha\beta}^+(n) = \omega_{\alpha\beta}^+(n-1) + \frac{2}{T_s}\left[\theta_{\alpha\beta}^+(n) - \theta_{\alpha\beta}^+(n-1)\right]$ $\omega_r(n) = \left|\omega_{\alpha\beta}^+(n)\right| LPF(n)$

5. Experimental Validation

The aim of this section is to evaluate the performance of the proposed PLL by extensive experimental studies under distorted conditions. The experimental setup presented in Figure 11 consists of signal generator and digital signal processor (DSP) TMS320F28335 control board. Throughout the experimental studies, the voltage benchmark is 311 V and the grid fundamental frequency f_0 is 50 Hz. The PI parameters in control loop are: $k_p = 189.2$ $k_i = 9746$ and the sampling frequency f_s is 12.8 kHz. In addition, three phase DC offsets (0.1 p.u., -0.1 p.u. and 0.1 p.u.) caused by measurement devices are considered all the time in the rest experiments. The detailed distorted conditions performed in experiments are summarized as follows:

Condition I: An asymmetric voltage sags (0.1, 0.2 and 0.3 p.u.) at $t = 30$ ms.
Condition II: An asymmetric phase jumps ($10°$, $20°$, and $30°$) at $t = 40$ ms.
Condition III: 5th and 7th order harmonics (0.2 and 0.1 p.u.) emerge at $t = 50$ ms.
Condition IV: Grid frequency jumps from 50 to 55 Hz at $t = 60$ ms.
Condition V: Grid frequency ramp change occurs at $t = 100$ ms with ramp rate of 20 Hz/s.

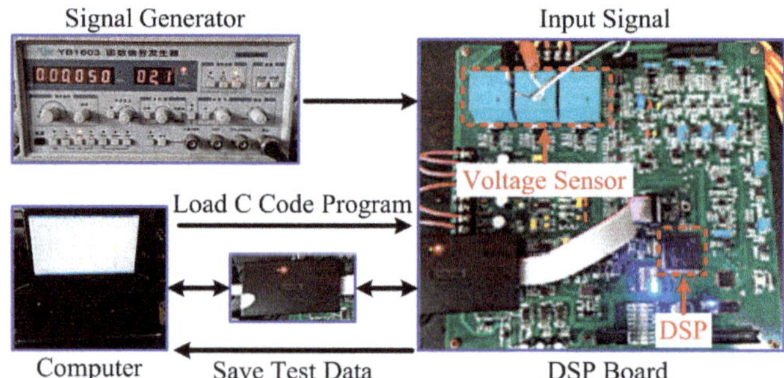

Figure 11. The experimental platform.

5.1. Experimental Results of the Proposed PLL under Distorted Conditions

The main variables of the proposed PLL are displayed: three-phase voltage (v_{abc}); FPSCs in two-phase stationary frame (v_α^+ and v_β^+); phase angle of fundamental positive sequence voltages (θ_p^+); calculated reference angular frequency (ω_r); detected grid frequency (f_m); estimated phase error ($\Delta\theta$); detected frequency error (Δf). The settling time is the time required for the response curve to reach and stay within certain range of 98% steady-state value for $\Delta\theta$ and Δf, respectively. Figure 12 shows the experimental results of the proposed PLL performed under voltage sag (condition I), phase jump (condition II), harmonics (condition III), and step-changed frequency (condition IV).

As can be seen in Figure 12, v_α^+ and v_β^+ give same amplitude and $\pi/2$ angle difference, which means the FPSCs are extracted accurately under conditions I–IV. As SGDFT needs one cycle to collect data, v_α^+ and v_β^+ become stable after one cycle when disturbances occur at $t = 30$ ms, $t = 40$ ms, $t = 50$ ms and $t = 60$ ms. Besides, it is worth noticing that θ_p^+ and v_α^+ reach the maximum simultaneously and f_m is in accordance with f_0 in steady-state regardless of the distorted conditions.

The performance of the proposed PLL under ramp-changed frequency (condition V) is also evaluated. The corresponding experimental results are shown in Figure 13. It can be observed that f_m tracks the ramp change of f_0 with a small error. In addition, f_m gives obvious overshoot at the start and end of ramp change, as SGDFT cannot give correct reference during its settling process. The detailed steady-state and dynamic performance indexes under conditions I–V are shown in the next section.

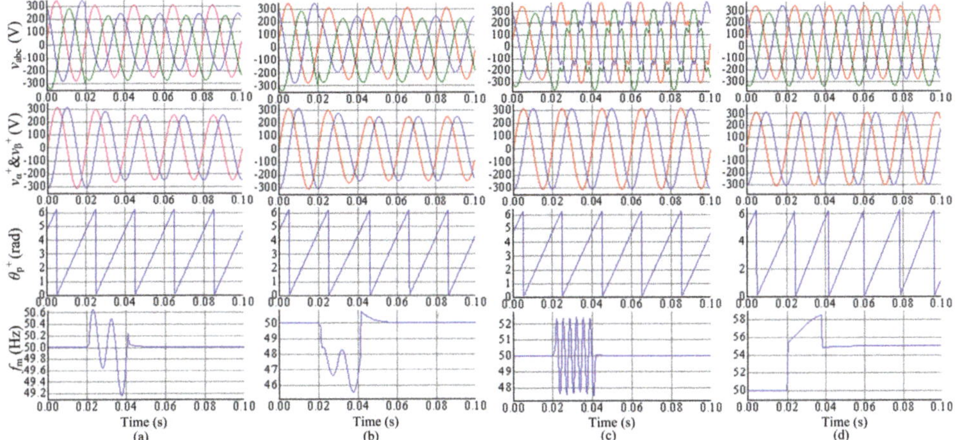

Figure 12. The experimental results under conditions I–IV. (**a**) Voltage sag; (**b**) Phase jump; (**c**) Harmonics; (**d**) Step-changed frequency.

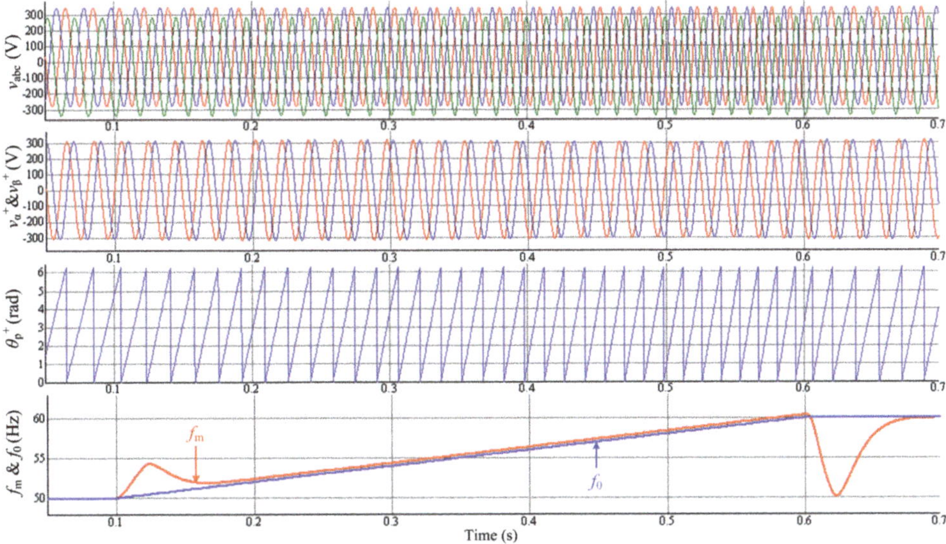

Figure 13. The experimental results under ramp-changed frequency.

5.2. Experimental Results Compared with Other Pre-Filtered PLLs

The effectiveness of the proposed PLL is further confirmed by comparing its performance with MCCF-PLL, DSOGI-PLL, PMAF-PLL and GDSC-PLL in [15,18,21,24], respectively. In order to allow a fair evaluation, the PI parameters of the above four PLLs are regulated according to the tunning methods in the corresponding articles. The PI parameters of the above four PLLs are: $k_{p1} = 141.1\ k_{i1} = 9952$, $k_{p2} = 222\ k_{i2} = 6169$, $k_{p3} = 390\ k_{i3} = 40{,}426$, and $k_{p4} = 266\ k_{i4} = 35{,}530$, respectively. The comparative tracking performance under conditions I–V are shown in Figures 14 and 15. Besides, the relevant data are summarized in Table 2.

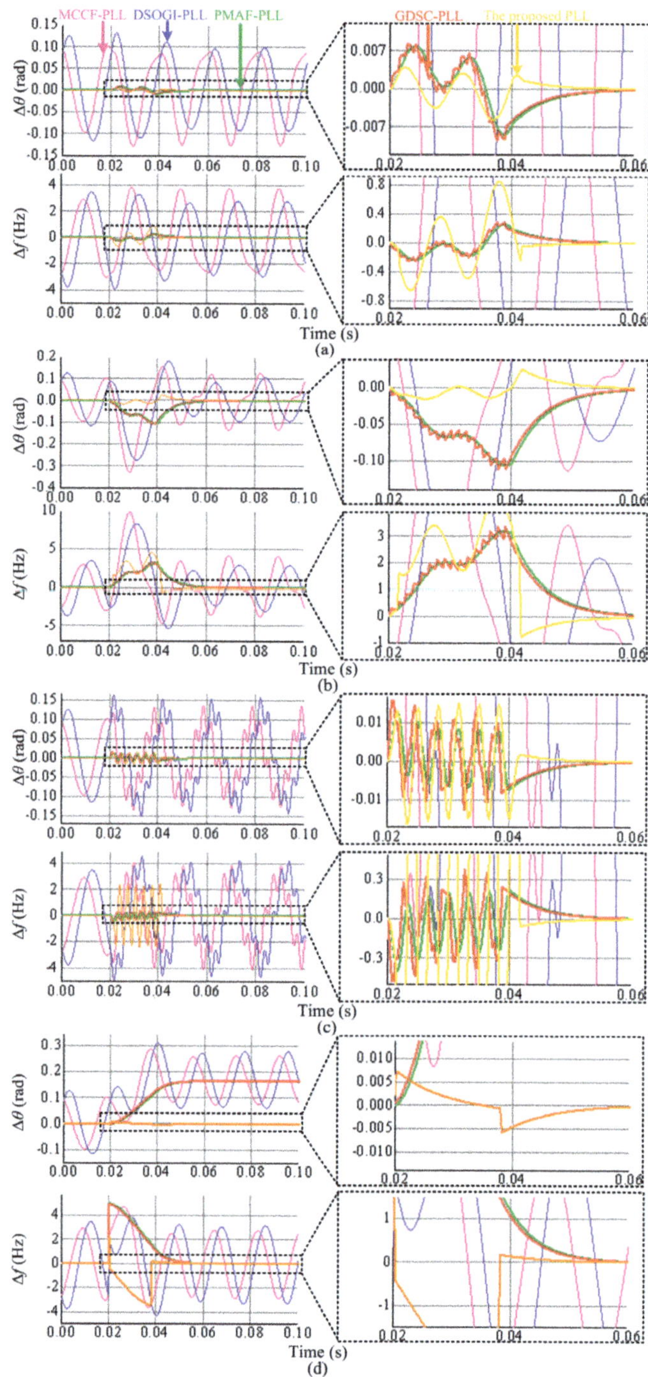

Figure 14. The comparative experimental results under conditions I–IV. (**a**) Voltage sag; (**b**) Phase jump; (**c**) Harmonics; (**d**) Step-changed frequency.

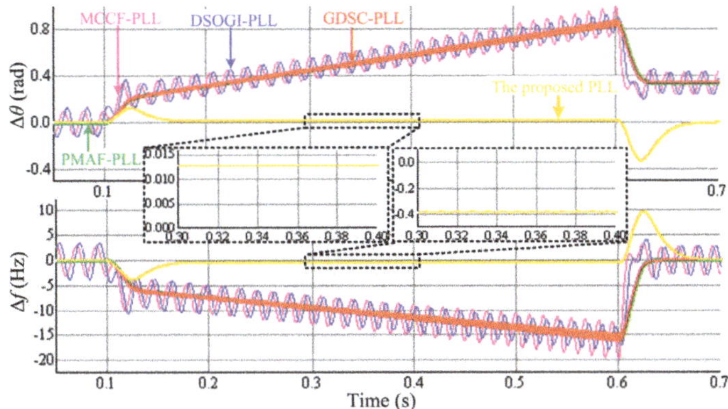

Figure 15. The comparative experimental results under ramp-changed frequency.

Table 2. Dynamic performance index.

Performance Index	Conditions	MCCF-PLL		DSOGI-PLL		PMAF-PLL		GDSC-PLL		Proposed PLL	
		$\Delta\theta$	Δf	$\Delta\theta$	Δf	$\Delta\theta$	Δf	$\Delta\theta$	Δf	$\Delta\theta$	Δf
Settling Time (ms)	I	≈30	≈33	≈40	≈40	≈36	≈31	≈35	≈30	≈25	≈23
	II	≈36	≈38	≈41	≈42	≈40	≈40	≈39	≈39	≈30	≈30
	III	≈30	≈35	≈38	≈39	≈34	≈34	≈33	≈33	≈30	≈28
	IV	≈31	≈32	≈38	≈40	≈41	≈36	≈40	≈35	≈35	≈25
	V	-	-	-	-	-	-	-	-	≈50	≈50
Overshoot (rad, Hz)	I	≈0.14	≈4	≈0.14	≈4	≈0.01	≈0.3	≈0.01	≈0.3	≈0.006	≈0.9
	II	≈0.32	≈10	≈0.28	≈7	≈0.11	≈3.1	≈0.11	≈3.1	≈0.03	≈4.5
	III	≈0.14	≈4	≈0.16	≈4.1	≈0.01	≈0.31	≈0.012	≈0.31	≈0.012	≈2.1
	IV	≈0.29	≈4.1	≈0.31	≈4.1	≈0.18	≈5	≈0.18	≈5	≈0.006	≈3.8
	V	-	-	-	-	-	-	-	-	≈0.18	≈4.5
Steady-state value (rad, Hz)	I	≈0.13	≈3.9	≈0.1	≈2.9	0	0	0	0	0	0
	II	≈0.13	≈4	≈0.1	≈3.5	0	0	0	0	0	0
	III	≈0.14	≈4	≈0.15	≈4.5	0	0	0	0	0	0
	IV	≈0.28	≈3	≈0.29	≈3.5	≈0.17	0	≈0.17	0	0	0
	V	-	-	-	-	-	-	-	-	≈0.013	≈0.39

DC offsets are considered in all conditions in Table 2.

It can be seen from Figure 14 that only MCCF-PLL and DSOGI-PLL contain obvious fundamental component in $\Delta\theta$ and Δf as their pre-filters are not effective for DC offsets. Also, it is clear that $\Delta\theta$ and Δf of MCCF-PLL and DSOGI-PLL are strongly distorted under heavily polluted condition due to their limited filtering characteristic for harmonics. As shown in Figure 14a–c, PMAF-PLL, GDSC-PLL and the proposed PLL have satisfactory disturbance rejection ability for DC offsets, voltage sag, phase jump and harmonics. However, PMAF-PLL and GDSC-PLL give obvious $\Delta\theta$ steady-state errors in Figure 14d. Moreover, in Figure 15 MCCF-PLL, DSOGI-PLL, PMAF-PLL and GDSC-PLL cannot track ramp-changed frequency accurately as their reference angular frequency is not updated with input signal's frequency. It is clear that the proposed PLL is effective under DC offsets, voltage sag, phase jump, harmonics, step- and ramp-changed frequency.

In Table 2, the performance indexes of the first four PLLs under condition V is not provided as they cannot track the ramp-changed frequency stably. Moreover, the steady-state values of MCCF-PLL and DSOGI-PLL refer to the peak value because they both contain fundamental components.

As summarized in Table 2, DSOGI-PLL gives the maximal settling time overall. The settling time of PMAF-PLL and GDSC-PLL are almost same. They are slightly smaller than DSOGI-PLL. MCCF-PLL tracks relatively faster than the previous PLLs. The proposed PLL gives the least settling time of $\Delta\theta$

and Δf under conditions I–IV and it needs about 2.5 cycles to track the ramp-changed frequency as SGDFT needs one cycle to collect data when the periodical signals change.

MCCF-PLL and DSOGI-PLL give obvious overshoots due to the existed DC offsets. The $\Delta\theta$ overshoots of PMAF-PLL and GDSC-PLL are smaller than the first two PLLs. It is worth noticing that the proposed PLL gives the smallest $\Delta\theta$ overshoot but it cause a slight increase in the overshoot of Δf under conditions I–IV. In addition, the proposed PLL gives acceptable overshoot in $\Delta\theta$ and Δf under condition V.

MCCF-PLL and DSOGI-PLL have obvious fundamental components of $\Delta\theta$ and Δf under steady-state condition, while the other three PLLs can obtain zero steady-state error of $\Delta\theta$ and Δf under conditions I–IV except for $\Delta\theta$ of PMAF-PLL and GDSC-PLL under condition IV. Moreover, the proposed PLL gives the smallest steady-state values of $\Delta\theta$ and Δf.

We can conclude from the experimental results in Figures 12–15 and Table 2 that, (1) Like PMAF-PLL and GDSC-PLL, the proposed PLL has good disturbance rejection capability of DC offsets; (2) Only the proposed PLL can solve the ramp-changed frequency problem due to its improved SCP; (3) The proposed PLL can obtain the least settling time of phase and frequency under conditions I–IV; (4) The proposed PLL gives the least overshoot of phase and the third small overshoot of frequency under conditions I–IV; (5) The proposed PLL almost obtains zero stead-state error of phase and frequency.

6. Conclusions

This paper presents an efficient PLL based on SGDFT filter and the improved SCP. SGDFT filter is employed to enhance separation accuracy of FPSC under distorted conditions. Lagrange-interpolation method is applied to remove the adverse effect of the fractional delay when the sampling number is not integer. The improved SCP is employed to promise precise phase estimation and enables the PLL tracking reference frequency rapidly. Comparative experimental results demonstrate that the proposed PLL can achieve zero steady-state error in phase and frequency with a rapid speed compared with the other four PLLs. Meanwhile, it has satisfactory disturbance rejection capability under unbalanced voltages, harmonics, step-and ramp-changed frequency and DC offsets.

Acknowledgments: This work was supported in part by the national Natural Science Foundation of China (NSFC) under Grant 61233008 and 51520105011, and in part by the Special Project of Hunan Province of China under Grant 2015GK1002 and 2015RS4022.

Author Contributions: Yijia Cao proposed the original idea, Jiaqi Yu carried out the main research tasks, Yong Xu and Yong Li carried out the experiments. Jiaqi Yu and Jingrong Yu wrote the full manuscript and supervised the experiments.

Conflicts of Interest: The authors declare no conflict of interest.

1. Hadjidemetriou, L.; Kyriakides, E.; Yang, Y.; Blaabjerg, F. A synchronization method for single-phase grid-tied inverters. *IEEE Trans. Power Electron.* **2016**, *31*, 2139–2149. [CrossRef]
2. Zhang, G.Q.; Wang, G.L.; Xu, D.G.; Zhao, N.N. Adaline network based PLL for position sensorless interior permanent magnet synchronous motor drives. *IEEE Trans. Power Electron.* **2016**, *31*, 1450–1460. [CrossRef]
3. Li, Y.; Liu, Q.Y.; Hu, S.J.; Liu, F.; Cao, Y.J.; Luo, L.F.; Rehtanz, C. A virtual impedance comprehensive control strategy for the controllably inductive power filtering system. *IEEE Trans. Power. Electron.* **2017**, *32*, 920–926. [CrossRef]
4. Li, Y.; Peng, Y.J.; Liu, F.; Sidorov, D.; Liang, C.G.; Luo, L.F.; Cao, Y.J. A controllably inductive filtering method with transformer-integrated linear reactor for power quality improvement of shipboard power system. *IEEE Trans. Power Deliv.* **2016**, *99*, 1–9. [CrossRef]
5. Zhang, D.L.; Wang, Y.J.; Hu, J.B.; Ma, S.C.; He, Q.; Guo, Q. Impacts of PLL on the DFIG-based WTG's electromechanical response under transient conditions: Analysis and modeling. *CSEE J. Power Energy Syst.* **2016**, *2*, 30–39. [CrossRef]

6. Golestan, S.; Freijedo, F.D.; Guerrero, J.M. A systematic approach to design high-order phase-locked loops. *IEEE Trans. Power Electron.* **2016**, *30*, 4013–4019. [CrossRef]
7. Kaura, V.; Blasko, V. Operation of phase locked loop system under distorted utility conditions. *IEEE Trans. Ind.* **1997**, *33*, 58–63. [CrossRef]
8. Blaabjerg, F.; Teodorescu, R.; Liserre, M.; Timbus, A.V. Overview of control and grid synchronization for distributed power generation systems. *IEEE Trans. Ind. Electron.* **2006**, *53*, 1398–1409. [CrossRef]
9. Subramanian, C.; Kanagaraj, R. Rapid tracking of grid variables using prefiltered synchronous reference frame PLL. *IEEE Trans. Instrum. Meas.* **2015**, *64*, 1826–1836. [CrossRef]
10. Lee, K.J.; Lee, J.P.; Shin, D.; Yoo, D.W.; Kim, H.J. A novel grid synchronization PLL method based on adaptive low-pass notch filter for grid-connected PCS. *IEEE Trans. Ind. Electron.* **2014**, *61*, 292–301. [CrossRef]
11. Wang, L.; Jiang, Q.; Hong, L.C. A novel three-phase software phase-locked loop based on frequency-locked loop and initial phase angle detection phase-locked loop. In Proceedings of the IECON 2012 38th Annual Conference on IEEE Industrial Electronics Society, Montreal, QC, Canada, 25–28 October 2012; pp. 150–155.
12. Kanjiya, P.; Khadkikar, V.; El Moursi, M.S. A novel type-1 frequency-locked loop for fast detection of frequency and phase with improved stability margins. *IEEE Trans. Power Electron.* **2016**, *31*, 2550–2561. [CrossRef]
13. Wang, Y.F.; Li, Y.W. Analysis and digital implementation of cascaded delayed-signal-cancellation PLL. *IEEE Trans. Power Electron.* **2011**, *26*, 1067–1080. [CrossRef]
14. Carugati, I.; Maestri, S.; Donato, P.G.; Carrica, D.; Benedetti, M. Variable sampling period filter PLL for distorted three-phase systems. *IEEE Trans. Power Electron.* **2012**, *27*, 321–330. [CrossRef]
15. Guo, X.Q.; Wu, W.Y.; Chen, Z. Multiple-complex coefficient-filter-based phase-locked loop and synchronization technique for three-phase grid-interfaced converters in distributed utility networks. *IEEE Trans. Power Electron.* **2011**, *58*, 1194–1204. [CrossRef]
16. Li, W.W.; Ruan, X.B.; Bao, C.L.; Pan, D.H.; Wang, X.H. Grid synchronization systems of three-phase grid-connected power converters: A complex-vector-filter perspective. *IEEE Trans. Ind. Electron.* **2014**, *61*, 1855–1870. [CrossRef]
17. Ohori, K.; Hattori, N.; Funaki, T. Phase-locked loop using complex-coefficient filters for grid-connected inverter. *Electr. Eng. Jpn.* **2014**, *189*, 52–60. [CrossRef]
18. Rodríguez, P.; Teodorescu, R.; Candela, I.; Timbus, A.V.; Blaabjerg, M.L.F. New Positive-sequence Voltage Detector for Grid Synchronization of Power inverters under Faulty Grid Conditions. In Proceedings of the 2006 37th IEEE Power Electronics Specialists Conference, Jeju, Korea, 18–22 June 2006; pp. 1–7.
19. Rodriguez, P.; Luna, A.; Candela, I.; Mujal, R.; Teodorescu, R.; Blaabjerg, F. Multiresonant frequency-locked loop for grid synchronization of power converters under distorted grid conditions. *IEEE Trans. Ind. Electron.* **2011**, *58*, 127–138. [CrossRef]
20. Xiao, F.; Dong, L.; Li, L.; Liao, X. A frequency-fixed SOGI-based PLL for single-phase grid-connected converters. *IEEE Trans. Power Electron.* **2017**, *32*, 1713–1719. [CrossRef]
21. Golestan, S.; Guerrero, J.M.; Vidal, A. PLL with MAF-based prefiltering stage: Small-signal modeling and performance enhancement. *IEEE Trans. Power Electron.* **2016**, *31*, 4013–4019. [CrossRef]
22. Robles, E.; Pou, J.; Ceballos, S.; Zaragoza, J.; Martin, J.L.; Ibanez, P. Frequency-adaptive stationary-reference-frame grid voltage sequence detector for distributed generation systems. *IEEE Trans. Ind. Electron.* **2011**, *58*, 4275–4287. [CrossRef]
23. Robles, E.; Ceballos, S.; Pou, J.; Martín, J.L.; Zaragoza, J.; Ibañez, P. Variable-frequency grid-sequence detector based on a quasi-ideal low-pass filter stage and a phase-locked loop. *IEEE Trans. Power Electron.* **2010**, *25*, 2552–2563. [CrossRef]
24. Neves, F.A.S.; Cavalcanti, M.C.; de Souza, H.E.P.; Bradaschia, F.; Bueno, E.J.; Rizo, M. A generalized delayed signal cancellation method for detecting fundamental-frequency positive-sequence three-phase signals. *IEEE Trans. Power Deliv.* **2010**, *25*, 1816–1825. [CrossRef]
25. Wang, Y.F. Grid synchronization PLL based on cascaded delayed signal cancellation. *IEEE Trans. Ind. Electron.* **2013**, *60*, 645–658. [CrossRef]
26. Batista, Y.N.; de Souza, H.E.P.; Neves, F.A.S.; Filho, R.F.D.; Bradaschia, F. Variable-structure generalized delayed signal cancellation PLL to improve convergence time. *IEEE Trans. Ind. Electron.* **2015**, *62*, 7146–7150. [CrossRef]

27. Neves, F.A.S.; de Souza, H.E.P.; Cavalcanti, M.C.; Bradaschia, F.; Bueno, E.J. Digital filters for fast harmonic sequence component separation of unbalanced and distorted three-phase signals. *IEEE Trans. Ind. Electron.* **2012**, *59*, 3847–3859. [CrossRef]

28. Liccardo, F.; Marino, P.; Raimondo, G. Robust and fast three-phase PLL tracking system. *IEEE Trans. Ind. Electron.* **2011**, *58*, 221–231. [CrossRef]

29. Rani, B.I.; Aravind, C.K.; Ilango, G.S.; Nagamani, C. A three phase PLL with a dynamic feed forward frequency estimator for synchronization of grid connected converters under wide frequency variations. *Int. J. Electr. Power Energy Syst.* **2012**, *41*, 63–70. [CrossRef]

30. Jacobsen, E.; Lyons, R. The sliding DFT. *IEEE Signal Process. Mag.* **2003**, *20*, 74–80.

31. Laakso, T.I.; Valimaki, V.; Karjalainen, M. Splitting the unit delay [FIR/all pass filters design]. *IEEE Signal Process. Mag.* **1996**, *13*, 30–60. [CrossRef]

32. Golestan, S.; Monfared, M.; Freijedo, F.D.; Guerrero, J.M. Dynamics assessment of advanced single-phase PLL structures. *IEEE Trans. Ind. Electron.* **2013**, *60*, 2167–2176. [CrossRef]

33. Yepes, A.G.; Freijedo, F.D.; Doval-Gandoy, J. Effects of discretization methods on the performance of resonant controllers. *IEEE Trans. Power Electron.* **2010**, *25*, 1692–1712. [CrossRef]

Article

Evaluation of Harmonics Impact on Digital Relays

Kinan Wannous * and Petr Toman

Department of Electrical Power Engineering, Brno University of Technology, Technicka 12,
61600 Brno, Czech Republic; toman@feec.vutbr.cz
* Correspondence: xwanno00@stud.feec.vutbr.cz; Tel.: +420-774-649-490

Received: 28 February 2018; Accepted: 4 April 2018; Published: 11 April 2018

Abstract: This paper presents the concept of the impact of harmonic distortion on a digital protection relay. The aim is to verify and determine the reasons of a mal-trip or failure to trip the protection relays; the suggested solution of the harmonic distortion is explained by a mathematical model in the Matlab Simulink programming environment. The digital relays have been tested under harmonic distortions in order to verify the function of the relays algorithm under abnormal conditions. The comparison between the protection relay algorithm under abnormal conditions and a mathematical model in the Matlab Simulink programming environment based on injected harmonics of high values is provided. The test is separated into different levels; the first level is based on the harmonic effect of an individual harmonic and mixed harmonics. The test includes the effect of the harmonics in the location of the fault point into distance protection zones. This paper is a new proposal in the signal processing of power quality disturbances using Matlab Simulink and the power quality impact on the measurements of the power system quantities; the test simulates the function of protection in power systems in terms of calculating the current and voltage values of short circuits and their faults. The paper includes several tests: frequency variations and decomposition of voltage waveforms with Fourier transforms (model) and commercial relay, the effect of the power factor on the location of fault points, the relation between the tripping time and the total harmonic distortion (THD) levels in a commercial relay, and a comparison of the THD capture between the commercial relay and the model.

Keywords: protection relay; Simulink; Matlab; Omicron CMC 256plus; power quality; enerlyzer; comtrade; distance protection

---

1. Introduction

In electrical engineering, the protective relay is a relay device designed to trip a circuit breaker when a fault is detected, and has the ability to measure the power system quantities through the internal logic of a microprocessor. Digital relays have become more efficient and functional, especially for each of the following processes: the digital relay features accurate methods to calculate the voltage, current measurements, and other electrical quantities, and has become a communication standard for electrical Substation Automation Systems (SAS). Digital relays include multi-protection functions such as distance protection, overcurrent protection, under-voltage protection, etc. In addition, there are many measurements that can be done using the microprocessor, such as internal/external fault diagnosis, fault measurements, zero current sequences, and disturbance recording. Additional functions of the digital relays, such as monitoring, metering, setting groups, fault recorder communication, and reports, have no direct relation to the protective elements. The hierarchy structure of power system automation comprises an electrical protection relay, control, measurement and monitoring, and data communications. Power system automation is a system that is integrated into the various components connected to the power network. The numerical relay is a focal concept of the power system automation for protecting the equipment and limiting the damage. The system's components have better communication with each other; the information is exchanged via dozens of communication protocols, this concept can be characterized by the fact that one sensor is enough to obtain and collect information through the network instead of a sensor per component in the power system. Power system automation has several levels to integrate between the power substations and the substation supervisory system (SCADA). It defines both the information model and services used for communication between the Intelligent Electronic Devices (IEDs) in a substation.

Some studies presented the practical test of the harmonic influence on electromechanical and microprocessor relays, the test implemented the current signal accompanied by the total harmonic distortion on relays, these studies found that the mixed-harmonics influence is minor on the protective relays, conversely, the influence of pure signal above the fundamental frequency found a significant effect on protection relays function [1–5]. The evaluation influence of the power quality on the proactive relays presented using the simulation models (Mathcad software) [6–9], these studies focused on the advantages and distances of the relay algorithm. The harmonic distortions in the power system and the associated problems caused by non-linear loads were briefly discussed. A review related to various methodologies for detecting and measuring harmonics based on this review of a new hybrid method for detecting and measuring harmonics is introduced in these references [10–12]. Few studies explore the research associated with quantifying the cost of reliability and power quality. Various power quality disturbances are investigated and possible methods of quantifying both the effect and cost are presented [13–15], moreover, it aims at analyzing and probing the influences of harmonics to differential relays. It analyzes and compares the mathematic models which are constructed by using EMTP and the test results [16,17]. It presents the impact of the fault in the power line based on the short circuit and abnormal condition, comparing the fault tripping time on the distance relay with two simulation scenarios developed using matlab environment [18].

This paper explains the signal processing of power quality disturbances using Matlab 2016, MathWorks, Natick, MA, USA) Simulink, especially the power quality impact on the measurements of the power system quantities; the test simulates the function of protection in power systems in terms of calculating the current and voltage values of short circuits and their faults. The model includes a number of blocks that process the signals for the current and voltage coming from the simulator side. First, the current and voltage signals are filtered through a low-pass filter which removes the high-frequency components from these signals. After the passage of the signals through the filter, there is the second level that converts the analog signals into digital signals through the processing signal within four stages. The first stage is the sampling process of the original signal, taking into account the sampling frequency which is determined to be 80 samples/cycle according to IEC standard 61850 (4000 Hz per second for systems operating at 50 Hz frequency). For this block, the input signals are analog and plus generator signals. The signal is then inserted into a quantizer that converts the sinusoidal signal into a digital signal. Meanwhile, the signal is filtered through a digital filter. The parameters of this filter are adjusted to match the previous filter. The last stage is the calculation relating to the signal itself, which includes the amplitude angle using the Fourier transformation or an arms calculation. The model characteristics are calculated from the parameters of the component with the intention to protect the relay, such as the transmission line, the transformer, and the generator, etc.

2. Impact of Harmonics on Protection Relays

The main reasons to study the power quality as follows:

- Intelligent electrical devices (IED) have become more accurate in power quality measurements and less tolerant with higher frequency components than the nominal frequency of these devices. The major risk of harmonics in power systems is mal-operation of the protective relays and the thermoelectric effect accompanied by these harmonics.

The reason to study the impact of harmonics on protective relays is the existence of harmonics in current and voltage signals which can cause a mal-trip of the relay when no fault happens or the system fails to trip when there is a fault in the power system, especially when a harmonic can be noticed during faulty performance. The relay algorithm is an important factor to define the harmonic impact. There is a difference between the higher values of relay tripping thresholds and the lower values of normal operation. A large difference will cause a higher risk of failure to trip and mal operation. The power systems quantities are converted to the digital form that provides easy implementation of digital processing signal (DPS) and analysis the power quality.

Usually, the harmonic can be determined by the load characteristics; moreover, the harmonic in the current has a more intense impact than the harmonic in the voltage.

In the power system, the harmonics might reduce the power quality and cause a number of problems, such as overloads in distribution systems because of an increase in the *rms* current value, and harmonics can also cause a shorter lifespan of generators, transformers, motors, and other power system components. On the other hand, the sensitive loads can be affected by harmonic distortion.

Harmonic Phenomena

The total harmonic distortion (%THD) is a percentage of the *rms* of the fundamental component. The percentage current total harmonic distortion is

$$THD_i = \frac{\sqrt{\sum_{k=2}^{\infty} I_{h,rms}^2}}{I_{rms}} \times 100\%, \tag{1}$$

where $I_{h,rms}$ is the current harmonic amplitude of order h (i.e., the h-th harmonic) and I_{rms} is the current amplitude of the normal frequency and harmonic components:

$$I_{rms} = \sqrt{\sum_{k=2}^{\infty} I_{h,rms}^2}, \tag{2}$$

This subsequently leads to the formula:

$$I_{rms} = I_1 \sqrt{1 + THD_i^2} \tag{3}$$

The total power factor

$$PF = \frac{P}{V_{1,rms} I_{1,rms} \sqrt{1 + \left(\frac{THD_i}{100\%}\right)^2}} \tag{4}$$

The percentage voltage total harmonic distortion is

$$THD_v = \frac{\sqrt{\sum_{k=2}^{\infty} V_{h,rms}^2}}{V_{rms}} \times 100\%, \tag{5}$$

where $V_{h,rms}$ is the voltage harmonic amplitude of order h (i.e., the hth harmonic) and V_{rms} is the voltage amplitude of the normal frequency and harmonic components:

$$V_{rms} = \sqrt{\sum_{k=1}^{\infty} V_{h,rms}^2}, \tag{6}$$

The European energy organizations define the power quality in three parameters [2]:

- Commercial quality: it refers to the services which provide by the electrical distribution companies. Voltage quality: it refers to the total harmonic distortion, frequency variation and voltage distortion can cause mal-operation. Levels of continuity of supply depend on the measurements which can be collected at all voltage level.

The IEC (International Electrotechnical Commission) is an organization for standardization comprising all national electrotechnical committees.

There are three European standards defining the limitation of harmonic currents injected to the power system. They define the limits of the harmonic component of the input current which may be produced by the equipment:

- IEC 61000-3-2:2014 Electromagnetic compatibility (EMC)—Part 3-2: Limits—Limits for harmonic current emissions (equipment input current $\leq$16 A per phase). This standard is applicable to the electrical equipment which required an input current up to 16 A.
- IEC 61000-4-7: Testing and measurement techniques—General guide on harmonics and inter-harmonics measurements and instrumentation, for power supply systems and equipment connected thereto. The new requirement launched for measuring harmonic should get 10 cycles of the fundamental current. Furthermore, no gap and no overlapping between the successive measuring windows are allowed.
- IEEE 519: The standard provides limitations for a harmonic distortion (a limitation on the harmonic voltage distortion provided by the distributor to the customers, a limitation on the harmonic current distortion superimposed to the system by a customer).
- Individual Voltage Distortion limits for the bus voltage level of 69 kV and below should be not be exceeded by 3% and the total voltage distortion THD should be not be exceeded by 5%. Moreover, according to IEEE 519, there are more conditions with regard to the effects of harmonics on protective relays (electromechanical, static, and digital relays).
- It is important to bear in mind that the impact of the power quality on the protection relays can cause incorrect trips and relay mal-operation, while an incorrect tripping time occurrence depends on the power frequency variations and harmonic distortion [7,16].

3. Description of Mathematical Model

The model is a suggested method of processing the voltage and current signals similarly to the process that takes place in relays, precisely, for the frequency variations and the number of samples per cycle. To ensure the validity of the test for this model, a comparison was made between the output of the model and the output of the physical relay located in the laboratory. The simulator generated signals for the voltage and current of a single phase fault (a short circuit) with the accompanying of these signals. These signals have been sent to the physical relay and have been recorded and changed the format to Comtrade (Common Format for Transient Data Exchange for power systems); after that the signals have been sent to the model; see Figure 1. Comtrade is a file format for status data related to a transient power system and storing signals. In addition to blocks that simulate these physical elements, the model also contains a display and calculation blocks for graphic representation of the results of simulated scenarios. The model presents the simulation and modeling of communication-based digital relay using Matlab and the model tested under abnormal conditions (short circuit) and under various fault types. The behavior of the model can be monitored and compared with the real protection relay by using the Enerlyzer in Omicron which can able to record voltage and current signals. The model reads the recorded signals by using a comtrade reader and the model offers the analysis of signals as shown in Figure 2.

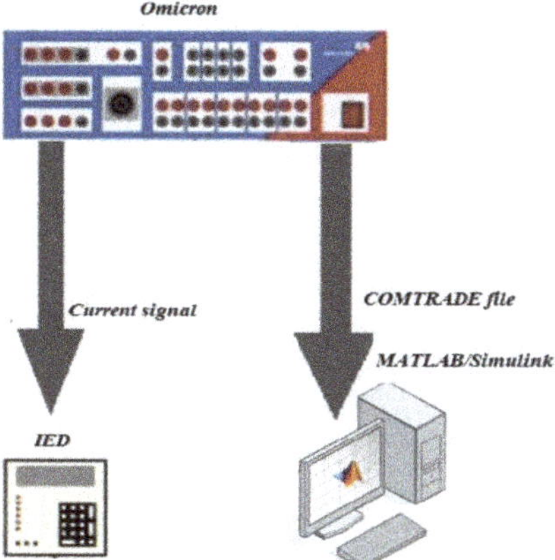

Figure 1. The test structure.

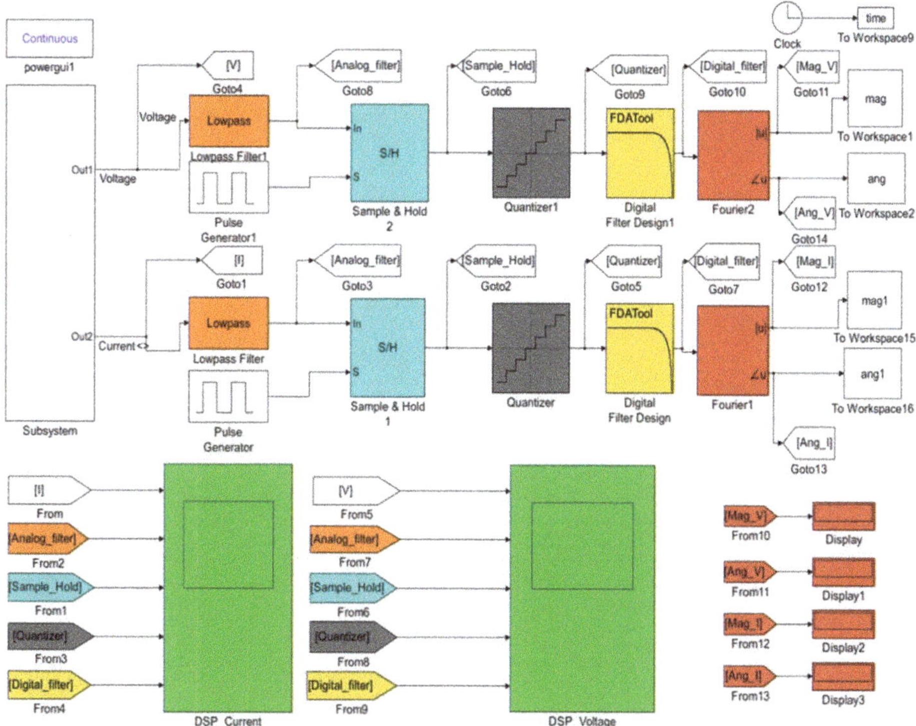

Figure 2. Model for current/voltage signal processing (DSP).

The model was made in the Matlab Simulink programming environment (Figure 2) using elements of the Sim-Power-Systems library. The first step is to get the current and voltage signals from the current and voltage transformer sides and apply some functions to them. The digital processing signal is the process of modifying a signal to improve the performance of the relay, to eliminate the high-frequency components, and to avoid the phenomenon of aliasing from a fault signal; low pass anti-aliasing analog filters with suitable cut-off frequency are used. Holding and sampling the signal is the second block of the signal process which converts the analog signal to the sample. The quantizer converts the smooth signal into a stair step output. Fast Fourier Transform (FFT) is a faster version of the Discrete Fourier Transform (DFT) which is used to find the fundamental frequency and higher frequencies contained in the input signal [6,7].

3.1. The Low Pass Filter Block

The low pass filter block can filter every channel of the input signal separately using the given design specification as shown in Figure 3 and the Table 1. A detailed description and the possibilities of determining equivalent parameters are thoroughly discussed in [9,19].

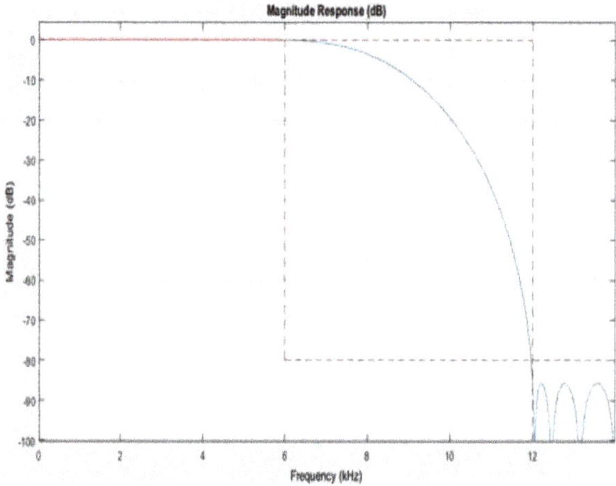

Figure 3. Low pass filter block characteristics.

Table 1. Parameters of low pass filter block.

Parameter	Value
Passband edge frequency (Hz)	6000
Stopband edge frequency (Hz)	12,000
Maximum passband ripple (dB)	0.1
Minimum stopband attenuation (dB)	80
Input sample rate (Hz)	28,000 *

* Sampling frequency (3 kHz, 9 kHz, 28 kHz).

The output signal from the Finite impulse response (FIR) filter can be described by Equation (1):

$$y[n] = b_0 x[n] + b_1 x[n-1] + \cdots + b_N x[n-N] = \sum_{i=0}^{N} b_i . x[n-i], \tag{7}$$

where:

$x[n]$ Refers to the current or voltage signals,
$y[n]$ Refers to the filtering signal,
N Refers to the filter order,
b_i is the value of the impulse response at the ith instant for $0 \leq i \leq N$ of an Nth-order FIR filter.

3.2. Sample and Hold

The sample and hold block gets the input signal when it receives a trigger event at the trigger port. The block holds the output signal until the next triggering event occurs.

- When the trigger input rises from a negative value to a positive value, the block starts to acquire the input signal.
- When the trigger input drops from a positive value to a negative value, the block starts to acquire the input signal.

3.3. Pulse Generator

The pulse generator block is the trigger input of the sample and hold block. It generates square wave pulses at regular intervals. The block waveform parameters (amplitude, pulse width, period, and phase delay) determine the shape of the output waveform.

3.4. Fourier Block

The Fourier block offers the calculation the amplitude and phase of the input signal (current and voltage), total harmonic distortion. The block offers analysis of the signal components as a percentage of the fundamental signal.

Recall that a signal $f(t)$ can be expressed by a Fourier series of the form

$$f(t) = \frac{a_0}{2} + \sum_{n=1}^{\infty} a_n cos(nwt) + b_n sin(nwt) \tag{8}$$

where n represents the rank of the harmonics. ($n = 1$ corresponds to the fundamental component.) The magnitude and phase of the selected harmonic component are calculated by these equations:

$$|H_n| = \sqrt{a_n^2 + b_n^2} \tag{9}$$

where

$$a_n = \frac{2}{T} \int_{t-T}^{t} f(t)cos(nwt)dt$$

$$b_n = \frac{2}{T} \int_{t-T}^{t} f(t)sin(nwt)dt$$

$$T = \frac{1}{f_1}$$

f_1: Fundamental frequency.

3.5. Phase Locked Loop (PLL) System

This block is used to synchronize a variable frequency sinusoidal signal. Meanwhile, this model is used to determine the frequency and the fundamental component of the signal phase angle which can be used to track the frequency and phase of the sinusoidal signal by using an internal frequency oscillator. The control system changes the internal frequency to keep the phase difference set to 0, as shown in Figure 4.

The model discusses two cases of Power Frequency variations when the power frequency gradually increases from 50 Hz to 52 Hz and returns to the normal frequency after a period of time; see Figure 5. The model compares the conventional way of calculating the frequency variation and the proposed way using a closed circuit. The proposed method is to track the frequency and amplitude of the input wave by using the frequency oscillator. In Figure 6, in contrast, the frequency decreases by 2 Hz gradually to 48 Hz.

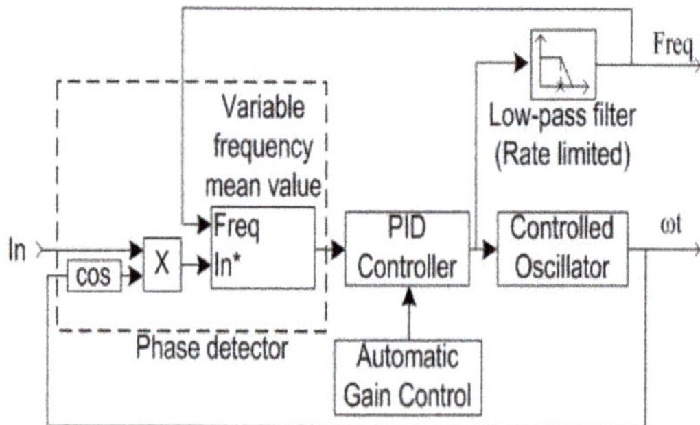

Figure 4. Phase locked loop (PLL) system. (PID: proportional–integral–derivative controller).

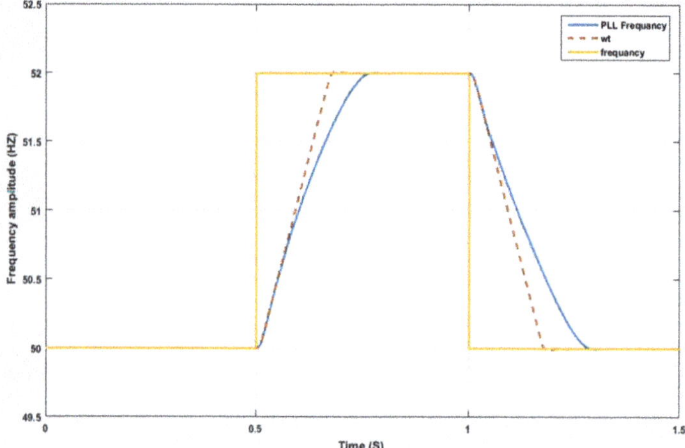

Figure 5. Frequency variation measurements between 50 and 52 Hz.

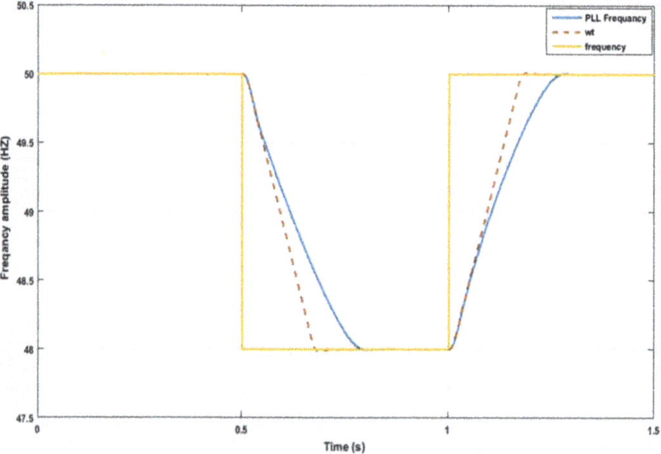

Figure 6. Frequency variation measurements between 48 and 50 Hz.

4. Relay Report and Simulink Result

Fault points were used in the tests, and a single phase with ground (SLG) was chosen; the fault points were included in the harmonic components. The points were located near the border of zone 1 and zone 2. The characteristic zones of the distance relay and the chosen points can be seen in Figure 7. As it is known, most digital relays use quadrilateral characteristics due to the many advantages which those characteristics can provide. The configuration and settings of the relay should be done in the first place; after that, exporting of these settings to CMC 256 is important to evaluate the relay functions. CMC 256 provides different platforms to test the relay with its zones setting, transmission line parameters, and time delay. Moreover, an advanced distance test displays the amplitude and phase of the three phase currents and voltages determined by the fault type and the fault location [11]. The distance relay is a universal short-circuit protection. Its mode of operation is based on the measurement and evaluation of the short-circuit impedance which, in the classic case, is proportional to the distance to the fault. Its tripping time is approximately one to two cycles (20 to 40 ms at 50 Hz) in the first zone for faults within the first 80% to 90% of the length. In the second zone, the tripping time depends on the settings which are usually 300–500 ms [20].

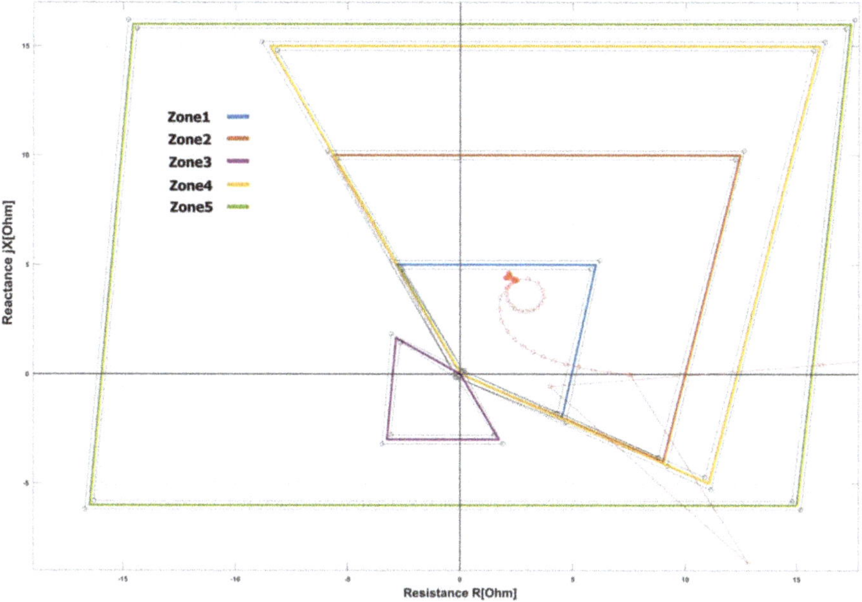

Figure 7. Characteristic zones of a distance relay and fault points.

The parameters of the derived model are based on the physical relay, as shown in Figure 7, which presents the characteristic zones of distance relay designed in the matlab environment.

The Testing Conditions

A harmonics test allows for creating a voltage and current signals with three states: pre-signal refers to the signal with the fundamental frequency and pre-signal time can be determined according to the test conditions, signal refers to the signal accompanied by the harmonic components, post signal refers to the signal with the fundamental frequency as shown in Table 2. Harmonic can be added to the voltage an current signals individually or mixed. The test offers the file export as comtrade formate and playback.

Table 2. Harmonic testing conditions Omicron.

Voltage	Current	Signal Definition	Triger Condition	Harmonic Input
VA	IA	Pre-signal time	Active high	% of fundamental
VB	IB	Signal	Active low	Absolute
VC	IC	Post-signal		
		Measured trip time		

The distorted voltage and the current waveform can be composed of harmonic components; however, the distorted waveform can be decomposed into a fundamental sinusoidal waveform at nominal and harmonic frequencies. The decomposition of distorted waveforms can be done by Fourier transform, as shown in Figure 8. An evaluation of the relay function and harmonics impact can be done in this test. The test is divided into different levels. The first level is based on the harmonic effect of mixed harmonics and individual harmonics. Both are used to test the protection operation and the Simulink model, considering that protection is IED and Matlab is a Simulink model. Figure 8 shows three mixed harmonics (2nd, 4th, 6th) of the fundamental signal. The discrete Fourier transform is used to calculate the THD in the Matlab model. The DFT Spectrum of a one-cycle (20 ms) voltage signal was taken, applying the DFT to the first 20 ms of sampled signal results in the line spectrum at discrete frequencies 50, 100, 150 Hz, etc. Figure 8 shows the magnitude values of the harmonics by using Equation (8). The figure shows that the 50 Hz component dominates; it is already visible from the original signal. The voltage signal contains significant components at even harmonics of 100, 200, and 300 Hz. Figure 9 shows decomposed voltage waveform with harmonic distortions of three harmonics (3rd, 5th, 7th) using the Fourier transform/ Matlab window. The harmonic analysis was tested and measured using a Matlab model.

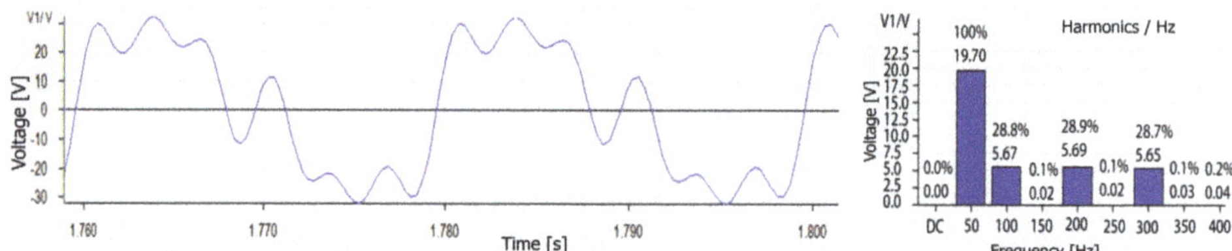

Figure 8. Decomposed voltage waveform with Fourier transform/Matlab window.

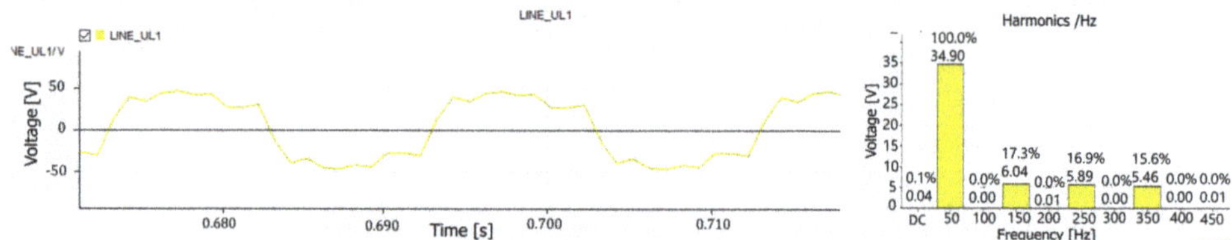

Figure 9. Decomposed voltage waveform with Fourier transform/Matlab window.

The test was conducted at five different fault points and a total harmonic distortion was added to each of these tests. The total harmonic distortion on the current wave added at two points (the arrow direction up) is called an overreach, and the total harmonic distortion was added to the voltage wave at three points, as shown in Figure 10. The total harmonic distortion was added as a percentage of the fundamental signal as follows: 10%, 20%, 30%, 40%, and 50%, and were colored in the following colors, respectively (violet, blue, yellow, red, and green).

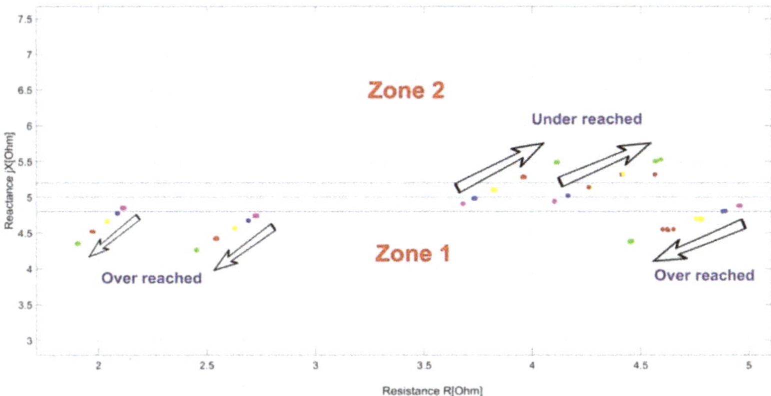

Figure 10. Quadrilateral characteristic and measured fault impedance locus.

Overreach and underreach of protection relays are common problems in power systems; they cause a mal-operation of the protection relays and it is impossible to detect the fault in the correct zone. The overreach of the point of the fault can cause the protection relay to send a mal-trip signal, which means that the relay instead of a trip in a delayed time zone 2 will send the tripping signal in zone 1, which is not desired; as shown in Figure 10. A harmonic distortion can change the power factor which leads to a change in the measured impedance lower than the actual value. The overreach can be noticed from a distorted voltage waveform, as shown in Figure 10, due to the lagging power factor. Conversely, an underreach of the point of fault can cause the change of protection relay and the decision to send a mal-trip signal; it means that the relay instead of a trip at a delayed time zone 1 will send a tripping signal at zone 2, which is not desired, as shown in Figure 10. A harmonic distortion can change the power factor, which leads to a change in the measured impedance higher than the actual value. The underreach can be noticed from a distorted voltage waveform, as shown in Figure 10, due to the leading power factor. The overreach of the distance relays should be avoided mostly in the first zone.

The test can illustrate harmonic distortion. A single-phase fault operation of a distance digital relay was applied and, as a result, the relay made a mal-operation decision. The analysis has been summarized from the integral disturbance recorder in the relay that has an area of memory. The relay can store all events in a disturbance recorder. Figure 11 shows a single phase to earth fault on the transmission line in a power system with a high harmonic content. It is assumed that three harmonics (3rd, 5th, and 7th) exist. The distance protection relay shows the instantaneous value of the three-phase voltage with a high harmonic content once a single phase with earth occurs.

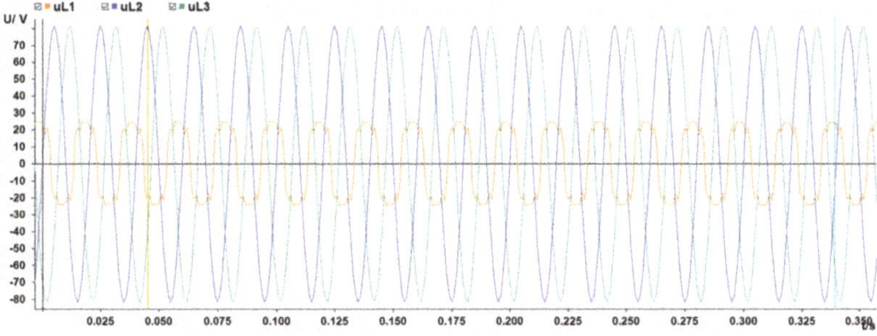

Figure 11. Voltage waveforms during single-phase fault (IED).

5. Distance Relay: Tripping Time vs. THD Level

Based on previous reports [21,22], the total harmonic distortion was measured at 30 different grids in the Czech Republic at different intervals. The measurements and evaluation of the harmonic level were made by the E.ON Distribution in the Czech Republic, operated by the company E.ON Czech Republic. More than 1000 MW of distributed energy sources (DES) are connected to this network, mainly from photovoltaic (PV) sources; see Table 3.

Table 3. Evaluation of the voltage harmonic from Grids.

Grid	3rd	5th	7th	15th
1	0.80%	1.60%	0.80%	0.10%
2	0.20%	0.80%	1.40%	0.10%
3	1.90%	2.20%	1.60%	0.42%
4	1.00%	1.80%	1.20%	0.38%
5	0.60%	1.70%	0.80%	0.02%
6	0.40%	2.20%	1.90%	0.50%
7	0.70%	1.90%	1.40%	0.18%
8	0.70%	1.80%	1.80%	0.17%
9	0.40%	1.60%	1.20%	0.08%
10	1.00%	1.60%	1.00%	0.26%
11	0.80%	1.80%	1.60%	0.30%
12	0.40%	1.40%	1.60%	0.10%
13	1.40%	1.40%	2.00%	0.41%
14	0.80%	1.80%	1.60%	0.20%
15	1.00%	1.80%	0.80%	0.24%
16	1.60%	1.80%	1.40%	0.36%
17	1.00%	2.60%	1.60%	0.08%
18	0.60%	1.00%	1.20%	0.19%
19	0.60%	0.80%	1.70%	0.16%
20	1.00%	1.80%	1.80%	0.42%
21	0.60%	1.70%	1.20%	0.40%
22	0.40%	1.00%	1.80%	0.12%
23	2.50%	2.40%	2.10%	0.60%
24	2.40%	2.00%	1.80%	0.60%
25	1.60%	2.40%	1.80%	0.48%
26	0.40%	1.00%	3.20%	0.18%
27	1.00%	1.00%	1.80%	0.18%
28	1.00%	2.00%	1.40%	0.32%
29	0.60%	0.60%	2.40%	0.12%
30	0.40%	1.00%	2.10%	0.08%

According to the total harmonics of the measured points in grids, the distance relay was tested. Its mode of operation is based on the measurement and evaluation of the short-circuit impedance which in the classic case is proportional to the distance to the fault.

This test explains how a different magnitude of the total harmonic distortion can influence the relay's operation and the tripping time of most digital protection relays used. A fault was located at zone 1 near the border zone 1–2 border (as shown in Figure 12: the tripping time for fault without distortion was 23 ms (average of 10 measurements); the tests were performed with zone 1 tripping time set to 0 ms and zone 2 tripping time set to 1000 ms. It is possible to see how the harmonic distortion can affect the distance relay's accuracy and assessment of where the fault took place. During harmonic distortion, a portion of the current is missing so an excessively large impedance is measured. In our measurements on the distance protection, the zone reach is reduced. It is acceptable for near faults, because the distance to the zone limit is long [13,16].

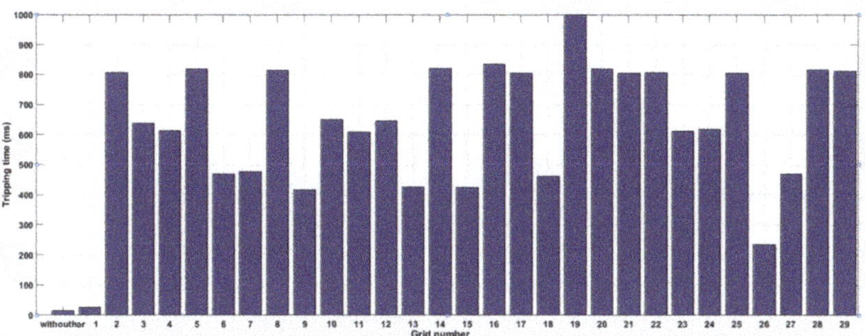

Figure 12. Distance relay: tripping time (seconds) vs. THD level of grids.

For faults close to the zone limit, an underreach is not permitted; the relay will trip in the second zone with a time delay [11]. Figure 12 shows the relation between the tripping time and the THD of the voltage. In this test, the THD does not exceed 4.5%. Note that the tripping time for the distance protection varies according to the added harmonic value.

Figure 13 presents the harmonic influence on the tripping time of the physical distance relay, where the current and voltage signals contain harmonics at a high level. The test was repeated five times and the average tripping time is presented. The type of fault is a single phase with the ground and tripping time without added harmonics of 1 s. The *rms* current and *rms* voltage should be constant during the test. For example, when the voltage and current signals contain the second harmonic, the tripping time of the distance protection is not constant, and the tripping time starts to change from 1 s to 1.6 s, meaning that the relay algorithm calculated the impedance in the third zone when the harmonic level was 10–40% of the current and voltage signals. Moreover, when the harmonic level was 50% of the current and voltage signals, the tripping time is changed to 3 s and the distance protection decision wrongly calculated the fault in zone 4.

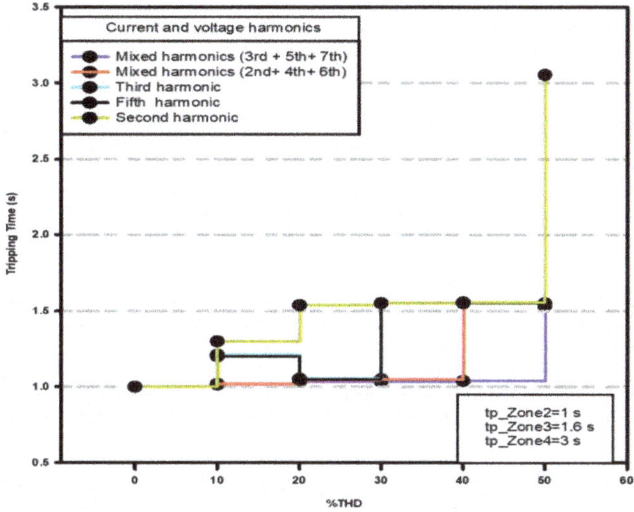

Figure 13. Distance relay: tripping time (seconds) vs. %THD level.

6. Comparison of Total Harmonic Measurement between Physical Relay and Model

The test involves a calculation of performance indicators concerning the level of harmonics through a commercial relay. This chapter contains a large number of measurements. The interval in the relay is 10 cycles in a 50-Hz system according to IEC 61000-4-30 and the model. The harmonics were added as the percentage of the current and voltage signals.

- Testing Conditions

To achieve a comparison between the THD in the commercial relay and the model, EnerLyzer is used to control the measuring features of the CMC test sets. It runs as a stand-alone test module. It has four modes of operation: a multimeter mode, a transient recording mode, a harmonic analysis mode, and a trend recording mode. It calculates the harmonic analysis of all configured inputs (up to 64 harmonics) and displays it in a bar graph and in a tabular format.

- Total Harmonic Distortion Detection in Physical Relay and Matlab Model

Through the test, the results show that the commercial relay of the harmonic capture ratio is lower than the original harmonic value. The total harmonic distortion is 10%, 20%, 30%, 40%, and 50% of the current signal according to the relay report. Figure 14 shows the harmonic measurements in a commercial relay. The second, third, and fourth harmonics were added as well as the three harmonics combined (2nd, 4th, 6th) and were added to the 3rd, 5th, and 7th harmonics. We can conclude that the commercial relay measures the THD with a difference of up to 35%, especially when there are three harmonics combined in the input signal, as shown in Figure 14. The digital relays start function when abnormal conditions occurred as faults. Abnormal events are accompanied by harmonics which are combined with the current and voltage signals.

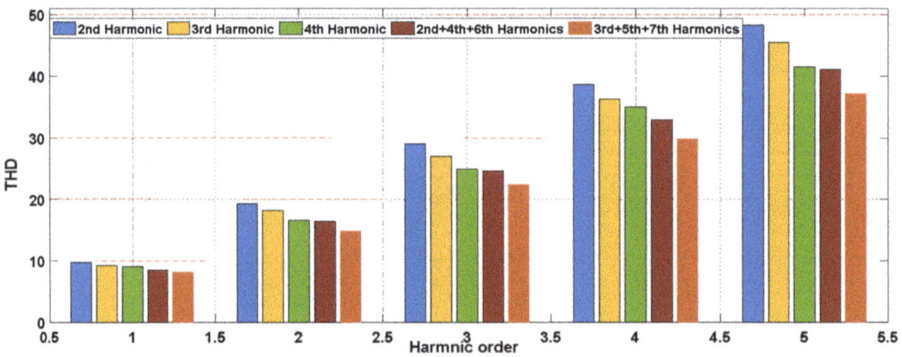

Figure 14. Commercial relay measurement of THD.

Figure 14 shows the measurement of THD in a commercial relay. A harmonic measurement evaluates the error of calculation, and the calculation method described above applies to the steady state fault conditions. The measurements of the third harmonic showed that the error of calculation in the commercial relay increased according to the harmonic percentage of the signal. The error of the calculation of the third harmonic is ca. 7% when the percentage of harmonic is 0–10%. After that, the error of the calculation of the third harmonic is stabilized to 10% when the percentage of harmonic is 10–50%. The highest error of the calculation of the THD can be found in mixed harmonics, as shown in Tables 4 and 5. The THD for mixed 3rd, 5th, and 7th harmonics is ca. 10–20% when the harmonic content is 0–10%. After that, the error of the calculation of THD for mixed 3rd, 5th, and 7th harmonics is stabilized to 33% when the percentage of harmonic is 10–50%.

Table 4. Data from distance relay's tests with fault near to the border from zone 1 to zone 2.

THD Level	Tripping Time (ms)					
	Test 1	Test 2	Test 3	Test 4	Test 5	Average
0.00%	0	20	30	20	1	14.2
1.63%	20	21	20	50	20	26.2
1.68%	1005	1000	26	1000	1000	806.2
1.96%	1000	170	1000	1000	15	637.0
1.97%	30	1000	1005	25	1005	613.0
1.98%	1005	1000	1005	80	1000	818.0
2.04%	1000	130	180	30	1005	469.0
2.10%	291	70	20	1000	1000	476.2
2.15%	1000	1000	70	1000	1000	814.0
2.17%	1000	1000	50	15	15	416.0
2.20%	1000	1000	1000	40	210	650.0
2.22%	20	1001	20	1000	1000	608.2
2.30%	1000	160	1000	1005	60	645.0
2.36%	80	20	1000	30	1000	426.0
2.41%	100	1000	1000	1000	1000	820.0
2.47%	60	40	20	1000	1005	425.0
2.55%	1006	1000	160	1000	1000	833.2
2.56%	1001	1000	20	1005	1000	805.2
2.65%	1000	80	160	70	1000	462.0
2.66%	1000	1000	1000	1000	1000	1000.0
2.77%	1000	1000	90	1005	1000	819.0
2.81%	1000	1000	20	1000	1000	804.0
2.84%	1000	30	1000	1001	1000	806.2
2.98%	1000	1000	40	15	1000	611.0
3.21%	1000	46	40	1000	1000	617.2
3.34%	1000	1000	26	1000	1000	805.2
3.38%	50	1000	70	20	30	234.0
3.43%	1000	1000	20	280	45	469.0
3.66%	70	1000	1000	1000	1005	815.0
4.10%	1000	1000	1000	1000	50	810.0

Table 5. The error of calculation THD for commercial relay.

THD	10%	20%	30%	40%	50%
2nd harmonic	2.04	3.62	3.45	3.62	3.52
3rd harmonic	7.526	9.89	10.29	10.19	10.13
4th harmonic	9.89	20.48	20.48	14.28	20.48
2nd + 4th + 6th harmonics	17.56	21.95	21.95	21.58	21.65
3rd + 5th + 7th harmonics	21.95	35.13	34.53	34.22	34.77

Figure 15 shows the measurement of THD in the model. The harmonic measurements evaluate the error of calculation. The measurements of the third harmonic show that the error of the calculation in the model is increasing according to the harmonic percentage of the signal and the error of the calculation of the third harmonic is ca. 1% when the harmonic percentage is 0–10%. After that, the error of the calculation of the third harmonic is stabilized to 2% when the harmonic percentage is 10–50%. The highest error in the calculation of THD can be found in mixed harmonics, as shown in Figure 15 and in Table 6; the THD for the 3rd + 5th + 7th harmonics is around 1–2% when the harmonic percentage is 0–10%. After that, the error in the calculation of the THD for the 3rd + 5th + 7th harmonics is stabilized to 3% when the harmonic percentage is 10–50%.

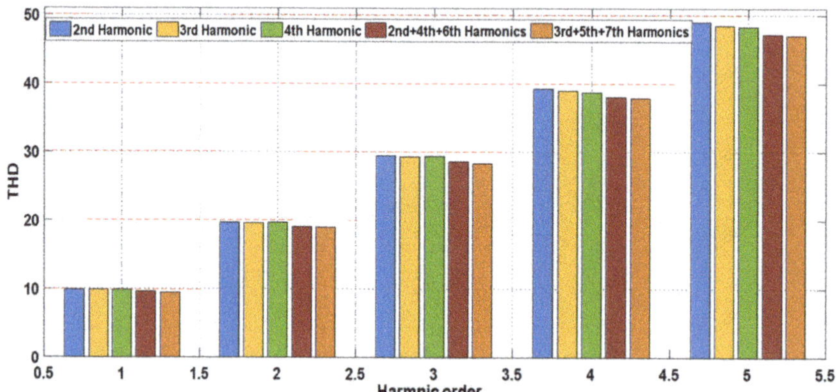

Figure 15. Model measurements of THD.

Table 6. The error of calculation THD for Matlab model.

THD	10%	20%	30%	40%	50%
Har2	1	1.52	1.69	2.04	2.04
Har3	2.045	2.38	2.74	2.827	3.092
Har4	1.01	1.522	2.739	3.359	3.519
Har246	4.16	4.712	4.89	5.26	5.932
Har357	5.266	5.263	6	6.1	6.38

Because the model implements the voltage and current signals, it is able to measure higher THD than the physical relay, as shown in Figure 15. The model captures harmonics with accuracy of 90–95%. In the case of the individual harmonics, however, the physical relay captures with accuracy of 80–85%, as shown in Figure 14. Similarly, mixed harmonics are inserted in the physical relay accompanied by the fault current and voltage signals. The physical relay captures mixed harmonics with accuracy of 65–70%, however the model captures mixed harmonics with accuracy of 85–90% , as shown in Figures 14 and 15.

The model provides a filter to reduce the harmonic distortion; this filter is built up from passive RLC components. Their values are computed using the specified nominal reactive power, tuning frequency, and quality factor. The filter has been implemented to mitigate the total harmonic distortion of the current and voltage. In case of an abnormal condition (a short circuit), the simulation implements a fault (a single-phase with the ground) from 0.1 to 0.15 s. Conversely, the steady state has been implemented during the period from (0 to 0.1) s and (0.15 to 0.2) s. During the implementation of the simulation, the delay to start the calculation at the very beginning takes 0.02 s or 1 cycle. The steady state of the model takes place under normal operation and the calculation of voltage total harmonic distortion (VTHD) and current total harmonic distortion (ITHD) are implemented. Abnormal operation begins at 0.1 s and lasts for 0.05 s (2.5 cycles), which is accompanied by increasing the fault current and decreasing the voltage.

Figure 16 shows the *rms* current calculation during the steady state and fault (short circuit) between phase B and the ground. The THD filter block offers the possibility to mitigate the harmonics, and the THD filter is designed to reduce the 5th, 7th, 11th, and 13th) harmonics, as it known that the physical relay has the ability to remove harmonics up to the 5th.

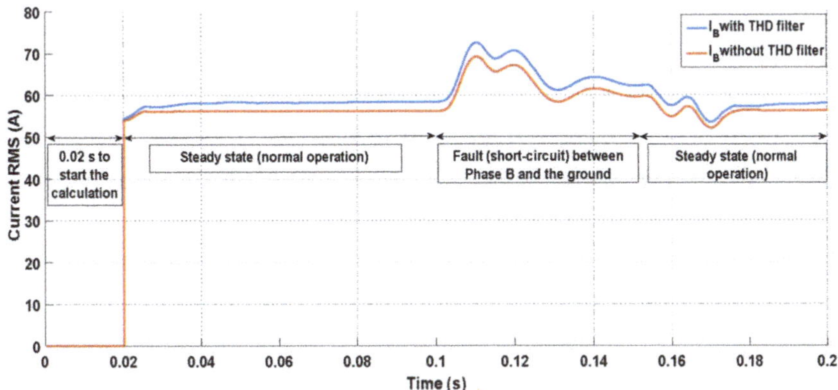

Figure 16. Compare *rms* current calculation using THD filter & without THD filter.

Figure 17 shows the comparison between the *rms* measured voltage with and without a THD filter. The figure confirms the assumption that the *rms* voltage with a THD filter is lower than the *rms* voltage without a THD filter.

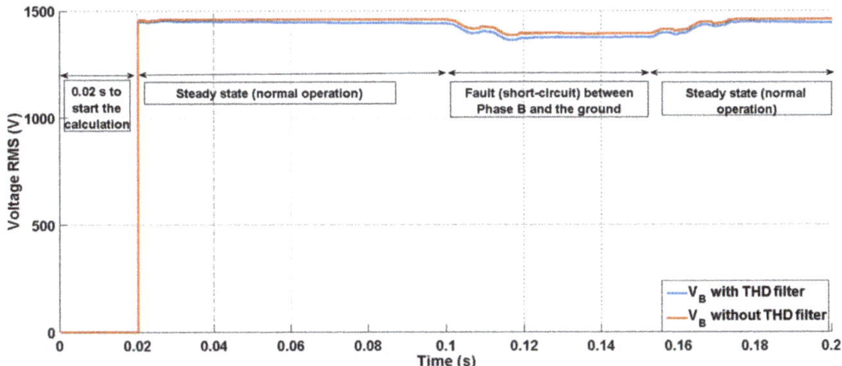

Figure 17. Compare rmsvoltage calculation using THD filter & without THD filter.

Figure 18 shows the computed total harmonic distortion (THD) of the current signal. The THD is defined as the *rms* value of the total harmonic content of the signal divided by the *rms* value of its fundamental signal. For example, for currents, the THD is defined as

$$THD = \frac{I_H}{I_F} \tag{10}$$

$$I_H = \sqrt{I_2{}^2 + I_3{}^2 + \cdots + I_n{}^2}, \tag{11}$$

In: *rms* value of the harmonic *n*,
I_F: *rms* value of the fundamental current.

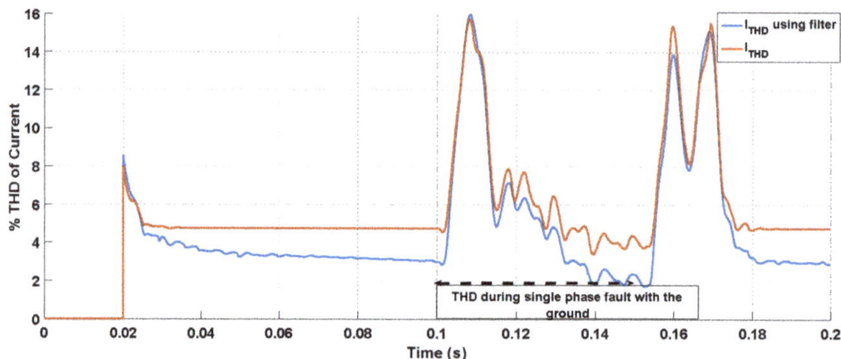

Figure 18. Compare %THD of current calculation using THD filter & without THD filter.

In Figure 18, when the simulation performs a normal condition, the I_{THD} has decreased accordingly by 1% to 2% when the filter has been implemented, during the short circuit the I_{THD} has decreased accordingly by 0.4% to 1.7%.

In Figure 19, when the simulation was run under normal conditions, the VTHD decreased accordingly by 5% to 7%. When a filter was implemented during the short circuit, the ITHD decreased accordingly by 2% to 5%.

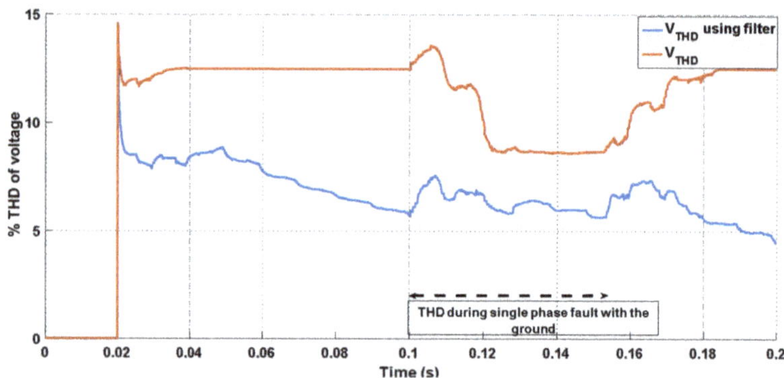

Figure 19. Comparison of the %THD of voltage calculation with and without THD filter.

7. Conclusions

This paper explains part of the digital signal processing in a power system. Moreover, the paper provides different methods to compare the relay algorithm which can be used in a power system based on the impact of harmonics once they are injected in high values. The implementation of the test requires analyzing the occurrence of events in the power system. Each event contained in the input signals can be imported to Matlab via a Comtrade reader which reads the selected event. Digital relays are limited because they can only respond to changes in the magnitude of the fundamental current or voltage. Regarding overcurrent relays, a low level of harmonic distortion may not affect their operation. However, concerning distance relays, while the relay's ability to find faults away from zone's limit may still be reliable, when it comes to faults located near the limit of the zone, there is a possibility for the distance relay to be misguided as to the location of the fault.

Protective relays implement different techniques to measure the current and voltage. The microprocessor relays use a digital filter to take out the fundamental component. Filtering techniques were developed to accommodate a wide variety of harmonic influences. Anti-aliasing provides the ability to remove the frequencies higher than the Nyquist frequency; filter techniques should be implemented to reduce the harmonic level from the power system measurements. The THD filter implemented in this paper can mitigate the THD of the current and voltage. The calculations of the THD during abnormal and normal conditions showed that the voltage harmonics were reduced by 2–5% and the current harmonics were reduced by 0.4–1.7%.

Acknowledgments: This research work has been carried out in the Centre for Research and Utilization of Renewable Energy (CVVOZE). Authors gratefully acknowledge financial support from the Ministry of Education, Youth and Sports of the Czech Republic under NPU I programme (project No. LO1210).

Author Contributions: Kinan Wannous performed the measurements, data analysis and created the simulation model. Petr Toman consulted the results of the analysis. All of the authors have been involved in writing the manuscript.

Conflicts of Interest: The authors declare no conflict of interest.

References

1. Al-Musawi, L.; Waye, A.; Yu, W.; Al-Mutawaly, N. The effects of waveform distortion on power protection relays. In Proceedings of the International Protection Testing Symposium, Feldkirch, Austria, 13–14 October 2015.

2. Wang, F.; Bollen, M. Classification of Component Switching Transients in the Viewpoint of Protective Relays. Ph.D. Thesis, Chalmers University of Technology, Göteborg, Sweden, 2003.

3. Paithankar, Y.G. Line protection with distance relays. In *Book Transmission Network Protection: Theory and Practice*, 6th ed.; Dekker, M., Ed.; Dekker: New York, NY, USA; Basel, Switzerland; Hong Kong, China, 1998; pp. 296–321.

4. Zocholl, S.E.; Benmouyal, G. How microprocessor relays respond to harmonics, saturation, and other wave distortions. In Proceedings of the 24th Annual Western Protective Relay Conference, Spokane, WA, USA, 21 October 1997.

5. Achleitner, G.; Fickert, L.; Obkircher, C.; Sakulin, M.; IFEA, T.; AG, Ö.I.B. Earth fault distance protection. In Proceedings of the 20th International Conference and Exhibition on Electricity Distribution, Prague, Czech Republic, 8–11 June 2009; pp. 1–4.

6. Cease, T.W.; Kunsman, S.A.; Apostolov, A.; Boyle, J.R.; Carroll, P.; Hart, D.; Johnson, G.; Kobet, G.; Nagpal, M.; Narendra, K.; et al. *Protective Relaying and Power Quality*; IEEE PSRC Working Group Report; IEEE: Piscataway, NJ, USA, 2003; pp. 1–65.

7. Pietkiewicz, A.; Melly, S. *Harmonic Filter White Paper*; White Paper; In Electric Power Quality; Springer: Dordrecht, The Netherlands, 2008.

8. Schweitzer, E.O.; Hou, D. Filtering for protective relays. In Proceedings of the IEEE WESCANEX 93 Communications, Computers and Power in the Modern Environment, Saskatoon, SK, Canada, 17–18 May 1993; pp. 15–23.

9. Kumar, B. Design of Harmonic Filters for Renewable Energy Applications. Master's Thesis, Gotland University, Visby, Sweden, 2011.

10. Durdhavale, S.R.; Ahire, D.D. A review of harmonics detection and measurement in power system. *Int. J. Comput. Appl.* **2016**, *143*, 42–45.

11. Hodder, S.; Kasztenny, B.; Fischer, N.; Xia, Y. Low second-harmonic content in transformer inrush currents—Analysis and practical solutions for protection security. In Proceedings of the 67th Annual Conference for Protective Relay Engineers, College Station, TX, USA, 31 March–3 April 2014; pp. 1–20.

12. Kim, G.; Lee, H. A study on IEC 61850 based communication for intelligent electronic devices. In Proceedings of the 9th Russian-Korean International Symposium on Science and Technology, Novosibirsk, Russia, 26 June–2 July 2005.

13. Leao, P.S.; Barroso, G.C.; Melo, N.X.; Sampaio, R.F.; Barbosa, J.A.; Antunes, F.L. Numerical relay: Influenced by and accessing the power quality. In *Book Power Quality*, 1st ed.; Eberhard, A., Ed.; InTech: Rijeka, Croatia, 2011; pp. 213–236, ISBN 978-953-307-180-0.

14. Elphick, S.; Ciufo, P.; Smith, V.; Perera, S. Summary of the economic impacts of power quality on consumers. In Proceedings of the Australasian Universities Power Engineering Conference (AUPEC), Wollongong, Australia, 27–30 September 2015; pp. 1–6.

15. Daut, I.; Hasan, S.; Taib, S. Magnetizing current, harmonic content and power factor as the indicators of transformer core saturation. *J. Clean Energy Technol.* **2013**, *1*, 304–307. [CrossRef]

16. Ho, J.M.; Liu, C.C. The effects of harmonics on differential relay for a transformer. In *IEE Conference Publication*; Institution of Electrical Engineers: London, UK, 2001; pp. 2–34.

17. Rafajdus, P.; Bracinik, P.; Hrabovcova, P.; Saitz, J.; Altus, J.; Höger, M.; Pyrhönen, J. *Examination of Instrument Transformers for Their Employment in New Fault Location Method*; ADE: Bologna, Italy, 2012; pp. 4585–4595.

18. Cintula, B.; Eleschova, Z.; Belan, A.; Volcko, V.; Konicek, M.; Kovac, M. Impact of fault location on transient stability of synchronous generator. In Proceedings of the 2014 15th International Scientific Conference on Electric Power Engineering (EPE), Brno, Czech Republic, 12–14 May 2014; pp. 17–21.

19. Alkandari, A.; Soliman, S. Measurement of a power system nominal voltage, frequency and voltage flicker parameters. *Int. J. Electr. Power Energy Syst.* **2009**, *31*, 295–301. [CrossRef]

20. Apostolov, A.; Vandiver, B. Maintenance testing of multifunctional distance protection IEDs. In Proceedings of the Transmission and Distribution Conference and Exposition, New Orleans, LA, USA, 19–22 April 2010; pp. 1–6.

21. Kaspirek, M.; Mikulas, L.; Mezera, D. Analysis of voltage quality parameters in LV distribution grids with connected distributed energy sources. *CIRED-Open Access Proc. J.* **2017**, *2017*, 513–516. [CrossRef]

22. Kaspirek, M.; Mikulas, L.; Mezera, D.; Prochazka, K.; Santarius, P.; Krejci, P. Analysis of voltage quality parameters in MV distribution grid. *CIRED-Open Access Proc. J.* **2017**, *2017*, 517–521. [CrossRef]

Article

Robustness Improvement of Superconducting Magnetic Energy Storage System in Microgrids Using an Energy Shaping Passivity-Based Control Strategy

Rui Hou [1,2], Huihui Song [2], Thai-Thanh Nguyen [1], Yanbin Qu [2,* and Hak-Man Kim [1,*

[1] Department of Electrical Engineering, Incheon National University, (Songdo-dong)119 Academy-ro, Yeonsu-gu, Incheon 22012, Korea; houruihit@163.com (R.H.); ntthanh@inu.ac.kr (T.-T.N.)

[2] School of Information and Electrical Engineering, Harbin Institute of Technology at Weihai, No.2 in Wenhua Xi Road, Weihai 264209, China; songhh@hitwh.edu.cn

* Correspondence: quyanbin@hit.edu.cn (Y.Q.); hmkim@inu.ac.kr (H.-M.K.);
Tel.: +86-189-6317-2108 (Y.Q.); +82-32-835-8769 (H.-M.K.);
Fax: +86-631-568-7028 (Y.Q.); +82-32-835-0773 (H.-M.K.)

Academic Editor: José Gabriel Oliveira Pinto
Received: 15 February 2017; Accepted: 9 May 2017; Published: 11 May 2017

Abstract: Superconducting magnetic energy storage (SMES) systems, in which the proportional-integral (PI) method is usually used to control the SMESs, have been used in microgrids for improving the control performance. However, the robustness of PI-based SMES controllers may be unsatisfactory due to the high nonlinearity and coupling of the SMES system. In this study, the energy shaping passivity (ESP)-based control strategy, which is a novel nonlinear control based on the methodology of interconnection and damping assignment (IDA), is proposed for robustness improvement of SMES systems. A step-by-step design of the ESP-based method considering the robustness of SMES systems is presented. A comparative analysis of the performance between ESP-based and PI control strategies is shown. Simulation and experimental results prove that the ESP-based strategy achieves the stronger robustness toward the system parameter uncertainties than the conventional PI control. Besides, the use of ESP-based control method can reduce the eddy current losses of a SMES system due to the significant reduction of 2nd and 3rd harmonics of superconducting coil DC current.

Keywords: microgrid (MG); renewable power generation; superconducting magnetic energy storage (SMES); energy shaping passivity (ESP)-based control

---

1. Introduction

The fossil energy crisis has expedited the development of renewable power generation (RPG) [1]. The microgrid (MG), an important part of the future smart grid, can flexibly organize the RPG integration into the power system [2–5]. Typical RPGs such as wind and photovoltaic generation have high intermittent and fluctuating natures, which deteriorate the stability of the power system [6]. Consequently, energy storage systems (ESSs) are essential in the MG to sustain the output power, voltage, and frequency [7–9]. The superconducting magnetic energy storage (SMES) system is a promising ESS in MG applications due to its significant life cycle and power density advantages [10–12]. The SMES system has been used in MG for the power smoothing, load following, power oscillation damping, power factor correction, and dynamic voltage improvement [13]. In this study, SMES is applied in MG for the power smoothing in the grid-connected (GC) mode as well as the frequency and voltage regulation in the islanded mode.

The power conditioning system (PCS) is used to transfer the energy between the superconducting coil and the AC grid. The power switch of the PCS can be the half-controlled device thyristor or the fully-controlled device such as insulated gate bipolar transistor (IGBT). The PCS topology is divided into two types: voltage source converter (VSC), and current source converter (CSC) [2]. In this study, and IGBT switched VSC-based topology is adopted for SMES due to the mature technology and wide applications.

The control performance of SMES mainly relies on a PCS control strategy [14]. The VSC-based SMES system has relatively high nonlinear and coupling characteristic because the PCS of a SMES system includes the DC/DC chopper and VSC. Robustness should be an important consideration in any SMES system to maintain its stability due to its high nonlinear and coupling characteristics [15,16]. In the real implementation, the model mismatch is inevitable due to the parameter drift and uncertainty [17,18]. In some cases, the structural parameters of the SMES converter may actively improve the certain performance such as the output impedance of the converter (the switching ripple filters). A controller with poor robustness might substantially undermine the performance in the SMES applications [18,19] whereas a controller with strong robustness can maintain satisfactory performance and prevent the redesign process. Conventionally, the proportional–integral (PI) strategy is employed to control the PCS of SMES. However, the PI-based SMES system might present an unsatisfactory robustness [20] because the SMES and MG systems are both complex nonlinear and strongly coupling systems [21]. Since the partially linearized model of the SMES system is usually used to design the PI controller, the PI method can only obtain the local stability and the respectable performance near the equilibrium point [20]. When the operation points deviate from the balance point, the conventional PI method might achieve poor performance, and even become unstable. To solve the problem, in [22], the authors have proposed an H_2/H_∞-based robust PI control method for the SMES system. However, this approach employs the particle swarm optimization (PSO) strategy to adjust the PI parameters, which increases the computation and complexity of the SMES control system significantly.

Nonlinear control methods such as hysteresis, sliding, and fuzzy control methods, which can improve the robustness of the system compared with the conventional PI method [23], have been studied in the converter control areas [24–26]. Hysteresis control is simple to implement with good transient performance and robustness, but the unfixed switching frequency may increase the difficulty of the filter design. Sliding control is insensitive to the parameter variation and disturbance, but nonetheless, chattering problems will deteriorate the control performance. Fuzzy control can be applied to the complicated system to enhance the control robustness. However, it presents low steady performance, and the design is not systematic. Due to their respective defects, these methods are subject to limitations when applied in a SMES control system. Recently, an advanced nonlinear control strategy, the so-called energy shaping passivity (ESP)-based control method, has drawn the interest of many researchers [27–29]. The property of passivity, a special case of the system's dissipation characteristics, means that the energy injected from the outside is always less than the energy accumulated in the system. The passivity is an important concept in control theory because it corresponds to the Lyapunov stability of the system. The passivity-based control (PBC) is a class of the advanced control using an expected energy storage function and the damping injection, which guarantees the system's passivity to achieve control stability. Unlike the conventional control concept which focuses on the signal, the PBC considers the controller design from the viewpoint of the energy which is the essential attribute of the physical system. Therefore, some desired control properties can be achieved easily such as the global stability, strong robustness, etc. Compared with the "classical" PBC which is usually called the Lyapunov control, the ESP-based control is proved to be superior because it retains the Lagrangian structure of the system [30]. Energy shaping (ES) is the power configuration process to stabilize the closed-loop control system at the expected equilibrium point. Interconnection and damping assignment (IDA) strategy, which have low control complexity and outstanding performance, can be considered as an effective implementation of ES [31]. In the IDA methodology, the structure of the expected system is firstly described by the port-controlled Hamiltonian (PCH) equations. Then the energy of the closed-loop system is configured to match the structure of the scheme. Finally, the control input can be achieved by solving the resulting simultaneous equations. The advantage to ESP-based method over the other nonlinear regulation approaches lies in the full consideration of the system interconnection features, which results in a straightforward controller implementation [32]. In contrast to the traditional PI control, the ESP-based approach provides better transient control performance. Moreover, the ESP-based approach can remain stable and robust over a wider operation range than PI control [33]. Consequently, the ESP-based control method can achieve superior robustness performance in SMES application in comparison with the conventional PI method.

The ESP-based method has been studied in the areas of mechanics, astronautics, robotics, and power electronics [33–40]. Recently, the ESP-based strategy has been studied for power converter control applications such as AC/DC converters [36], motor drive converters [37], battery charging converters [38], GC inverters [39], and active power filters (APFs) [40]. Although the ESP-based controllers are designed and can get good control effect, there are still some limitations in these studies. References [36–39] only focused on the control performance of ESP-based strategy itself and did not compare the performance of ESP-based method with that of some other approaches. Besides, the conditions of parameter variations, which are very common in the practical application, were not examined in [37–39]. Additionally, the PI regulator proposed in [40] for the comparison has an unclear design basis and a weak control effect.

So far, no ESP-based strategy has been proposed for the SMES control. In this study, the ESP-based method is implemented to control the PCS of the SMES for robustness improvement of the SMES system. In particular, the robustness to the uncertainties of the converter parameters is considered in this paper. Furthermore, a step-by-step ESP-based design method including the robustness considerations for the SMES system is proposed, which might provide a useful reference for other ESP-based researches. The IDA strategy and integral operation incorporate organically to ensure the outstanding performance even under the model mismatch circumstances. The robustness of the ESP-based SMES controller is compared with that of the conventional PI controller. Besides, the DC current in the superconducting coil is also compared between ESP-based and PI control system to evaluate the efficiency of ESP-based SMES control system. The remainder of this paper is organized as follows: Section 2 presents a stepwise ESP-based regulator design strategy which is used for the current control loop of SMES system with the consideration of the robustness issue. Section 3 provides the example MG system and the overall regulation scheme of SMES system for MG application. Section 4 offers the simulation results of the ESP-based method and compares the robustness of the ESP-based and PI methods. Section 4 also provides some discussions of the simulation results. Section 5 presents the experimental results. Finally, Section 6 concludes this study.

2. Design of the Current Control Loop for SMES System Using ESP-Based method

Figure 1 depicts the SMES topology utilized in this study. The superconducting coil is firstly linked to the DC capacitor by a bidirectional DC/DC chopper and then interfaced to the AC MG by a VSC. This topology is well known as VSC-based SMES. VSC can independently control the real and reactive powers transfer.

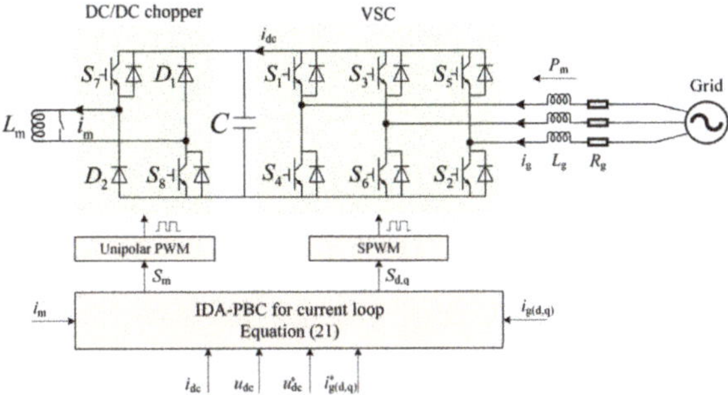

Figure 1. The topology of the SMES system.

In Figure 1, S stands for the IGBTs, and D is the power diode. It is assumed that the direction flowing into the SMES is positive both for the SMES current i_g and power P_m. The superconducting coil charges and discharges through the bidirectional DC/DC converter to maintain a stable DC capacity voltage. When the DC voltage is controlled to be constant, energy releasing from the SMES is the sum of system losses and the energy injected into the grid.

In this study, the ESP-based method used in the current loop of the SMES system is proposed to contain both DC/DC and VSC converters for the robustness improvement. As shown in Figure 1, the ESP-based control directly calculates the duty ratio of the converter to control the SMES. The detailed ESP-based design process including the robustness considerations for SMES system consists of the following four steps.

2.1. The Establishment of Mathematical Model

The design of ESP-based control method requires the mathematical model of the VSC-based SMES system. To facilitate the analysis and control, the VSC model is represented in the two-phase synchronous rotating coordinate (d-q frame). Then, the mathematical model of SMES system depicted in Figure 1 is given as in Equation (1):

$$\begin{cases} L_g \frac{di_{gd}}{dt} = -R_g i_{gd} + \omega_0 L_g i_{gq} - s_d u_{dc} + u_{gd} \\ L_g \frac{di_{gq}}{dt} = -R_g i_{gq} - \omega_0 L_g i_{gd} - s_q u_{dc} + u_{gq} \\ C \frac{du_{dc}}{dt} = i_{dc} - s_m i_m \\ L_m \frac{di_m}{dt} = s_m u_{dc} \end{cases} \tag{1}$$

where s represents the duty cycle of the system. In particular, s_d and s_q are for the VSC control, and s_m is for the DC/DC chopper control. To simplify the analysis, s_m is defined as in (2):

$$s_m = s_7 + s_8 - 1 \tag{2}$$

Meanwhile, in the ESP-based strategy, the PCH model of the system should be established as the following standard form:

$$\begin{cases} \dot{x} = [J(x) - R(x)]\nabla H(x) + g(x)u(x) \\ y(x) = g^T(x)\nabla H(x) \end{cases} \tag{3}$$

where $R(x)$ describes the damping properties of the dissipative system, and it should meet $R(x) = R^T(x) > 0$; $J(x)$ and $g(x)$ can depict the interconnect characteristics of the physical system. $H(x)$ is the energy function that corresponds to the storage energy of the system.

By using (1) to (3), the PCH model of the VSC-based SMES system is given as below:

$$\begin{cases} x = [L_g i_{gd} \quad L_g i_{gq} \quad C u_{dc} \quad L_m i_m]^T \\ u = [s_d \quad s_q \quad u_{gd} \quad u_{gq} \quad i_{dc} \quad 0]^T \\ H = \frac{1}{2} L_g i_{gd}^2 + \frac{1}{2} L_g i_{gq}^2 + \frac{1}{2} C u_{dc}^2 + \frac{1}{2} L_m i_m^2 \\ y = \nabla H = [i_{gd} \quad i_{gq} \quad u_{dc} \quad i_m]^T \end{cases} \tag{4}$$

$$J = \begin{bmatrix} 0 & \omega_0 L_g & 0 & 0 \\ -\omega_0 L_g & 0 & 0 & 0 \\ 0 & 0 & 0 & -s_m \\ 0 & 0 & s_m & 0 \end{bmatrix}, R = \begin{bmatrix} R_g & 0 & 0 & 0 \\ 0 & R_g & 0 & 0 \\ 0 & 0 & 0 & 0 \\ 0 & 0 & 0 & 0 \end{bmatrix}, g = \begin{bmatrix} -u_{dc} & 0 & 1 & 0 & 0 & 0 \\ 0 & -u_{dc} & 0 & 1 & 0 & 0 \\ 0 & 0 & 0 & 0 & 1 & 0 \\ 0 & 0 & 0 & 0 & 0 & 0 \end{bmatrix} \tag{5}$$

2.2. The Determination of the Equilibrium Points

The IDA energy-shaping method is employed in this study because it is easy to implement and the control performance is satisfactory. The structure of closed-loop SMES control system should be configured in the following PCH equation:

$$\begin{cases} \dot{x} = [J_d(x) - R_d(x)]\nabla H_d(x) \\ J_d(x) = J(x) + J_a(x) = -J_d^T(x) \\ R_d(x) = R(x) + R_a(x) = R_d^T(x) > 0 \\ H_d(x) = H(x) + H_a(x) \end{cases} \tag{6}$$

The matrices with the subscript *d* are the desired parameters in the feedback control system, whereas the matrices with the subscript *a* mean the functions awaiting to be determined, which correspond to the parts introduced by the feedback regulation. For example, $H_a(x)$ describes the energy introduced by the regulation. Based upon IDA strategy, the expected Hamilton energy function $H_d(x)$ should adapt the configured structure of the closed-loop control system with the minimum value at the equilibrium point x^*. Furthermore, the balance positions of the closed-loop control system can be gained by:

$$\dot{x}\big|_{x=x^*} = 0 \Rightarrow \frac{\partial H_d(x^*)}{\partial x} = 0 \tag{7}$$

Based on the instantaneous power equations in the d-q frame, the current references of VSC can be obtained:

$$\begin{cases} i_{gd}^* = \frac{\frac{2}{3}(u_{gd}P^* + u_{gq}Q^*)}{u_{gd}^2 + u_{gq}^2} \\ i_{gq}^* = \frac{\frac{2}{3}(u_{gq}P^* - u_{gd}Q^*)}{u_{gd}^2 + u_{gq}^2} \end{cases} \tag{8}$$

The DC voltage should be controlled stably at the reference u_{dc}^* to maintain the proper operation of VSC. u_{dc}^* should be high enough to ensure the tracking ability of the VSC. Meanwhile, the selection of u_{dc}^* should also consider the withstand voltage capability of the IGBTs.

2.3. The Configuration of Energy Shaping and Stability Analysis

In the closed-loop control system, the desired Hamilton energy function which should meet the Equation (7) to ensure the convergence of balance points is given in Equation (9):

$$\begin{cases} H_d = \frac{1}{2}L_g(i_{gd} - i_{gd}^*)^2 + \frac{1}{2}L_g(i_{gq} - i_{gq}^*)^2 + \frac{1}{2}C(u_{dc} - u_{dc}^*)^2 + \frac{1}{2}L_m(i_m - i_m^*)^2 \\ \nabla H_d = [i_{gd} - i_{gd}^* \quad i_{gq} - i_{gq}^* \quad u_{dc} - u_{dc}^* \quad i_m - i_m^*]^T \\ \quad\quad \nabla H_a = [-i_{gd}^* \quad -i_{gq}^* \quad -u_{dc}^* \quad -i_m^*]^T \end{cases} \tag{9}$$

Energy shaping is the process to make the closed-loop system reach the expected energy H_d by controlling the injected energy H_a. By combining the (3) and (6), it can be obtained by:

$$(J_d - R_d)(-\nabla H_a) = (J_a - R_a)\nabla H - gu \tag{10}$$

Equation (10) is a significant constraint in the implementation of the ESP-based method. By choosing suitable indefinite matrices J_a and R_a, the expected duty ratio *s* for PCS in the closed-loop control system can be achieved by solving Equation (10). With the control law obtained by Equation (10), the energy of the closed-loop system will converge to the expected H_d value. The system will be stable at the balance points, and the energy shaping is achieved.

The design method for indefinite matrices in the ESP-based method is provided in [30]. For the purpose of facilitating the design and decreasing the complexity of the ESP-based controller, here the indefinite matrices are determined as:

$$
J_a = \begin{bmatrix} 0 & 0 & 0 & 0 \\ 0 & 0 & 0 & 0 \\ 0 & 0 & 0 & 0 \\ 0 & 0 & 0 & 0 \end{bmatrix}, R_a = \begin{bmatrix} r & 0 & 0 & 0 \\ 0 & r & 0 & 0 \\ 0 & 0 & r_1 & 0 \\ 0 & 0 & 0 & r_2 \end{bmatrix}
\tag{11}
$$

By substituting Equations (5), (6), (9) and (11) into (10), the intermediate variable i_m^* can be eliminated. The following equation can be acquired:

$$
\begin{cases}
s_d = [-R_g i_{gd}^* + r(i_{gd} - i_{gd}^*) + \omega_0 L_g i_{gq}^* + u_{gd}]/u_{dc} \\
s_q = [-R_g i_{gq}^* + r(i_{gq} - i_{gq}^*) - \omega_0 L_g i_{gd}^* + u_{gq}]/u_{dc} \\
s_m = -\left[r_2 i_m + \sqrt{r_2^2 i_m^2 + 4r_2 u_{dc}^*(r_1 u_{dc} - r_1 u_{dc}^* + i_{dc})}\right]/2u_{dc}^*
\end{cases}
\tag{12}
$$

Stability is an essential consideration for the regulator design. The Lyapunov second theorem is convenient to check the stability of the nonlinear system like ESP-based SMES. H_d can be chosen as the Lyapunov function because it describes the energy features of the closed-loop control system with the nonnegative characteristic and just achieves the zero value at the balance position. This choice can substantially decrease the complexity of stability evaluation, which is a significant advantage for the ESP-based control strategy. Based on the Lyapunov criterion, the stability of the controller can be judged by the derivative of H_d showed in Equation (13):

$$
\frac{dH_d}{dt} = [\nabla H_d]^T \dot{x} = [\nabla H_d]^T (J_d - R_d)\nabla H_d = -[\nabla H_d]^T R_d \nabla H_d \leq 0
\tag{13}
$$

The derivation process of Equation (13) employs two mathematical conditions. Firstly, J_d has the antisymmetric feature. Therefore, it can be known $[\nabla H_d]^T J_d \nabla H_d \equiv 0$. Secondly, R_d has the nonnegative and symmetric characteristics. From Equation (13), the derivative of H_d is non-positive and just get the zero value at the balance position. Consequently, the equilibrium position gains the asymptotical stability. Moreover, it can be easily concluded that if $\|x\| \to \infty$, $H_d \to \infty$. The system can acquire a large-scale asymptotic stability near the balance position, which is another outstanding merit of the ESP-based control.

2.4. Introducing the Integrators

The ESP-based method configures of the structure and energy of the system with the aid of the PCH model. Parameter variation and the external noise may introduce the steady-state performance problem to the ESP-based method such as the existence of the steady error. For the purpose of dealing with this issue, the ESP-based control can introduce the integrators in the IDA strategy to achieve the outstanding static performance even in the situation of the model mismatch and the interference [41]. The duty ratios are functioning to the SMES converter regulation. The original location of s_m lies in the interconnection matrix J. It should be transferred to the input vector u in the new PCH model for the SMES system as shown below:

$$
\begin{cases}
\dot{x}_n = (J_n - R_n)\nabla H_n + g_1 u_1 + g_2 u_2 \\
y_1 = g_1^T \nabla H_n \\
y_2 = g_2^T \nabla H_n
\end{cases}
\tag{14}
$$

$$J_n = \begin{bmatrix} 0 & \omega_0 L_g & 0 \\ -\omega_0 L_g & 0 & 0 \\ 0 & 0 & 0 \end{bmatrix}, R_n = \begin{bmatrix} R_g & 0 & 0 \\ 0 & R_g & 0 \\ 0 & 0 & 0 \end{bmatrix}, g_1 = \begin{bmatrix} 0 & 0 & 1 & 0 & 0 \\ 0 & 0 & 0 & 1 & 0 \\ 0 & 0 & 0 & 0 & 1 \end{bmatrix},$$

$$g_2 = \begin{bmatrix} -u_{dc} & 0 & 0 \\ 0 & -u_{dc} & 0 \\ 0 & 0 & -i_m \end{bmatrix}, J_{an} = \begin{bmatrix} 0 & 0 & 0 \\ 0 & 0 & 0 \\ 0 & 0 & 0 \end{bmatrix}, R_{an} = \begin{bmatrix} r & 0 & 0 \\ 0 & r & 0 \\ 0 & 0 & r_1 \end{bmatrix} \tag{15}$$

$$\begin{cases} x_n = \begin{bmatrix} L_g i_{gd} & L_g i_{gq} & C u_{dc} \end{bmatrix}^T, u_1 = \begin{bmatrix} 0 & 0 & u_{gd} & u_{gq} & i_{dc} \end{bmatrix}^T, u_2 = \begin{bmatrix} s_d & s_q & s_m \end{bmatrix}^T \\ \nabla H_n = \begin{bmatrix} i_{gd} & i_{gq} & u_{dc} \end{bmatrix}^T, \nabla H_{dn} = \begin{bmatrix} i_{gd} - i_{gd}^* & i_{gq} - i_{gq}^* & u_{dc} - u_{dc}^* \end{bmatrix}^T \\ y_1 = \begin{bmatrix} 0 & 0 & i_{gd} & i_{gq} & u_{dc} \end{bmatrix}^T, y_2 = \begin{bmatrix} -i_{gd} u_{dc} & -i_{gq} u_{dc} & -i_m u_{dc} \end{bmatrix}^T \end{cases} \tag{16}$$

The subscript n means new. It is worth noting that the state variable $L_m i_m$ is ignored in the new PCH model. The reason is that the expected controlled variables are i_{gd}, i_{gq}, and u_{dc}. i_m is only the intermediate variable in the model and can be omitted in the integral combination.

It can be noticed in Equation (16) that ∇H_{dn} represents the differences between the actual value and the expected value for the control variables. Defining $y_{2d} = g_2^T \nabla H_{dn}$, a new part can be introduced as:

$$\begin{cases} Z = -K_I \int y_{2d} dt = -K_I \int g_2^T \nabla H_{dn} dt \\ K_I = \begin{bmatrix} K_{Idq} & 0 & 0 \\ 0 & K_{Idq} & 0 \\ 0 & 0 & K_{Idc} \end{bmatrix} \end{cases} \tag{17}$$

where K_I is the integral coefficient matrix, and Z is an introduced integrator array.

When $\dot{Z} = -K_I g_2^T \nabla H_{dn} \to 0$, the static error will be eliminated even under the parameter uncertainty situations. New parts introduced in the model are defined as:

$$\begin{cases} H_{dnZ} = \dfrac{Z^T K_I^{-1} Z}{2} \\ H_T(x, Z) = H_{dn}(x) + H_{dnZ}(Z) \end{cases} \tag{18}$$

The ESP-based SMES PCH model with the integrator extension can be built as in Equation (19):

$$\begin{cases} \begin{bmatrix} \dot{x}_n \\ \dot{Z} \end{bmatrix} = \begin{bmatrix} J_{dn} - R_{dn} & K_I g_2 \\ -K_I g_2^T & 0 \end{bmatrix} \begin{bmatrix} \frac{\partial H_T(x,Z)}{\partial x} \\ \frac{\partial H_T(x,Z)}{\partial Z} \end{bmatrix} \\ \frac{\partial H_T(x,Z)}{\partial x} = \nabla H_{dn} \\ \frac{\partial H_T(x,Z)}{\partial Z} = K_I^{-1} Z \end{cases} \tag{19}$$

Reference [41] has proposed and proved the integral stability theorem for the IDA strategy. Based on this theorem, the integrator extension of IDA methodology in the PCH system will not change any stability characteristic of the control system. Furthermore, the steady errors can be eliminated by:

$$\begin{bmatrix} s_{dn} \\ s_{qn} \\ s_{mn} \end{bmatrix} = \begin{bmatrix} s_d \\ s_q \\ s_m \end{bmatrix} + Z = \begin{bmatrix} s_d \\ s_q \\ s_m \end{bmatrix} - K_I \int g_2^T \nabla H_{dn} dt \tag{20}$$

By taking the related variables into Equation (20), it can be finally acquired:

$$\begin{cases} s_d = [-R_g i_{gd}^* + r(i_{gd} - i_{gd}^*) + \omega_0 L_g i_{gq}^* + u_{gd}]/u_{dc} + K_{Idq} \int u_{dc}(i_{gd} - i_{gd}^*) dt \\ s_q = [-R_g i_{gq}^* + r(i_{gq} - i_{gq}^*) - \omega_0 L_g i_{gd}^* + u_{gq}]/u_{dc} + K_{Idq} \int u_{dc}(i_{gq} - i_{gq}^*) dt \\ s_m = -[r_2 i_m + \sqrt{r_2^2 i_m^2 + 4 r_2 u_{dc}^*(r_1 u_{dc} - r_1 u_{dc}^* + i_{dc})}]/2 u_{dc}^* + K_{Idc} \int i_m(u_{dc} - u_{dc}^*) dt \end{cases} \tag{21}$$

The determination of K_I just needs to consider the magnitude the steady error. Equation (21) is the control law in the ESP-based method. The duty ratio of the converter can be directly calculated by the sampling value of the related variables. s_d and s_q are used to control the VSC, while s_m is used to regulate the DC chopper. It can be concluded from Equation (21) that the ESP-based control strategy is simply implemented due to the direct calculation.

3. Test Microgrid System and SMES Control System for Microgrid Application

In this study, an AC MG consisting of the wind generation (WG), DG, load, and SMES is employed as an example. The constitution and parameters of the example MG system for the test are illustrated in Figure 2. The rated line voltage RMS value is 380 V and the rated frequency is 60 Hz. A constant-speed wind turbine generator (WTG) is employed in the simulation with the expected output power of 300 kW. The capacity of the DG is 200 kVA. The 500 kW resistor load is considered in the simulation. The inductance value of the superconductive coil is 1.5 H, and the initial coil current is 1000 A. The forward direction of the primary power flow in MG system is also provided in Figure 2. P_M has the reversible power flow characteristics, and the charge is determined to be forward.

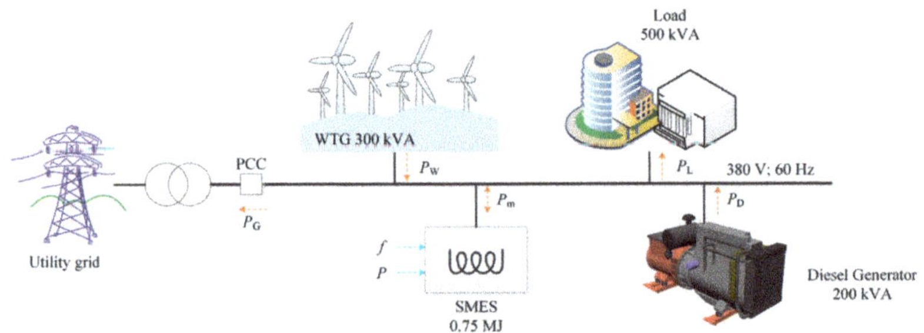

Figure 2. The test MG system.

Figure 3 provides the control diagram of ESP-based SMES system. In the GC mode, the SMES real power reference is calculated by $P_g^* = P_E - P_A$ whereas the reactive power reference is set constantly. SMES can stabilize the power fluctuation of the WG and achieve the steady power flow at the point of common coupling (PCC). The power references of SMES are obtained by using the first-order low pass filter to smooth the fluctuations in WG. In the islanded mode, an additional frequency-real power (*f-p*) PI regulator generates the real power reference of the SMES. Another outer voltage-reactive power (*v-q*) PI regulator calculates the reactive power reference of the SMES. The designed ESP-based current control strategy ensures that the SMES power can be controlled to follow the given trajectory. Meanwhile, the ESP-based strategy can stabilize the DC voltage at the given value to guarantee the normal operation of the PCS. It is noteworthy that the power tracking control is adopted without the extra power feedback in this study. Although this extra feedback can improve the precision of the power regulation, it is a tough to tune the parameters of the multi-loop PI controllers with sufficient stability. The structure of the ESP-based control system is quite different from that of the conventional PI-based control system. In the PI control system, VSC utilizes one PI controller for the power control, and the DC chopper needs another different PI regulator to maintain the DC voltage. Instead, the ESP-based method considers the SMES converters as a whole and integrates the control of VSC and DC-DC chopper together. Therefore a better overall performance it can be anticipated for the ESP-based control strategy. Finally, space vector PWM (SVPWM) technology is employed in VSC control to increase the DC voltage utilization. The unipolar PWM method is used in the DC chopper to reduce the harmonics.

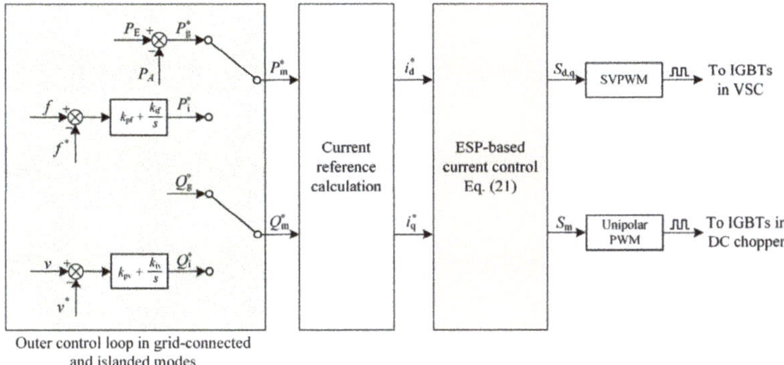

Figure 3. Overall control diagram of the SMES system.

For the purpose of showing the merits of the proposed ESP-based method, the performance of the ESP-based control is compared with that of the conventional PI current control strategy in the example SMES system. In this part, the design of the conventional PI control system for comparison is briefly provided. The PI current controller for the VSC is used to regulate the output power of the SMES. The classic current feedforward decoupling PI control which is commonly employed in VSC control area is adopted in this study. The structure of the controller is depicted in Figure 4.

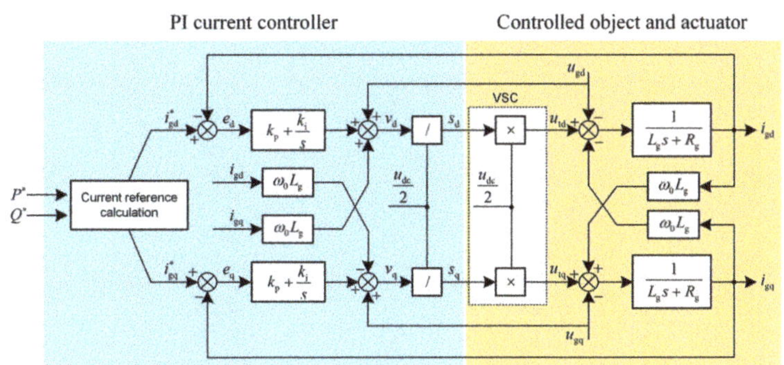

Figure 4. The structure of the PI controller for VSC.

As shown in Figure 4, the grid voltage is fed forward to restrain the disturbance in the control loop. The feedforward decoupling operation eliminates the control coupling between d and q components. The detailed design strategy of this controller is presented in [42]. The simplified control block diagram of the current loop is illustrated in Figure 5.

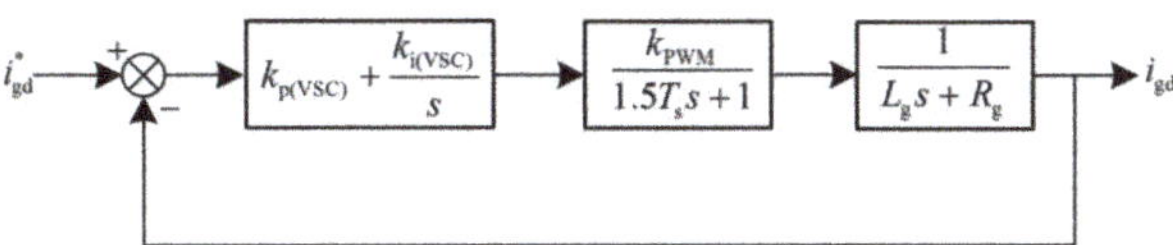

Figure 5. The simplified control block diagram of the PI current loop.

In Figure 5, the function of the PWM and sampling delay are combined into one inertia link with the time constant of $1.5T_s$. According to [42], the adverse pole $-R_g/L_g$ can be directly eliminated by the

PI current controller to ensure the tracking ability of the system. By this design method, the closed-loop transfer function of the control system is a typical second-order oscillation link. Based on the optimal design method for the second-order oscillation system, the damping coefficient is determined as $\xi_{VSC} = 0.707$ to achieve the satisfactory overall performance. Accordingly, the implementation equations for PI current controller are as follows:

$$k_{p(VSC)} = \frac{L_g}{3T_s}, \quad k_{i(VSC)} = \frac{R_g}{3T_s} \tag{22}$$

When the parameters of the system such as L_g and R_g vary, the performance of the PI control system may deteriorate because the position of the adverse pole changes and the PI controller cannot eliminate it. Meanwhile, the PI controller for the DC/DC chopper is used to control the DC voltage. The detailed design method of the PI voltage controller is proposed in [43]. The closed-loop transfer function of the DC voltage loop is as follows:

$$\begin{cases} \Phi(s) = \frac{u_{dc}}{u_{dc}^*} = \frac{\omega_n^2(1+T_{i(dc)}s)}{s^2+2\xi_{dc}\omega_n s+\omega_n^2} \\ \omega_n = \sqrt{\frac{k_{p(dc)}}{T_{i(dc)}C}}, \quad \xi_{dc} = 0.5\sqrt{\frac{T_{i(dc)}k_{p(dc)}}{C}} \end{cases} \tag{23}$$

As is shown in Equation (23), the closed-loop system is also a second-order oscillation link. In the equation, ω_n is the natural oscillation frequency; $T_{i(dc)}$ is the integral time constant in the PI voltage controller; ξ_{dc} is the damping ratio of the system. Considering the characteristic and demand of the DC voltage control, $T_{i(dc)}$ is chosen to be 16 ms, and ξ_{dc} is determined to be 2 [43].

The additional f-p and v-q PI controllers are designed by the classic Ziegler-Nichols parameter tuning method. The primary simulation parameters are provided in Table 1.

Table 1. The primary simulation parameters.

Parameter Type	Parameter Value
Sampling/switching frequency	10 kHz/10 kHz
DC voltage reference	750 V
C	32,000 μF
L_g, R_g	1 mH, 1.1 mΩ
L_m	1.5 H
P_E	500 kW
$k_{p(vsc)}, k_{i(vsc)}$	3.333, 3.667
$k_{p(dc)}, k_{i(dc)}$	32, 2000
k_{pf}, k_{if} [1]	139.5, 3799
k_{pv}, k_{iv} [1]	4, 2
K_{Idq}, K_{Idc}	3, 0.35
r, r_1, r_2	1500, 3000, 3000

[1] The power unit is W or Var.

It should be noted that the state of charge (SOC) is a significant factor in the energy management studies which is considered for a long-term operation. However, this study focuses on the control problem in SMES which deals with the short-term operation. It is assumed that the energy managing of the SMES system has been designed appropriately. Accordingly, the SOC issue of the SMES can be neglected in our study.

4. Simulation Results and Discussion

In the simulation, the constant step-size and ode3 arithmetic is employed. Figure 6 illustrates the power of each primary element in the example MG test system. The whole simulation period lasts 16 s. At the time point of 0.5 s, PCS converters of SMES is put into operation. At the time point of 5 s,

MG transits from GC mode to the islanded mode. A load with $\Delta P_L = 40$ kW is switched into the MG at the time point of 8 s and disconnected at the time point of 12 s.

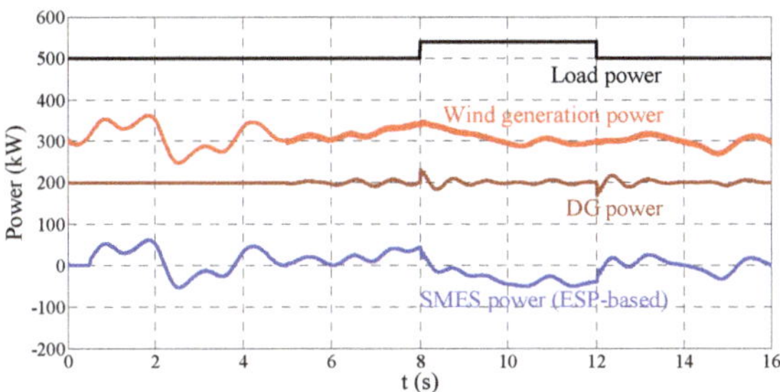

Figure 6. The power of each primary element in the example test MG system.

4.1. Simulation in the GC Mode

In the GC mode, SMES is employed to stabilize the WG power fluctuation, which can be seen in Figure 7. Ideally, the constant power flow at the PCC can be achieved. In this simulation, the MG is in the GC mode before 5 s.

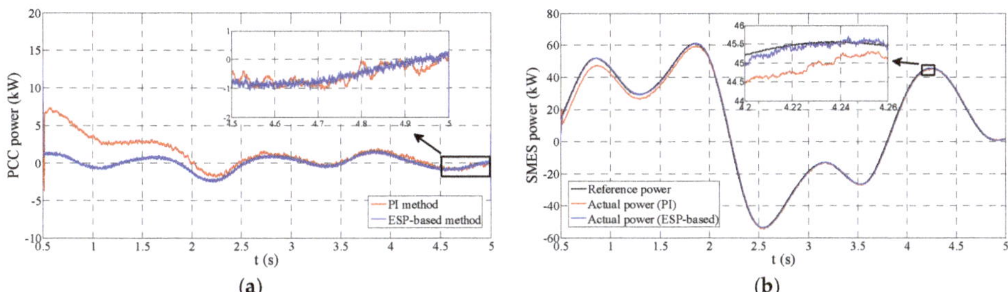

Figure 7. Performance under the grid-connected (GC) mode (L_g = 1 mH, R_g = 1.1 mΩ): (**a**) the performance on stabilizing the grid-side power; and (**b**) the performance on SMES power tracking.

Figure 7a demonstrates the performance comparison of smoothing PCC power between PI and ESP-based control strategies under the situation of L_g = 1 mH and R_g = 1.1 mΩ. Figure 7b provides a comparison of the SMES power tracking performance between these two approaches. It can be seen that both PI and ESP-based approaches can stabilize the PCC power flow and follow the power reference well. However, some visible differences between them still deserve the concern. Firstly, when the PCS of SMES is put into operation, the PI-based system has a visible 7 kW overshoot and an around 2 s regulation period as shown in Figure 7a. Accordingly, we can notice in Figure 7b that before 2 s the SMES power by the PI control has a distinct difference compared to the power reference and needs a regulation period to remove the error. The primary reason for this problem is that the start of the PCS is a disturbance for the control system, and the conventional PI control strategy has a relatively poor disturbance rejection performance. Conversely, the ESP-based control system practically avoids this disturbance regulation process, which can be seen in Figure 7. Hence, the ESP-based control has stronger disturbance rejection ability in comparison to the PI method. Secondly, the steady-state performance of PI control strategy is not as satisfactory as that of the ESP-based control strategy. From the enlarged figure in Figure 7b, it can be noticed that the ESP-based control strategy provides a smaller tracking error. Accordingly, the stabilized PCC power by the PI control strategy is not steady as that by the ESP-based control strategy. Due to the better tracking ability, the ESP-based method can reduce the low-frequency harmonic of the system and has a lower ripple in the smoothed power compared to the PI control, which can be seen in the enlarged figure of Figure 7a.

Figure 8 provides the DC voltage regulation performance. It can be found that the ESP-based control strategy has much superior dynamic performance in contrast to the PI control. More specifically, the PI control strategy has a 50 V overshoot and nearly 0.06 s regulation period. Instead, ESP-based control method turns stable in the very short time without any overshoot.

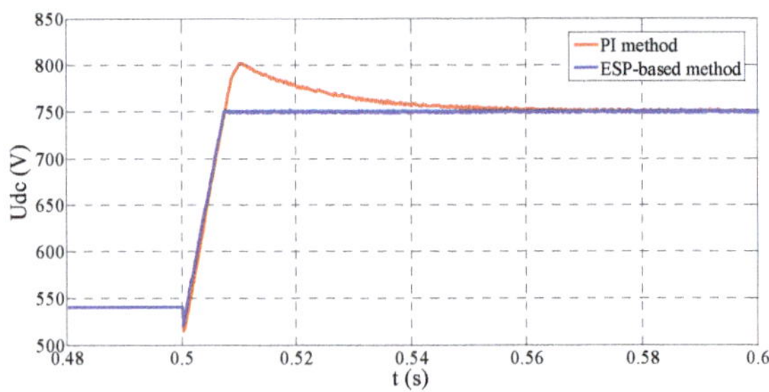

Figure 8. The performance on DC voltage regulation.

Robustness is a critical consideration for SMES control system design because the model mismatch is unavoidable in any real system. In this study, the robustness to the parameters variation of L_g and R_g are considered. The reason for this selection mainly relies on two aspects. On the one hand, the actual values of these two parameters are usually unknown and varying with the service time. Nevertheless, this kind of parameter change is often in the small range. On the other hand, these parameters could be modified deliberately to enhance the performance. This type of parameter uncertainty may be on a large scale. Hence the parameter uncertainty considered in this study is in a considerable region. If the controller can maintain the reliable performance and stability in this condition, it will greatly facilitate the design and reform of the system.

Figure 9 shows the comparison of the GC mode performance between PI and ESP-based control strategies under the condition of L_g = 3 mH, R_g = 0.1 Ω. while Figure 10 provides the comparison of the situation of L_g = 4 mH and R_g = 0.2 Ω. It is obvious in the figures that the performance of these two approaches is quite different under the two test situations. The ESP-based control strategy keeps the good performance on both the grid power stabilizing and the reference power tracking. In contrast, the PI control strategy presents an inferior regulation performance. There is apparent reference power tracking error during most of the time. As a result, the PCC power stabilized by the PI control has a much bigger fluctuation than that by the ESP-based control. The following aspects can explain this performance difference. On the one hand, due to the enlargement of connection impedance, the tracking ability of the PCS converter decreases correspondingly. Power tracking for PCS significantly increases the difficulty and requires an outstanding tracking ability for the controller. On the other hand, PI parameters are designed based on the situation of L_g = 1 mH and R_g = 1.1 mΩ. Under the new circumstances, the PI control parameters are not suitable, and the controller gets a poor tracking capability. To the contrary, the ESP-based method maintains an outstanding tracking performance and presents a better robustness to parameter uncertainty. Also, we can find the performance of PI control method is especially bad near 2.5 s. Around this time, the SMES power strongly deviates from the reference power, which can be seen in Figures 9b and 10b. Accordingly, the smoothed PCC power gets a large peak error as shown in Figures 9a and 10a. The reason for this problem is probably that the absolute value of reference power is high and increases very steeply during this time. The PI method has serious trouble with this situation due to the insufficient tracking ability.

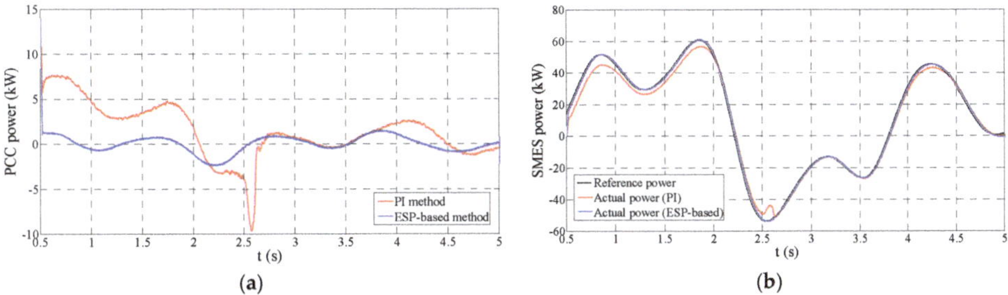

Figure 9. Performance under the grid-connected (GC) mode ($L_g = 3$ mH, $R_g = 0.1\ \Omega$): (**a**) The performance on stabilizing the grid-side power; and (**b**) The performance on SMES power tracking.

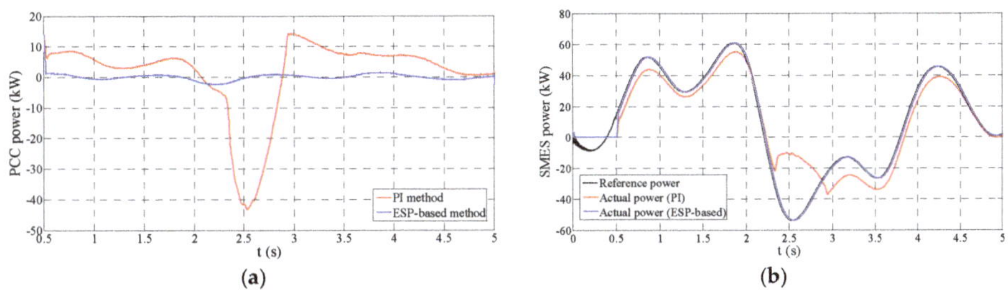

Figure 10. Performance under the grid-connected (GC) mode ($L_g = 4$ mH, $R_g = 0.2\ \Omega$): (**a**) The performance on stabilizing the grid-side power; and (**b**) The performance on SMES power tracking.

4.2. Simulation in the Islanded Mode

At the time point of 5 s, MG is switched to the islanded mode. The SMES system takes part in the frequency and magnitude regulation for the MG voltage. Figure 11 provides the regulation performance under the circumstance of $L_g = 1$ mH and $R_g = 1.1$ mΩ.

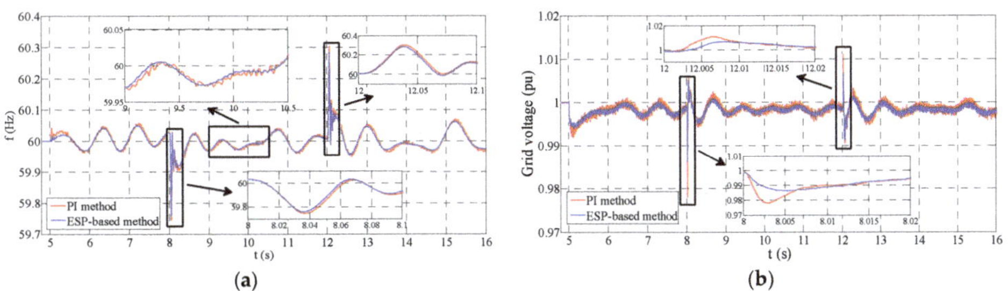

Figure 11. Performance under the islanded mode ($L_g = 1$ mH, $R_g = 1.1$ mΩ): (**a**) Frequency control performance; and (**b**) Magnitude control performance.

Figure 11a,b show the comparison of the control performance in the islanded mode between PI and ESP-based control approaches, respectively. The regulation performance of the PI and ESP-based control methods is similar because the external *f-p* and *v-q* PI controllers are the same. Both PI and ESP-based methods can follow the reference well in this parameter situation. However, there are still some performance differences between these two approaches. On the one hand, the ESP-based control strategy presents a superior dynamic performance in comparison to the PI method. During the mode transition and load change periods, the ESP-based control has a smaller overshoot in both frequency and magnitude regulations. On the other hand, the ESP-based control method has a smaller frequency and voltage ripple.

Figure 12 provides the performance comparison under the situation of $L_g = 3$ mH and $R_g = 0.1\ \Omega$. It can be found that the performance of PI and ESP-based control methods has a sharp distinction. In Figure 12a, it is noticed that the ESP-based control provides a much smaller frequency overshoot during the islanded mode. Also, it can be seen that the frequency ripple by the ESP-based method is much lower than that by the PI method. Similarly, the voltage overshoot and ripple by the ESP-based control is also much smaller than those by the PI method. The main reason for these differences is that the tracking ability of the PI method is awful in this parameter situation as mentioned before.

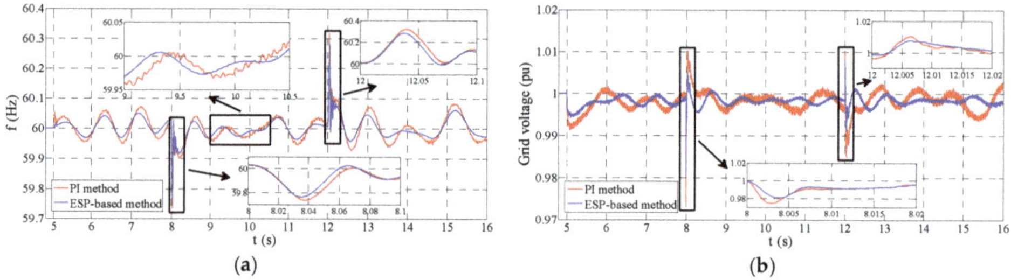

Figure 12. Performance under the islanded mode ($L_g = 3$ mH, $R_g = 0.1\ \Omega$): (**a**) frequency control performance; and (**b**) magnitude control performance.

Figure 13 shows the frequency regulation performance under the situation of $L_g = 4$ mH and $R_g = 0.2\ \Omega$. It can be seen that the PI method becomes unstable, whereas the ESP-based method maintains the proper performance in this situation. The poor tracking ability can explain this stability problem for the PI method. When the power reference is large and fast-changing, the frequency regulation of the PI control method gets unstable due to the insufficient tracking capability under this parameter situation. It is concluded that the ESP-based method has a larger stability region than PI method under the situation of parameter uncertainty.

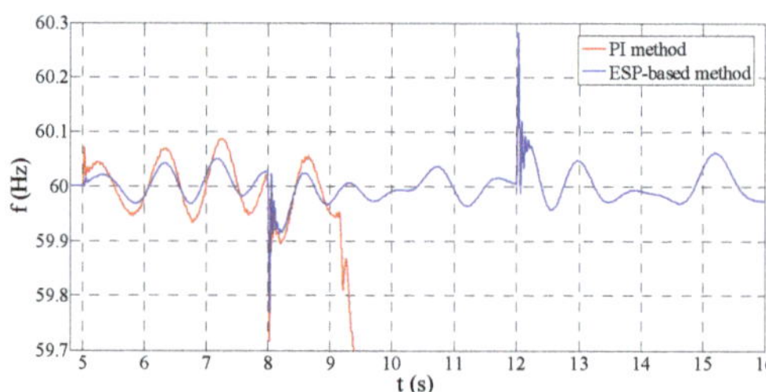

Figure 13. Frequency control performance under the islanded mode ($L_g = 4$ mH, $R_g = 0.2\ \Omega$).

Figure 14 compares the harmonic of the grid voltage controlled by the PI and ESP-based control methods.

The results show that the typical harmonic in the grid voltage by the ESP-based method is smaller than that by the PI method due to the superior tracking ability. When the system parameters change, the harmonic performance of the ESP-based method is much better than that of the PI control.

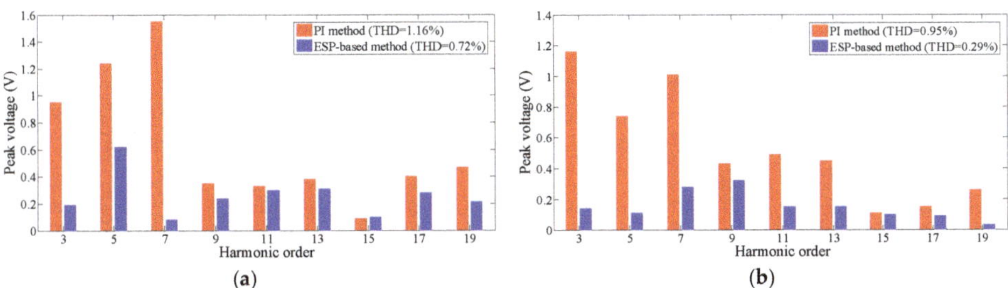

Figure 14. Comparison of the harmonic of the grid voltage: (**a**) L_g = 1 mH, R_g = 1.1 mΩ; and (**b**) L_g = 3 mH, R_g = 0.1 Ω.

4.3. The Harmonic Analysis of the Superconducting Coil Current

The eddy current loss in the SMES system is an important consideration, for it greatly influences the efficiency of the system. Meanwhile, the eddy current loss has a significant relationship with the harmonic of the DC current in the superconducting coil, especially the low-order harmonic (2nd and 3rd) [44]. In this part, the comparison of the DC current harmonic between PI and ESP-based methods is provided to estimate the eddy current loss conditions.

It can be easily found in Figure 15a that the DC current by the ESP-based method is much more sinusoidal than that by the PI method. We also provide the low-order harmonic analysis in Figure 15b. The harmonic of the DC current is substantially reduced by the ESP-based method due to the outstanding tracking ability. Therefore, the ESP-based method can be estimated to achieve the eddy current loss reduction and the system efficiency improvement.

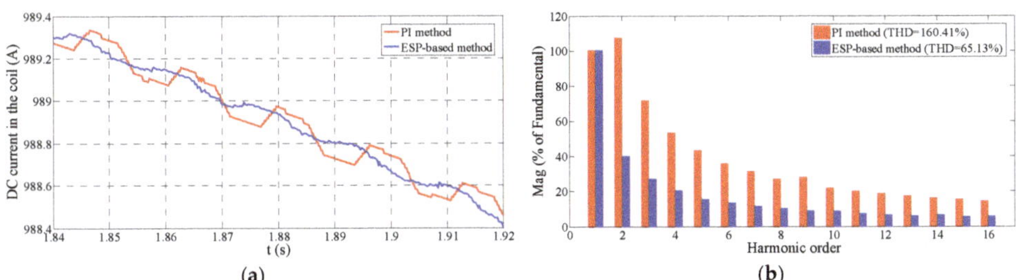

Figure 15. The DC current in the superconducting coil (L_g = 1 mH, R_g = 1.1 mΩ): (**a**) the current waveforms; and (**b**) harmonic analysis.

4.4. Discussion

In this part, some discussions based on the simulation results are provided. Firstly, in order to well explain the difference in simulation results between PI and ESP-based method, the frequency response of the current control loops are tested by the control design toolbox in Matlab. The amplitude of the input test current is 10 A, and the frequency variation range of the input is from 10 Hz to 200 Hz.

Figures 16 and 17 provide the frequency response results of the PI and ESP-based current controllers under two different parameter situations. Under the expected parameters, both the PI and ESP-based control present a satisfactory frequency response, which proves the correctness of the controller design in this study. The magnitude gain range is within ±1 dB, and the phase shift is within ±10 degree. Both PI and ESP-based control can achieve good tracking performance under this situation. When the parameters change, the magnitude gain of the ESP-based control is still in the region of ±1 dB. Nevertheless, the maximum phase shift increases to be −35 degree which is still acceptable due to the enlargement of the output impedance of the converter. In contrast, both the magnitude and phase characteristics of the PI control deteriorate seriously under this situation. From the frequency response analysis, it can be concluded that the ESP-based control can remain the good tracking capability when parameters change, while the parameters affect the tracking ability of the PI control significantly. This conclusion also can well explain the previous simulation results.

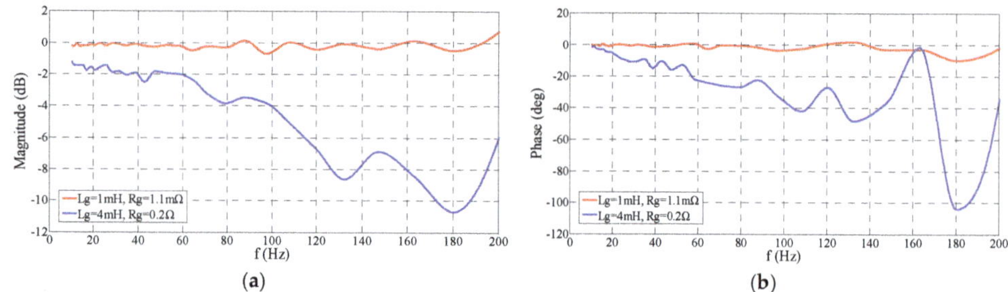

Figure 16. The frequency response of PI control: (**a**) magnitude-frequency characteristics; and (**b**) phase-frequency characteristics.

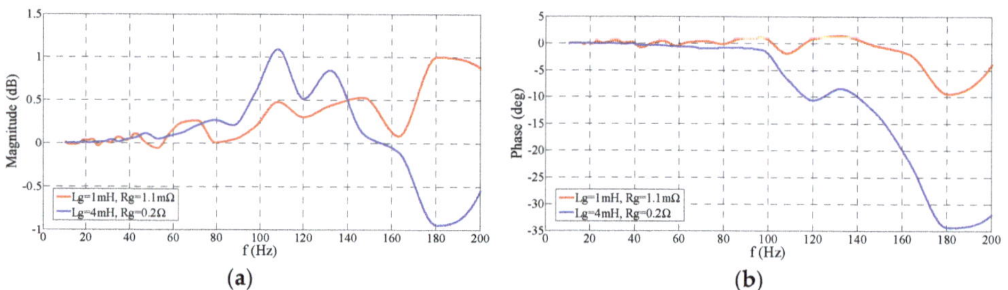

Figure 17. The frequency response of ESP-based control: (**a**) magnitude-frequency characteristics; and (**b**) phase-frequency characteristics.

Secondly, in order to verify the model adopted to design the ESP-based controller, a brief sensitivity analysis [45] is provided in this part. The sensitivity analysis factor (SAF) which is defined in Equation (24) is introduced to evaluate the sensitivity of the controller to system parameters:

$$\text{SAF} = \frac{\Delta A/A}{\Delta F/F} \tag{24}$$

In Equation (24), $\Delta A/A$ represents the change rate of evaluating indicator; $\Delta F/F$ means the change ratio of the uncertain system parameter. The positive/negative signs of the SAF reveal that the evaluating indicator changes with uncertain system parameter at the same/opposite direction. In the GC mode, SMES is used to smooth the WG power. Hence, the peak-to-peak power ($P_{\text{p-p}}$) of the PCC power is adopted as the evaluating indicator.

The magnitude of $P_{\text{p-p}}$ can evaluate the steady degree and the ripple level of the smoothed power. Table 2 provides SAF performance when the system parameters change. Considering that some parameters in the power system are prone to change, the parameter variations in the sensitivity analysis consider not only the structural parameters of SMES but also the related parameters of power grid. It can be found that the performance is not sensitive to the most of the parameter variation, which proves the effectiveness of the adopted model and the designed ESP-based controller. The performance is a little sensitive to the grid voltage and the WG power fluctuation. However, $P_{\text{p-p}}$ is still small enough to guarantee the good performance.

Table 2. The SAF performance of ESP-based control in the GC mode.

Parameters	Change Rate (%)	$P_{\text{p-p}}$ (kW)	SAF
No	0	4.163	/
L_{m}	10	4.248	0.204
i_{m} (initial)	10	4.188	0.060
C	10	4.185	0.052
u_{dc}^*	10	4.108	−0.133
L_{g}	100	3.953	−0.050
R_{g}	100	4.246	0.020
ΔP_{W} [1]	50	5.988	0.876
P_{L}	50	4.107	−0.027
u_{g}	5	4.45	1.379
f	0.5	4.156	−0.336

[1] ΔP_{W} means the fluctuation of the WG power.

In the islanded mode, the frequency overshoot ($\Delta f = f - 60$) when the load changes (at the time point of 8 s) is employed to be the evaluation indicator in the sensitivity analysis. It can be seen from Table 3 that the most of the parameters have little effect on the transient performance of the controller.

Table 3. The SAF performance of ESP-based control in the islanded mode.

Parameters	Change Rate (%)	Δf (Hz)	SAF
No	0	−0.272	/
L_{m}	10	−0.2726	0.022
i_{m} (initial)	10	−0.2713	−0.026
C	10	−0.274	0.073
u_{dc}^*	10	−0.2722	0.007
L_{g}	100	−0.2703	−0.006
R_{g}	100	−0.2735	0.006
ΔP_{W}	50	−0.277	0.035
ΔP_{L} [1]	50	−0.349	0.566

[1] The positive/negative ΔP_{L} means the load connecting/disconnecting to MG.

The load change ΔP_{L} influences the transient performance of the system to some extent. If the tolerance limit of Δf is determined as ± 0.5 Hz, the permitted load change region is approximately ± 140 kW which is obtained by the simulation test.

5. Experimental Verification

In order to further verify the feasibilities of the proposed control method, the real-time simulation based on the hardware-in-the-loop simulation (HILS) using OP5600 and OP8665 (OPAL-RT Technologies, Montreal, QC, Canada) is implemented. The HILS system used for the test of proposed control method is depicted in Figure 18.

As shown in Figure 18, the HILS system can simulate the complex MG and the converter of SMES by the programming of the software RT-LAB (Version 11, OPAL-RT Technologies, Montreal, QC, Canada). The OP8665 including the digital signal processor (DSP) TMS320F28335 implements the proposed control method by sampling the analog signal from the OP5600 and generates the PWM commands for the converter of SMES. All of the parameters in the experiment are the same with those in the previous simulation to facilitate the comparison. Nevertheless, the WG power is extended periodically every 16 s in contrast to Figure 6.

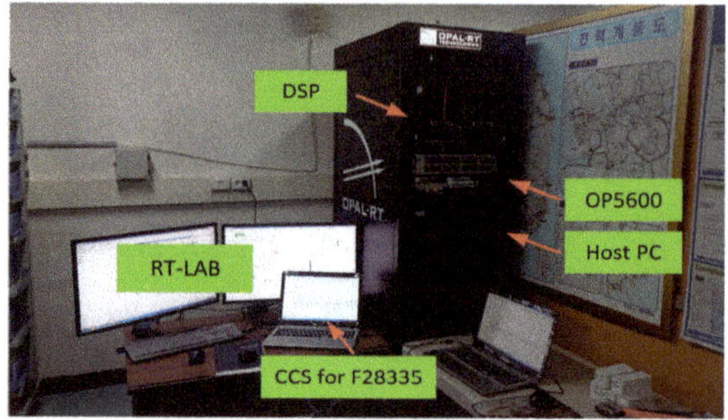

Figure 18. The HILS system used for the test of proposed control method.

Figure 19a compares the DC voltage performance between the PI and ESP-based methods. The DC voltage by the ESP-based method has a much better transient performance without any overshoot than PI control. Figure 19b provides the comparison of the WG power smoothing performance under the expected parameter. The performance of PI and ESP-based control methods are both satisfactory under this situation. It can be noticed that the switching ripple of ESP-based control mainly lies in the high-frequency region, which is easily absorbed by the additional LC or LCL filter. Figure 19c,d demonstrate the performance comparison under the situation of the parameter change. Figure 20a,b provide the SMES power tracking performance under these two conditions. The experimental results are quite similar with the previous simulation results. Both of the power smoothing and the power tracking performance are poor by the PI control method when parameters change. In contrast, the performance of ESP-based control method is still satisfactory under the situation of the parameter variation. The reasons for the performance difference have been analyzed in Section 4.

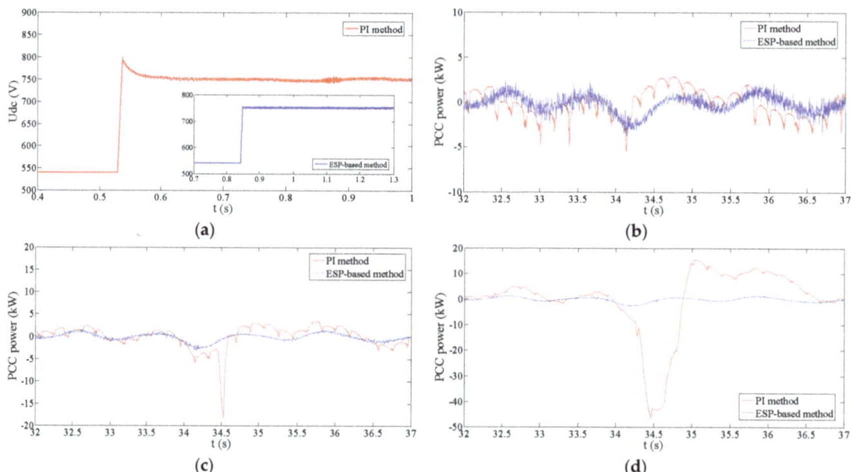

Figure 19. The DC voltage and WG power smoothing performance: (**a**) DC voltage; (**b**) PCC power ($L_g = 1$ mH, $R_g = 1.1$ mΩ); (**c**) PCC power ($L_g = 3$ mH, $R_g = 0.1$ Ω); and (**d**) PCC power ($L_g = 4$ mH, $R_g = 0.2$ Ω).

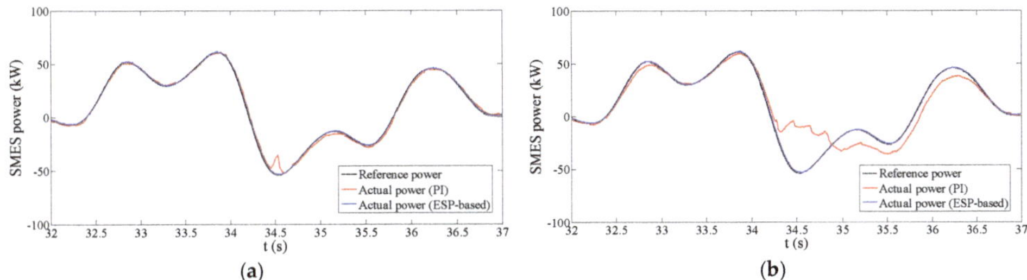

Figure 20. SMES power tracking performance: (**a**) when $L_g = 3$ mH, $R_g = 0.1$ Ω; and (**b**) when $L_g = 4$ mH, $R_g = 0.2$ Ω).

Figure 21 provides the frequency and voltage regulation performance in the islanded mode without the parameter variation. It can be found that under the expected parameter situation the performance of the PI and ESP-based methods are quite similar. The reason is that their outer f-p or v-q controllers are the same, and in this parameter situation, both PI and ESP-based methods have the good tracking ability. Figure 22 shows the regulation performance when $L_g = 4$ mH and $R_g = 0.2$ Ω. Under this situation, the PI-based control system becomes unstable due to the poor tracking capability. To the contrary, the ESP-based method remains the good regulation performance when parameters change.

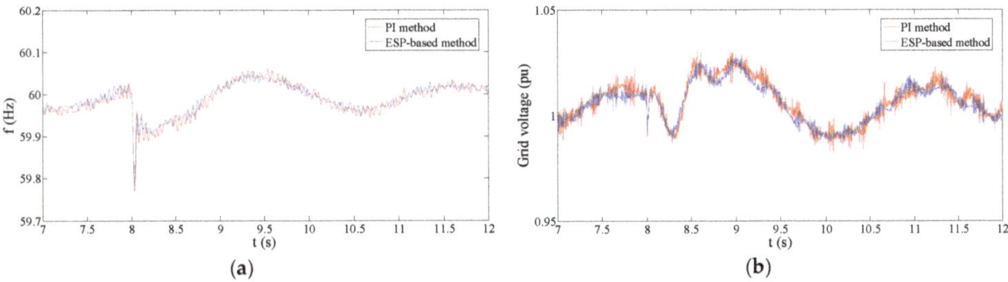

Figure 21. Experimental performance of the islanded mode ($L_g = 1$ mH, $R_g = 1.1$ mΩ): (**a**) MG frequency; and (**b**) MG voltage.

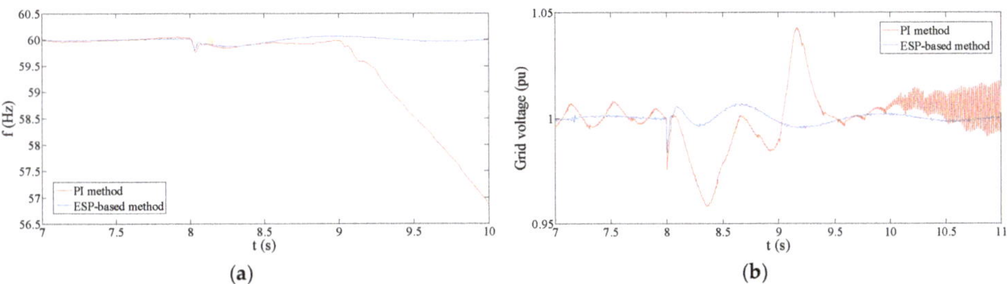

Figure 22. Experimental performance of the islanded mode ($L_g = 4$ mH, $R_g = 0.2$ Ω): (**a**) MG frequency; and (**b**) MG voltage.

6. Conclusions

In this study, an ESP-based control strategy, which can integrate the VSC and DC chopper regulation together, has been proposed for the robustness improvement of SMES systems. A stepwise ESP-based design strategy in the SMES system has been presented, which consists of four steps:

building the mathematical model, establishing the balance positions, energy shaping, and introducing the integrators. The designed ESP-based regulator is easily realized in the programming. Experimental verification by using the HILS system with DSP was performed in this study to show the validity of the proposed controller.

A comparison between the ESP-based and PI strategies is presented to show the merits of the proposed ESP-based control. The parameters of the connection impedance of the PCS converter were changed to test the robustness of the controllers. Both the PI and ESP-based control methods could follow the power reference well under the expected parameter situation. Nevertheless, ESP-based control avoids the disturbance rejection process and achieves a flatter PCC power flow under the GC mode. Moreover, the ESP-based control provides a smaller overshoot and ripple in contrast to the PI control method under the islanded mode. When the system parameters change, the ESP-based control method could achieve a much superior performance than PI control. In particular, when the reference power was high in value and varied steeply simultaneously, the SMES power controlled by the PI controller would deviate from the reference due to the insufficient tracking ability. This problem also caused a weak stability region for the PI method when parameters vary. It can be concluded from the simulation results that the ESP-based control strategy has a superior robustness to the parameter variation in comparison with the conventional PI control strategy. Besides, the ESP-based method significantly reduced the harmonic of DC current compared with the PI method, which results in the reduction of eddy current losses compared to the PI control method.

The SMES with the designed ESP-based control method might provide better overall performance compared with the conventional PI-based SMES. On the one hand, the good robustness of ESP-based SMES could maintain the satisfactory performance even when converter parameters change, and this could avoid the redesign process in some cases. On the other hand, ESP-based SMES can provide better dynamic property for improving the transient performance of the power system.

Acknowledgments: This research was supported by Post-Doctor Research Program (2016) through Incheon National University (INU), Incheon, Republic of Korea.

Author Contributions: The paper was a collaborative effort between the authors. The authors contributed collectively to the theoretical analysis, modeling, simulation, and manuscript preparation.

Conflicts of Interest: The authors declare no conflict of interest.

Nomenclature

Abbreviations

SMES	Superconducting magnetic energy storage
ESP	Energy shaping passivity
IDA	Interconnection and damping assignment
PCS	Power conditioning system
RPG	Renewable power generation
MG	Microgrid
VSC	Voltage source converter
CSC	Current source converter
PCH	Port-controlled Hamiltonian
GC	Grid-connected
DG	Diesel generator
WG	Wind generation
SOC	State of charge

Variables, Parameters, and Constants

P_W, P_G, P_m, P_D, P_L	The power of WG/utility grid/SMES/DG/load
L_g, L_m	The inductance of the filter inductor/the superconducting coil
R_g	The total resistance of the filter inductor and IGBTs (lumped parameter)
C	The DC bus capacitance of VSC
f, v	Frequency/voltage
f^*, v^*	Frequency/voltage reference
p, q	Real/reactive power
i_{dc}, i_m	The current flowing into the DC capacitor/the superconducting coil
P_E, P_A	The expected/actual power of WG
P_m^*, Q_m^*	The real/reactive power reference of SMES
P_g^*, Q_g^*	The real/reactive power reference of SMES under the GC mode
P_i^*, Q_i^*	The real/reactive power reference of SMES under the islanded mode
$k_{p(vsc)}, k_{i(vsc)}$	The proportional/integral parameter in the PI controller for VSC
$k_{p(dc)}, k_{i(dc)}$	The proportional/integral parameter in the PI controller for DC-DC chopper
k_{pf}, k_{if}	The proportional/integral parameter in additional f-p regulator
k_{pv}, k_{iv}	The proportional/integral parameter in additional v-q regulator
K_{Idq}, K_{Idc}	The integral parameters in ESP-based controller
$i_{g(d,q)}^*, i_{g(d,q)}$	The reference/actual current of SMES (in d-q axis)
ω_0	Fundamental angle frequency
u_{dc}^*, u_{dc}	The reference/actual DC voltage of VSC
u_{gd}, u_{gq}	Grid voltage in d/q axis
$s_{d,q}, s_m$	The duty cycle of the VSC (in d-q axis)/the DC chopper
$u(x), y(x)$	The input/output vector of the PCH model
$J(x), g(x)$	The interconnection matrices
$J_d(x), R_d(x)$	Expected interconnection/dissipation matrix
$H(x), H_d(x)$	The original/expected Hamilton energy function
$J_a(x), R_a(x)$	The interconnection/dissipation matrix to be decided
$H_a(x)$	Hamilton energy function to be decided

1. Kim, Y.S.; Kim, E.S.; Moon, S.I. Distributed generation control method for active power sharing and self-frequency recovery in an islanded microgrid. *IEEE Trans. Power Syst.* **2017**, *32*, 544–551. [CrossRef]
2. Han, H.; Hou, X.; Yang, J.; Wu, J.; Su, M.; Guerrero, J.M. Review of power sharing control strategies for islanding operation of AC microgrids. *IEEE Trans. Smart Grid* **2016**, *7*, 200–215. [CrossRef]
3. Lotfi, H.; Khodaei, A. AC versus DC microgrid planning. *IEEE Trans. Smart Grid* **2017**, *8*, 296–304. [CrossRef]
4. Hou, X.; Sun, Y.; Yuan, W.; Han, H.; Zhong, C.; Guerrero, J.M. Conventional P-ω/Q-V droop control in highly resistive line of low-voltage converter-based AC microgrid. *Energies* **2016**, *9*, 943. [CrossRef]
5. Sreekumar, P.; Khadkikar, V. Direct control of the inverter impedance to achieve controllable harmonic sharing in the islanded microgrid. *IEEE Trans. Ind. Electron.* **2017**, *64*, 827–837. [CrossRef]
6. Liu, Y.; Hou, X.; Wang, X.; Lin, C.; Guerrero, J.M. A coordinated control for photovoltaic generators and energy storages in low-voltage AC/DC hybrid microgrids under islanded mode. *Energies* **2016**, *9*, 651. [CrossRef]
7. Lian, B.; Sims, A.; Yu, D.; Wang, C.; Dunn, R.W. Optimizing LiFePO4 battery energy storage systems for frequency response in the UK system. *IEEE Trans. Sustain. Energy* **2017**, *8*, 385–394. [CrossRef]
8. Guo, L.; Liu, W.; Li, X.; Liu, Y.; Jiao, B.; Wang, W.; Wang, C.; Li, F. Energy management system for stand-alone wind-powered-desalination microgrid. *IEEE Trans. Smart Grid* **2016**, *7*, 1079–1087. [CrossRef]
9. Tang, X.; Hu, X.; Li, N.; Deng, W.; Zhang, G. A novel frequency and voltage control method for islanded microgrid based on multienergy storages. *IEEE Trans. Smart Grid* **2016**, *7*, 410–419. [CrossRef]
10. Tang, Y.; Mu, C.; He, H. SMES-based damping controller design using fuzzy-grhdp considering transmission delay. *IEEE Trans. Appl. Supercond.* **2016**, *26*, 5701206. [CrossRef]

11. Zheng, Z.; Xiao, X.; Li, C.; Chen, Z.; Zhang, Y. Performance evaluation of SMES system for initial and steady voltage sag compensations. *IEEE Trans. Appl. Supercond.* **2016**, *26*, 5701105. [CrossRef]

12. Song, M.; Shi, J.; Liu, Y.; Xu, Y.; Hu, N.; Tang, Y.; Ren, L.; Li, J. 100 kJ/50 kW HTS SMES for micro-grid. *IEEE Trans. Appl. Supercond.* **2015**, *25*, 5700506. [CrossRef]

13. Ali, M.H.; Wu, B.; Dougal, R.A. An overview of SMES applications in power and energy systems. *IEEE Trans. Sustain. Energy* **2010**, *1*, 38–47. [CrossRef]

14. Molina, M.G.; Mercado, P.E. Power flow stabilization and control of microgrid with wind generation by superconducting magnetic energy storage. *IEEE Trans. Power Electron.* **2011**, *26*, 910–922. [CrossRef]

15. Wan, Y.; Zhao, J. H∞ control of single-machine infinite bus power systems with superconducting magnetic energy storage based on energy-shaping and backstepping. *IET Control Theory Appl.* **2013**, *7*, 757–764. [CrossRef]

16. Dechanupaprittha, S.; Sakamoto, N.; Hongesombut, K.; Watanabe, M.; Mitani, Y.; Ngamroo, I. Design and analysis of robust SMES controller for stability enhancement of interconnected power system taking coil size into consideration. *IEEE Trans. Appl. Supercond.* **2009**, *19*, 2019–2022. [CrossRef]

17. Davari, M.; Mohamed, Y.A.R.I. Dynamics and robust control of a grid-connected VSC in multiterminal DC grids considering the instantaneous power of DC-and AC-side filters and DC grid uncertainty. *IEEE Trans. Power Electron.* **2016**, *31*, 1942–1958. [CrossRef]

18. Ngamroo, I.; Supriyadi, A.C.; Dechanupaprittha, S.; Mitani, Y. Stabilization of tie-line power oscillations by robust SMES in interconnected power system with large wind farms. In Proceedings of the 2009 Transmission & Distribution Conference & Exposition: Asia and Pacific (T&D Asia), Seoul, Korea, 26–30 October 2009; pp. 1–4.

19. Iqbal, M.; Bhatti, A.I.; Ayubi, S.I.; Khan, Q. Robust parameter estimation of nonlinear systems using sliding-mode differentiator observer. *IEEE Trans. Ind. Electron.* **2011**, *58*, 680–689. [CrossRef]

20. Liu, F.; Mei, S.; Xia, D.; Ma, Y.; Jiang, X.; Lu, Q. Experimental evaluation of nonlinear robust control for SMES to improve the transient stability of power systems. *IEEE Trans. Energy Convers.* **2004**, *19*, 774–782. [CrossRef]

21. Tan, Y.L.; Wang, Y. Stability enhancement using SMES and robust nonlinear control. In Proceedings of the 1998 International Conference on Energy Management and Power Delivery (EMPD), Singapore, 3–5 March 1998; pp. 171–176.

22. Vachirasricirikul, S.; Ngamroo, I. Improved H 2/H∞ control-based robust PI controller design of SMES for suppression of power fluctuation in microgrid. In Proceedings of the 2014 International Electrical Engineering Congress (iEECON), Pattaya, Thailand, 19–21 March 2014; pp. 1–4.

23. Lu, Q.; Sun, Y.; Mei, S. *Nonlinear Control Systems and Power System Dynamics*; Kluwer: Norwell, MA, USA, 2001; pp. 343–372.

24. Yi, H.; Zhuo, F.; Wang, F.; Wang, Z. A digital hysteresis current controller for three-level neural-point-clamped inverter with mixed-levels and prediction-based sampling. *IEEE Trans. Power Electron.* **2016**, *31*, 3945–3957. [CrossRef]

25. Flores-Bahamonde, F.; Valderrama-Blavi, H.; Bosque-Moncusi, J.M.; García, G.; Martínez-Salamero, L. Using the sliding-mode control approach for analysis and design of the boost inverter. *IET Power Electron.* **2016**, *9*, 1625–1634. [CrossRef]

26. Tao, C.-W.; Wang, C.-M.; Chang, C.-W. A design of a DC–AC inverter using a modified ZVS-PWM auxiliary commutation pole and a DSP-based PID-like fuzzy control. *IEEE Trans. Ind. Electron.* **2016**, *63*, 397–405. [CrossRef]

27. Nageshrao, S.P.; Lopes, G.A.D.; Jeltsema, D.; Babuška, R. Port-hamiltonian systems in adaptive and learning control: A survey. *IEEE Trans. Autom. Control* **2016**, *61*, 1223–1238. [CrossRef]

28. Schöberl, M.; Siuka, A. On casimir functionals for infinite-dimensional port-hamiltonian control systems. *IEEE Trans. Autom. Control* **2013**, *58*, 1823–1828. [CrossRef]

29. Dirksz, D.A.; Scherpen, J.M.A. Structure preserving adaptive control of port-hamiltonian systems. *IEEE Trans. Autom. Control* **2012**, *57*, 2880–2885. [CrossRef]

30. Ortega, R.; Garcəa-Canseco, E. Interconnection and damping assignment passivity-based control: A survey. *Eur. J. Control* **2004**, *10*, 432–450. [CrossRef]

31. Nunna, K.; Sassano, M.; Astolfi, A. Constructive interconnection and damping assignment for port-controlled Hamiltonian systems. *IEEE Trans. Autom. Control* **2015**, *60*, 2350–2361. [CrossRef]

32. Kwasinski, A.; Krein, P.T. Passivity-based control of buck converters with constant-power loads. In Proceedings of the 2007 IEEE Power Electronics Specialists Conference (PESC), Orlando, FL, USA, 17–21 June 2007; pp. 259–265.
33. Zeng, J.; Zhang, Z.; Qiao, W. An interconnection and damping assignment passivity-based controller for a DC–DC boost converter with a constant power load. *IEEE Trans. Ind. Appl.* **2014**, *50*, 2314–2322. [CrossRef]
34. Qiu, J.; Hu, C.; Yang, S. PMSM Hamiltonian energy shaping control with parameters self-tuning PID control. In Proceedings of the 2015 34th Chinese Control Conference (CCC), Hangzhou, China, 28–30 July 2015; pp. 4506–4511.
35. Tang, Y.; Yu, H.; Zou, Z. Hamiltonian modeling and energy-shaping control of three-phase AC/DC voltage-source converters. In Proceedings of the 2008 IEEE International Conference on Automation and Logistics (ICAL), Qingdao, China, 1–3 September 2008; pp. 591–595.
36. Yu, H.; Teng, Z.; Yu, J.; Zang, Y. Energy-shaping and passivity-based control of three-phase PWM rectifiers. In Proceedings of the 2012 10th World Congress on Intelligent Control and Automation (WCICA), Beijing, China, 6–8 July 2012; pp. 2844–2848.
37. Santos, G.V.; Cupertino, A.F.; Mendes, V.F.; Seleme, S.I. Interconnection and damping assignment passivity-based control of a PMSG based wind turbine for maximum power tracking. In Proceedings of the 2015 IEEE 24th International Symposium on Industrial Electronics (ISIE), Rio de Janeiro, Brazil, 3–5 June 2015; pp. 306–311.
38. Serra, F.M.; De Angelo, C.H. IDA-PBC control of a single-phase battery charger for electric vehicles with unity power factor. In Proceedings of the 2016 IEEE Conference on Control Applications (CCA), Buenos Aires, Argentina, 19–22 September 2016; pp. 261–266.
39. Serra, F.M.; Angelo, C.H.D.; Forchetti, D.G. IDA-PBC control of a three-phase front-end converter. In Proceedings of the 2012 38th Annual Conference on IEEE Industrial Electronics Society (IECON), Montreal, QC, Canada, 25–28 October 2012; pp. 5203–5208.
40. Liang, Z.; Xie, Z.; Wei, X.; Zhang, H. Computer control design of active power filter based on the energy shaping control principle. In Proceedings of the 2009 WRI World Congress on Computer Science and Information Engineering, Los Angeles, CA, USA, 31 March–2 April 2009; pp. 196–200.
41. Donaire, A.; Junco, S. On the addition of integral action to port-controlled Hamiltonian systems. *Automatica* **2009**, *45*, 1910–1916. [CrossRef]
42. Yazdani, A.; Iravani, R. *Voltage-Sourced Converters in Power Systems: Modeling, Control, and Applications*; John Wiley & Sons: Hoboken, NJ, USA, 2010; pp. 217–226.
43. Casadei, D.; Grandi, G.; Reggiani, U.; Serra, G. Analysis of a power conditioning system for superconducting magnetic energy storage (SMES). In Proceedings of the 1998 IEEE International Symposium on Industrial Electronics (ISIE), Pertoria, South Africa, 7–10 July 1998; pp. 546–551.
44. Nguyen, T.T.; Yoo, H.J.; Kim, H.M. Applying model predictive control to SMES system in microgrids for eddy current losses reduction. *IEEE Trans. Appl. Supercond.* **2016**, *26*, 5400405. [CrossRef]
45. Hamby, D.M. A review of techniques for parameter sensitivity analysis of environmental models. *Environ. Monit. Assess.* **1994**, *32*, 135–154. [CrossRef] [PubMed]

Article

A Time-Efficient Approach for Modelling and Simulation of Aggregated Multiple Photovoltaic Microinverters

Sungmin Park [1,*], Weiqiang Chen [2], Ali M. Bazzi [2] and Sung-Yeul Park [2]

[1]   Department of Electronic & Electrical Engineering, Hongik University, Sejong 30016, Korea
[2]   Department of Electrical & Computer Engineering, University of Connecticut, Storrs, CT 06269, USA;
    weiqiang.chen@uconn.edu (W.C.); bazzi@uconn.edu (A.M.B.); supark@engr.uconn.edu (S.-Y.P.)
*   Correspondence: smpark@hongik.ac.kr; Tel.: +82-44-860-2510

Academic Editor: José Gabriel Oliveira Pinto
Received: 14 February 2017; Accepted: 29 March 2017; Published: 31 March 2017

Abstract: This paper presents a time-efficient modeling and simulation strategy for aggregated microinverters in large-scale photovoltaic systems. As photovoltaic microinverter systems are typically comprised of multiple power electronic converters, a suitable modeling and simulation strategy that can be used for rapid prototyping is required. Dynamic models incorporating switching action may induce significant computational burdens and long simulation durations. This paper introduces a single-matrix-form approach using the average model of a basic microinverter with two power stages consisting of a dc-dc and dc-ac converter. The proposed methodology using a common or intermediate source between two average models of cascaded converters to find the overall average model is introduced and is applicable to many other converter topologies and combinations. It provides better flexibility and simplicity when investigating various power topologies in system-level studies of microinverter and other power electronic systems. A 200 W prototype microinverter is tested to verify the proposed average and dynamic models. Furthermore, MATLAB/Simulink (2010a, Mathworks, Natick, MA, USA) is used to show the improved simulation speed and maintained accuracy of the multiple microinverter configurations when the derived average model is compared to a dynamic switching simulation model.

Keywords: computer simulation; modeling; microinverter; photovoltaic systems; state-space model

---

1. Introduction

Photovoltaic (PV) power is becoming more prevalent as its cost has become more competitive with traditional power sources and it is especially attractive for microgrids. As PV systems become more popular, higher efficiency and reliability are of increasing importance. Researchers have developed many effective designs and methodologies that address several essential aspects to improve the performance of PV systems. Modified or optimized power electronic energy conversion topologies have gained significant interest since power stage configuration has significant impact on the system: For example, recent work has shown that microinverters can yield significant reliability enhancement in PV systems [1]. Another example is shown in [2] where different targets, such as efficiency, reliability and cost, were considered to optimize the design of synchronous rectification (SR) boost topology. A third example is a new cascaded active-front-end converter is proposed in [3] to increase system reliability and decrease cost. A single-stage PV system [4] and a modified dual-stage inverter [5] along with maximum power point tracking (MPPT), grid synchronization, and harmonic reduction, have been implemented. Among all these designs, selecting proper converter topologies and components and evaluating their performance, system modelling is critical to address higher efficiency and reliability needs [6,7]. Advanced control has also been of interest enhance maximum power point tracking (MPPT) in PV systems [8–10]. For example, recent work shown in [11] focused on distributed maximum power point tracking (DMPPT) converter design methodology to deal with the problem of variation in temperature, irradiance and shadowing. Other efficiency and power quality enhancement methods include those that utilize an energy storage device [12] or shunt active power filter [13] on the bases of conventional inverter operation. PV systems can be classified per their connection method between the PV modules and the power conditioning system (PCS). In a conventional string configuration, shown in Figure 1a, which can also be connected with several parallel strings, several series PV modules deliver electrical power to the grid and local ac loads through a central PCS. An example is Nano Grid integration [14], which utilizes a central PCS to control multiple of small distributed generation clusters. However, the central PCS configuration may cause mismatch losses of arrays due to differences in manufacturing, temperature, shading, and degradation conditions among the PV modules, resulting in a less efficient PV system. Also, failure of the PCS affects the reliability of the whole system. On the other hand, the microinverter configuration shown in Figure 1b, also referred to as the module-integrated converter (MIC), uses individual small PCSs mounted on each PV module, allowing a simple "plug and play" installation and more localized control such as independent maximum power point tracking (MPPT) at the individual PV module scale [15–18]. Compared to the centralized PCS configuration, this system is expected to be more reliable with higher energy yield, which justifies its minor cost increase. Comparison based on experimental testing of PCSs is provided in [19], but due to the variations of operating conditions, experimental testing is not convenient or cost effective for one-to-one comparisons between various PCSs and under different conditions.

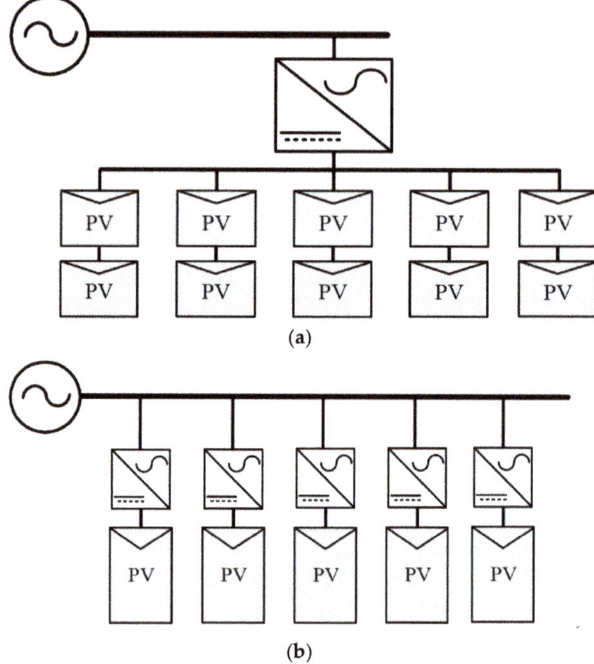

Figure 1. PV system configurations. (**a**) Conventional string type; (**b**) multiple microinverters type.

Modeling and simulation has thus become essential especially in order to choose the proper topology, select appropriate circuit component types and values, evaluate circuit performance, and complete a system design [20,21]. There are several methods to build power electronics models and they usually involve switching power devices and passive power elements such as inductors, capacitors, and resistors. Dynamic models have been used to analyze the performance of an inverter [5], but dynamic models including switching actions may not be suitable for multiple-microinverter aggregated system simulations despite their simple implementation and accurate transient responses. The increase in the number of microinverters in a system simulation yields significant computational burdens and long simulation times [22]. Another approach is to use the average pulse width modulation (PWM) switch model replacing the switches in the dynamic models with time-averaged models represented by voltage and current sources [23]. With the average modeling method, some simulation accuracy is lost but the resulting simulation run time and setup time can be significantly reduced. The state-space average model is employed to ascertain a set of equations describing the system behavior over one switching period, which aids designers in understanding the physical relationship between control parameters and converter states [20]. Using an average model, the transfer function of the system can also be obtained, and larger simulation step sizes can be utilized with minimal loss of accuracy which leads to a faster simulation time. Consequently the state-space average model will be the most competitive modeling method in simulation studies for aggregated multiple microinverters.

A number of papers have examined the state-space averaging model for various power converter topologies in different operating modes [24–33]. The proper analytical averaging model for discontinuous conduction mode (DCM) operation in the dc-dc converter has been studied in [24], which contains implicitly elements to generate a base model applicable to both fixed and variable frequency operations. Many efforts have also been made to develop adequate seamless mode transitions from DCM to continuous conduction mode (CCM) and vice versa in simulation studies of dc-dc converters [25,26]. Furthermore, the issue of parasitic components on these modeling methods is investigated in [27]. Recently the conventional converter configurations such as the boost, buck and fly-back converter tend to be combined and integrated with other power electronics circuits for high efficiency converters and thus their proper averaging model have been required. The boost converter with a voltage multiplier cell has been analyzed in [28] to derive its average model which is complex and requires the use of advanced techniques due to the resonant circuit. Not only small-signal model approaches, but also large-signal model approaches have been conducted to investigate large signal behaviors and capabilities in multiple dc-dc [29] or dc-ac [30] converters where a generalized state-space averaging method employing the Fourier series with time-dependent coefficients. However, these models cannot predict the complete dynamic behavior of these systems. While most papers focus on dc-dc converter modeling, a small number of papers have been presented on average modeling of dc-ac converters to approximate their behavior in grid-connected power electronics such as static synchronous compensators (STATCOM), active power filters [31] and PV applications [32]. It should be noted that references [24–33] have all addressed single-stage power converters and inverters, but average modeling of multiple-stage converters such as microinverters is still open to research along with the adequate simulation strategy that is necessary to improve simulation speed and accuracy of multiple cascaded converters, such as multiple PV microinverters in a microgrid.

The primary objectives of this paper are to introduce the simulation strategy for aggregated multiple microinverters by employing state-space average modeling of multi-stage power converters represented by a single-matrix-form (SMF). SMF can be derived by using the intermediate source model to link to converters with established average models, and thus this approach can facilitate integration of existing average models for cascaded, parallel, and many topologies. Since the proposed simulation strategy is based on the state-space average modeling technique which allows investigating small-signal behavior of the systems, it can capture important dynamics of PV microinverters in smaller power systems such as microgrids even under aggregated multiple converters, but switching transients that slow down simulations are over-ridden. The main advantages of this proposed approach are: (1) achieving a faster simulation time in research on aggregated multiple microinverter systems compared to dynamic switching models of converters; (2) providing better flexibility with easily interchangeable converter models; (3) understanding the relationship of state variables between multi-stage converters; and (4) simple extension to other power electronics conversion systems with multiple-stage configurations. It is important to note that the main purpose of this paper is to introduce this intermediate source methodology in multiple-converter systems which can be extended to various topologies and applications, especially in the area of modeling and control of microgrids to capture finer dynamics with faster simulations that mask switching transients.

This paper starts with descriptions of the proposed simulation strategy and state-space average modeling of microinverters in Section 2. Model validation is carried out by comparing waveforms from simulation results in MATLAB/Simulink with experimental results from a 200 W prototype microinverter board in Section 3. These results are compared in open-loop mode to test the "plant" dynamics using the proposed model, dynamic simulation model, and experiments. The general control strategy for microinverters in PV applications is explained in Section 4, and then the simulation times between the proposed modeling approach and the dynamic model are compared in Section 5. Finally, Section 6 concludes the paper.

2. Proposed Simulation Strategy and State-Space Average Model with a Single Matrix-Form

It is common for PV microinverters to have multiple-stage configurations consisting of dc-dc and dc-ac stages [15,16,34]. The dc-dc converter provides MPPT at the PV panel terminals while the dc-ac converter delivers PV power to the grid or local ac loads in the second stage as shown in Figure 2. The proposed approach to simulate the multiple microinverter system is shown in Figure 3. Several models are required for a PV system simulation—PV module model, power converter model, grid model and local load model. PV cell models are very common in the literature, especially those utilizing the current source and inverse diode configuration [35]. Using this model, a 200 W PV panel model is developed under different irradiance conditions. As for the load model, a simple resistor model is used as the passive Local Load Model in Figure 3 since it yields simpler SMF. If reactive power needs to be consumed or generated by local loads, the resistive load model can be replaced with other inductive and capacitive loads or an RLC combination. The grid model encompasses a stiff single-phase voltage source with fixed voltage and frequency characteristics in series with the grid impedance consisting of a resistor and inductor. By using this model, it is possible to simulate all grid disturbances such as sag, swell, and interruptions, and adjust the voltage source to include voltage harmonics. Details of the converter dc-dc, dc-ac, and combined converter models are presented in this section.

While this paper presents examples of specific dc-dc and dc-ac stages, adjusting the matrices in the average model of each stage to reflect other converter topologies is possible. The methodology proposed in this paper for integrating two average models of dc-dc and dc-ac stages as shown in Figure 4 can thus be followed. A SMF integrated with the dc-dc converter and dc-ac converter models is used for better flexibility when other topologies are used as it directly links the state variables between both energy conversion stages.

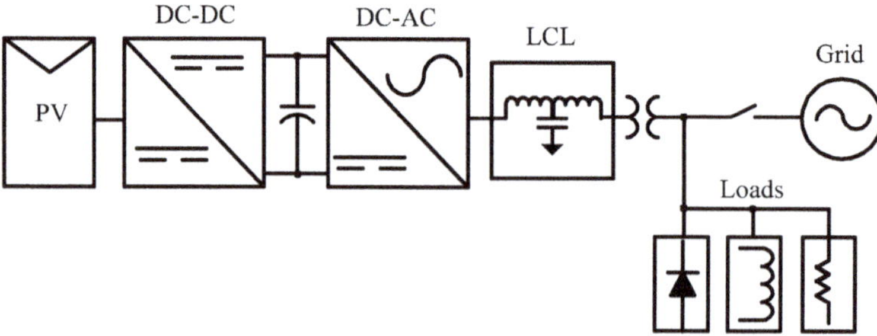

Figure 2. General PCS structure for converting PV power.

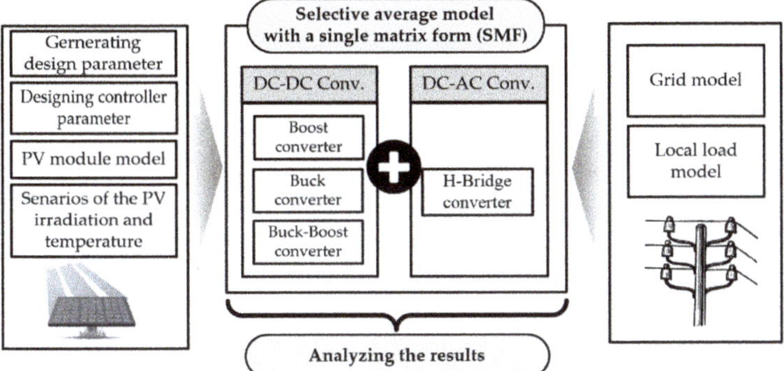

Figure 3. Proposed modeling approach for the multiple microinverters system.

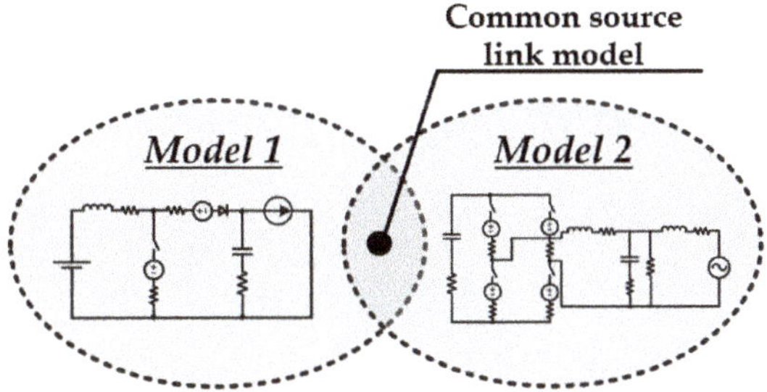

Figure 4. Multi-stage converter model with single matrix form.

2.1. DC-DC Converter Model

State-space average modeling is employed to obtain a set of differential equations applicable to a selected converter topology [20]. The resulting model is expected to combine the dc-dc converter in CCM and dc-ac converter. Those equations are capable of describing the system behavior over one switching period. It is also desirable to include all parasitic effects in the state-space average model to predict the dynamic behavior and frequency response of the microinverter accurately [27,36]. The state space equation of the converter in CCM can be expressed as:

$$\dot{\mathbf{x}}(t) = (q(t)\mathbf{A}_1 + (1 - q(t))\mathbf{A}_2)\mathbf{x}(t) + (q(t)\mathbf{B}_1 + (1 - q(t))\mathbf{B}_2)\mathbf{u}(t) \tag{1}$$

where $q(t)$ is the switching function corresponding to the power device's on/off states, A_k and B_k are the system matrices where $k = 1$ or 2 depending on the switch status, and $\mathbf{u}(t)$ is the input vector. The low-frequency components of state variables such as the inductor currents, capacitor voltages and the output duty can be modeled by averaging over the switching period T_s and can be defined as:

$$\bar{\mathbf{x}}(t) = \frac{1}{T_s}\int_t^{t+T_s} \mathbf{x}(\tau)d\tau \tag{2}$$

where the bar symbol denotes the so-called fast average or true average of a state variable $\mathbf{x}(t)$. Using (2), the state average equation can be expressed as:

$$\dot{\bar{\mathbf{x}}}(t) = (d(t)\mathbf{A}_1 + (1 - d(t))\mathbf{A}_2)\bar{\mathbf{x}}(t) + (d(t)\mathbf{B}_1 + (1 - d(t))\mathbf{B}_2)\bar{\mathbf{u}}(t) \tag{3}$$

where $d(t)$ is the duty cycle function.

As an example, Figure 5 depicts the grid-connected microinverter, consisting of a non-ideal boost converter and H-bridge inverter. To obtain the state-space average model with a SMF, these converters can be considered separately as shown in Figures 6 and 7. The circuit equations of these converters for turn-on and turn-off periods can be derived by applying Kirchoff's voltage and current laws, and then the system matrices according to the switching status can be obtained. From the dc-dc boost converter shown in Figure 6, the state-space average model can be defined as:

$$
\begin{aligned}
\dot{\bar{x}}_d(t) &= \mathbf{A}_d\bar{x}_d(t) + \mathbf{B}_d\bar{u}_d(t) \\
&= (D_{dc}(t)\mathbf{A}_{d1} + (1 - D_{dc}(t))\mathbf{A}_{d2})\bar{x}_d(t) \\
&\quad + (D_{dc}(t)\mathbf{B}_{d1} + (1 - D_{dc}(t))\mathbf{B}_{d2})\bar{u}_d(t)
\end{aligned}
\tag{4}
$$

where $\mathbf{x}_d = [i_{pv}\ v_{Cdc}]^\mathrm{T}$, $\mathbf{u}_d = [v_{pv}\ v_m\ v_d\ i_{dc}]^\mathrm{T}$ and i_{pv} is the dc inductor current, v_{Cdc} is the link capacitor voltage, v_{pv} is the output voltage of the PV module, v_m is the drain-to-source voltage of the boost switch, v_d is the forward voltage of the diode, i_{dc} is the current in the dc-bus and D_{dc} is the duty ratio of the boost converter. Matrices $\mathbf{A}_{d1}$, $\mathbf{A}_{d2}$, $\mathbf{B}_{d1}$ and $\mathbf{B}_{d2}$ when the switch is on and off are presented in Appendix A. Using these matrices and (4), matrices $\mathbf{A}_d$ and $\mathbf{B}_d$ presented with averaged values during one-sample time are:

$$
\mathbf{A}_d =
\begin{bmatrix}
-\dfrac{R_{Ldc} + D_{dc}R_{Mdc} + (R_{Cdc} + R_d)(1 - D_{dc})}{L_{dc}} & -\dfrac{1 - D_{dc}}{L_{dc}} \\[2ex]
\dfrac{1 - D_{dc}}{C_{dc}} & 0
\end{bmatrix}
\tag{5}
$$

$$
\mathbf{B}_d =
\begin{bmatrix}
\dfrac{1}{L_{dc}} & -\dfrac{D_{dc}}{L_{dc}} & -\dfrac{1 - D_{dc}}{L_{dc}} & \dfrac{(1 - D_{dc})R_{Cdc}}{L_{dc}} \\[2ex]
0 & 0 & 0 & -\dfrac{1}{C_{dc}}
\end{bmatrix}
\tag{6}
$$

where L_{dc} is the boost inductor value, C_{dc} is the dc link capacitor value, R_{Ldc} and R_{Cdc} are the parasitic resistors of the passive and active components, R_{Mdc} and R_d are the switch on-resistance, respectively.

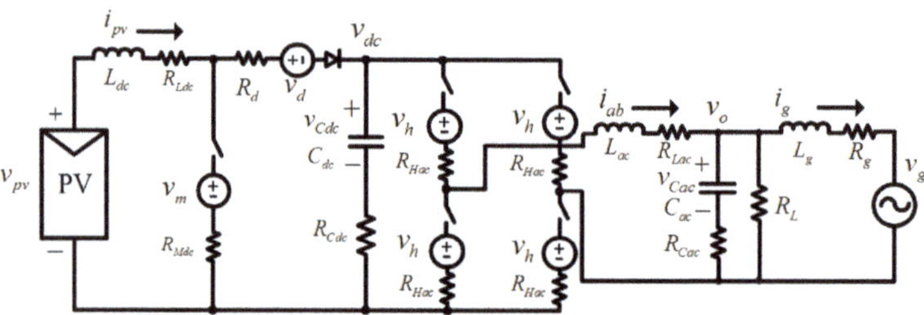

Figure 5. Circuit model for deriving the average model.

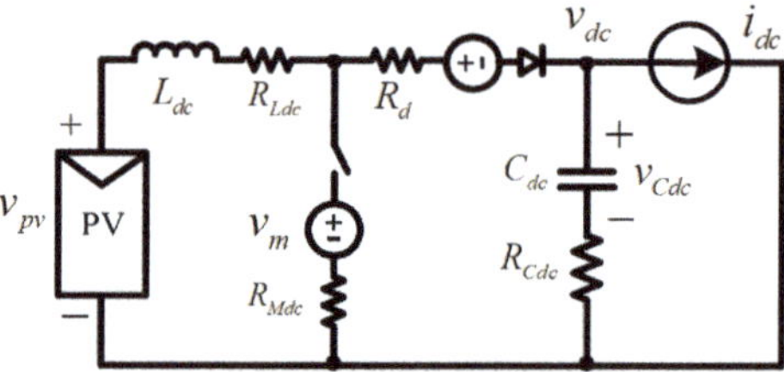

Figure 6. Circuit model for the dc-dc converter.

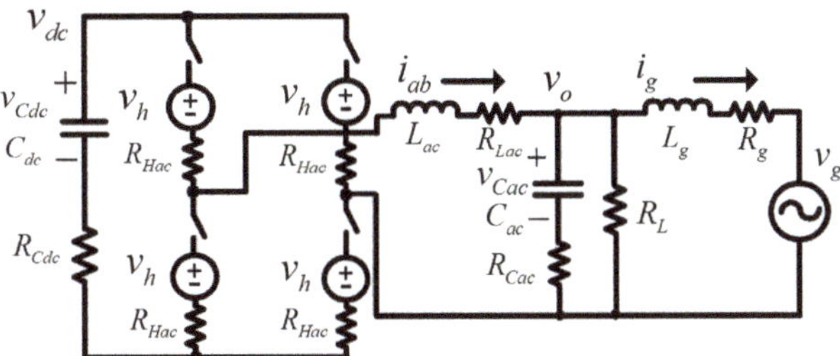

Figure 7. Circuit model for the dc-ac converter.

2.2. DC-AC Converter Model

In a procedure similar to that discussed in the previous section, the average model of the H-bridge converter shown in Figure 7 is defined as:

$$
\begin{aligned}
\dot{\bar{\mathbf{x}}}_a(t) &= \mathbf{A}_a \bar{\mathbf{x}}_a(t) + \mathbf{B}_a \bar{\mathbf{u}}_a(t) \\
&= (D_{ac}(t)\mathbf{A}_{a1} + (1 - D_{ac}(t))\mathbf{A}_{a2})\bar{\mathbf{x}}_a(t) \\
&\quad + (D_{ac}(t)\mathbf{B}_{a1} + (1 - D_{ac}(t))\mathbf{B}_{a2})\bar{\mathbf{u}}_a(t)
\end{aligned}
\tag{7}
$$

where $\mathbf{x}_a = [i_{ab}\ v_{Cac}\ i_g]^{\mathrm{T}}$, $\mathbf{u}_a = [v_{Cdc}\ v_h\ v_g]^{\mathrm{T}}$, i_{ab} is the ac inductor current, v_{Cac} is the capacitor voltage in the LC filter, i_g is the grid current, v_h is the drain-to-source voltage of the H-bridge switch, v_g is the grid voltage and D_{ac} is the duty ratio of the H-bridge converter.

It is worthwhile to mention that the relationship between the average duty and modulation index (M) of the H-bridge converter can be expressed as $D_{ac}(t) = 0.5 + \text{M}\cdot\sin(\omega t)$ where ω is the grid frequency since the average duty is between 0 and 1. Matrices $\mathbf{A}_{a1}$, $\mathbf{A}_{a2}$, $\mathbf{B}_{a1}$ and $\mathbf{B}_{a2}$ are shown in the Appendix A. Using these matrices and (7), matrices A_a and B_a for averaged values during one-sample time are given by:

$$
\mathbf{A}_a =
\begin{bmatrix}
-\left(\dfrac{2R_{Hac} + R_{Lac} + R_{Cac}\Psi + R_{Cdc}}{L_{ac}} \right) & -\dfrac{\Psi}{L_{ac}} & \dfrac{R_{Cac}\Psi}{L_{ac}} \\[2ex]
\dfrac{\Psi}{C_{ac}} & -\dfrac{\Psi}{C_{ac}R_L} & -\dfrac{\Psi}{C_{ac}} \\[2ex]
\dfrac{R_{Cac}\Psi}{L_g} & \dfrac{\Psi}{L_g} & -\dfrac{\Psi}{L_g}\left(R_{Cac} + R_g + \dfrac{R_{Cac}R_g}{R_L} \right)
\end{bmatrix}
\tag{8}
$$

where $\Psi = \dfrac{R_L}{R_L + R_{Cac}}$, and:

$$
\mathbf{B}_a =
\begin{bmatrix}
\dfrac{(2D_{ac}-1)}{L_{ac}} & -\dfrac{2}{L_{ac}} & 0 \\[2ex]
0 & 0 & 0 \\[2ex]
0 & 0 & -\dfrac{1}{L_g}
\end{bmatrix}
\tag{9}
$$

where, L_{ac} is the ac inductor value, C_{ac} is the capacitor value in the LC filter, R_{Lac} and R_{Cac}, are the parasitic resistors of the passive components, R_L is the local load resistance, and L_g and R_g reflect the grid impedance, respectively.

2.3. Combined Average Model

As the next step, using the relationship between the dc-bus current and output current, the two state-space average models obtained in (4) and (7) can be combined into SMF since the common source in the previously derived two state-space average model is the dc-bus current. Thus, the dc-bus current can be represented by ac current and ac duty in the H-bridge converter as:

$$\bar{i}_{dc} = (2D_{ac} - 1)\bar{i}_{ab} \tag{10}$$

It is notable in (10) that the common source model causes the dc-bus current to include ac ripples that are twice the ac output frequency, which is also reflected to dc-bus voltage. Moreover, the dc voltage v_{dc} and the output ac voltage v_o can be expressed as:

$$\bar{v}_{dc} = R_{Cdc}\bar{i}_{pv} + \bar{v}_{Cdc} - R_{Cdc}\bar{i}_{ab} \tag{11}$$

$$\bar{v}_o = \left(\frac{R_L R_{Cac}}{R_L + R_{Cac}}\right)\bar{i}_{ab} + \left(\frac{R_L}{R_L + R_{Cac}}\right)\bar{v}_{Cac} - \left(\frac{R_L R_{Cac}}{R_L + R_{Cac}}\right)\bar{i}_g \tag{12}$$

The SMF state-space representation of the dc-dc and dc-ac converters in the circuits of Figure 5 is:

$$\frac{d\bar{\mathbf{x}}(t)}{dt} = \mathbf{A}\bar{\mathbf{x}}(t) + \mathbf{B}\bar{\mathbf{u}}(t) \tag{13}$$

$$\bar{\mathbf{y}}(t) = \mathbf{C}\bar{\mathbf{x}}(t) \tag{14}$$

The state variables **x**, **u** and **y** in Figure 5, for grid connected mode are defined as:

$$\bar{\mathbf{x}} = \begin{bmatrix} \bar{i}_{pv} & \bar{v}_{Cdc} & \bar{i}_{ab} & \bar{v}_{Cac} & \bar{i}_g \end{bmatrix}^T \tag{15}$$

$$\bar{\mathbf{u}} = \begin{bmatrix} \bar{v}_{pv} & \bar{v}_m & \bar{v}_d & \bar{v}_h & \bar{v}_g \end{bmatrix}^T \tag{16}$$

$$\bar{\mathbf{y}} = \begin{bmatrix} \bar{i}_{pv} & \bar{v}_{dc} & \bar{i}_{ab} & \bar{v}_o & \bar{i}_g \end{bmatrix}^T \tag{17}$$

Finally, using (5)–(12), matrices for the state, input, and output variables including the boost converter and H-bridge converter can be derived as:

$$\mathbf{A} = \begin{bmatrix} & \mathbf{A}_d & & (1-D_{dc})(2D_{ac}-1)R_{Cdc}/L_{dc} & 0 & 0 \\ & & & -(2D_{ac}-1)/C_{dc} & 0 & 0 \\ 0 & (2D_{ac}-1)/L_{ac} & & & & \\ 0 & 0 & & \mathbf{A}_a & & \\ 0 & 0 & & & & \end{bmatrix} \tag{18}$$

$$\mathbf{B} = \begin{bmatrix} 1/L_{dc} & -D_{dc}/L_{dc} & -(1-D_{dc})/L_{dc} & 0 & 0 \\ 0 & 0 & 0 & 0 & 0 \\ 0 & 0 & 0 & -2/L_{ac} & 0 \\ 0 & 0 & 0 & 0 & 0 \\ 0 & 0 & 0 & 0 & -1/L_g \end{bmatrix} \tag{19}$$

$$\mathbf{C} = \begin{bmatrix} 1 & 0 & 0 & 0 & 0 \\ R_{Cdc} & 1 & -R_{Cdc} & 0 & 0 \\ 0 & 0 & 1 & 0 & 0 \\ 0 & 0 & R_{Cac}\Psi & \Psi & -R_{Cac}\Psi \\ 0 & 0 & 0 & 0 & 1 \end{bmatrix} \tag{20}$$

Moreover, the state-space average model for stand-alone mode can be obtained in (18)–(20) by setting the grid impedance (L_g and R_g) as infinity and the grid voltage (v_g) as zero.

3. Model Validation

The state-space average model of the microinverter in SMF obtained in the previous section can be validated by simulating the model and comparing waveforms with a dynamic switching based model, as well as experimental testing. A 200 W proto-type microinverter board is used for experimental tests. Since long input wires are used for the experimental test in the dc-dc converter side, extra input resistance R_{in} is added to R_{Ldc} in the simulation. For experimental validation, an open-loop control scenario with resistive load is considered.

This validation procedure is intended to validate the plant dynamics in the average model, dynamic switching model, and experiments without control effects where the plant is the microinverter power stage. The main system parameters for simulation and experimental testing are summarized in Table 1. The input voltage is 30 V to mimic that of a solar PV panel. Figure 8 shows the dynamic behavior of the dc inductor current, dc voltage, and ac output current waveforms under open-loop duty disturbances of the dc-dc converter while the modulation index of the H-bridge converter remains constant at 0.935 in both MATLAB simulation models. At 0.35 ms, a 1% step change from 0.800 to 0.792 in duties is applied. Simulation results show excellent correspondence between the proposed average model based on (18)–(20) and the dynamic switching model where the current and voltage values from established average model are in the middle of the dynamic model switching ripple. An experiment was carried out to ensure that the simulated models (average and dynamic) match a real setup under the same test conditions where IRFP4332PbF and MUR840G are used for power devices. As shown in Figure 9 and Table 2, experimental results are in agreement with simulation results in Figure 8, with the exception of the settling time which is sensitive to various experimental set-up characteristics such as the printed circuit board and line parasitic elements.

Table 1. Simulation and experimental parameters for model validation.

Parameters	Value	Parameters	Value
v_{pv}	25~40 V	L_{dc}	2.63 mH
v_g	110 Vrms	R_{Ldc}	0.15 Ω
V_d	0.975 V	L_{ac}	1.3 mH
V_m	0.2 V	R_{Lac}	0.075 Ω
V_h	0.2 V	C_{dc}	680 μF
R_{in}	0.2 Ω	R_{Cdc}	0.03 Ω
R_{Mdc}	0.029 Ω	C_{ac}	1 μF
R_d	0.02 Ω	R_{Cac}	0.01 Ω
R_{Hac}	0.029 Ω	L_g	3 mH
R_L	62.5 Ω	R_g	0.01 Ω

Table 2. Experimental results for model validation.

Parameters	Simulation Results	Experimental Results	Errors (%)
v_{dc} (D_{dc} = 80.0%)	141 V	138 V	2.1%
v_{dc} (D_{dc} = 79.2%)	134 V	132 V	1.5%
i_{pv} (D_{dc} = 80.0%)	4.2 A	4.1 A	2.4%
i_{pv} (D_{dc} = 79.2%)	3.8 A	3.6 A	5.3%
i_{ab} (D_{dc} = 80.0%)	1.39 Arms	1.32 Arms	5.0%
i_{ab} (D_{dc} = 79.2%)	1.31 Arms	1.26 Arms	3.8%
Settling time (v_{dc})	0.07 s	0.09 s	−28.6%
Undershoot (v_{dc})	132 V	127 V	3.8%
Settling time (i_{pv})	0.07 s	0.09 s	−28.6%
Undershoot (i_{pv})	1.9 A	1.8 A	5.3%

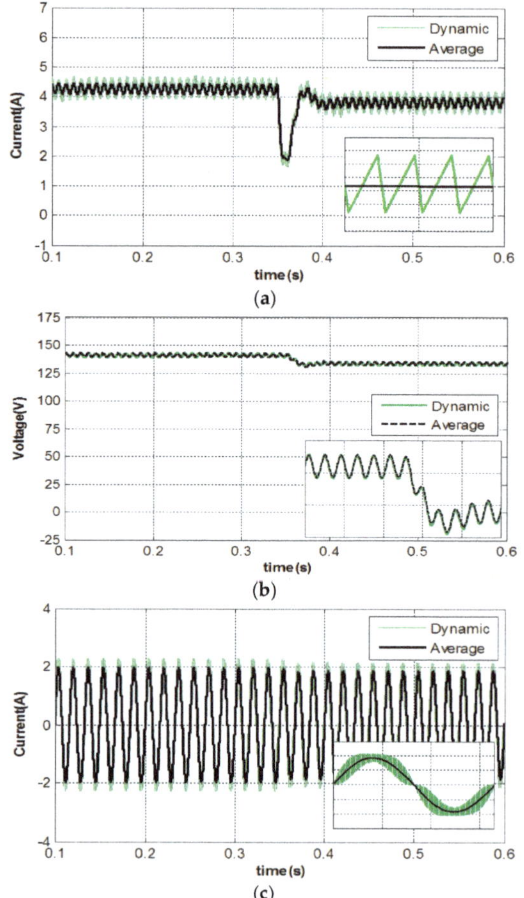

Figure 8. Simulation results for a single microinverter under open-loop control with resistive load. (**a**) PV current; (**b**) dc-bus voltage; (**c**) ac output current.

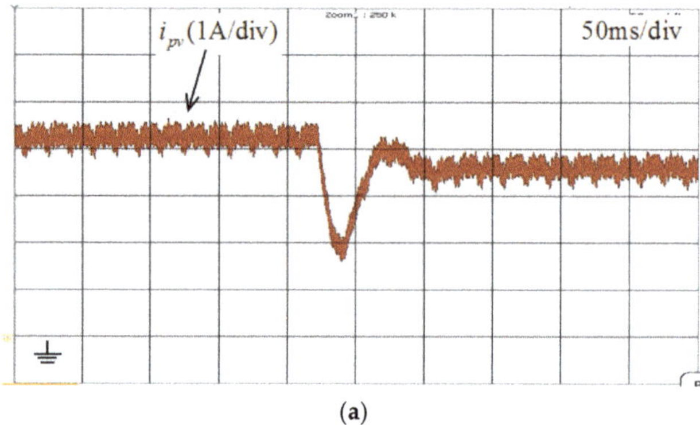

(**a**)

Figure 9. *Cont.*

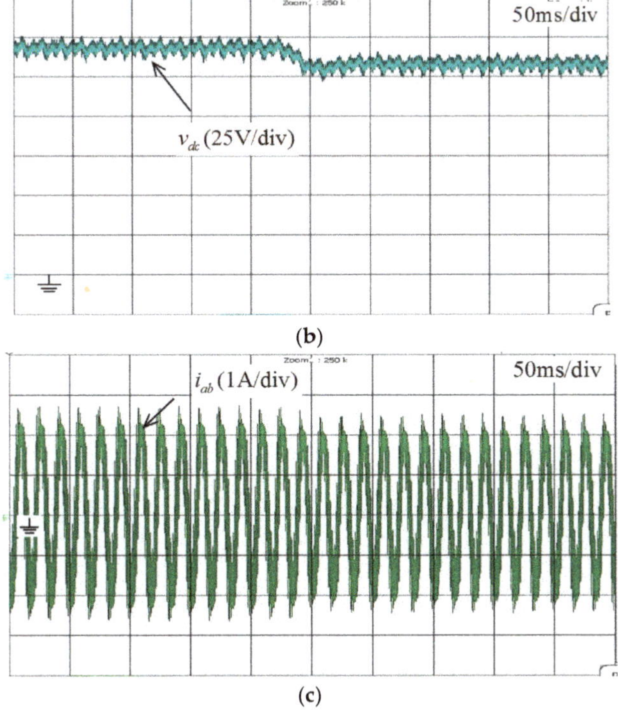

Figure 9. Experimental results for a single microinverter under open-loop control with resistive load. (**a**) PV current; (**b**) dc-bus voltage; (**c**) ac output current.

4. Control Strategy for PV Microinverter

After validating the proposed model with dynamic simulations and experimental tests, it is desired to study the effect of integrating multiple PV microinverters in a grid-connected large-scale system. For this purpose, two control block diagrams for the dc-dc and dc-ac converters are developed as shown in Figure 10. Cascaded voltage regulators with current regulators are used to maintain the PV bus voltage and dc link voltage, while maintaining operation at the maximum power point in the I-V curve of the PV modules by using MPPT algorithms [37,38] and delivering the generated PV power to the grid or the local ac load with a sinusoidal current waveform in phase with the grid voltage [36,39,40] (assume that the direction of the output current is going into the grid).

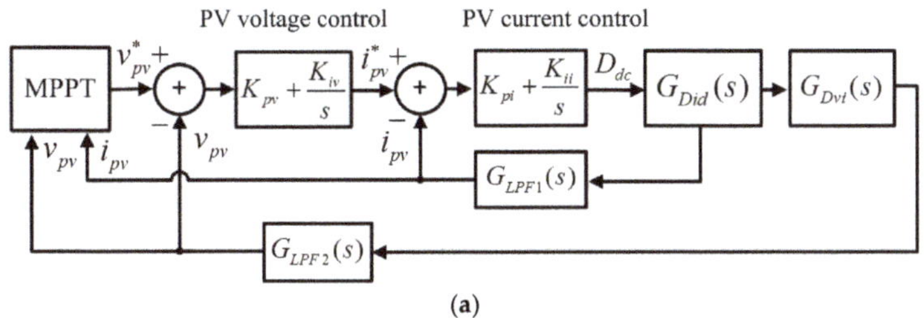

Figure 10. *Cont.*

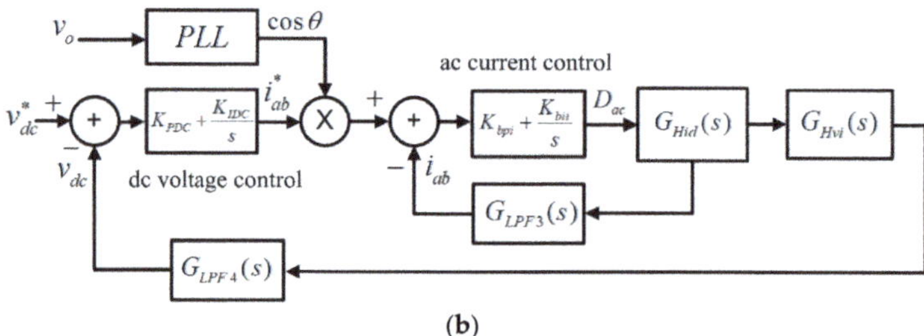

(**b**)

Figure 10. Control block diagram for the microinverter. (**a**) dc-dc converter; (**b**) dc-ac converter.

Figure 11 shows simulation results with closed-loop control. The PV voltage is changed from 30 V to 40 V with a current reference of 5 A while the dc-ac converter controls the dc-bus voltage with a reference of 200 V. Waveforms from the established average model match the averaged values of the switching model as expected.

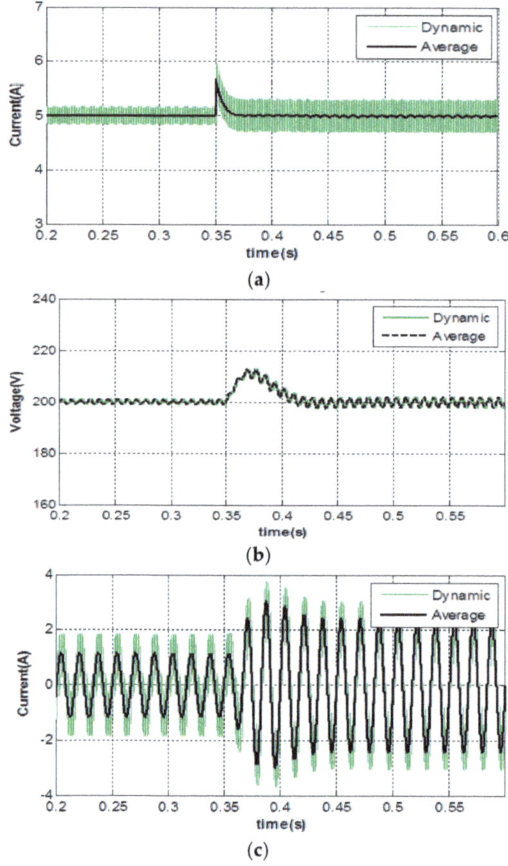

Figure 11. Simulation results with closed-loop control. (**a**) PV currents; (**b**) dc-bus voltage; (**c**) ac output current.

5. Aggregated Microinverter in PV Applications

In order to validate the fast simulation time of the proposed modeling approach in a multiple microinverter configuration, a simulation is carried out in MATLAB/Simulink with 20 parallel microinverters at 200 W per PV panel for a total power output of 4 kW. Figures 12 and 13 show a high-level block diagram of the simulated system and the PV module model emulating PV-TD195HA6 from Mitsubishi electric corporation (Tokyo, Japan) [41], respectively. The grid voltage is at 110 Vrms, 60 Hz, and the local loads are zero ($R_L \to \infty$). This means that all generated power from the PV modules is sent to the grid.

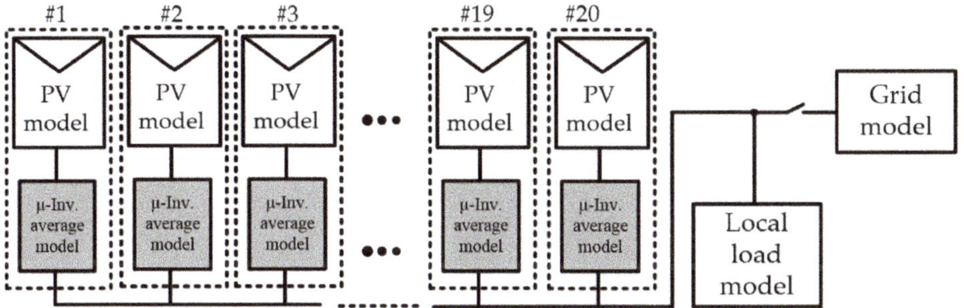

Figure 12. Simulation structure for aggregated microinverters.

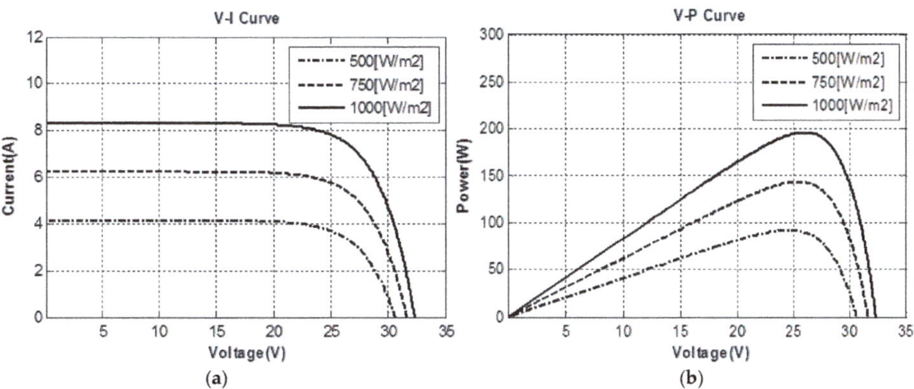

Figure 13. Electrical characteristics of the simulated PV modules under different irradiance conditions. (a) V-I curve; (b) V-P curve.

Partial shading is applied to some panels in order to demonstrate the simulation flexibility. Figure 14a,b show the current waveforms of the second and third microinverters among the 20 inverters where these panels have irradiance values of 900 W/m^2 and 800 W/m^2, respectively, and the irradiance of the 18 other microinverters is 1000 W/m^2. Another waveform shown in Figure 14a,b is that of #1 which is at 1000 W/m^2. Figure 14c shows the total grid current from the 20 microinverters. Resulting waveforms from using the dynamic switching model under the same conditions are shown in Figure 15. As expected, taking the average of waveforms generated using the dynamic model eliminates switching effects and the results match those in Figure 15 from the proposed average model.

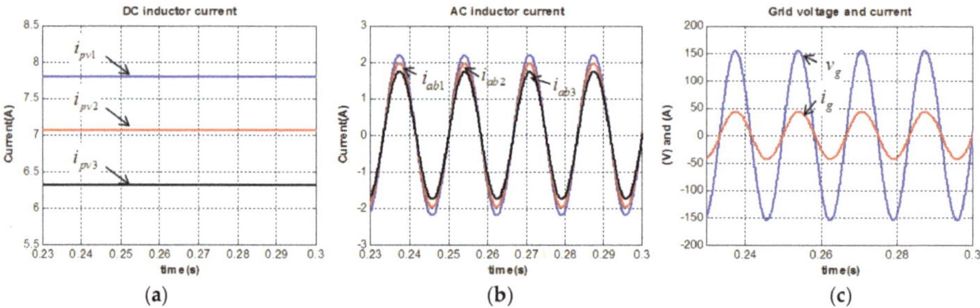

Figure 14. Simulation waveforms using the state-space average model. (**a**) PV currents; (**b**) the output currents; (**c**) grid voltage and current.

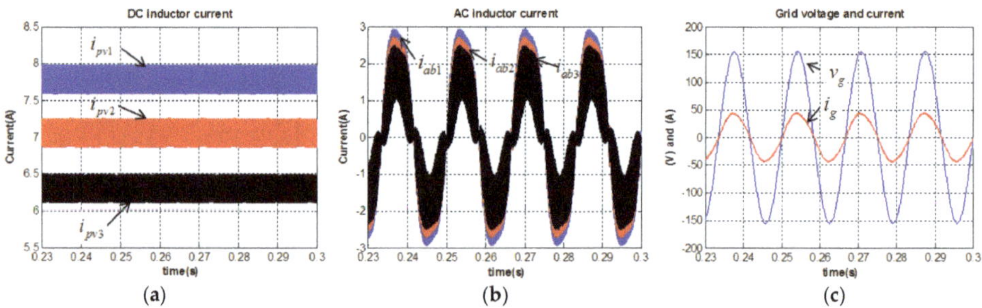

Figure 15. Simulation waveforms using the dynamic switching model. (**a**) PV currents; (**b**) the output currents; (**c**) grid voltage and current.

Generally, the simulation runtime is highly dependent on the computer's performance and specifications, where the simulation tool is run, and solver options. In this paper, a computer (Thinkpad T420S, Lenovo) with an Intel core i5 processor, 16 GB memory, and MATLAB 2010(a) with fixed-time step and Ode4 (Runge-Kutta) are utilized as a computational medium for simulations. Simulation runtime is compared for both average and dynamic models as shown in Figure 16.

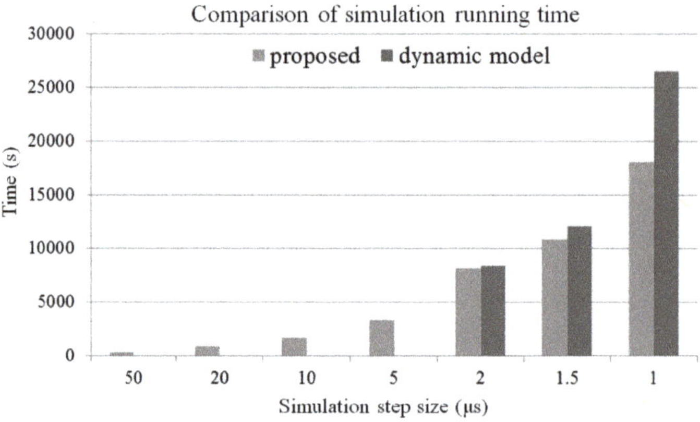

Figure 16. Comparison of the simulation runtime

The proposed method allows the use of a smaller step size compared to the dynamic model, thus reducing the total simulation runtime. Note that time steps ≥ 5.0 µs resulted in erratic results in the dynamic model while the average model still performed well and all signals in the simulation model were as expected.

6. Conclusions

As photovoltaic microinverter systems are typically comprised of multiple power electronic converters, a suitable modeling and simulation strategy for rapid prototyping is required. The paper has presented a time-efficient modeling and simulation strategy in multiple converter systems which can be extended to various topologies and applications, especially in the area of modeling and control of microgrids to capture finer dynamics with faster simulations. The simulation model represented by a single-matrix-form using a common or intermediate source between two average models of cascaded converters to find the overall average model was derived by employing a basic microinverter composed of a boost stage and an H-bridge. Results show that the proposed state-space average model matches experiments and dynamic simulations. The proposed model also provides significant reduction in simulation runtime and simulation set-up time with aggregated microinverters compared to the dynamic model. Larger step sizes were shown to be possible when using the average model to achieve both accurate and fast simulation convergence. Finally, the advantages of the proposed modeling and approach can be summarized as follows: (1) achieving a faster simulation time in research on aggregated multiple microinverter systems compared to dynamic switching models of converters; (2) providing better flexibility with easily interchangeable converter models; (3) understanding the relationship of state variables between multi-stage converters; and (4) simple extension to other power electronics conversion systems with multiple-stage configurations. It is expected that the proposed model will have valuable use in very large-scale PV farm simulations and can be extended for other cascaded power electronic topologies.

Acknowledgments: This research was supported by Basic Science Research Program through the National Research Foundation of Korea (NRF) funded by the Ministry of Science, ICT & Future Planning (2017R1C1B2008200).

Author Contributions: Sungmin Park developed the proposed modeling approach and validated it through simulation results. Weiqiang Chen performed the experiments and analyzed the data. Ali M. Bazzi conceived the main concept of the modelling approach. Sung-Yeul Park guided and revised the manuscript. All authors contributed to the writing of the manuscript, and approved the final manuscript.

Conflicts of Interest: The authors declare no conflict of interest.

Abbreviations

PV	photovoltaic
PCS	power conditioning system
MIC	module-integrated converter
MPPT	maximum power point tracking
PWM	pulse width modulation
CCM	continuous conduction mode
DCM	discontinuous conduction mode
SMF	single-matrix-form

Appendix A

In the dc-dc boost converter shown in Figure 6, when the switch turns on, the dc inductor current and capacitor voltage as state variables can be obtained as:

$$\frac{di_{pv}}{dt} = -\frac{1}{L_{dc}}\left(R_{Ldc} + R_{Mdc}\right) + \frac{1}{L_{dc}}v_{pv} - \frac{1}{L_{dc}}v_m \tag{A1}$$

$$\frac{dv_{Cdc}}{dt} = -\frac{1}{C_{dc}}i_{dc} \tag{A2}$$

When the switch turns off, the inductor current and capacitor voltage as natural state variables can also be obtained as:

$$\frac{di_{pv}}{dt} = -\frac{1}{L_{dc}}(R_{Ldc} + R_d + R_{Cdc})i_{pv} - \frac{1}{L_{dc}}v_{Cdc} - \frac{1}{L_{dc}}v_d + \frac{1}{L_{dc}}v_{pv} + \frac{R_{Cdc}}{L_{dc}}i_{dc} \tag{A3}$$

$$\frac{dv_{Cdc}}{dt} = \frac{1}{C_{dc}}i_{pv} - \frac{1}{C_{dc}}i_{dc} \tag{A4}$$

Thus, the matrix $\mathbf{A_{d1}}$, $\mathbf{A_{d2}}$, $\mathbf{B_{d1}}$ and $\mathbf{B_{d2}}$ for the dc-dc converter is given by:

$$\mathbf{A}_{d1} = \begin{bmatrix} -\frac{(R_{Ldc}+R_{Mdc})}{L_{dc}} & 0 \\ 0 & 0 \end{bmatrix} \tag{A5}$$

$$\mathbf{B}_{d1} = \begin{bmatrix} \frac{1}{L_{dc}} & -\frac{1}{L_{dc}} & 0 & 0 \\ 0 & 0 & 0 & -\frac{1}{C_{dc}} \end{bmatrix} \tag{A6}$$

$$\mathbf{A}_{d2} = \begin{bmatrix} -\frac{(R_{Ldc}+R_{Cdc}+R_d)}{L_{dc}} & -\frac{1}{L_{dc}} \\ \frac{1}{C_{dc}} & 0 \end{bmatrix} \tag{A7}$$

$$\mathbf{B}_{d2} = \begin{bmatrix} \frac{1}{L_{dc}} & 0 & -\frac{1}{L_{dc}} & \frac{R_{Cdc}}{L_{dc}} \\ 0 & 0 & 0 & -\frac{1}{C_{dc}} \end{bmatrix} \tag{A8}$$

From (4) and (A5)–(A8), the state-space averaging model for the dc-dc boost converter over a switching period can be written as:

$$\begin{bmatrix} \dot{i}_{pv} \\ \dot{v}_{Cdc} \end{bmatrix} = \begin{bmatrix} -\frac{R_{Ldc}+D_{dc}R_{Mdc}+(R_{Cdc}+R_d)(1-D_{dc})}{L_{dc}} & -\frac{1-D_{dc}}{L_{dc}} \\ \frac{1-D_{dc}}{C_{dc}} & 0 \end{bmatrix} \begin{bmatrix} i_{pv} \\ v_{Cdc} \end{bmatrix} + \begin{bmatrix} \frac{1}{L_{dc}} & -\frac{D_{dc}}{L_{dc}} & -\frac{1-D_{dc}}{L_{dc}} & \frac{(1-D_{dc})R_{Cdc}}{L_{dc}} \\ 0 & 0 & 0 & -\frac{1}{C_{dc}} \end{bmatrix} \begin{bmatrix} v_{pv} \\ v_m \\ v_d \\ i_{dc} \end{bmatrix} \tag{A9}$$

Similarly, as for the H-bridge converter shown in Figure 7, when the switches S1, S4 of the H-bridge converter turn on, the ac inductor current and capacitor voltage in the LC filter as state variables can be obtained as:

$$\frac{di_{ab}}{dt} = \frac{-2R_{Hac}-R_{Lac}-\frac{R_{Cac}R_L}{R_{Cac}+R_L}-R_{Cdc}}{L_{ac}}i_{ab} - \frac{1}{L_{ac}}\left(\frac{R_L}{R_L+R_{Cac}}\right)v_{Cac} + \frac{1}{L_{ac}}\left(\frac{R_LR_{Cac}}{R_L+R_{Cac}}\right)i_g - \frac{2}{L_{ac}}v_h + \frac{1}{L_{ac}}v_{Cdc} \tag{A10}$$

$$\frac{dv_{Cac}}{dt} = \frac{1}{C_{ac}}\left(\frac{R_L}{R_L+R_{Cac}}\right)i_{ab} - \frac{1}{(R_L+R_{Cac})C_{ac}}v_{Cac} - \frac{R_L}{(R_L+R_{Cac})C_{ac}}i_g \tag{A11}$$

$$\frac{di_g}{dt} = \frac{1}{L_g}\left(\frac{R_LR_{Cac}}{R_L+R_{Cac}}\right)i_{ab} + \frac{R_L}{(R_L+R_{Cac})L_g}v_{Cac} - \frac{R_LR_{Cac}+R_LR_g+R_{Cac}R_g}{(R_L+R_{Cac})L_g}i_g - \frac{1}{L_g}v_g \tag{A12}$$

When the switches S2, S3 of the H-bridge converter turn on, the ac inductor current and capacitor voltage in the LC filter as natural state variables can be obtained as:

$$\frac{di_{ab}}{dt} = \frac{-2R_{Hac}-R_{Lac}-\frac{R_{Cac}R_L}{R_{Cac}+R_L}-R_{Cdc}}{L_{ac}}i_{ab} - \frac{1}{L_{ac}}\left(\frac{R_L}{R_L+R_{Cac}}\right)v_{Cac} + \frac{1}{L_{ac}}\left(\frac{R_LR_{Cac}}{R_L+R_{Cac}}\right)i_g - \frac{2}{L_{ac}}v_h - \frac{1}{L_{ac}}v_{Cdc} \tag{A13}$$

$$\frac{dv_{Cac}}{dt} = \frac{1}{C_{ac}}\left(\frac{R_L}{R_L+R_{Cac}}\right)i_{ab} - \frac{1}{(R_L+R_{Cac})C_{ac}}v_{Cac} - \frac{R_L}{(R_L+R_{Cac})C_{ac}}i_g \tag{A14}$$

$$\frac{di_g}{dt} = \frac{1}{L_g}\left(\frac{R_LR_{Cac}}{R_L+R_{Cac}}\right)i_{ab} + \frac{R_L}{(R_L+R_{Cac})L_g}v_{Cac} - \frac{R_LR_{Cac}+R_LR_g+R_{Cac}R_g}{(R_L+R_{Cac})L_g}i_g - \frac{1}{L_g}v_g \tag{A15}$$

Thus, the matrix $\mathbf{A_{a1}}$, $\mathbf{A_{a2}}$, $\mathbf{B_{a1}}$ and $\mathbf{B_{a2}}$ for the dc-ac converter is given by:

$$\mathbf{A}_{a1} = \begin{bmatrix} -\frac{\left(\begin{array}{c}2R_{Hac}+R_{Lac}\\+R_{Cac}\Psi+R_{Cdc}\end{array}\right)}{L_{ac}} & -\frac{\Psi}{L_{ac}} & \frac{R_{Cac}\Psi}{L_{ac}} \\ \frac{\Psi}{C_{ac}} & -\frac{\Psi}{C_{ac}R_L} & -\frac{\Psi}{C_{ac}} \\ \frac{R_{Cac}\Psi}{L_g} & \frac{\Psi}{L_g} & -\frac{\Psi}{L_g}\left(\begin{array}{c}R_{Cac}+R_g\\+\frac{R_{Cac}R_g}{R_L}\end{array}\right) \end{bmatrix} \tag{A16}$$

$$\mathbf{B}_{a1} = \begin{bmatrix} \frac{1}{L_{ac}} & -\frac{2}{L_{ac}} & 0 \\ 0 & 0 & 0 \\ 0 & 0 & -\frac{1}{L_g} \end{bmatrix} \tag{A17}$$

$$\mathbf{A}_{a2} = \begin{bmatrix} -\dfrac{\left(\begin{array}{c} 2R_{Hac} + R_{Lac} \\ +R_{Cac}\Psi + R_{Cdc} \end{array}\right)}{L_{ac}} & -\dfrac{\Psi}{L_{ac}} & \dfrac{R_{Cac}\Psi}{L_{ac}} \\ \dfrac{\Psi}{C_{ac}} & -\dfrac{\Psi}{C_{ac}R_L} & -\dfrac{\Psi}{C_{ac}} \\ \dfrac{R_{Cac}\Psi}{L_g} & \dfrac{\Psi}{L_g} & -\dfrac{\Psi}{L_g}\left(\begin{array}{c} R_{Cac} + R_g \\ +\dfrac{R_{Cac}R_g}{R_L} \end{array}\right) \end{bmatrix} \tag{A18}$$

$$\mathbf{B}_{a2} = \begin{bmatrix} -\frac{1}{L_{ac}} & -\frac{2}{L_{ac}} & 0 \\ 0 & 0 & 0 \\ 0 & 0 & -\frac{1}{L_g} \end{bmatrix} \tag{A19}$$

From (7) and (A16)–(A19), the state-space averaging model for the dc-ac converter can be written as:

$$\begin{bmatrix} \dot{i}_{ab} \\ \dot{v}_{Cac} \\ \dot{i}_g \end{bmatrix} = \begin{bmatrix} \dfrac{-2R_{Hac}-R_{Lac}-\frac{R_{Cac}R_L}{R_{Cac}+R_L}-R_{Cdc}}{L_{ac}} & -\dfrac{1}{L_{ac}}\left(\dfrac{R_L}{R_L+R_{Cac}}\right) & \dfrac{1}{L_{ac}}\left(\dfrac{R_L R_{Cac}}{R_L+R_{Cac}}\right) \\ \dfrac{1}{C_{ac}}\left(\dfrac{R_L}{R_L+R_{Cac}}\right) & -\dfrac{1}{C_{ac}}\left(\dfrac{1}{R_L+R_{Cac}}\right) & -\dfrac{1}{C_{ac}}\left(\dfrac{R_L}{R_L+R_{Cac}}\right) \\ \dfrac{1}{L_g}\left(\dfrac{R_L R_{Cac}}{R_L+R_{Cac}}\right) & \dfrac{1}{L_g}\left(\dfrac{R_L}{R_L+R_{Cac}}\right) & -\dfrac{1}{L_g}\left(\dfrac{R_L R_{Cac}+R_L R_g+R_{Cac}R_g}{R_L+R_{Cac}}\right) \end{bmatrix} \begin{bmatrix} i_{ab} \\ v_{Cac} \\ i_g \end{bmatrix}$$
$$+ \begin{bmatrix} \dfrac{(2D_{ac}-1)}{L_{ac}} & -\dfrac{2}{L_{ac}} & 0 \\ 0 & 0 & 0 \\ 0 & 0 & -\dfrac{1}{L_g} \end{bmatrix} \begin{bmatrix} v_{Cdc} \\ v_h \\ v_g \end{bmatrix} \tag{A20}$$

1. Scholten, D.M.; Ertugrul, N.; Soong, W.L. Micro-inverters in small scale PV systems: A review and future directions. In Proceedings of the Australasian Universities Power Engineering Conference (AUPEC), Hobart, Australia, 29 September–3 October 2013.
2. Adinolfi, G.; Graditi, G.; Siano, P.; Piccolo, A. Multi-Objective Optimal Design of Photovoltaic Synchronous Boost Converters Assessing Efficiency, Reliability and Cost Savings. *IEEE Trans. Ind. Inform.* **2015**, *11*, 1038–1048. [CrossRef]
3. Liu, C.; Sun, P.; Lai, J.S.; Ji, Y.; Wang, M.; Chen, C.L.; Cai, G. Cascade dual-boost/buck active-front-end converter for intelligent universal transformer. *IEEE Trans. Ind. Electron.* **2012**, *59*, 4671–4680.
4. Libo, W.; Zhengming, Z.; Jianzheng, L. A Single-Stage Three-Phase Grid-Connected Photovoltaic System with Modified MPPT Method and Reactive Power Compensation. *IEEE Trans. Energy Convers.* **2007**, *22*, 881–886. [CrossRef]
5. Martins, D.C. Analysis of a three-phase grid-connected PV power system using a modified dual-stage inverter. *ISRN Renew. Energy* **2013**, *2013*, 406312. [CrossRef]
6. Koran, A.; LaBella, T.; Lai, J. High efficiency photovoltaic source simulator with fast response time for solar power conditioning systems evaluation. *IEEE Trans. Power Electron.* **2014**, *29*, 1285–1297. [CrossRef]
7. Seyedmahmoudian, M.; Mekhilef, S.; Rahmani, R.; Yusof, R.; Renani, E.T. Analytical modeling of partially shaded photovoltaic systems. *Energies* **2013**, *6*, 128–144. [CrossRef]
8. Seyedmahmoudian, M.; Horan, B.; Rahmani, R.; Maung Than Oo, A.; Stojcevski, A. Efficient photovoltaic system maximum power point tracking using a new technique. *Energies* **2016**, *9*, 147. [CrossRef]
9. Aurilio, G.; Balato, M.; Graditi, G.; Landi, C.; Luiso, M.; Vitelli, M. Fast hybrid MPPT technique for photovoltaic applications: Numerical and Experimental validation. *Adv. Power Electron.* **2014**, *2014*, 125918. [CrossRef]
10. Graditi, G.; Adinolfi, G.; Femia, N.; Vitelli, M. Comparative analysis of synchronous rectification boost and diode rectification boost converter for DMPPT applications. In Proceedings of the 2011 IEEE International Symposium on Industrial Electronics (ISIE), Gdansk, Poland, 27–30 June 2011; pp. 1000–1005.
11. Graditi, G.; Adinolfi, G.; Tina, G.M. Photovoltaic optimizer boost converters: Temperature influence and electro-thermal design. *Appl. Energy* **2014**, *115*, 140–150. [CrossRef]

12. Liu, Y.; Hu, Z.; Zhang, Z.; Suo, J.; Liang, K. Research on the three-phase photovoltaic grid-connected control strategy. *Energy Power Eng.* **2013**, *5*, 31–35. [CrossRef]

13. Bojoi, R.I.; Limongi, L.R.; Roiu, D.; Tenconi, A. Enhanced power quality control strategy for single-phase inverters in distributed generation systems. *IEEE Trans. Power Electron.* **2011**, *26*, 798–806. [CrossRef]

14. Cecati, C.; Khalid, H.; Tinari, M.; Adinolfi, G.; Graditi, G. Hybrid DC Nano grid with renewable energy sources and modular DC/DC LLC converter building block. *IET Power Electron.* **2016**, *10*, 536–544. [CrossRef]

15. Hu, H.; Harb, S.; Kutkut, N.; Batarseh, I.; Shen, Z.J. A review of power decoupling techniques for microinverters with three different decoupling capacitor locations in PV systems. *IEEE Trans. Power Electron.* **2013**, *28*, 2711–2726. [CrossRef]

16. Kim, K.A.; Lertburapa, S.; Xu, C.; Krein, P.T. Efficiency and cost trade-offs for designing module integrated converter photovoltaic systems. In Proceedings of the IEEE Power and Energy Conference at Illinois, Champaign, IL, USA, 24–25 February 2012; pp. 1–7.

17. Chen, B.; Gu, B.; Zhang, L.; Zahid, Z.U.; Lai, J.-S.; Liao, Z.; Ho, R. A high-efficiency MOSFET transformer less inverter for nonisolated microinverter applications. *IEEE Trans. Power Electron.* **2015**, *30*, 3610–3622. [CrossRef]

18. Lin, F.J.; Chiang, H.C.; Chang, J.K. Modeling and controller design of PV micro inverter without using electrolytic capacitors and input current sensors. *Energies* **2016**, *9*, 993. [CrossRef]

19. Spertino, F.; Graditi, G. Power conditioning units in grid-connected photovoltaic systems: A comparison with different technologies and wide range of power ratings. *Sol. Energy* **2014**, *108*, 219–229. [CrossRef]

20. Middlebrook, R.D. Small-signal modeling of pulse-width modulated switched-mode power converters. *Proc. IEEE* **1988**, *76*, 343–354. [CrossRef]

21. Li, W.; Bélanger, J. An equivalent circuit method for modelling and simulation of modular multilevel converters in real-time HIL test bench. *IEEE Trans. Power Deliv.* **2016**, *31*, 2401–2409. [CrossRef]

22. Abramovitz, A. An approach to average modeling and simulation of switch-mode systems. *IEEE Trans. Educ.* **2011**, *54*, 509–517. [CrossRef]

23. Dijk, E.; Spruijt, J.N.; O'Sullivan, D.M.; Klaassens, J.B. PWM-switch modeling of DC-DC converters. *IEEE Trans. Power Electron.* **1995**, *10*, 659–665. [CrossRef]

24. Sun, J.; Mitchell, D.M.; Greuel, M.F.; Krein, P.T.; Bass, R.M. Averaged modeling of PWM converters operating in discontinuous conduction mode. *IEEE Trans. Power Electron.* **2001**, *16*, 482–492.

25. Suntio, T. Unified average and small-signal modeling of direct-on-time control. *IEEE Trans. Ind. Electron.* **2006**, *53*, 287–295. [CrossRef]

26. Davoudi, A.; Jatskevich, J.; Rybel, T. Numerical state-space average-value modeling of PWM DC-DC converters operating in DCM and CCM. *IEEE Trans. Power Electron.* **2006**, *21*, 1003–1012. [CrossRef]

27. Davoudi, A.; Jatskevich, J. Parasitics realization in state-space average-value modeling of PWM DC-DC converters using an equal area method. *IEEE Trans. Circuits Syst. I Reg. Pap.* **2007**, *54*, 1960–1967. [CrossRef]

28. Dupont, F.H.; Rech, C.; Gules, R.; Pinheiro, J.R. Reduced-order model and control approach for the boost converter with a voltage multiplier cell. *IEEE Trans. Power Electron.* **2013**, *28*, 3395–3404. [CrossRef]

29. Emadi, A. Modeling and analysis of multiconverter DC power electronic systems using the generalized state-space averaging method. *IEEE Trans. Ind. Electron.* **2004**, *51*, 661–668. [CrossRef]

30. Emadi, A. Modeling of power electronic loads in AC distribution systems using the generalized state-space averaging method. *IEEE Trans. Ind. Electron.* **2004**, *51*, 992–1000. [CrossRef]

31. Bina, M.Y.; Bhat, A. Averaging technique for the modeling of STATCOM and active filters. *IEEE Trans. Power Electron.* **2008**, *23*, 723–734. [CrossRef]

32. Figueres, E.; Garcera, G.; Sandia, J.; Gonzalez-Espin, F.; Rubio, J.C. Sensitivity study of the dynamics of three-phase photovoltaic inverters with an LCL grid filter. *IEEE Trans. Ind. Electron.* **2009**, *56*, 706–717. [CrossRef]

33. Krein, P.T.; Bentsman, J.; Bass, R.; Lesieutre, B.C. On the use of averaging for the analysis of power electronic systems. *IEEE Trans. Power Electron.* **1990**, *5*, 182–190. [CrossRef]

34. Meneses, D.; Blaabjerg, F.; García, O.; Cobos, J.A. Review and comparison of step-up transformerless topologies for photovoltaic AC-module application. *IEEE Trans. Power Electron.* **2013**, *28*, 2649–2663. [CrossRef]

35. Rodriguez, C.; Amaratunga, G.A.J. Analytic solution to the photo-voltaic maximum power point problem. *IEEE Trans. Circuits Syst. I Reg. Pap.* **2007**, *54*, 2054–2060. [CrossRef]

36. Park, S.M.; Bazzi, A.; Park, S.Y.; Chen, W. A time-efficient modeling and simulation strategy for aggregated multiple microinverters in large-scale PV systems. In Proceedings of the IEEE Applied Power Electronic Conference, Fort Worth, TX, USA, 16–20 March 2014.
37. Subudhi, B.; Pradhan, R. A comparative study on maximum power point tracking techniques for photovoltaic power systems. *IEEE Trans. Sustain. Energy* **2013**, *4*, 89–98. [CrossRef]
38. Mastromauro, R.A.; Liserre, M.; Dell'Aquila, A. Control issues in single-stage photovoltaic systems: MPPT, current and voltage control. *IEEE Trans. Ind. Inform.* **2012**, *8*, 241–254. [CrossRef]
39. Blaabjerg, F.; Teodorescu, R.; Liserre, M.; Timbus, A.V. Overview of control and grid synchronization for distributed power generation systems. *IEEE Trans. Ind. Electron.* **2006**, *53*, 1398–1409. [CrossRef]
40. Ma, L.; Sun, K.; Teodorescu, R.; Guerrero, J.M.; Jin, X. An integrated multifunction DC/DC converter for PV generation systems. In Proceedings of the IEEE International Symposium on Industrial Electronics, Bari, Italy, 4–7 July 2010.
41. Mitsubishi Electric Corporation. Available online: http://www.mitsubishielectric.com/bu/solar/pv_modules/pdf/L-175-0-B8492-A.pdf (accessed on 14 February 2017).

Article

Application of a Continuous Particle Swarm Optimization (CPSO) for the Optimal Coordination of Overcurrent Relays Considering a Penalty Method

Abdul Wadood [†], Chang-Hwan Kim [†], Tahir Khurshiad, Saeid Gholami Farkoush and Sang-Bong Rhee *

Department of Electrical Engineering, Yeungnam University, 280 Daehak-Ro, Gyeongsan, Gyeongsangbuk-do 38541, Korea; wadood@ynu.ac.kr (A.W.); kranz@ynu.ac.kr (C.-H.K.); tahir@ynu.ac.kr (T.K.); saeid_gholami@ynu.ac.kr (S.G.F.)
* Correspondence: rrsd@yu.ac.kr
† These authors contributed equally to this work.

Received: 28 February 2018; Accepted: 29 March 2018; Published: 9 April 2018

Abstract: In an electrical power system, the coordination of the overcurrent relays plays an important role in protecting the electrical system by providing primary as well as backup protection. To reduce power outages, the coordination between these relays should be kept at the optimum value to minimize the total operating time and ensure that the least damage occurs under fault conditions. It is also imperative to ensure that the relay setting does not create an unintentional operation and consecutive sympathy trips. In a power system protection coordination problem, the objective function to be optimized is the sum of the total operating time of all main relays. In this paper, the coordination of overcurrent relays in a ring fed distribution system is formulated as an optimization problem. Coordination is performed using proposed continuous particle swarm optimization. In order to enhance and improve the quality of this solution a local search algorithm (LSA) is implanted into the original particle swarm algorithm (PSO) and, in addition to the constraints, these are amalgamated into the fitness function via the penalty method. The results achieved from the continuous particle swarm optimization algorithm (CPSO) are compared with other evolutionary optimization algorithms (EA) and this comparison showed that the proposed scheme is competent in dealing with the relevant problems. From further analyzing the obtained results, it was found that the continuous particle swarm approach provides the most globally optimum solution.

Keywords: continuous particle swarm optimization (CPSO); overcurrent relay coordination (OCR); time multiplier setting (TMS); power system protection

---

1. Introduction

In an electric power system, the overcurrent relays provide both primary and backup protection to maintain the system reliable and healthy, and to ensure minimum exposure to the healthy portion of the system. In a transmission system, sometimes this type of relay is used as backup protection when deploying distance protection as the primary protection. The overcurrent relays are a useful choice for telecommunication networks, industries, and consumers in terms of offering fast protection and from an economic point of view. Once the overcurrent relays fulfil the requirements of reliability, sensitivity, and selectivity they can operate quickly and without mal-operating issues by isolating the faulty portion of the system with the help of circuit breakers [1]. The main aim for the coordination of relays is to set the relays so that the whole system receives both the primary and backup protection if the level of load and fault current are known. Therefore, accurate coordination of the overcurrent relays is necessary. Coordination amongst these OCRs should be maintained at the optimum value to reduce the overall operating time and ensure that the minimum number of power outages occur during fault conditions. Hence, the coordination of OCRs is formulated as a minimization problem [2,3]. In the research undertaken by [4], a technical survey was presented for the optimal coordination of time overcurrent relays. In the past, different optimization algorithms have been investigated in order to deal with the problem of optimum coordination of relays. A linear programming technique was applied in the study by [5]. In experiments described in [6], a random search method was used. In [7], an evolutionary algorithm was applied for the first time to deal with the relay coordination problem. In various papers [8–13], different versions of a genetic algorithm have been applied to improve the convergence characteristics of genetic algorithms overall. A different version of the particle swarm optimization technique has also been suggested in order to achieve the optimum values for relay coordination [14–19]. In the research of [20], five different versions of the Modified Differential Evolution Algorithm (MDE) were proposed to solve the coordination problem in order to figure out the best performance of the MDE with respect to other algorithms. The artificial bee colony technique was utilized in research by [21]. In [22], a hybrid evolutionary algorithm based on tabu search was used for optimum relay coordination. Reference [23] suggested that the coordination of the overcurrent relay is formulated as a mixed integer nonlinear program by employing a population-based heuristic search algorithm, which regards optimization process as a search of optimal solution by a seeker population. In [24], to improve and enhance the quality of solution the chaos theory is incorporated to the conventional firefly algorithm to figure out the coordination problem. The coordination problem is solved using different metaheuristic method is [25]. Reference [26] suggests that the coordination of the overcurrent relay is formulated as a nonlinear program by employing a group search optimization algorithm. In the paper by [27], the hybridized symbiotic organism search method is used to deal with the directional overcurrent relay problem. The authors found the optimal coordination of directional overcurrent relays by using a firefly algorithm [28]. In the paper by [29], distributed system protection coordination is investigated based on directional overcurrent protection with an inverse time characteristic. Reference [30] designed a protection scheme for a distribution system considering different modes of operation. In [31], a robust optimization strategy was proposed for the protection of micro-grids using microprocessor-based relays. The major deficiency of the previously proposed methods, including both the mathematical and evolutionary approaches, is the possibility of convergence to values which may not be a global optimum but rather are stuck at a local optimum. To solve this issue, a continuous particle swarm optimization (CPSO) is examined in this study for the optimum coordination of overcurrent relays, and is compared with continuous genetic algorithms (CGA), genetic algorithms (GA), the dual and two phase simplex methods (DSM, TPSM), fire fly algorithms (FA), and chaotic firefly algorithms (CFA).

This paper proposes that a continuous particle swarm optimization (CPSO), using the penalty method, can achieve the optimum coordination of overcurrent relays. To enhance and improve the quality of the solutions the CPSO scheme is assimilated, thereby preventing the search from becoming stuck in local minima. The proposed algorithm has a high search capability and convergence speed as compared to other evolutionary techniques, and these characteristics make the population member of the CPSO more discriminative in finding the optimal solution than other evolutionary techniques. To the best of the authors' knowledge the CPSO has not previously been implemented for the optimization of the overcurrent relay coordination problem, investigation into which is presented in this paper. The main aim of this paper is to find the optimal values of the Time Multiplier Setting (TMS) to minimize the operating time of overcurrent relays under several constraints, such as relay setting and backup constraints.

2. Formulation of the Overcurrent Relay Problem

In a multi- or single-source loop system the coordination of the directional overcurrent relays is formulated as an optimization problem. However, the coordination problem has an objective function and constraints that should satisfy the distinct constraints:

$$min f = \sum_{j=1}^{n} w_j T_{j,k} \tag{1}$$

where the parameters w_i and T_i are the weight and operation of the relays. For all the relays the value $w_i = 1$. Therefore, the characteristic curve for the operating relay R_i can be selected from the IEC standards and could be defined as follows:

$$T_{op} = TMS_i \left(\frac{\alpha}{\left(\frac{If_j}{Ip_j} \right)^k - 1} \right) \tag{2}$$

where α and k are constant parameters which define the relay characteristic and are assumed as $\alpha = 0.14$ and $k = 0.02$ for a normal inverse type relay. The variables TMS_i and Ip_j are the Time Multiplier Setting and pickup current of the ith relay, while If_i is the fault current flowing through relay R_i:

$$PSM = \frac{I_{fj}}{Ip_j} \tag{3}$$

where Ip_j is the primary pickup current and PSM stands for the Plug Setting Multiplier:

$$T_{op} = TMS_i \left(\frac{\alpha}{(PSM)^k - 1} \right) \tag{4}$$

The above problem, as represented in Equation (4), is a nonlinear problem in nature. By taking the plug setting of the relay as fixed, and the operating time of the relays, and a linear function of the TMS, the coordination can be expressed as linear programming. In linear programming only the TMS is continuous while the rest of the parameters are constant, so Equation (4) becomes:

$$T_{op} = a_\rho (TMS_i) \tag{5}$$

where:

$$a_\rho = \frac{\alpha}{(PSM)^k - 1} \tag{6}$$

Hence the objective function can be formulated as:

$$min f = \sum_{i=1}^{n} a_\rho (TMS_i) \tag{7}$$

Constraints

The objective of minimizing the total operating times of the relays should be achieved under two types of constraints; the constraints of the relay setting parameters and coordination constraints.

The first type consists of the boundaries of the TMS, whereas the second type is pertinent to the coordination of the primary and backup relays. The boundary on the relay setting parameters imposes constraints (8) on the choice of relay parameters:

$$TMS_i^{min} \leq TMS_i \leq TMS_i^{max} \tag{8}$$

The second type of constraint is pertinent to the adjustment of the operating time of the primary and backup relays. Since the fault would be sensed by both the primary and the backup relays simultaneously, in order to avoid mal-operation the coordination time interval (CTI) should be taken into account in the tripping action. The CTI includes the sum of the operating time of the circuit breaker (CB) associated with the primary relay, the overshoot time of the backup relay, and an appropriate safety margin. Therefore, according to Figure 1, the backup relay R_j should operate later than the primary relay R_i [32]. This is critical for satisfying the requirement for selectivity of the primary and backup relays. The coordination constraint is defined as follows:

$$T_j \geq T_i + \text{CTI} \tag{9}$$

where T_i and T_j are the operating times of the primary and backup relays, respectively, for a fault occurring in front of the primary relay. The value of the CTI could vary from 0.2 to 0.5 s, depending upon different circumstances and factors.

Figure 1. A single end radial distribution system.

3. Continuous Particle Swarm Optimization

PSO is one of the EA techniques that is, in basic terms, inspired by the swarm behavior associated with fish schooling and bird flocking [33–37]. The task of the PSO algorithm is to control the agents or particle population, and these agents or particles are called a "swarm". Each particle serves as the possible result of the objective function under consideration. The particles in the population can memorize the current position with respect to the objective function, and the best position and velocity as visited during its fish flying tour or bird flocking tour in the group will be referred to as the "personal best position" (p_{best}). The tour will find the best position among all the possible solutions and this is referred to as the "global best position" (g_{best}). Some features of the continuous particle swarm optimization are found in the literature [38–40]. In the literature survey, PSO in its standard form has been widely used for unconstrained optimization projects. In this paper two modifications have been added to the authentic PSO algorithm; the penalty method and the initialization of PSO with a local search. As CPSO basically solves the unconstrained optimization problem, to convert the relay coordination problem into an unconstrained optimization problem a new objective function is defined via the penalty method. It is probable that the PSO executes such a bearded exploration that it generates immature results, which is an insufficient solution. To produce a more satisfactory solution, it is necessary to insert a local search algorithm into the original PSO. In this paper, the author inserted a local search alongside the global best position vector. The CPSO method proposed for the coordination of the overcurrent relay problem deals with each particle position on three key vectors; velocity (v_i), position (x_i) and open facility (y_i), where v_i expresses the ith velocity vector in the swarm, x_i represents the ith position vector in swarm, and y_i expresses the opening facilities determined based on the position vector (x_i). For N number of facility problems, each particle contains N number of dimensions so the position vector x_i approaches the continuous value for N facilities, $x_i = [x_{i1}, x_{i2}, \ldots, x_{in}]$, although it does not describe a candidate solution to calculate the total cost. To create a candidate solution, the position vector is reciprocated to a binary variable, $y_i \leftarrow x_i$. Specifically, a discrete set is formed from the continuous set for generating a candidate solution. The fitness of the ith particle is calculated with the help of the open facility vector (y_i). The personal best fitness value of the ith particle p_i is expressed by f_i^{bp}. At the beginning the personal best vector is computerized with the position vector ($p_i = x_i$), where p_i is the position vector and the fitness values of the personal bests are equal to the fitness of the positions, $fi^k = f(x_i{}^k)$. Then, the best particle in the whole swarm with respect to the fitness value is selected with the named global best and expressed as g_i. The global best, $fb^k = f(y \leftarrow g)$ can be achieved by finding the best of the personal bests over the whole swarm, $fi^k = min\{f(x_i{}^k)\}$, with its corresponding position vector xg which is to be used for $g = xg$ and $yg = y$ where yg express the y_i vector of the global best. Then, the velocity of the individual particle is updated based on its personal best and the global best in the following way (10):

$$v_{ik}^{(t+1)} = (w.v_{ik}^t + c_1 r_1(p_{ik}^t - x_{ik}^t) + c_2 r_2(g_k^t - x_{ik}^t)) \tag{10}$$

where w, c_1 and c_2 are the inertia weight and learning factors, also known as the social and cognitive parameters respectively, while r_1 and r_2 are random numbers with limits between [0, 1]. The job of w is to control the influence of the preceding velocity on the present one. The next step is to update the positions that are given as follows:

$$x_{ik}^{(t+1)} = x_{ik}^t + v_{ik}^{t+1} \tag{11}$$

Algorithm 1 scale equations to the same size as the rest of the text

---

1. Set parameter w_{min}, w_{max}, c_1, c_2 and r_1, r_2 of PSO
2. Initialize population of particles as having positions X and velocities V
3. Set iteration $k = 1$
4. Calculate fitness of particles $F_i^k = f(x_i^k)$ $\forall i$ and find the index of the best particle b
5. Select $Pbest_i^k = x_i^k, \forall i$ and $Gbest^k = x_b^k$
6. $w = w_{max} - k \times (w_{max} - w_{min})/\text{Maxite}$
7. Update velocity and position of particles
$v_{ik}^{(t+1)} = (w.v_{ik}^t + c_1 r_1(p_{ik}^t - x_{ik}^t) + c_2 r_2(g_k^t - x_{ik}^t))$ $x_{ik}^{(t+1)} = x_{ik}^t + v_{ik}^{t+1}$
8. Update $Pbest$ population
9. If $F_i^{k+1} < F_i^k$ then $Pbest_i^{k+1} = x_i^{k+1}$
Or else
$Pbest_i^{k+1} = Pbest_i^k$
10. If $Fb_{b1}^{k+1} < F_b^k$ then
$Gbestb^{k+1} = Pbest_{b1}^{k+1}$ and set $b = b1$
Or else
$Gbestb^{k+1} = Pbest^k$
11. If $K < \text{Maxite}$ then $K = K + 1$ and go to step 6
12. End while
13. End PSO-LS
Or go to step 14
14. Display optimum solution as $Gbest^k$

After getting the updated position values of all the particles, if the prearranged meeting condition is not fulfilled the corresponding open facility vectors are resolved with their fitness values to start a new repetition, as the PSO produces a premature and unsatisfactory solution as a result of a rough search. In this regard there is a need to implant a local search algorithm (LSA) into the PSO to produce more satisfactory solutions. At the end of each iteration of the PSO the global best that is found is adopted as the initial solution by the LSA. The Flow chart of CPSO is shown in Figure 2, and the pseudocode of the proposed algorithm (CPSO) is also given above in algorithm 1 [39–41].

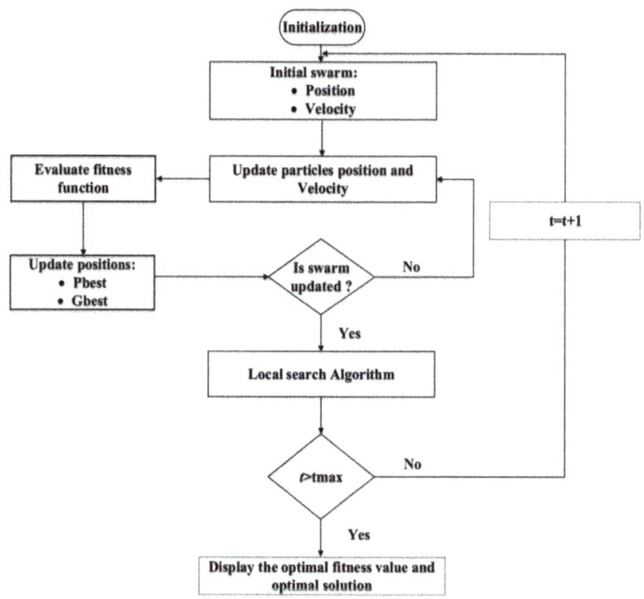

Figure 2. Flow chart of continuous particle swarm optimization (CPSO).

4. Results and Discussion

To study the continuous particle swarm optimization algorithm, four case studies have been considered. The system details of all case studies can be found in earlier works [9,24,42,43].

4.1. Case I

In this case a single end fed system with four overcurrent relay is used, as shown in Figure 3. The relays R_1 and R_4 are non-directional while relays R_2 and R_3 have a directional feature. Two faults are taken into consideration: A and B. At bus 2 the maximum load current, including overload, is 600 A. The current transformer (CT) and plug setting ratio for each relay is 300:1 and 1, respectively. The maximum fault current is 4000 A. For each relay the minimum operating time (MOP) is 0.1 s. The primary and backup relation of the relays is shown in Table 1. Table 2 provides the detail of the a_ρ constant and current seen by the relays for different fault points. In this case the total number of constraints is six; four constraints emerge as a result of the boundaries of the relay operation and two constraints emerge as a result of the coordination condition. The TMS range is 0.025–1.2. The CTI is 0.3 s. The TMSs of all four relays are x_1–x_4. The optimal operations of the relays obtained by the proposed algorithm are given in Table 3, which also provides the comparative results of the proposed algorithm with a previously published algorithm described in the literature. According to Table 3, the proposed algorithm is a better solution for the current case.

Table 1. Primary and backup relationships of the relays for Case I.

Fault Point	Primary Relay	Backup Relay
A	2	4
B	3	1

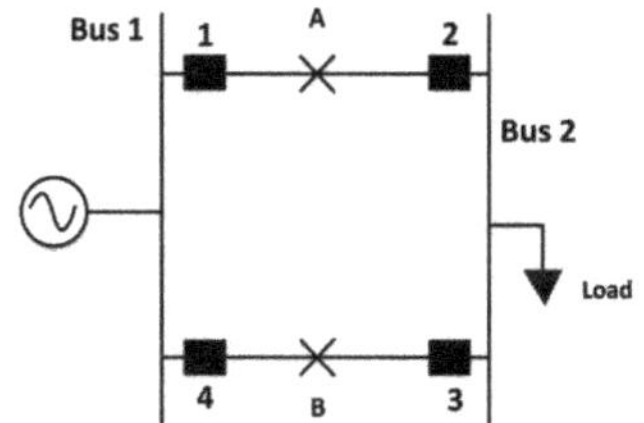

Figure 3. A single end system with parallel feeders.

Table 2. a_ρ Constants and relay currents for Case I.

Fault Point		1	2	3	4
A	I_{relay}	10	3.33	-	3.33
	a_ρ	2.97	5.749	-	5.749
B	I_{relay}	3.33	-	3.33	10
	a_ρ	5.749	-	5.749	2.97

- Indicates the fault is not seen by the relay.

The objective function for minimization can be stated as:

$$z = 8.764x_1 + 5.749x_2 + 5.749x_3 + 8.764x_4 \tag{12}$$

The constraints that emerge because of the MOPs of the relays are:

$$2.97x_1 \geq 0.1 \tag{13}$$

$$5.749x_2 \geq 0.1 \tag{14}$$

$$5.749x_3 \geq 0.1 \tag{15}$$

$$2.97x_4 \geq 0.1 \tag{16}$$

The constraints explained by Equations (14) and (15) violate the constraints of the minimum value of the TMS. The minimum limit on the TMS is 0.025. Hence, these constraints are rewritten as:

$$x_2 \geq 0.025 \tag{17}$$

$$x_3 \geq 0.025 \tag{18}$$

The constraints that emerge because of coordination are:

$$5.749x_4 - 5.749x_2 \geq 0.3 \tag{19}$$

$$5.749x_1 - 5.749x_3 \geq 0.3 \tag{20}$$

The objective function was solved using a continuous particle swarm algorithm. In each case study the number of iterations and population size are both taken to be 300, the minimum and maximum inertia weights are 0.4 and 0.9, and the acceleration factor (c_1, c_2) is 2. In addition, r_1 and r_2 are between [0, 1]. As can be seen in Table 3, the proposed method works and it performs better as compared to other evolutionary techniques. The proposed algorithm gives an optimal solution and lower total operating time ($\sum T_{op}$) and can solve the overcurrent relay problem faster and in a superior way. The values of the TMSs obtained are found to satisfy all the constraints. They give the minimum operating time of the relays for any fault location and also ensure proper coordination. Figure 4 depicts the graphical representation of the optimized Time Multiplier Setting in the literature.

Table 3. Comparison of the optimized TMS by the proposed method with the literature for Case I.

TMS	GA 1 [42]	GA 2 [42]	SM [42]	DSM [43]	CPSO
TMS 1	0.081	0.168	0.07718	0.15	0.078
TMS 2	0.025	0.0250	0.0250	0.041	0.0250
TMS 3	0.025	0.0250	0.0250	0.041	0.0250
TMS 4	0.081	0.168	0.07718	0.15	0.078
$T_{op}\, z$ (s)	**1.70**	**3.23**	**1.64**	**3.09**	**1.65**

GA 1: Solution 1 using GA; GA 2: Solution 2 using GA.

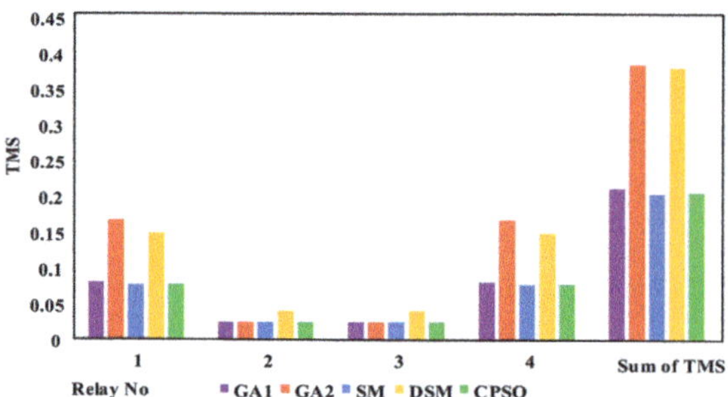

Figure 4. Graphical representation of the optimized TMS compared with the literature for Case I.

The convergence characteristic graph for the total operating time (TOP) obtained for Case 1 during the simulation is shown in Figure 5, demonstrating that the convergence is faster and achieved a better value for the objective function (z) in fewer iterations. The total net gain in time achieved by the proposed algorithm is shown in Table 4, which demonstrates the superiority and advantages of CPSO over the techniques mentioned in the literature.

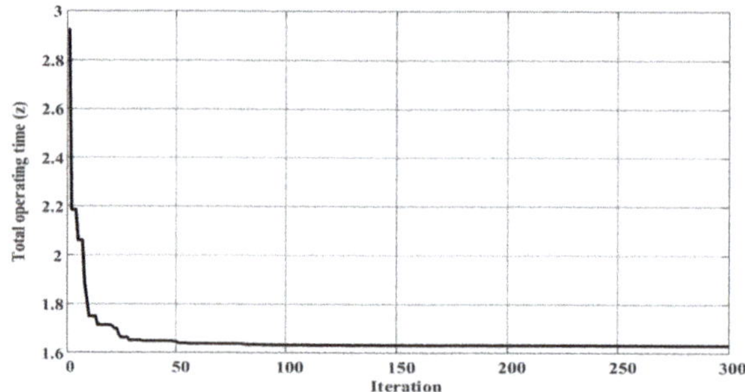

Figure 5. Convergence characteristic graph for Case I.

Table 4. Comparison of the total net gain in time achieved by the proposed algorithm with the literature for Case I.

Net Gain	CPSO/GA	CPSO/GA	CPSO/DSM
$\sum \Delta(t)s$	0.05	1.58	1.44

4.2. Case II

In this case a parallel distribution system that is fed from a single end with five overcurrent relays is shown in Figure 6. Five different fault points were considered with a negligible load current as compared to the fault current. The primary and backup relation of the relays for five different fault points are given in Table 5. The plug setting and CT ratios are illustrated in Table 6. The a_ρ constants and current seen by the relays for the different fault locations are given in Table 7. In this case there are nine constraints in total; five of these constraints arise as a result of boundaries of the relay operation and the other four constraints emerge as a result of the coordination condition. The MOP of each relay is 0.1 s and the CTI is 0.2 s. The TMSs of all the relays is x_1–x_5.

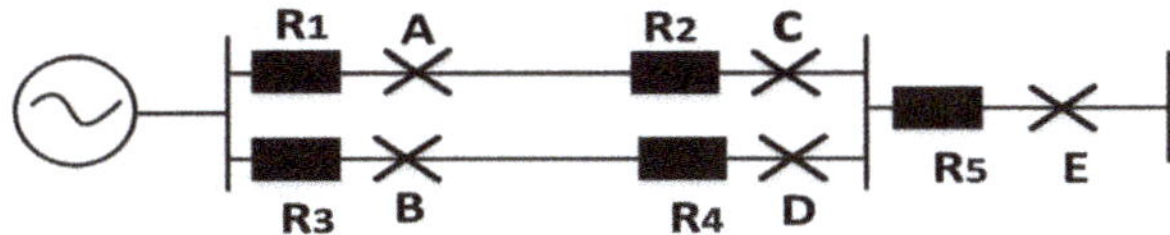

Figure 6. A single end fed parallel feeder distribution system.

Table 5. Primary and backup relationships of the relays for Case II.

Fault Point	Primary Relay	Backup Relay
A	1	-
B	3	-
C	1, 2	-, 3
D	3, 4	-, 1
E	5	1, 3

- Indicates no back up relay.

Table 6. CT ratios and plug settings of the relays for Case II.

Relay	CT Ratio	Plug Setting
1	300/1	1
2	300/1	1
3	300/1	1
4	300/1	1
5	100/1	1

Table 7. a_ρ Constants and relay currents for Case II.

Fault Point		Relay				
		1	2	3	4	5
A	I_{relay}	42.34	-	-	-	-
	a_ρ	1.799	-	-	-	-
B	I_{relay}	-	42.34	-	-	-
	a_ρ	-	1.799	-	-	-
C	I_{relay}	4.876	4.876	4.876	-	-
	a_ρ	4.348	4.348	4.348	-	-
D	I_{relay}	4.876	-	4.876	4.876	-
	a_ρ	4.348	-	4.348	4.348	-
E	I_{relay}	4.876	-	4.876	-	29.25
	a_ρ	4.348	-	4.348	-	2.004

- Indicates the fault is not seen by the relay.

The optimization problem can be stated as:

$$z = 14.843x_1 + 6.147x_2 + 13.044x_3 + 4.348x_4 + 2.004x_5 \tag{21}$$

The constraints that emerge because of the MOP of the relays are:

$$1.799x_1 \geq 0.1 \tag{22}$$

$$4.348x_2 \geq 0.1 \tag{23}$$

$$1.799x_3 \geq 0.1 \tag{24}$$

$$4.348x_4 \geq 0.1 \tag{25}$$

$$2.004x_5 \geq 0.1 \tag{26}$$

The constraints that emerge because of coordination are:

$$4.348x_3 - 4.348x_2 \geq 0.2 \tag{27}$$

$$4.348x_1 - 4.348x_4 \geq 0.2 \tag{28}$$

$$4.348x_1 - 2.004x_5 \geq 0.2 \tag{29}$$

$$4.348x_3 - 2.004x_5 \geq 0.2 \tag{30}$$

The objective function was solved using the proposed algorithm and keeping the same parameters. Table 8 provides the results of the proposed method for this case and a comparison with previous works, respectively. In this case no miscoordination or violations were found. All the relays will initiate operation at a minimum operating time while maintaining coordination. The time required by the relay R_1 to initiate its operation is lowest for a fault at point A (0.214 s) and will require extra time for a fault at point D (0.43 s) and E (0.3 s). Hence, at fault point A the relay R_1 will operate first while at fault points D and E, relays R_4 and R_5 should operate first. If the expected relay fails to activate, then relay R_1 should take over the tripping action. The graphical representation of the optimized TMS is shown in Figure 7, and demonstrates that the TMS is optimized up to the optimum value. Figure 8 shows the convergence characteristic graph obtained during the simulation. According to Tables 8 and 9, the proposed method finds a better solution for this case.

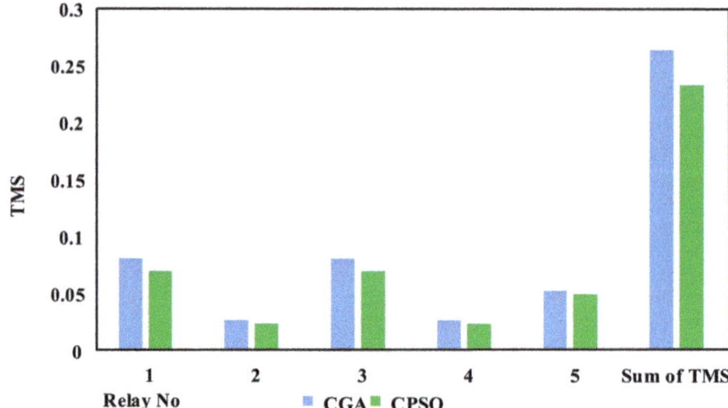

Figure 7. Graphical representation of the optimized TMS compared with the literature for Case II.

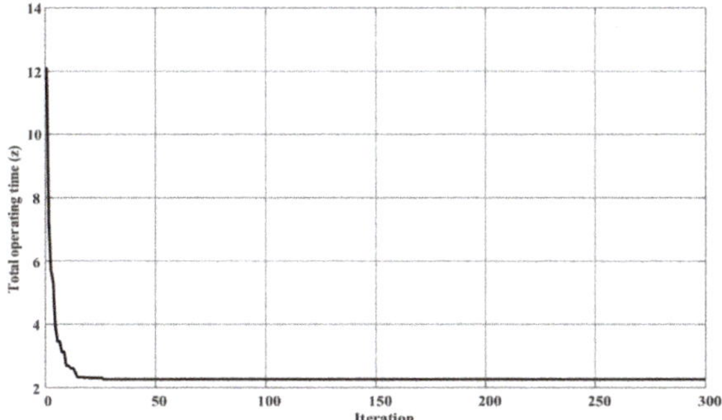

Figure 8. Convergence characteristic graph for Case II.

Table 8. Comparison of the proposed method with the literature for Case II.

TMS	CGA [9]	CPSO
TMS 1	0.08	0.0690
TMS 2	0.026	0.0230
TMS 3	0.08	0.0690
TMS 4	0.026	0.0230
TMS 5	0.052	0.0499
T_{op} (z)	2.52	2.21

Table 9. Total net gain in time achieved by the proposed algorithm compared with the literature for Case II.

Net Gain	$\sum \Delta(t)s$
CPSO/CGA	0.31

4.3. Case III

A parallel distribution system that is fed from a single end with five overcurrent relays is shown in Figure 9. The current transfer ratio and plug setting of the relays are assumed to be 300:1 and 1, respectively. Two fault currents are imposed in the middle of the lines, i.e., A and B. For the fault at A backup protection will be provided by relay R_3 to relay R_2 while for the fault at B the backup will be provided by relay R_1 to R_4, and for the fault at C back up will be provided by R_1, R_3 to R_5. In this case, the total number of constraints is nine; five constraints arise as a result of the boundaries of the relay operation and four constraints emerge as a result of the coordination condition. The MOP of each relay is 0.1 s. The CTI is 0.2 s. The TMSs of all the relays is x_1–x_5. The currents seen by the relays and a_ρ constants for different fault locations are given in Table 10.

Table 10. a_ρ Constants and relay currents for Case II.

Fault Point		Relay				
		1	2	3	4	5
A	I_{relay}	9.059	3.019	3.019	-	-
	a_ρ	3.106	6.265	6.265	-	-
B	I_{relay}	3.019	-	9.059	3.019	-
	a_ρ	6.265	-	3.106	6.265	-
C	I_{relay}	4.875	-	4.875	-	29.25
	a_ρ	4.348	-	4.348	-	2.004

- Indicates no fault current seen by the relay.

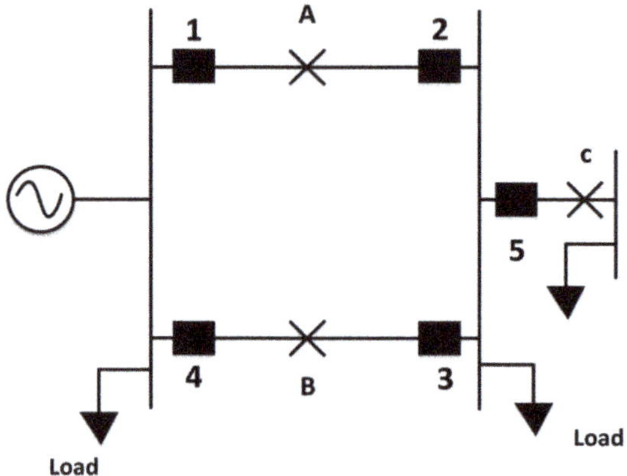

Figure 9. A single end parallel feeder distribution system.

In this case two optimization problems are derived from Table 10 for comparison with the other published techniques mentioned in the literature and can be stated as:

$$\min z = 13.719x_1 + 6.265x_2 + 13.719x_3 + 6.265x_4 + 2.004x_5 \tag{31}$$

$$\min z = 3.106x_1 + 6.265x_2 + 3.106x_3 + 6.265x_4 + 2.004x_5 \tag{32}$$

The constraints that emerge because of the MOP of the relays are:

$$3.106x_1 \geq 0.1 \tag{33}$$

$$6.265x_2 \geq 0.1 \tag{34}$$

$$3.106x_3 \geq 0.1 \tag{35}$$

$$6.265x_4 \geq 0.1 \tag{36}$$

$$2.004x_5 \geq 0.1 \tag{37}$$

The constraints that emerge because of coordination are:

$$6.265x_3 - 6.265x_2 \geq 0.2 \tag{38}$$

$$6.265x_1 - 6.265x_4 \geq 0.2 \tag{39}$$

$$4.348x_1 - 2.004x_5 \geq 0.2 \tag{40}$$

$$4.348x_3 - 2.004x_5 \geq 0.2 \tag{41}$$

The problem was solved using the CPSO algorithm. The optimized graphical representation and convergence characteristic graph for this case is shown in Figures 10–12, which demonstrate that the proposed method yields a faster convergence and a better solution for the objective function "z". Table 11 provides the comparative results of the proposed algorithm with a previous optimization algorithm explained in the literature, which ensures that for the fault at point A, relay R_1 is first to operate, whereas for the fault at point B relay R_4 will operate, and for the fault at point C the relay R_5 should get the first chance to operate. The total net gain in time achieved by the proposed algorithm is tabulated in Table 12.

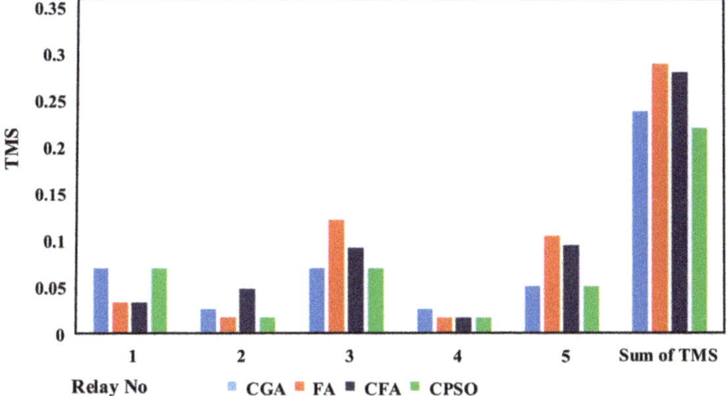

Figure 10. Graphical representation of the optimized TMS compared with the literature for Case III.

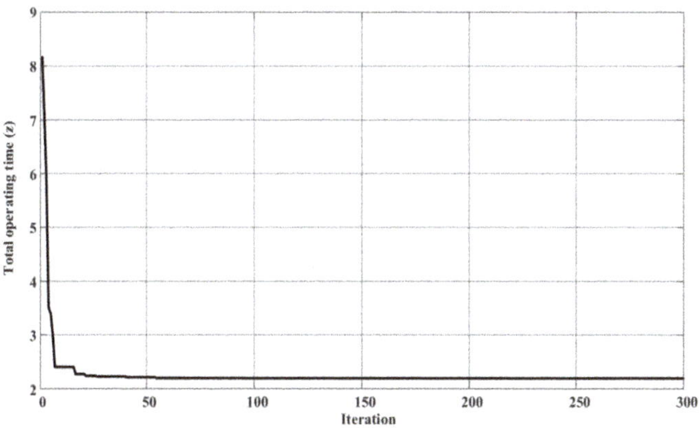

Figure 11. Convergence characteristic graph of the objective function z (31) for Case III.

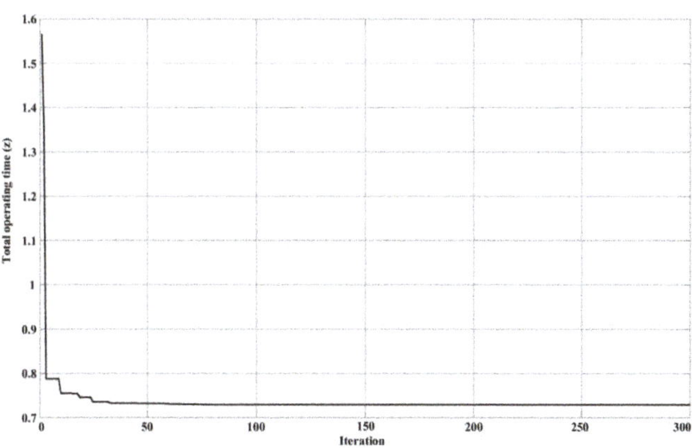

Figure 12. Convergence characteristic graph of the objective function z (32) for Case III.

Table 11. Comparison of the optimized TMS with the literature for Case III.

TMS	TPSM [44]	CPSO [1]	FA [24]	CFA [24]	CPSO [2]
TMS 1	0.069	0.069	0.032	0.032	0.069
TMS 2	0.025	0.0160	0.0160	0.047	0.0160
TMS 3	0.069	0.069	0.121	0.091	0.069
TMS 4	0.025	0.0160	0.0160	0.0160	0.0160
TMS 5	0.0499	0.0499	0.104	0.094	0.0499
T_{op} z (s)	2.27	2.17	1.73	1.63	0.7291

[1] For the objective function mentioned in Equation (31); [2] for the objective function mentioned in Equation (32).

Table 12. Comparison of the total net gain in time achieved by the proposed algorithm compared with the literature for Case III.

Net Gain	$\sum \Delta(t)s$
CPSO [1]/TPSM	0.10
CPSO [2]/FA	1.01
CPSO [2]/CFA	0.91

[1] For the objective function mentioned in Equation (31); [2] for the objective function mentioned in Equation (32).

4.4. Case IV

In this case, a multi-loop system with six overcurrent relays and with negligible line charging admittances is considered, as shown in Figure 13. A set of various primary and backup relays are designed which are subject to the locations of the various faults. These configurations are contingent on the path of the fault current in the different feeders. The line data of the system are shown in Table 13. Four different fault positions are considered. The primary and backup relationships of the relays for the four fault points are given in Table 14. The CT ratios and plug setting are illustrated in Table 15. The a_ρ constants currents seen by the relays for the different fault locations are given in Table 16. In this case study the total number of constraints is eleven; six constraints emerge as a result of the boundaries of the relay operation and five constraints emerge as a result of the coordination condition. The MOP of each relay is 0.1, while the range of the TMS is 0.025–1.2, except x_1 which is 0.027. The CTI is 0.3 s. The TMSs of all six relays are x_1–x_6. The optimal operation of the relays as achieved by the proposed algorithm is given in Table 17, which also provides the comparative results of the proposed algorithm with a previous optimization algorithm explained in the literature. According to Table 18, the proposed algorithm achieves a better solution for this case.

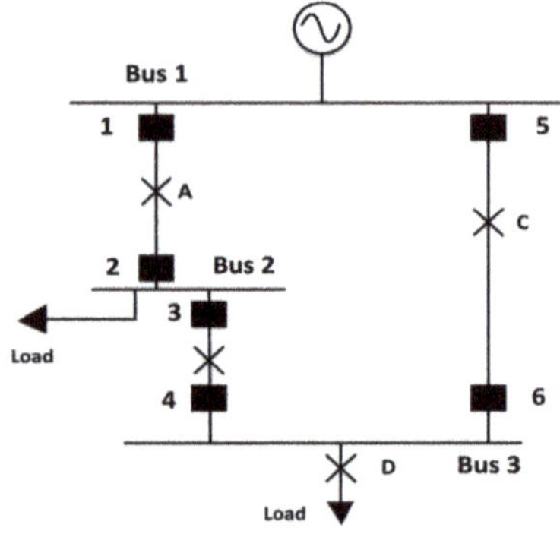

Figure 13. A multi-loop distribution system.

Table 13. Line data for Case IV.

Line	Impedance (Ω)
1-2	$0.08j1$
2-3	$0.08 + j1$
1-3	$0.16 + j2$

Table 14. Primary and backup relationships of the relays for Case IV.

Fault Point	Primary Relay	Backup Relay
A	1, 2	-, 4
B	3, 4	1, 5
C	5, 6	-, 3
D	3, 5	1, -

- Indicates no back up relay.

Table 15. CT ratios and plug settings of the relays for Case IV.

Relay	CT Ratio	Plug Setting
1	1000/1	1
2	300/1	1
3	1000/1	1
4	600/1	1
5	600/1	1
6	600/1	1

Table 16. a_ρ Constants and relay currents for Case IV.

Fault Point		Relay					
		1	2	3	4	5	6
A	I_{relay}	6.579	3.13	-	1.565	1.565	-
	a_ρ	3.646	6.065	-	15.55	15.55	-
B	I_{relay}	2.193	-	2.193	2.193	2.193	-
	a_ρ	8.844	-	8.844	8.844	8.844	-
C	I_{relay}	1.096	-	1.096	-	5.482	1.827
	a_ρ	75.91	-	75.91	-	4.044	11.539
D	I_{relay}	1.644	-	1.644	-	2.741	-
	a_ρ	13.99	-	13.99	-	6.872	-

- Indicates the fault is not seen by the relay.

Table 17. Comparison of the optimized TMS by the proposed method with the literature for Case IV.

TMS	CGA [9]	FA [24]	CFA [24]	CPSO
TMS 1	0.0765	0.027	0.027	0.0589
TMS 2	0.034	0.130	0.221	0.0250
TMS 3	0.0339	0.025	0.025	0.0250
TMS 4	0.036	0.025	0.025	0.0290
TMS 5	0.0711	0.489	0.363	0.0630
TMS 6	0.0294	0.0285	0.029	0.0250
$T_{op} z$ (s)	15.88	16.25	14.39	11.87

Table 18. Comparison of the proposed method results with the literature for Case IV.

Method	Objective Function
CGA [9]	15.88
FA [21]	16.25
CFA [21]	14.69
Proposed CPSO	11.87

The objective function for minimization can be defined as:

$$z = 102.4x_1 + 6.06x_2 + 98.75x_3 + 24.4x_4 + 35.31x_5 + 11.53x_6 \tag{42}$$

The constraints that emerge because of the MOP of the relays are:

$$3.646x_1 \geq 0.1 \tag{43}$$

$$6.055x_2 \geq 0.1 \tag{44}$$

$$8.844x_3 \geq 0.1 \tag{45}$$

$$8.844x_4 \geq 0.1 \tag{46}$$

$$4.044x_5 \geq 0.1 \tag{47}$$

$$11.539x_6 \geq 0.1 \tag{48}$$

The constraints explained by Equations (44)–(48) violate the constraints of the minimum value of the Time Multiplier Setting (TMS). All the TMSs should be greater than 0.025. Hence, these constraints are rewritten as:

$$x_2 \geq 0.025 \tag{49}$$

$$x_3 \geq 0.025 \tag{50}$$

$$x_4 \geq 0.025 \tag{51}$$

$$x_5 \geq 0.025 \tag{52}$$

$$x_6 \geq 0.025 \tag{53}$$

The constraints that emerge as a result of coordination are:

$$15.55x_4 - 6.065x_2 \geq 0.3 \tag{54}$$

$$8.844x_1 - 8.844x_3 \geq 0.3 \tag{55}$$

$$8.844x_5 - 8.844x_4 \geq 0.3 \tag{56}$$

$$75.91x_3 - 11.53x_6 \geq 0.3 \tag{57}$$

$$13.998x_1 - 13.998x_3 \geq 0.3 \tag{58}$$

The objective function was solved using a continuous particle swarm algorithm. As can be seen in Tables 17 and 18, the proposed method achieves a satisfactory solution as compared to other methods. The values shown in Table 17 prove that the relays will operate in the minimum possible time for a fault at any point in the system, and will also maintain coordination. The time taken by relay 1 to operate is the minimum possible time for the fault at point A, while it will take maximum time for the fault at point C. This is desirable, because for the fault at point A, relay 1 is the first to operate, whereas for the fault at point C, relay 6 should be the first to operate. If relay 6 fails to operate then relay 3 should take over the tripping action, and if relay 3 also fails to operate only then should relay 1 take over the tripping action. The proposed algorithm gives an optimal solution and optimized total operating time up to the optimum value. Figure 14 depicts a graphical representation of the optimized Time Multiplier Setting compared with the literature. The convergence characteristic for the total operating time obtained for Case IV during the simulation is shown in Figure 15, which demonstrates that the convergence is faster and achieved a better value for the objective function in fewer iterations. The total net gain in time achieved by the proposed algorithm is shown in Table 19, demonstrating the superiority and advantages of CPSO over the techniques mentioned in the literature.

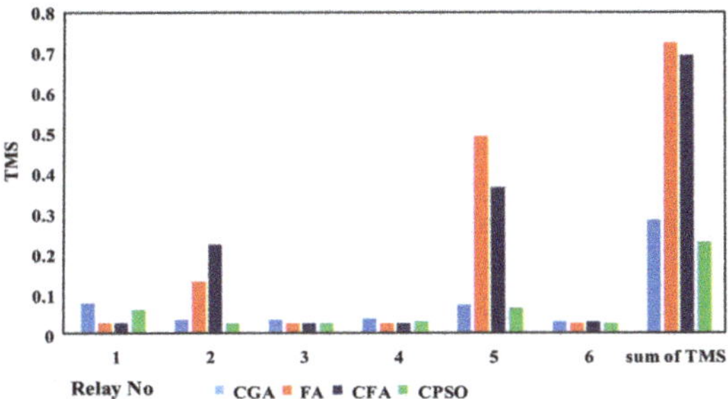

Figure 14. Graphical representation of the optimized TMS compared with the literature for Case IV.

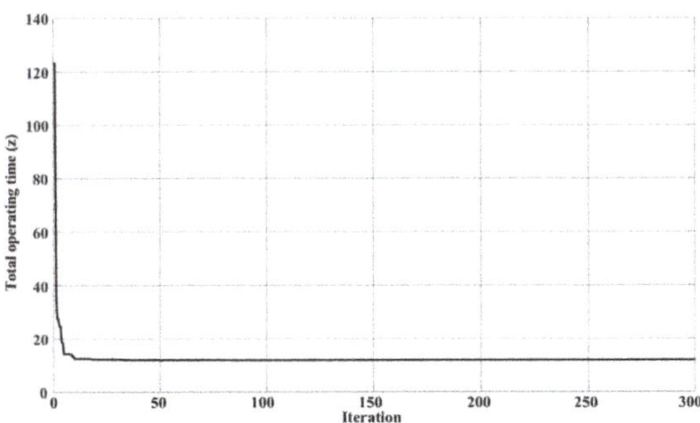

Figure 15. Convergence characteristic graph for Case IV.

Table 19. Comparison of the total net gain in time achieved by the proposed algorithm compared with the literature for Case IV.

Net Gain	$\sum \Delta(t)s$
CPSO/CGA	3.242
CPSO/FA	4.348
CPSO/CFA	2.82

5. Discussion

The CPSO algorithm was used to evaluate the overcurrent relay coordination problem. The proposed algorithm has a high search capability and convergence speed as compared to other optimization techniques, and these characteristics make the population member of the CPSO more discriminative in finding the optimal solution than that of other optimization techniques. The case studies presented in this paper have also been evaluated by GA, CGA, FA, CFA, DSM, SM, and TPSM optimization algorithms, with several different initial conditions and parameter values as shown in the literature, and an improved optimal solution was observed from the proposed CPSO algorithm compared to these other algorithm options. The optimum relay coordination problem is basically a highly constrained optimization problem. As CPSO can solve constrained and unconstrained optimization problems, the relay coordination problem has been converted into an unconstrained optimization problem by defining a new objective function (using the penalty method) and by using the boundaries on the TMS (and boundaries on the relay operating time) as the limits of the variables. A systematic procedure for converting a relay coordination problem into an optimization problem has been developed in this paper. A program has been developed in MATLAB for finding the optimum time coordination of relays using the CPSO method. The program can be used for setting the optimum time coordination of relays in a system with any number of relays and any number of primary-backup relationships. The TMS and total operating time of relays obtained for all case studies by the proposed CPSO algorithm ensured that the relays will activate in the minimum possible amount of time for a fault at any point in the system. However, if the number of relays is increased the nature of the highly constrained problem becomes more distinct. Therefore, an accurate optimum relay coordination minimizes the total operating time as well as reduces and limits the damage produced by the fault. Unwanted tripping of the circuit breaker can also be bypassed by this method. The convergence characteristic graphs obtained during simulations show that the convergence is faster and obtains a superior solution for the fitness function "z" in fewer iterations. The CPSO algorithm is superior to the GA, DSM, TPSM, FA, CFA, and CGA algorithms, as shown in Tables 4, 9, 12 and 19. The CPSO algorithm gains 1.58 s and 1.44 s over the GA and DSM algorithms in Case I, and although this may appear insufficient it should be noted that it is a very small system. In Case II the CPSO algorithm gains 31 ms over the CGA algorithm. In Case III the CPSO gives an advantage of 1 ms over the TPSM and 1.01 s and 0.91 s over the FA and CFA algorithms, respectively. In Case IV the CPSO algorithm gives an advantage of 3.24 s, 4.34 s, and 2.82 s over the CGA, FA and CFA algorithms. For Case IV this advantage is sufficient given that it is a very small system, as it can be clearly seen from Tables 4, 9, 12 and 19, and from Figure 5, Figure 8, Figure 11, Figure 12, and Figure 14 that the proposed method is superior to the recent published techniques mentioned in the literature in term of the quality of the solution, convergence, and minimizing the objective function to the optimum value. The proposed method additionally addressed the weaknesses of the previous methods.

6. Conclusions

This paper proposed a CPSO algorithm based on inspirited swarm behavior associated with fish schooling and bird flocking. The overcurrent relay coordination problem was pursued using the CPSO algorithm for the various test systems. The prolificacy of the CPSO algorithm has been determined and tested on various single end multi-loop distribution systems, by analyzing its superiority compared with GA, SM, DSM, TPSM, CGA, FA, and CFA published techniques. The simulation results of the CPSO algorithm efficiently minimize all four models of the problem. The efficiency of the CPSO can be observed from the minimum function evaluations required by the algorithm to reach the optimum as compared to the CGA, FA, CFA, GA, and TPSM algorithms. The CPSO contributes a new approach for clarification as one of its distinctions is the generous field of research considering the characterization of fish schooling and bird flocking. The simulation results acknowledge the supremacy of the proposed CPSO algorithm in solving the overcurrent relay coordination problem. In future work, this algorithm can be extended to solve overcurrent relay problems of higher buses and complex power systems.

Acknowledgments: This research was supported by Korea Electric Power Corporation, grant number (R17XA05-38).

Author Contributions: Abdul Wadood, Chang-Hwan Kim, Tahir Khurshiad, and Saeid Gholami Farkoush contributed to the literature review of the optimum coordination of overcurrent relay, then developed the first discussions about the different parameters of the proposed methodology and conducted the simulations and analysis of results together with Sang-Bong Rhee. Sang -Bong Rhee guided the investigation and supervised this work. Kyu-Ho kim and Namhun Cho gave suggestions and guidance for the research. All of the authors read and approved the final manuscript.

Conflicts of Interest: The authors declare no conflict of interest.

References

1. Blackburn, J. *Protective Relaying, Principles and Applications*; Marcel Dekker Inc.: New York, NY, USA, 1987.
2. Urdaneta, A.J.; Nadira, R.; Jimenez, L.G.P. Optimal coordination of directional overcurrent relays in interconnected power systems. *IEEE Trans. Power Deliv.* **1988**, *3*, 903–911. [CrossRef]
3. Perez, L.G.; Urdaneta, A.J. Optimal coordination of directional overcurrent relays considering definite time backup relaying. *IEEE Trans. Power Deliv.* **1999**, *14*, 1276–1284. [CrossRef]
4. Birla, D.; Maheshwari, R.P.; Gupta, H.O. Time-overcurrent relay coordination: A review. *Int. J. Emerg. Electr. Power Syst.* **2005**, *2*. [CrossRef]
5. Chattopadhyay, B.; Sachdev, M.S.; Sidhu, T.S. An on-line relay coordination algorithm for adaptive protection using linear programming technique. *IEEE Trans. Power Deliv.* **1996**, *11*, 165–173. [CrossRef]
6. Birla, D.; Maheshwari, R.P.; Gupta, H.O.; Deep, K.; Thakur, M. Application of random search technique in directional overcurrent relay coordination. *Int. J. Emerg. Electr. Power Syst.* **2006**, *7*. [CrossRef]
7. So, C.W.; Li, K.K. Time coordination method for power system protection by evolutionary algorithm. *IEEE Trans. Ind. Appl.* **2000**, *36*, 1235–1240. [CrossRef]
8. Koochaki, A.; Asadi, M.R.; Mahmoodan, M.; Naghizadeh, R.A. Optimal overcurrent relays coordination using genetic algorithm. In Proceedings of the 11th International Conference on Optimization of Electrical and Electronic Equipment, Brasov, Romania, 22–24 May 2008.
9. Bedekar, P.P.; Bhide, S.R. Optimum coordination of overcurrent relay timing using continuous genetic algorithm. *Expert Syst. Appl.* **2011**, *38*, 11286–11292. [CrossRef]
10. So, C.W.; Li, K.K.; Lai, K.T.; Fung, K.Y. Application of genetic algorithm to overcurrent relay grading coordination. In Proceedings of the Sixth International Conference on Developments in Power System Protection, Nottingham, UK, 25–27 March 1997; pp. 283–287.
11. Bedekar, P.P.; Bhide, S.R. Optimum coordination of directional overcurrent relays using the hybrid GA-NLP approach. *IEEE Trans. Power Deliv.* **2011**, *26*, 109–119. [CrossRef]
12. Razavi, F.; Abyaneh, H.A.; Al-Dabbagh, M.; Mohammadi, R.; Torkaman, H. A new comprehensive genetic algorithm method for optimal overcurrent relays coordination. *Electr. Power Syst. Res.* **2008**, *78*, 713–720. [CrossRef]
13. Noghabi, A.S.; Sadeh, J.; Mashhadi, H.R. Considering different network topologies in optimal overcurrent relay coordination using a hybrid GA. *IEEE Trans. Power Deliv.* **2009**, *24*, 1857–1863. [CrossRef]
14. Rathinam, A.; Sattianadan, D.; Vijayakumar, K. Optimal coordination of directional overcurrent relays using particle swarm optimization technique. *Int. J. Comput. Appl.* **2010**, *10*, 43–47. [CrossRef]
15. Zeineldin, H.H.; El-Saadany, E.F.; Salama, M.M.A. Optimal coordination of overcurrent relays using a modified particle swarm optimization. *Electr. Power Syst. Res.* **2006**, *76*, 988–995. [CrossRef]
16. Mansour, M.M.; Mekhamer, S.F.; El-Kharbawe, N. A modified particle swarm optimizer for the coordination of directional overcurrent relays. *IEEE Trans. Power Deliv.* **2007**, *22*, 1400–1410. [CrossRef]
17. Motlagh, M.; Hadi, S.; Mazlumi, K. Optimal Overcurrent Relay Coordination Using Optimized Objective Function. *ISRN Power Eng.* **2014**, *2014*. [CrossRef]
18. Liu, A.; Yang, M.-T. A new hybrid nelder-mead particle swarm optimization for coordination optimization of directional overcurrent relays. *Math. Probl. Eng.* **2012**, *2012*. [CrossRef]
19. Yang, M.-T.; Liu, A. Applying hybrid PSO to optimize directional overcurrent relay coordination in variable network topologies. *J. Appl. Math.* **2013**, *2013*. [CrossRef]
20. Thangaraj, R.; Pant, M.; Deep, K. Optimal coordination of over-current relays using modified differential evolution algorithms. *Eng. Appl. Artif. Intell.* **2010**, *23*, 820–829. [CrossRef]
21. Uthitsunthorn, D.; Pao-La-Or, P.; Kulworawanichpong, T. Optimal overcurrent relay coordination using artificial bees colony algorithm. In Proceedings of the 2011 8th International Conference on Electrical Engineering/Electronics, Computer, Telecommunications and Information Technology (ECTI-CON), Khon Kaen, Thailand, 17–19 May 2011.
22. Xu, C.; Zou, X.; Yuan, R.; Wu, C. Optimal coordination of protection relays using new hybrid evolutionary algorithm. In Proceedings of the 2008 IEEE Congress on Evolutionary Computation (IEEE World Congress on Computational Intelligence), Hong Kong, China, 1–6 June 2008.

23. Amraee, T. Coordination of directional overcurrent relays using seeker algorithm. *IEEE Trans. Power Deliv.* **2012**, *27*, 1415–1422. [CrossRef]

24. Gokhale, S.S.; Kale, V.S. An application of a tent map initiated Chaotic Firefly algorithm for optimal overcurrent relay coordination. *Int. J. Electr. Power Energy Syst.* **2016**, *78*, 336–342. [CrossRef]

25. Alam, M.N.; Das, B.; Pant, V. A comparative study of metaheuristic optimization approaches for directional overcurrent relays coordination. *Electr. Power Syst. Res.* **2015**, *128*, 39–52. [CrossRef]

26. Alipour, M.; Teimourzadeh, S.; Seyedi, H. Improved group search optimization algorithm for coordination of directional overcurrent relays. *Swarm Evol. Comput.* **2015**, *23*, 40–49. [CrossRef]

27. Sulaiman, M.; Ahmad, A.; Khan, A.; Muhammad, S. Hybridized Symbiotic Organism Search Algorithm for the Optimal Operation of Directional Overcurrent Relays. *Complexity* **2018**, *2018*. [CrossRef]

28. Sulaiman, M.; Muhammad, H.; Khan, A. Improved Solutions for the Optimal Coordination of DOCRs Using Firefly Algorithm. *Complexity* **2018**, *2018*. [CrossRef]

29. Ehrenberger, J.; Švec, J. Directional Overcurrent Relays Coordination Problems in Distributed Generation Systems. *Energies* **2017**, *10*, 1452. [CrossRef]

30. Ates, Y.; Boynuegri, A.; Uzunoglu, M.; Nadar, A.; Yumurtacı, R.; Erdinc, O.; Paterakis, N.; Catalão, J. Adaptive protection scheme for a distribution system considering grid-connected and islanded modes of operation. *Energies* **2016**, *9*, 378. [CrossRef]

31. Núñez-Mata, O.; Palma-Behnke, R.; Valencia, F.; Jiménez-Estévez, P.M.G. Adaptive Protection System for Microgrids Based on a Robust Optimization Strategy. *Energies* **2018**, *11*, 308. [CrossRef]

32. Wadood, A.; Kim, C.-H.; Farkoush, S.G.; Rhee, S.B. An Adaptive Protective Coordination Scheme for Distribution System Using Digital Overcurrent Relays. In Proceedings of the Korean Institute of Illuminating and Electrical Installation Engineers, Gangwon, Korea, 30 August 2017; p. 53.

33. Kennedy, J.; Eberhart, R. Particle swarm optimization (PSO). In Proceedings of the IEEE International Conference on Neural Networks, Perth, Australia, 27 November–1 December 1995; pp. 1942–1948.

34. Shi, Y.; Eberhart, R. A modified particle swarm optimizer. In Proceedings of the 1998 IEEE International Conference on Evolutionary Computation Proceedings, IEEE World Congress on Computational Intelligence, Anchorage, AK, USA, 4–9 May 1998; Cat. No.98TH8360.

35. Eberhart, R.C.; Hu, X. Human tremor analysis using particle swarm optimization. In Proceedings of the 1999 Congress on Evolutionary Computation, Washington, DC, USA, 6–9 July 1999; Volume 3.

36. Park, J.B.; Lee, K.S.; Shin, J.R.; Lee, K.Y. A Particle swarm optimization for economic dispatch with non-smooth cost functions. *IEEE Trans. Power Syst.* **2005**, *20*, 34–42. [CrossRef]

37. Eberhart, R.; Kennedy, J. A new optimizer using particle swarm theory. In Proceedings of the Sixth International Symposium on Micro Machine and Human Science, Nagoya, Japan, 4–6 October 1995.

38. Sevkli, M.; Guner, A.R. A continuous particle swarm optimization algorithm for uncapacitated facility location problem. In *Ant Colony Optimization and Swarm Intelligence*; Springer: Berlin/Heidelberg, Germany, 2006.

39. Emara, H.M.; Fattah, H.A.A. Continuous swarm optimization technique with stability analysis. In Proceedings of the American Control Conference, Boston, MA, USA, 30 June–2 July 2004.

40. Khoshahval, F.; Zolfaghari, A.; Minuchehr, H.; Sadighi, M.; Norouzi, A. PWR fuel management optimization using continuous particle swarm intelligence. *Ann. Nucl. Energy* **2010**, *37*, 1263–1271. [CrossRef]

41. Eberhart, R.C.; Shi, Y. Comparing inertia weights and constriction factors in particle swarm optimization. In Proceedings of the 2000 Congress on Evolutionary Computation, La Jolla, CA, USA, 16–19 July 2000.

42. Bedekar, P.P.; Bhide, S.R.; Kale, V.S. Optimum coordination of overcurrent relay timing using simplex method. *Electr. Power Compon. Syst.* **2010**, *38*, 1175–1193. [CrossRef]

43. Bedekar, P.P.; Bhide, S.R.; Kale, V.S. Optimum coordination of overcurrent relays in distribution system using dual simplex method. In Proceedings of the 2009 2nd International Conference on Emerging Trends in Engineering and Technology (ICETET), Nagpur, India, 16–18 December 2009.

44. Bedekar, P.P.; Bhide, S.R.; Kale, V.S. Optimum time coordination of overcurrent relays using two phase simplex method. *World Acad. Sci. Eng. Technol.* **2009**, *28*, 1110–1114.

Article

A New Method to Monitor the Primary Neutral Integrity in Multi-Grounded Neutral Systems

Xiangmin Xie [1], Yuanyuan Sun [1,2,*], Xun Long [3] and Bingwei Zhang [1]

[1] School of Electrical Engineering, Shandong University, Jinan 250061, China; xxm922@163.com (X.X.); jnzhangbw@126.com (B.Z.)
[2] Key Laboratory of Power System Intelligent Dispatch and Control Ministry of Education, Shandong University, Jinan 20061, China
[3] FortisAlberta, Calgary T3K6E8, AB, Canada; Xun.Long@ieee.org
* Correspondence: sunyy@sdu.edu.cn; Tel.: +86-139-5316-4875

Academic Editor: José Gabriel Oliveira Pinto
Received: 14 December 2016; Accepted: 13 March 2017; Published: 17 March 2017

Abstract: In the three-phase four-wire system, there are usually multiple grounding points in the primary neutral line due to safety and economic considerations. The rising "neutral to earth voltage (NEV)" caused by a broken primary neutral can threaten the safety of nearby facilities and humans; therefore, the integrity of the primary neutral conductor is of vital importance for the multi-grounded neutral (MGN) system. In this paper, a new passive method is proposed to monitor the integrity of the primary neutral line in the MGN system. The method is based on the measured voltage and current data at the service transformer, and there is no need to install any signal generators. Therefore, it causes no disturbance to the utility and customer. In the paper, the equivalent analysis circuit is established and a new parameter is proposed to reflect the neutral condition. The value of the parameter is estimated based on the measured data, and then, the equivalent impedance of the primary neutral groundings can be obtained. On the other hand, the impedance value for the primary neutral under normal operating conditions can be estimated based on the derived analytical formulas. By comparing the monitored primary neutral impedance with its normal value, the broken neutral condition in the primary system can be detected. Different primary neutral broken cases are analyzed in the paper based on the Monte Carlo simulation. The results indicate that the integrity condition in the primary neutral can be accurately monitored by the proposed method.

Keywords: primary neutral integrity; multi-grounded neutral (MGN) system; neutral integrity detection; passive method

1. Introduction

Nowadays, the three-phase four-wire multi-grounded neutral (MGN) system has been widely adopted in many utilities due to its lower installation costs and higher sensitivity of fault protection compared with the three-phase three-wire system [1,2]. Since the primary neutral line is grounded through multiple points, the unbalanced current in the system will flow through the neutral into the earth, which can produce neutral to earth voltage (NEV) at the customer side. Especially, when the MGN network experiences a broken neutral or loose-connected neutral, more current will be forced into the ground, resulting in a rising NEV [3,4]. The NEV is very harmful to the safety of nearby facilities and humans [5,6]. Therefore, in order to guarantee that the MGN system works safely and efficiently, it is necessary for the utility to routinely monitor the neutral integrity condition of the MGN primary system.

Due to the complicated construction of the MGN system, the detection of the primary neutral integrity is not a trivial task. There is not even an accurate model for the multiple neutral grounding systems, and the impact of the primary neutral is either merged into the phase wires or directly neglected in order to simplify the simulation of an MGN system. From 2000, a full-scale model for the MGN distribution system was introduced in [1]. However, in this method, the multiple grounding electrodes were represented by only one node. Other research about the MGN system mainly focuses on the power flow analysis or ground fault analysis [7–11], and the research on the broken neutral integrity monitoring is relatively little. Some researchers have proposed the method of measuring the primary neutral impedance, such as "fall of potential"-based methods [12] and "staged fault"-based methods. However, for these methods, it is difficult to find a "true" ground to measure the voltage in order to calculate the primary neutral impedance. Moreover, once the soil conditions and some other conditions are changed, the potential electrode has to be relocated [13–15]. In [16], an active method based on the disturbance generated by a signal generator is proposed to reflect the neutral condition. This is generally a good detection method for the primary neutral condition, but usually, the customers are reluctant to install signal generators at their homes, which is needed for the active method.

In this paper, a new passive method to monitor the primary neutral integrity is proposed. Rather than directly measuring the primary neutral impedance, it is proposed to estimate the primary neutral condition based on the measured primary and secondary neutral current at the monitored transformer. There is no need for the method to install any devices or proxy loads to generate the disturbance signal; therefore, the method causes no disturbance to the utility and customer. In the paper, a parameter is first proposed to reflect the condition of the primary neutral. The actual value of the parameter can be estimated based on the measured neutral current. Then, the equivalent impedance of the MGN system in the primary side can be calculated. The theoretical value of the equivalent impedance can be calculated based on the system parameters under normal operating conditions. By comparing the monitored primary neutral impedance with its normal value, the condition of the multi-grounding neutral in the primary system can be determined.

The remainder of the paper is organized as follows: Section 2 illustrates the proposed passive method. Section 3 introduces the data selection criteria for calculating the proposed parameter and further summarizes the flowchart of the proposed method. Section 4 verifies the validity of the proposed method by using the Monte Carlo simulation. Finally, the conclusions of the paper are given in Section 5.

2. The Proposed Primary Neutral Monitoring Method

2.1. Establishment of the Equivalent Analysis Circuit

The single-line diagram of a typical three-phase four-wire system with multiple primary neutral groundings is shown in Figure 1. The neutral line in the primary system is grounded at multiple electrodes along its route. For the North American power system, there are usually at least four grounds per 1.6 km [17]. The primary neutral and secondary neutral are often tied at the service transformer grounding, as shown in Figure 1. Here, for simplification, only one phase of the distribution transformer is depicted. The loads in the transformer secondary side are represented by customer houses. Generally, one service transformer supplies dozens or hundreds of individual houses depending on the community size. At the transformer secondary side, the neutral is grounded via the grounding resistance at the service panels of the customer. In Figure 1, R_T represents the transformer grounding resistance, and R_{gc} represents the customer grounding resistance. Under normal operating conditions, the current flowing in the neutral line of the MGN system is mainly caused from the unbalanced utility loads and customer loads, and the flow of current is shown in Figure 1. In Figure 1, the green line represents the unbalanced current caused by the unbalanced customer loads, and the red line represents the unbalanced current caused by the unbalanced utility loads.

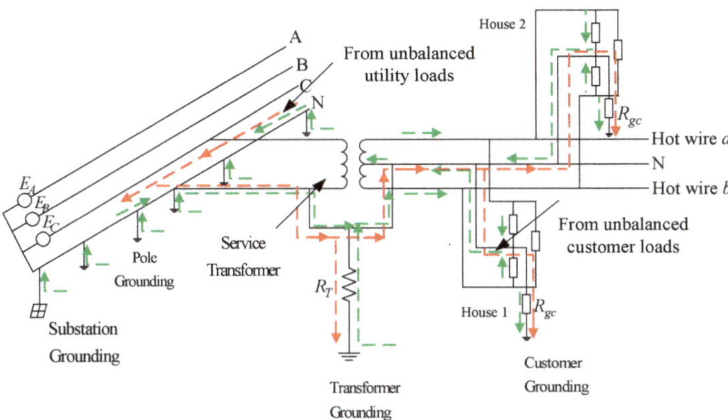

Figure 1. The single line diagram of the three-phase four-wire multi-grounded neutral system.

In order to analyze the neutral current at the transformer primary and secondary side, an equivalent circuit is first established, as shown in Figure 2. The transformer primary system is equivalent to be a Thévenin circuit represented by V_P and Z_S, where V_P is the phase to neutral voltage and Z_S is the equivalent impedance for the primary system. Z_{MGN} represents the equivalent impedance of the primary multiple groundings. The derivation process for the value of Z_{MGN} will be discussed in the next subsection. Z_{an} represents the equivalent customer loads between the hot wire a and neutral; Z_{bn} represents the equivalent customer loads between the hot wire b and neutral; Z_{ab} represents the equivalent customer loads between the hot wire a and b. Z_{sn} is the impedance of the neutral line in the transformer secondary side, and Z_{sp} is the distribution line impedance. In Figure 2, the current flowing in the primary neutral is labeled as I_{np}, and the current flows in the secondary neutral are labeled as I_{ns}. I_a, V_a and I_b, V_b represent the phase current and voltage of the secondary side. I_P represents the current in the transformer primary side, and I_{R_T} represents the grounding current of the transformer. The unbalanced current in the transformer secondary side is labeled as I_{ub_S} in Figure 2.

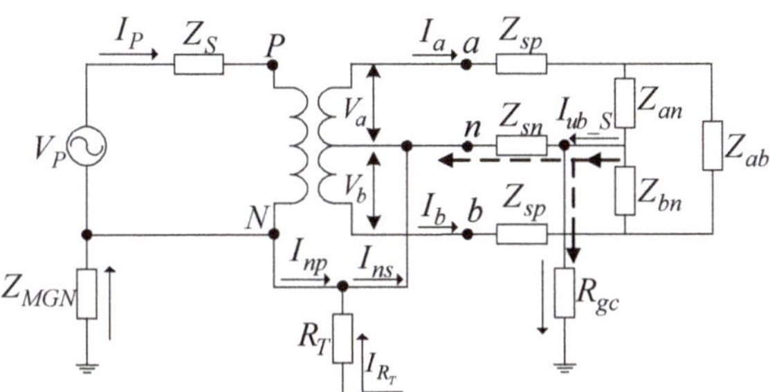

Figure 2. The equivalent analysis circuit for the multi-grounded neutral (MGN) system.

Suppose that the neutral current is totally caused by the customer, and the utility has no contribution. According to Figure 2, there are two circulation paths for this unbalanced current caused by the customer loads. One part of it flows through the secondary neutral line and then back to the customer loads. The other part flows into the ground through the customer grounding R_{gc} and then flows from the earth to the transformer grounding R_T and primary grounding Z_{MGN}. This neutral current further circulates back to the transformer secondary neutral.

According to the above descriptions, the equivalent circuit for the neutral current circulation can be simplified as shown in Figure 3. In the figure, the current source I_U is determined by the unbalanced voltage V_U and the equivalent load impedance Z_U, which is caused by the imbalance of the two single-phase loads (Z_{an} and Z_{bn}). The analytical formulas to calculate Z_U and I_U are given in Equations (1) and (2).

$$Z_U = Z_{an}//Z_{bn} = \frac{Z_{an} \times Z_{bn}}{Z_{an} + Z_{bn}} \tag{1}$$

$$I_U = V_U/Z_U \tag{2}$$

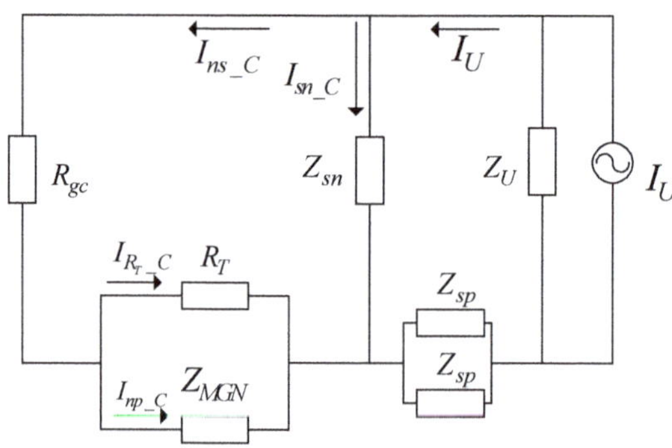

Figure 3. The equivalent analysis circuit for the unbalanced current totally caused by the customer.

V_U is the equivalent unbalanced voltage source, which can be calculated based on the transformer secondary rated voltage V_N.

$$V_U = \frac{Z_{bn} - Z_{an}}{Z_{an} + Z_{bn}} \times V_N \tag{3}$$

In Figure 3, I_{sn_C} represents the current flowing through the secondary neutral line, and I_{ns_C} represents the current flowing into the ground through the customer grounding impedance R_{gc}. The current I_{ns_C} is split into two parts as I_{RT_C} and I_{np_C}. The subscript C in each variable represents that the neutral currents are caused only by the customer. Based on Figure 3, it can be seen that the current flowing in the neutral of the primary system I_{np_C} can be calculated by Equation (4).

$$I_{np\_C} = \frac{R_T}{R_T + Z_{MGN}} I_{ns\_C} \tag{4}$$

Here, a parameter g is introduced, which is shown in Equation (5).

$$g = \frac{R_T}{R_T + Z_{MGN}} \tag{5}$$

It can be seen that the value of parameter g is determined by the transformer grounding impedance R_T and the primary neutral equivalent impedance Z_{MGN}. For an MGN system, the transformer grounding resistance R_T is usually constant. Therefore, the value of g is mainly determined by the equivalent neutral impedance Z_{MGN} of the primary system. If the primary system neutral is broken, the equivalent impedance Z_{MGN} should increase, and the value of parameter g should decrease. Therefore, the variation for the value of parameter g can be used to reflect the primary neutral condition.

2.2. Impedance Determination under Normal Operating Conditions

Under normal operating conditions, the value of transformer grounding impedance R_T can be estimated based on the substation database. The value of impedance Z_{MGN} can be determined based on the following procedures.

The primary neutral line of an MGN system is naturally a ladder network, as shown in Figure 4, where z_{pn} is the primary neutral conductor impedance between two grounding poles. R_{gn} represents the pole grounding resistance. $Z_{lad(k)}$ and $Z_{lad(k+1)}$ represent the equivalent impedance for the multiple grounding system with k or $(k+1)$ grounding poles, respectively. The equivalent impedance of a ladder network with $(k+1)$ grounding poles can be calculated according to Equation (6) [18].

$$Z_{lad(k+1)} = \frac{Z_{lad(k)}R_{gn}}{Z_{lad(k)} + R_{gn}} + z_{pn} \tag{6}$$

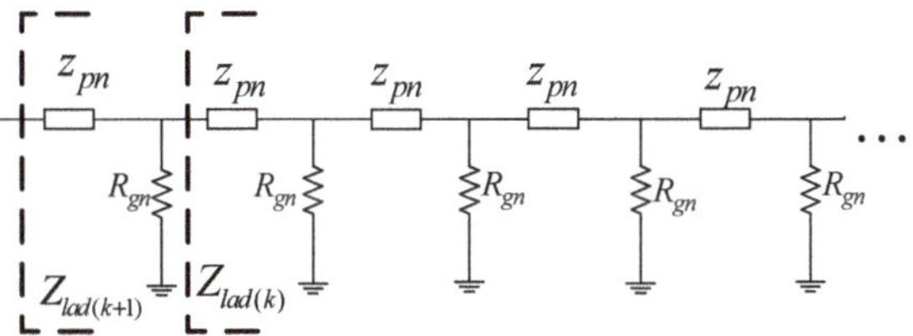

Figure 4. The equivalent circuit of multi-grounding neutral ladder network.

For the ladder network shown in Figure 4, a ladder network has the property that after certain more parallel networks are added, the change in the equivalent impedance becomes insignificant; therefore, when the number k is sufficiently large, $Z_{lad(k)} \approx Z_{lad(k+1)}$.

Then, Equation (6) can be expressed as Equation (7).

$$Z_{lad(k)} = \frac{Z_{lad(k)}R_{gn}}{Z_{lad(k)} + R_{gn}} + z_{pn} \tag{7}$$

After certain rearrangement of Equation (7), Equation (8) can be obtained:

$$\begin{aligned} Z_{lad}^2 - z_{pn}Z_{lad} - z_{pn}R_{gn} &= 0 \\ Z_{lad} &= \tfrac{1}{2}\left(z_{pn} \pm \sqrt{z_{pn}^2 + 4z_{pn}R_{gn}}\right) \end{aligned} \tag{8}$$

The subscript (k) is omitted for simplicity in Equation (8). The positive root of Z_{lad} is the equivalent impedance of the ladder network; thus, Equation (9) is derived.

$$Z_{lad} = \frac{1}{2}\left(z_{pn} + \sqrt{z_{pn}^2 + 4z_{pn}R_{gn}}\right) \tag{9}$$

Since the neutral conductor impedance z_{pn} is much smaller than R_{gn}, Equation (9) can be simplified to Equation (10).

$$Z_{lad} \approx \sqrt{z_{pn}R_{gn}} \tag{10}$$

The impedance calculation result based on Equation (10) is compared with the results from Equations (6) and (9) for an example system. The parameters for the MGN system are: $z_{pn} = (0.0427 + j0.0961)\,\Omega$ and $R_{gn} = 15\ \Omega$. The estimation results for the equivalent impedance of the ladder network are shown in Figure 5. It is apparent that the equivalent impedance of the ladder network reaches a constant value when the length of the ladder network is larger than 1000 m.

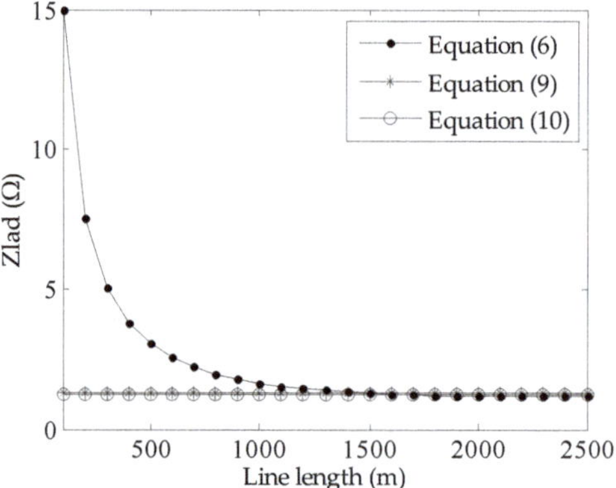

Figure 5. The system equivalent impedance versus the neutral line length.

The neutral system seen from the transformer primary node N (referring to Figure 2) can be equivalent to the circuit shown in Figure 6. It can be seen that, from the transformer connection point N, the multi-grounding primary system can be seen as two parallel ladder networks, labeled as Z_{lad1} and Z_{lad2}. The equivalent impedance of each network can be estimated by Equation (10). Additionally, the equivalent primary system impedance seen from the transformer connection point can be calculated by the parallel connection of Z_{lad1} and Z_{lad2} according to the analytical formula shown in Equation (11).

$$Z_{MGN} \approx Z_{lad1} // Z_{lad2} = \sqrt{z_{pn}R_{gn}} // \sqrt{z_{pn}R_{gn}} = \frac{1}{2}\sqrt{z_{pn}R_{gn}} \tag{11}$$

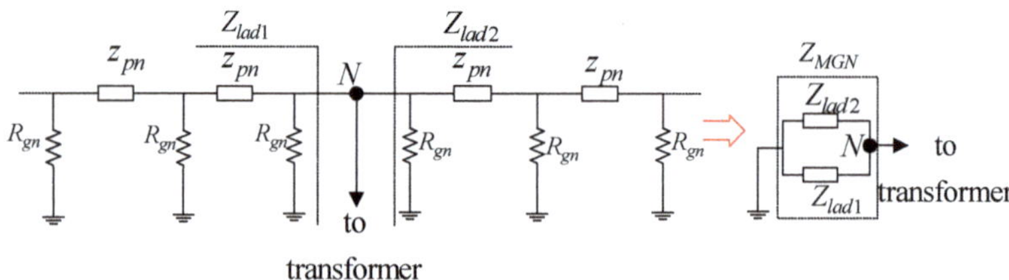

Figure 6. Estimation of Z_{MGN} under normal operating conditions.

2.3. Impedance Determination Based on the Measurement Data

Referring to the multi-grounding system shown in Figure 2, the neutral current in the primary system is caused simultaneously by both the utility and customer; therefore, the primary neutral current I_{np} can be divided into two parts, as shown in Equation (12).

$$I_{np} = I_{np\_U} + I_{np\_C} \tag{12}$$

I_{np_U} is the primary neutral current caused by the utility, and I_{np_C} is the primary neutral current caused by the customer. Similarly, the neutral current in the secondary system I_{ns} is also composed of two parts, I_{ns_U} and I_{ns_C}.

$$I_{ns} = I_{ns\_U} + I_{ns\_C} \tag{13}$$

Bring Equations (4) and (5) into Equations (12) and (13), the following relation can be derived.

$$I_{np} = I_{np\_U} + g(I_{ns} - I_{ns\_U}) \tag{14}$$

Based on the measurement data of two consecutive time instants t_1 and t_2, Equation (14) can be rewritten as Equations (15) and (16).

$$I_{np,1} = I_{np\_U,1} + g(I_{ns,1} - I_{ns\_U,1}) \tag{15}$$

$$I_{np,2} = I_{np\_U,2} + g(I_{ns,2} - I_{ns\_U,2}) \tag{16}$$

Subscripts 1 and 2 indicate that the data are for time instants t_1 and t_2, respectively. Subtracting Equation (15) from Equation (16), then Equation (17) can be obtained.

$$\Delta I_{np} = \Delta I_{np\_U} + g(\Delta I_{ns} - \Delta I_{ns\_U}) \tag{17}$$

Here, Δ is used to indicate the difference between the values of each variable at different time instants. From Equation (17), it can be seen that, if the utility causes no variation to the neutral current, ΔI_{np_U} and ΔI_{ns_U} will both equal zero. Equation (17) can be further simplified to Equation (18); therefore, parameter g can be calculated by Equation (19).

$$\Delta I_{np} = g \Delta I_{ns} \tag{18}$$

$$g = \frac{\Delta I_{np}}{\Delta I_{ns}} \tag{19}$$

Therefore, the variation of I_{np} and I_{ns} can be used to estimate the value of parameter g. Furthermore, based on the estimated parameter g, the primary neutral impedance Z_{MGN} can be estimated from Equation (20) according to Equation (5).

$$Z_{MGN} = \frac{1 - g}{g} \times R_T \tag{20}$$

2.4. Determination of the Primary Neutral Condition

Under normal operating conditions, the equivalent impedance Z_{MGN} of the primary neutral system can be calculated according to Equation (11). The actual value of Z_{MGN} can be estimated based on the procedures shown in Section 2.3. When the primary neutral line is loosened or broken, the equivalent impedance of the primary neutral conductor Z_{MGN} should increase. Therefore, by monitoring the value of Z_{MGN}, the condition of the primary neutral integrity can be known.

In the actual application, data acquisition devices can be installed at the service transformer, and the neutral currents I_{np} and I_{ns} can be monitored continuously. Therefore, parameter g can be estimated according to Equation (19) based on the variation of the neutral current. The transformer grounding impedance can be known based on the substation data, and then Z_{MGN} can be estimated by Equation (20).

A sensitivity coefficient SC is proposed to evaluate the discrepancy between the monitored Z_{MGN} and theoretical Z_{MGN}.

$$SC = \frac{|Z_{MGN\_m} - Z_{MGN\_t}|}{|Z_{MGN\_t}|} \times 100\% \tag{21}$$

where Z_{MGN_t} is the theoretical value of Z_{MGN} obtained from Equation (11), which represents the primary neutral impedance of a normal operating system; Z_{MGN_m} is the monitored value of Z_{MGN} based on Equation (20), which can reflect the actual condition of the primary neutral groundings. Under normal operating conditions, the monitored parameter Z_{MGN_m} should equal its theoretical value Z_{MGN_t}, and SC should be of a value nearly of zero. If there is a neutral broken problem happening at the primary side, the value of Z_{MGN_m} will increase. Referring to Figure 1, it is known that the nearer the neutral broken point to the monitored transformer, the larger Z_{MGN} and the larger SC. Therefore, the value of SC not only reflects the primary neutral condition, but also reflects the distance of the primary neutral broken point to the monitored service transformer.

3. Data Selection Criteria

The derivation of Equation (19) is based on the assumption that the neutral current variation is totally caused by the customer and the utility has no contribution. Actually, under most situations, the variation of neutral current is caused by both the utility and customer. Therefore, in order to use Equation (19) to estimate parameter g, the data corresponding to the variation of load should be first selected out.

Under normal operating conditions, the neutral current caused by the utility is mainly due to the unbalanced loads in the system. In order to guarantee that there is no contribution of the utility to the neutral current variation, the load at the primary side should remain constant. On the other hand, in order to select out the data that corresponds to the variation of customer load, there should be enough variation in the phase a and phase b loads of the monitored transformer. In summary, in order to select out the proper data, the following data selection criterion should be satisfied: (a) variations of the sum power consumed by customer side loads have to be small enough; therefore, the neutral current caused by the utility can be neglected; (b) the individual variations for the transformer secondary single phase power P_a and P_b should be large enough in order to guarantee that enough neutral current disturbances can be generated by the customer load. Moreover, the time interval between two consecutive sampled data should be short enough in order to further assure that the current variation caused by the primary side is small. The data that can satisfy the above criterion are selected out, and then, the time instants corresponding to these data are determined. Since the estimation of parameter g is based on the variation of neutral current in the primary and secondary neutral, the neutral current data corresponding to the above time instants are further selected out and used for the parameter estimation according to Equation (19).

In practice, many groups of neutral current ΔI_{np} and ΔI_{ns} can be selected out according to the above data selection criterion. Assume that there are N groups of neutral current that can meet the above requirements, such as $(\Delta I_{ns,1}, \Delta I_{np,1})$, $(\Delta I_{ns,2}, \Delta I_{np,2})$, $\dots$, $(\Delta I_{ns,i}, \Delta I_{np,i})$, $\dots$, $(\Delta I_{ns,N}, \Delta I_{np,N})$. According to the least square method, parameter g can be estimated based on the selected datasets as shown in Equation (22).

$$g = \frac{\sum_{i=1}^{N} (\Delta I_{np,i} - \Delta \overline{I}_{np})(\Delta I_{ns,i} - \Delta \overline{I}_{ns})}{\sum_{i=1}^{N} (\Delta I_{ns,i} - \Delta \overline{I}_{ns})^2} \tag{22}$$

where subscript i indicates the i-th selected data, and $\Delta \overline{I}_{np} = \sum_{i=1}^{N} \Delta I_{np,i}/N$, $\Delta \overline{I}_{ns} = \sum_{i=1}^{N} \Delta I_{ns,i}/N$.

After obtaining the value of parameter g, the equivalent impedance for the primary multi-grounding Z_{MGN_m} can be estimated, and the sensitivity coefficient SC can be determined according to Equation (21). Parameter Z_{MGN_m} can be estimated at a certain time interval based on the continuously monitored data, and thus, the value of SC can be updated for each time period. The broken or loosened problems happening in the primary neutral line can be reflected by the variation of SC. A threshold value (TV) to evaluate the variation of SC should be set beforehand based on the parameters of the power system. If SC is less than TV, this indicates that the condition of the primary neutral system is good. Otherwise, one should check the primary neutral condition near the monitored transformer.

To summarize, the estimation flowchart of the proposed method for the monitoring of the primary neutral integrity is presented in Figure 7.

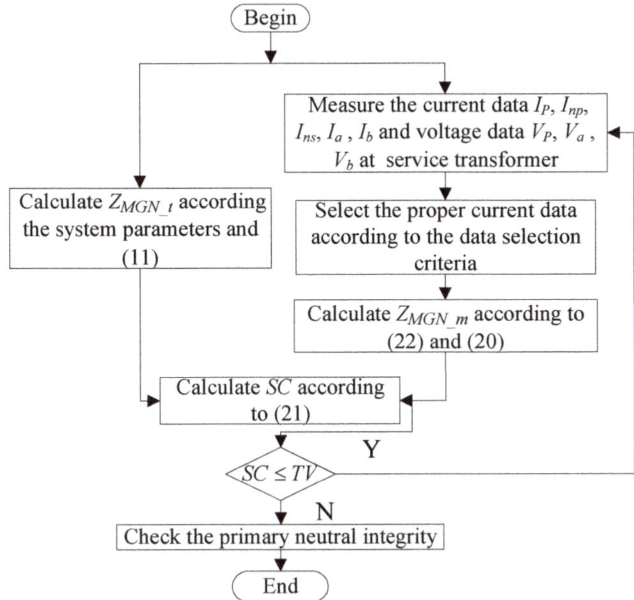

Figure 7. Flowchart of the proposed method for monitoring the primary neutral integrity.

4. Simulation Verification

In this section, the validity of the proposed method is verified based on the simulation studies. There are three parts for the verification. Firstly, the validity of using the variation of neutral current to calculate parameter g is verified based on the system shown in Figure 2. Secondly, Monte Carlo simulation is set up to imitate a real distribution network that contains five residential houses. The parameters g and Z_{MGN} are estimated, and SC is calculated based on the proposed method; and the results are compared with their theoretical values. Finally, in the third part, the broken primary neutral situations are simulated, and the proposed method is used to reflect the primary neutral condition.

4.1. Validity Verification of Parameter g

The simulation system is constructed as the equivalent circuit shown in Figure 2. The system is 60 Hz, and the capacity of the service transformer is 37.5 KVA. The rated voltage for the primary and secondary of the step-down transformer is 14.4 kV and 120 V, respectively. Other parameters of the system are listed in Table 1. The system is set up in MATLAB Simulink (The MathWorks, Natick, MA, USA).

Table 1. Equivalent circuit parameters for the studied system.

#	Parameters	Values
	Z_S (Ω)	0.00249
Primary System	Z_{MGN} (Ω)	0.4404
	R_T (Ω)	15
	R_{gc} (Ω)	1.0000
Secondary System	Z_{sn}, Z_{sp} (Ω)	0.0498, 0.0249
	Z_{an}, Z_{bn}, Z_{ab} (Ω)	20, 12, 10

Based on the analysis in Section 2, it is known that the value of parameter g is influenced by the primary system equivalent neutral impedance Z_{MGN} and the transformer grounding impedance R_T. In order to verify the proposed method for systems under different neutral integrity conditions, the following three scenarios are simulated.

Scenario 1: A normal operating condition of the system with $R_T = 15\ \Omega$ and $Z_{MGN} = 0.4404\ \Omega$.

Scenario 2: A broken primary neutral condition with $R_T = 15\ \Omega$ and $Z_{MGN} = 7.5\ \Omega$.

Scenario 3: A transformer neutral broken condition with $R_T = 150\ \Omega$ and $Z_{MGN} = 0.4404\ \Omega$.

The theoretical value of parameter g can be calculated according to Equation (5). Additionally, for the above three scenarios, the theoretical values of g are shown below.

$$\text{Scenario1} : g = \frac{R_T}{R_T + Z_{MGN}} = \frac{15}{15 + 0.4404} = 0.9715$$
$$\text{Scenario2} : g = \frac{R_T}{R_T + Z_{MGN}} = \frac{15}{15 + 7.5} = 0.6667$$
$$\text{Scenario3} : g = \frac{R_T}{R_T + Z_{MGN}} = \frac{150}{150 + 0.4404} = 0.9971$$

Under normal operating conditions, parameter g can be estimated based on the variations of neutral current I_{np} and I_{ns}. Additionally, the parameter calculation should be under the assumption that the neutral current variation is caused only by the customer. Therefore, by adjusting the individual load power in the two hot wires of the customer and at the same time maintaining the total load power to be constant, many cases have been generated. Due to space limitation, the results for four cases of each scenario are shown in Table 2. Based on the variation of the neutral current, parameter g can be calculated based on Equation (19), and the estimation results of g are compared with their theoretical value.

Table 2. Simulation data for the estimation of parameter g in three different scenarios.

#	Parameters	Case 1	Case 2	Case 3	Case 4
Equivalent customer loads	$Z_{an}\ (\Omega)$	17	18	19	20
	$Z_{bn}\ (\Omega)$	13.4	12.9	12.4	12
	$Z_{ab}\ (\Omega)$	10	10	10	10
Scenario 1	$I_{ns}\ (A)$	0.06277	0.08725	0.1113	0.1324
	$I_{np}\ (A)$	0.06098	0.08476	0.1081	0.1286
Scenario 2	$I_{ns}\ (A)$	0.01533	0.02131	0.02717	0.03233
	$I_{np}\ (A)$	0.01022	0.01420	0.01812	0.02156
Scenario 3	$I_{ns}\ (A)$	0.06230	0.08659	0.1104	0.1314
	$I_{np}\ (A)$	0.06212	0.08633	0.1101	0.1310

From Figure 8, it can be seen that the value of parameter g can correctly reflect the condition of the neutral grounding system. For example, in Scenario 2, when the primary neutral is broken, Z_{MGN} becomes larger, and the value of g should be less than its normal value. Correctly, the estimation result of g based on the variation of the neutral currents reflects this situation. In Scenario 3, when the transformer grounding is broken, R_T becomes larger, and the value of g should become larger than its normal value. Again, the estimated result correctly reflects this situation. Moreover, the estimation values of parameter g for all cases nearly overlap with their theoretical values. Therefore, analysis in this section reveals that the proposed parameter g can correctly reflect the neutral condition in the MGN system, and the method to estimate g based on the variation of neutral current is valid and accurate.

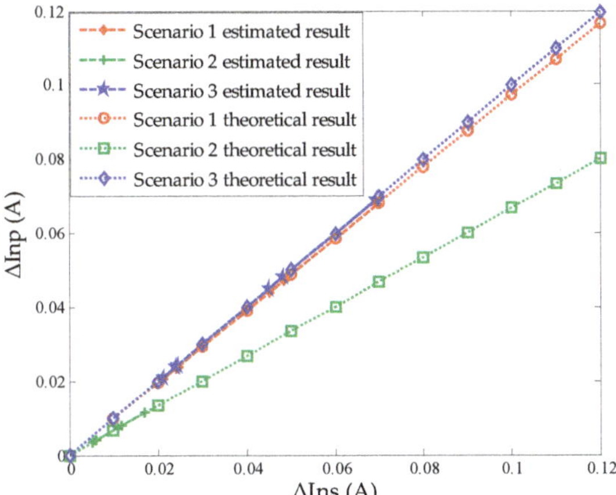

Figure 8. Results comparison for parameter g under three different scenarios.

4.2. Verification of the Proposed Method Based on a Typical MGN Network

A typical MGN network in North America is depicted in Figure 9. It contains a three-phase 25-kV MGN distribution system, a single-phase service transformer (14.4/0.12 kV) and five houses connected in parallel in the secondary side of the transformer. The span of primary neutral between two grounded points is 100 m [17]. The total length of the distribution line is 5 km. The distance between two houses is 20 m. Since the residential loads of each house are time-varying during a day, the Monte Carlo method is adopted to simulate this characteristic [19]. The Monte Carlo simulation is performed to provide a realistic one-hour load behavior for the studied system. Then, the proposed method can be verified based on the data provided by the Monte Carlo simulation.

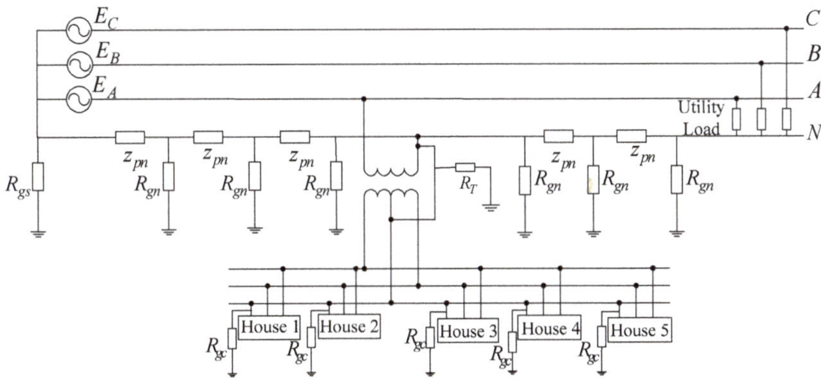

Figure 9. Single-line circuit for the typical MGN system.

Other parameters for the simulated system are listed in Table 3, where R_{gs} is the substation grounding resistance (less than 1 Ω), R_{gn} is the pole grounding resistance, Z_{line1} is the primary phase conductor impedance, z_{pn} is the primary neutral conductor impedance, R_T is the transformer grounding resistance, R_{gc} is the customer grounding resistance, Z_{line2} is the secondary phase conductor impedance and Z_{sn} is the secondary neutral conductor impedance.

Table 3. System parameters of the Monte Carlo simulation.

Parameter	Value
R_{gs} (Ω)	0.15
R_{gn} (Ω)	15
Z_{line1} (Ω/km)	0.2494 + 0.8782
z_{pn} (Ω)	0.04271 + j0.09609
R_T (Ω)	15
R_{gc} (Ω)	1
Z_{line2} (Ω/km)	0.2028 + j0.0936
Z_{sn} (Ω/km)	0.5500 + j0.3650

Under normal operating conditions, the equivalent neutral impedance for the above MGN system is estimated as follows.

$$Z_{MGN\_t} = \frac{1}{2}\sqrt{z_{pn}R_{gn}} = 0.5265 + j0.3422 \tag{23}$$

Therefore, the theoretical value of parameter g can be calculated according to Equation (24).

$$g_t = \frac{R_T}{R_T + Z_{MGN\_t}} = 0.9656 - j0.0213 \tag{24}$$

In the Monte Carlo simulation, the power flow of the system is calculated once every second, so there are 3600 sets of data for the one-hour simulation time period. The datasets of the transformer power are shown in Figure 10. P_a and P_b represent the phase a and phase b loads in the transformer secondary side, respectively. $(P_a + P_b)$ is the total load in this phase of the transformer. S_p represents the total three-phase power for the whole transformer. The quantitative data selection criterion for this system is summarized as shown in Equation (25).

$$\left\{ \begin{array}{l} \Delta(P_a + P_b)/(P_a + P_b) \leq 1\% \\ \Delta S_p/S_p \leq 1\% \\ \Delta P_a/P_a \geq 3\% \\ \Delta P_b/P_b \geq 3\% \end{array} \right. \tag{25}$$

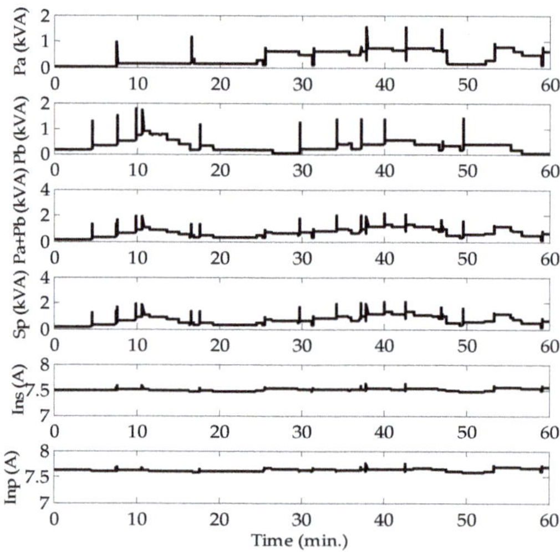

Figure 10. The powers and neutral currents of the transformer under normal operating conditions.

Based on the proposed data selection criteria, the proper power data are selected out, and the corresponding neutral current data are determined. The result of the estimated g is shown in Figure 11 according to Equation (22). Then, according to Equation (20), Z_{MGN_m} can be obtained, and the results of Z_{MGN_m} and SC are shown in Table 4.

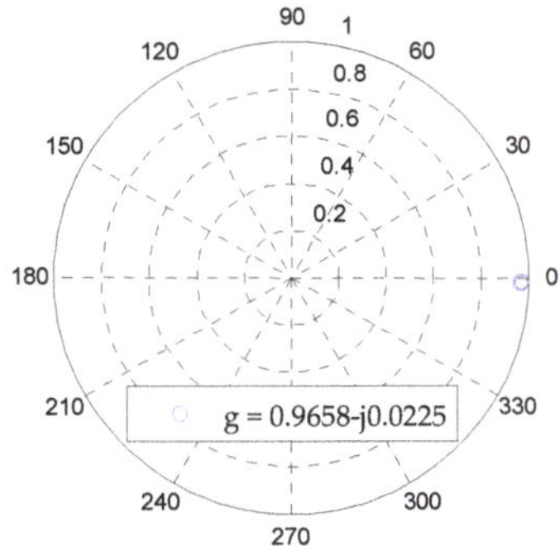

Figure 11. Results of the estimated g based on LSM.

Table 4. Estimation results of Z_{MGN_m} and SC.

Parameter	Value
Z_{MGN_m} (Ω)	$0.5227 + j0.3616$
Z_{MGN_t} (Ω)	$0.5265 + j0.3422$
SC (%)	3.15%

From the results, it can be seen that, under normal operating conditions, the monitored Z_{MGN_m} matches the theoretical Z_{MGN_t} very well, and the value of SC is very small. This means that the quantitative data selection criterion is proper and the proposed method is accurate to estimate the equivalent primary neutral impedance Z_{MGN} under normal operating conditions.

4.3. Monitoring of the Primary Neutral Broken Condition

In order to verify the effectiveness of the proposed method to monitor the neutral condition in the primary system, two different neutral broken cases are simulated. In the first case, the primary neutrals at both sides of the transformer are broken. In the second case, only the primary neutral is broken at one side of the transformer. Moreover, monitoring of the broken transformer grounding neutral is also simulated. Furthermore, in this part, the sensitivity of parameter Z_{MGN} to reflect the primary neutral condition is studied, and the threshold value of SC is proposed.

4.3.1. Broken Primary Neutral at Two Sides

Figure 12 shows the case that the primary neutrals are broken at both sides of the service transformer. The distance between the broken Point A and the monitored transformer is greater than 100 m and less than 200 m. The distance of broken Point B away from the service transformer is less than 100 m. Under this condition, the primary neutral system changes from a ladder network to a single point grounding system. Monte Carlo simulation is used to generate a realistic one-hour load behavior for the system. The powers and neutral currents at the transformer primary and secondary side are shown in Figure 13. The neutral current data that can meet the data selection criteria are selected out, and parameter g is estimated shown in Figure 14 according to Equation (21). The estimated value of parameter Z_{MGN} is shown in Table 5. The theoretical value Z_{MGN_t} under normal operating conditions can be used as a reference to check the estimated Z_{MGN_m} value. Then, the sensitivity coefficient *SC* can be calculated according to Equation (21), which is an indicator that can be used to reflect the primary neutral integrity.

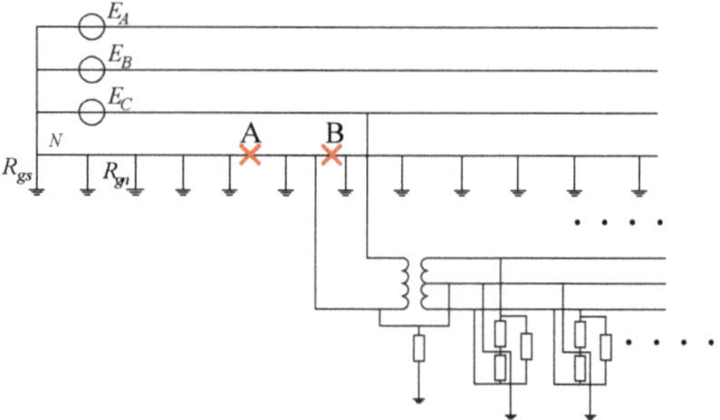

Figure 12. Two broken neutral points at both sides of the monitored service transformer.

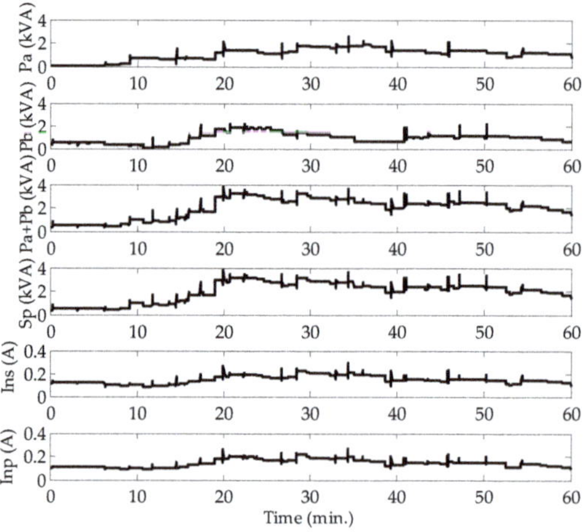

Figure 13. The powers and neutral currents of the transformer for the two-side broken neutral condition.

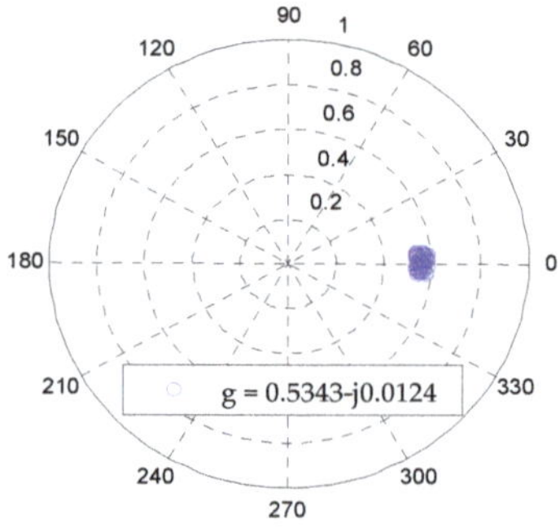

Figure 14. Results of the estimated *g* based on LSM (two broken points).

Table 5. Estimation results of Z_{MGN_m} and *SC*.

Parameter	Value
Z_{MGN_m} (Ω)	13.0590 + *j*0.6512
Z_{MGN_t} (Ω)	0.5265 + *j*0.3422
SC (%)	2002.1%

From Table 5, it can be seen that the value of SC reaches a very high value of 2002.1%, which is a good indicator that the primary neutral is broken near the monitored transformer.

4.3.2. Broken Primary Neutral at One Side

Figure 15 shows the condition that the primary neutral is broken at only one side of the service transformer. The distance of broken Point A is greater than 100 m and less than 200 m from the monitored transformer.

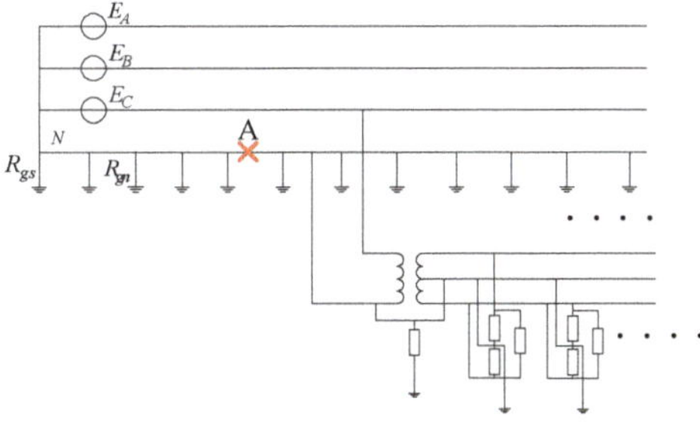

Figure 15. One broken neutral point near the monitored service transformer.

According to Equation (25), the proper neutral current data are selected out from Figure 16. The parameter *g* can be estimated from the selected data, and the result is shown in Figure 17. The theoretical values of Z_{MGN_t} and *SC* are given in Table 6.

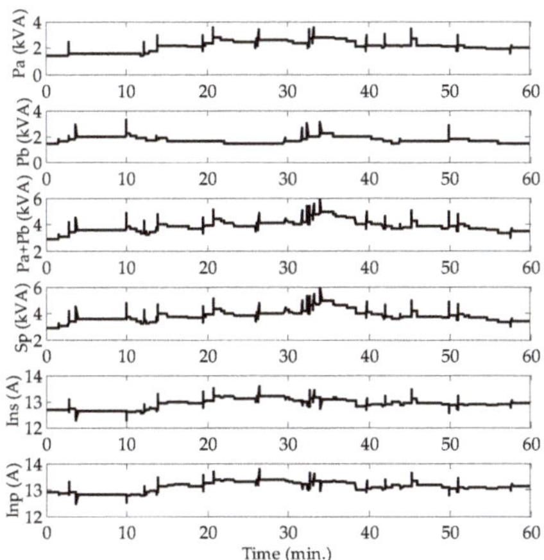

Figure 16. The powers and neutral currents of the transformer for the single-point neutral-broken condition.

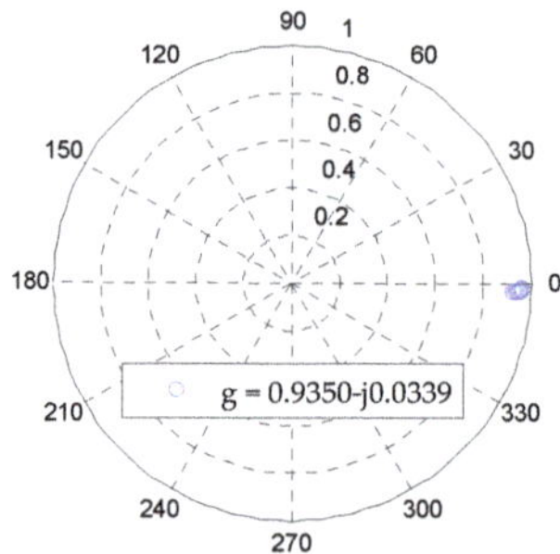

Figure 17. Results of the estimated g based on LSM (single broken point).

Table 6. Estimation results of Z_{MGN_m} and SC.

Parameter	Value
Z_{MGN_m} (Ω)	$1.0216 + j0.5826$
Z_{MGN_t} (Ω)	$0.5265 + j0.3422$
SC (%)	87.64%

Based on the results shown in Table 6, it can be seen that, when the primary neutral is broken at only one side of the transformer, the estimated parameter Z_{MGN_m} is larger than its normal value, and the sensitivity parameter SC can reach a value of 87.64%, which can correctly reflect the broken neutral condition in the primary system near this transformer. However, SC is smaller than the two-broken-point case; this is reasonable since the neutral broken at two sides of the transformer is more serious than the neutral broken at only one side of the transformer. By changing the distances between the neutral broken point and the service transformer, the relationship between the broken distance and parameter SC is analyzed, whose results are shown in Figure 18. Additionally, from the results, it can be seen that, when the broken neutral is close to the transformer, SC is large. As the distance between the broken neutral and the transformer increases, SC becomes smaller. This is reasonable since the sensitivity of SC will get lower for the broken neutral happening far away from the monitored transformer.

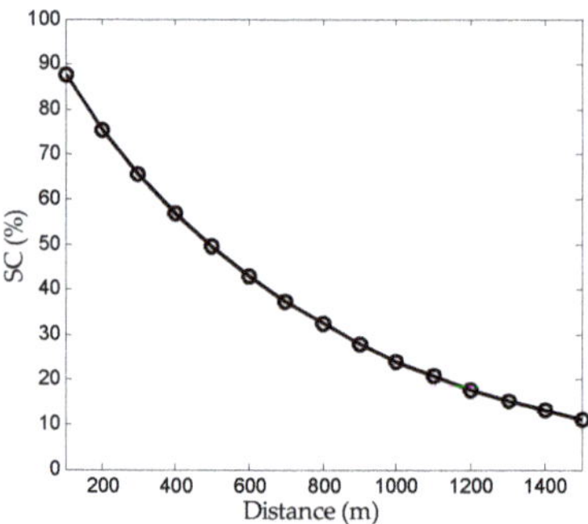

Figure 18. The relationship between the broken point distance and parameter SC.

Based on the analysis results, the threshold value of SC is suggested to be 10%, which also means that the monitoring distance of one service transformer is about 1500 m. Additionally, if the distance between two adjacent transformers is less than 3000 m, the primary neutral condition between these two transformers can be under good monitoring.

4.3.3. Broken Transformer Grounding

The proposed method can also be used to monitor the condition of the transformer grounding. Theoretically, if the transformer grounding is broken, the transformer grounding resistance will significantly increase compared with the normal situation. Therefore, based on Equation (5), it is known that the value of parameter g should increase. Thus, by comparing the value of estimated g and theoretical g, the neutral condition of the monitored transformer can be directly determined.

In this part, the situation of broken transformer grounding is simulated by setting the transformer neutral resistance to 150 ohms. Figure 19 shows the variation of transformer secondary and primary powers. Additionally, proper data can be selected out to calculate parameter g from Figure 19. The results of the estimated g based on the LSM method is $g = 0.9880 - j0.0025$.

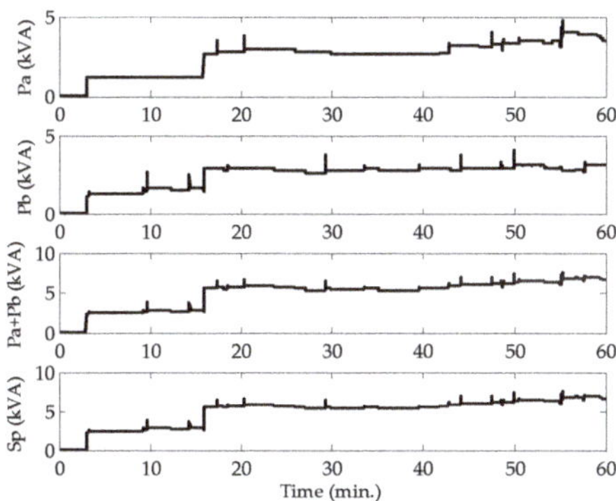

Figure 19. The secondary and primary side power for the transformer.

Therefore, the value of parameter g has increased compared with its normal value, as shown in Equation (24) ($g = 0.9656 - j0.0213$). Therefore, the result indicates that there is a broken problem in the transformer grounding.

5. Conclusions

A new method to monitor the integrity of the primary neutral in an MGN system is proposed in this paper. The method is based on the measurement data at the service transformer. Therefore, it does not need any signal generator and causes no disturbance to the utility and customer. The equivalent impedance for the primary neutral can be estimated based on the monitored neutral currents at the service transformer. By comparing the calculated value of the primary neutral impedance with its theoretical value under normal conditions, the integrity of the multiple grounding neutral in the primary side can be reflected. The method is verified based on an MGN distribution system using the Monte Carlo simulation. From the simulation results, it can be concluded that the proposed parameter can correctly detect the existence of a broken neutral in the primary system. Moreover, the method can also be used to reflect the broken condition in the transformer grounding.

Acknowledgments: This work was supported by the National Natural Science Foundation of China (51577108) and Young Scholars Program Shandong University (2016WLJH07). The authors appreciate Wilsun Xu at the University of Alberta for the discussion about this work.

Author Contributions: Yuanyuan Sun and Xun Long proposed the main idea of the method. Xiangmin Xie and Yuanyuan Sun established the model and validated the proposed method, including conducting the simulation study, doing the sensitivity analysis and writing the paper. Bingwei Zhang has actively contributed to finalizing the manuscript.

Conflicts of Interest: The authors declare no conflict of interest.

1. Chen, T.H.; Yang, W.C. Analysis of multi-grounded four-wire distribution systems considering the neutral grounding. *IEEE Trans. Power Deliv.* **2001**, *16*, 710–717. [CrossRef]
2. Parise, G.; Martirano, L.; Parise, L. Ecodesign of ever net-load microgrids. *IEEE Trans. Ind. Appl.* **2014**, *50*, 10–16. [CrossRef]
3. Bouford, J.D. The utility's perspective on calculating stray voltage. In Proceedings of the 2008 General Meeting of the IEEE-Power-and-Energy-Society, Pittsburgh, PA, USA, 20–24 July 2008; pp. 1–3.
4. Dorr, D.; McGranaghan, M.; Perry, C. Standardized measurements for elevated NEV concerns. In Proceedings of the IEEE/PES Transmission and Distribution Conference and Exhibition, Dallas, TX, USA, 21–26 May 2006; pp. 268–271.
5. Mitolo, M. Of electrical distribution systems with multiple grounded neutrals. *IEEE Trans. Ind. Appl.* **2010**, *46*, 1541–1546. [CrossRef]
6. Burke, J.; Marshal, M. Distribution system neutral grounding. In Proceedings of the 16th IEEE PES Transmission and Distribution Conference and Exposition, Atlanta, GA, USA, 28 October–2 November 2001; pp. 166–170.
7. Agudelo, L.; Ortiz, R.H.; Montoya, M.; Zapata, W.H.; Cardenas, C.; Echeverri, J.C.; Moreno, G. Transferred voltages in multigrounded systems. In Proceedings of the IEEE PES Transmission and Distribution Conference and Exposition-Latin America, Bogota, Colombia, 13–15 August 2008; pp. 1–5.
8. Ciric, R.M.; Ochoa, L.F.; Padilla-Feltrin, A.; Nouri, H. Fault analysis in four-wire distribution networks. *IEE Proc. Gener. Transm. Distrib.* **2005**, *152*, 977–982. [CrossRef]
9. Alam, M.J.E.; Muttaqi, K.M.; Sutanto, D. A three-phase power flow approach for integrated 3-wire MV and 4-wire multi-grounded LV networks with rooftop solar PV. *IEEE Trans. Power Syst.* **2013**, *28*, 1728–1737. [CrossRef]
10. Ciric, R.M.; Feltrin, A.P.; Ochoa, L.F. Power flow in four-wire distribution networks—General approach. *IEEE Trans. Power Syst.* **2003**, *18*, 1283–1290. [CrossRef]
11. Halevidis, C.; Koufakis, E. Power flow in PME distribution systems during an open neutral condition. *IEEE Trans. Power Syst.* **2013**, *28*, 1083–1092. [CrossRef]
12. Choi, J.; Ahn, Y.; Ryu, H.; Jung, G.; Han, B.; Kim, K. A new method of grounding performance evaluation of multigrounded power systems by ground current measurement. In Proceedings of the International Conference on Power System Technology (POWERCON 2004), Singapore, 21–24 November 2004; pp. 1144–1146.
13. Gustafson, R.; Pursley, R.; Albertson, V. Seasonal grounding resistance variations on distribution systems. *IEEE Trans. Power Deliv.* **1990**, *5*, 1013–1018. [CrossRef]
14. Long, X.; Dong, M.; Xu, W.; Li, Y.W. Online monitoring of substation grounding grid conditions using touch and step voltage sensors. *IEEE Trans. Smart Grid* **2012**, *3*, 761–769. [CrossRef]
15. Sunderman, W.G.; Dugan, R.; Dorr, D. The neutral-to-earth voltage (NEV) test case and distribution system analysis. In Proceedings of the 2008 General Meeting of the IEEE-Power-and-Energy-Society, Pittsburgh, PA, USA, 20–24 July 2008; pp. 1–7.
16. Long, X. Distribution System Condition Monitoring Using Active Disturbances. Ph.D. Thesis, University of Alberta, Edmonton, AB, Canada, 2013.
17. IEEE Standard. *National Electrical Safety Code C2-2012*; Institute of Electrical and Electronics Engineers: New York, NY, USA, 2012.
18. Acharya, J.R.; Wang, Y.; Xu, W. Temporary overvoltage and GPR characteristics of distribution feeders with multi-grounded neutral. *IEEE Trans. Power Deliv.* **2010**, *25*, 1036–1044. [CrossRef]
19. Torquato, R.; Shi, Q.; Xu, W.; Freitas, W. A Monte Carlo simulation platform for studying low voltage residential networks. *IEEE Trans. Smart Grid* **2014**, *5*, 2766–2776. [CrossRef]

Article

Nonlinear Robust Control for Low Voltage Direct-Current Residential Microgrids with Constant Power Loads

Martín-Antonio Rodríguez-Licea [1,*], Francisco-Javier Pérez-Pinal [2], Jose-Cruz Nuñez-Perez [3] and Carlos-Alonso Herrera-Ramirez [1]

1   CONACYT-Instituto Tecnológico de Celaya, Celaya, Guanajuato 38010, Mexico; carlos.herrera@itcelaya.edu.mx
2   Departamento de Electrónica, Instituto Tecnológico de Celaya, Celaya, Guanajuato 38010, Mexico; francisco.perez@itcelaya.edu.mx
3   Instituto Politécnico Nacional, IPN-CITEDI, Tijuana, Baja California 22435, Mexico; nunez@citedi.mx
*   Correspondence: martin.rodriguez@itcelaya.edu.mx

Received: 21 March 2018; Accepted: 27 April 2018; Published: 3 May 2018

Abstract: A Direct Current (DC) microgrid is a concept derived from a smart grid integrating DC renewable sources. The DC microgrids have three particularities: (1) integration of different power sources and local loads through a DC link; (2) on-site power source generation; and (3) alternating loads (on-off state). This kind of arrangement achieves high efficiency, reliability and versatility characteristics. The key device in the development of the DC microgrid is the power electronic converter (PEC), since it allows an efficient energy conversion between power sources and loads. However, alternating loads with strictly-controlled PECs can provide negative impedance behavior to the microgrid, acting as constant power loads (CPLs), such that the overall closed-loop system becomes unstable. Traditional CPL compensation techniques rely on a damping increment by the adaptation of the source or load voltage level, adding external circuitry or by using some advanced control technique. However, none of them provide a simple and general solution for the CPL problem when abrupt changes in parameters and/or in alternating loads/sources occur. This paper proposes a mathematical modeling and a robust control for the basic PECs dealing with CPLs in continuous conduction mode. In particular, the case of the low voltage residential DC microgrid with CPLs is taken as a benchmark. The proposed controller can be easily tuned for the desired response even by the non-expert. Basic converters with voltage mode control are taken as a basis to show the feasibility of this analysis, and experimental tests on a 100-W testbed include abrupt parameter changes such as input voltage.

Keywords: robust control; DC-DC converter; constant power load; low voltage direct-current residential microgrid

---

1. Introduction

A Distributed Generation System (DGS) can be understood as a flexible structure in the generation, transmission, distribution and storage of electric energy [1]. Nowadays, it is a common practice to use DGS in smart grids. A smart grid based on a DGS must have the following characteristics: (a) fast response to load transients; (b) energy storage capability, which feeds electricity back to the grid from the client. (c) self-control of power load demands; and (d) intensive use of small power generators based on renewable energy sources [2].

A relatively novel and important kind of smart grid is the DC microgrid. The DC microgrid is generally structured using DC renewable energy sources, for example fuel cells, photovoltaic panels, wind generators with a rectifier and energy storage systems (batteries) with an energy management system, among others. A particularity of the DC microgrid with respect to the smart grids is the integration of different power sources joined through a DC link and connected to local loads. In addition, the energy source can be generated on-site in such a way that the microgrid achieves high efficiency, reliability and versatility [3].

Today, there are numerous electronically-managed power loads that can be connected to the DC-link of a smart grid, such as televisions, electric ovens, coffee makers, LED lamps, computing equipment, among many others. However, it is necessary to mention that power electronics is the key device in the development of a DC microgrid, mainly due to the use of PECs, which allow an efficient power energy conversion between sources and loads. There are different kinds of PECs that can be connected to a DC microgrid in order to reduce, invert or increase the desired/nominal DC-link voltage, i.e., buck, boost, interleaved boost, positive buck/boost, forward, fly-back, half bridge and full bridge converter, to name a few [4].

However, for this paper, a low voltage DC microgrid for building/residential applications similar to the one proposed in [5] is considered as a benchmark. This arrangement is divided into three sections as in the following description. The first one includes high power elements for common building facilities (greater or equal to 538 V and 10 kW), including photovoltaic panels, wind generators, fuel cells, among others. The second one relies on the medium power elements, which are usually found in kitchen appliances and laundry rooms ($[230, 400]$ V, $[0.4, 10]$ kW). It is necessary to mention that the current loading can remain similar to the 230-V AC system. However, the voltage can be increased up to 400 V if a size reduction is needed. Finally are the low power loads, which include devices located in bedrooms, living rooms and outdoor areas ($[24, 48]$ V, less than 0.4 kW).

It is well known that in a closed-loop scenario, the load can act as a constant power load (CPL) for its source, producing a negative-rate varying resistance (known as negative impedance), which destabilizes the system. Since the theoretical analysis complexity increases with this characteristic, the development of different approaches to ensure the stability of the closed-loop system has been encouraged. Traditional CPL compensation techniques rely on a damping increment by adapting the voltage level of the source or the load, adding external circuits or by using advanced control techniques. An interesting review of these approaches was reported in [6], which classifies the CPLs into two groups, control approaches and auxiliary circuits/power buffer. In the first group, control techniques such as feedback linearization, sliding mode, pulse adjustment and pole placement were reviewed such that their main benefit is that they do not diminish the load performance. In the second group, the instability problem seems solved; however, an auxiliary circuit is needed, increasing the implementation cost and complexity; the overall system stability is omitted.

Another attractive survey can be found in [7] in which a review of the state of the art in the performance and usual properties of DC microgrids connected to CPLs, stability approaches and compensation methods were described. Recently, a novel discrete-time modeling and an active stabilizer for PWM converters was proposed in [8]. This proposal allows performing the stability analysis for the DC distribution systems with nonlinear characteristics. The authors reported that the proposed discrete-time method could identify slow-scale and fast-scale instabilities and also a measure of the sensitivity to some parameter change. The robustness of the proposal to filter the parameters' variation was also studied. Finally, this proposal was numerically- and experimentally-verified. Nevertheless, a formal analysis of the CPL case is not reported.

On the other hand, an active damping method was stated in [9]. In this paper, a variation of the CPL in the system by attaching a supercapacitor-based energy storage system was proposed. As a result, the CPL's instability effects can be decreased by virtually increasing the resistive loads. Numerical simulations and practical results were reported. The limitation of this proposal relies on the extra components needed such as a supercapacitor and a bidirectional converter; besides, the production of an energy consumption increase and a slow dynamics response are present.

In contrast, [10] present a sufficient condition for the local stability of DC linear and time-invariant (LTI) circuits with CPLs. A set of steps that should be met for the reported method was also proposed. In order to validate their proposal, two cases were analyzed: a single-port DC RLC circuit with a constant voltage input connected to an ideal CPL and a two-port DC circuit with a single constant voltage input connected to two CPLs, both modeled as LTI systems. Likewise, in [11], a sum of squares (SOS) technique for DC power networks with CPL was presented. In this paper, the region of attraction of the system was estimated by using the SOS and a semidefinite programming algorithm. Additionally, two methods were proposed and compared for the calculation of the region of attraction; it was stated that a reduction of the estimation was achieved by increasing the degree of a Lyapunov function. Furthermore, [12] reported a generalized method to algebraically analyze the stability of DC distribution systems. The authors derive necessary and sufficient conditions for stability by algebraically determining the system's eigenvalues. Furthermore, an insight into the parameter's relationship with the DGS stability was presented. Numerical simulations were reported to validate the proposed method, and unfortunately, practical results were not presented. Another interesting approach was reported in [13]. In this paper, a load side compensation technique was proposed by using a sliding mode (SM) approach. Additionally, a comparative study was performed with a PID controller. Besides the good dynamic response achieved by SM, only numerical simulations were reported. In [14], inner and outer control loops for higher order PECs, such as SEPIC, Cuk and Zeta, connected to a CPL were proposed. For the inner loop, a virtual resistance, connected in series with the input inductor, worked as an active damping stage. An integral controller was added in the outer loop with the main aim to compensate the output voltage steady state error. In this paper, the stability conditions for the equilibrium points and a validation given by numerical simulation were reported. Practical experiments for a high order PEC were also reported. In [15], the DC microgrid stability was analyzed for a known quantity of CPLs. In this paper, the problem was formulated as a robust stability problem of a polytopic uncertain linear system and a set of sufficient conditions to guarantee, at least locally, the robust stability. On the other hand, [16] investigated the DC microgrid with a resistive-inductive load in a CPL scheme using a droop control. A stability condition was discussed, and the corresponding stable region was obtained. Validation of the proposal was reported by numerical simulations. Unfortunately, practical results on a test bed were not reported. In [17], two controllers were proposed: the first an integer order and the second a fractional order, for sliding mode controllers in a DC/DC boost converter feeding a CPL on a typical DC microgrid scheme. Nonlinear sliding manifolds and stability were also researched, and numerical simulations to compare their performance were reported. Nevertheless, practical results were not reported. Further information can be found in [18]. Finally, in [19], the DC microgrid stability was analyzed for a known quantity of CPLs. The problem was formulated as a robust stability one for a polytopic uncertain linear system. A set of sufficient conditions to guarantee, at least locally, the robust stability was obtained. Several numerical simulations were reported to show the effectiveness and non-conservativeness of this proposal. Unfortunately, experiments were not performed.

From the previous paragraph, it can be noted that a large amount of research work has been done to solve the instability problem imposed by the CPLs and that many methods have been proposed to overcome the problem, but none of them provide a simple and more general solution for this issue. In this paper, the mathematical modeling, the robust control and the stability analysis for the basic PECs (buck, boost and buck-boost), even in a reconfigurable scheme dealing with CPLs and parameter variation in continuous conduction mode (CCM), are presented to gain knowledge about more complex configurations with CPLs. The reconfigurable converter presented in this paper is a kind of PEC that can change its structure while the activity is going on (on-the-fly), providing a high flexibility and increased operating range with few components; see, for example, [20–22].

The major contributions of the proposed approach are listed below:

- Robust stability against arbitrary (bounded) changes in source voltage, output power, inductance and capacitance on the desired operating-point.

- Flexibility for use with the basic converters and even in a reconfigurable scheme.
- Low implementation time and cost.
- Easy tuning for a certain desired response.

This paper is structured as follows. In Section 2, a unified model for the buck, boost and buck-boost PECs, in continuous conduction mode, voltage mode and current mode, is presented; then, the instability with CPLs is analyzed. In Section 3, a feedback linearization is used to build a linear parameter varying (LPV) system of the voltage-mode model. Robust stability against arbitrary (bounded) changes in source voltage and output power are analyzed so that the nonlinear CPL converter is stabilized on the desired operating point. In this section, a discussion is presented regarding the advantages and limitations of this scheme. After that, a Taylor series linearization is introduced for voltage and current mode; the latter in order to obtain the polytopic LPV representation for both cases. The advantages of the polytopic LPV as a natural representation of the CPL converter are highlighted. Subsequently, robust controllers are proposed in order to demonstrate analytically the stability of the polytopic systems despite that the CPLs are alternated/switched. It is necessary to mention that abrupt changes on CPL are a typical condition in renewable energy scenarios. In order to illustrate the robust stability of the proposed controller, even under arbitrary (bounded) changes, a 100-W testbed was implemented by using an inexpensive 100-W P-MOSFET and an 8-bit microchip microcontroller. Representative experimental tests with CPL, while an output voltage stabilization is reached, are presented in Section 4. Tests include parameter changes such as input voltage and CPL alternating. In this section, a brief comparison with other proposals is given. In the last section of this paper, final conclusions are given, which summarize the feasibility of the proposed model and control strategy to be robust in CPL scenarios and to be implemented on a low-cost platform.

2. Modeling

In this section, the models of the unified PEC for voltage-mode control and for current-mode control in continuous conduction mode are presented. That is, the mathematical models that are presented are valid for buck, boost and buck-boost converters. As reported in [7,23], the PEC dynamics for voltage and current control mode in discontinuous conduction mode with CPL are open-loop stable, and they are not within the scope of this article.

2.1. Voltage-Mode Modeling

In a microgrid scenario, several CPLs are switched depending on client necessities; each load can be modeled as a resistance with an additional current source parallel to its value of current, indirectly proportional to the voltage and directly proportional to the value of the constant power dissipated. A PEC provides adequate voltage/current levels to the CPL, but, since different voltage levels in the DC-link are needed for different kinds of CPLs, a PEC is attached to each DC-link in such a way that a cascaded configuration is structured. An example of such a configuration is shown in Figure 1. For instance, a buck, a boost or a buck-boost converter stands for the PEC in the most common scenarios. In order to consider any of the above PECs, even in an on-the-fly reconfiguration between them, the reconfigurable PEC (RPEC) from [24] is used. This means that the following analysis is valid for any of the mentioned configurations even if a reconfiguration is performed. Consider the RPEC in Figure 2; each mode/reconfiguration can be modeled as follows.

The overall power demand is provided by one or more power supplies; however, this scenario can be easily simplified from the point of view of each RPEC, considering that the power source is a time-varying voltage/current source and that the load is a CPL (orange block in the figure) with a piecewise constant level or even as a fast varying power level. In this study, the CPL is simplified by a time-varying resistance with known bounds (known and bounded CPL levels); for instance, with S1, S2 closed and S3 open, a buck converter is configured with a CPL in its output as shown in Figure 3. In this figure, the schematics of a buck-type PEC with varying parameters are shown.

The load resistance $R(t)$ is indirectly proportional to a constant power value and directly proportional to the square of the output voltage $V_0(t)$, and it can represent a set of CPLs. Since the objective is to maintain $V_0(t)$ at a set-point level V_x, a voltage/duty cycle representation is obtained as follows.

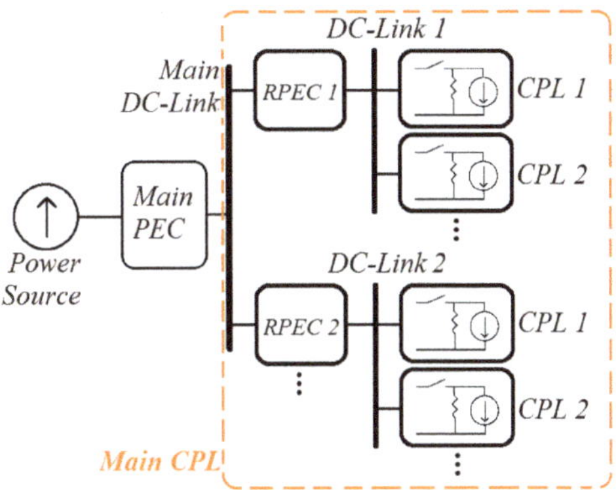

Figure 1. Illustrative block diagram of a microgrid. RPEC, reconfigurable power electronic converter.

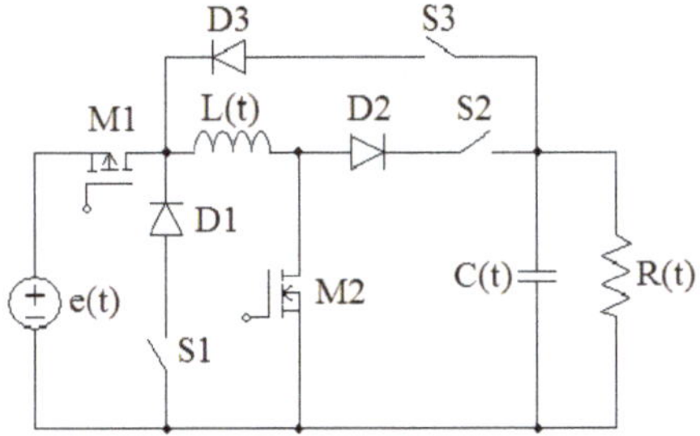

Figure 2. Reconfigurable PEC of this paper.

Consider the nominal values for $E(t) = e$, $L(t) = L$ and $C(t) = C$; note that later in this paper, a design range for the variation of these parameters will be considered for the stability analysis. Similar to [25], by comparing the particular dynamic systems' descriptions with respect to the MOSFET (M) state, with ideal components, a dynamic model is obtained:

$$L\frac{di}{dt} = u_d e - V_o \tag{1}$$

$$C\frac{dV_o}{dt} = i - \frac{P}{V_o} \tag{2}$$

where the time dependence is omitted for readability and $u_d \in \{0,1\}$ is a discrete function that represents the state of M with a 1 for closed and a 0 for open, in a high enough switching frequency operation. Redefining the control variable u_d as a sufficiently smooth function taking values in the compact interval $[0,1]$, the averaged dynamic description that predicts the behavior of the PEC with CPL is:

$$L\frac{di}{dt} = ue - V_o \tag{3}$$

$$C\frac{dV_o}{dt} = i - \frac{P}{V_o} \tag{4}$$

The C^∞ differentiability property of the above system allows replacing Equations (3) in (4) to obtain the voltage-mode PEC with the CPL mathematical description:

$$\dot{x}_1 = x_2 \tag{5}$$

$$\dot{x}_2 = \frac{e}{LC}u - \frac{1}{LC}x_1 + \frac{P}{Cx_1^2}x_2 \tag{6}$$

where $x_1 = V_o$ and $x_2 = \dot{V}_o$.

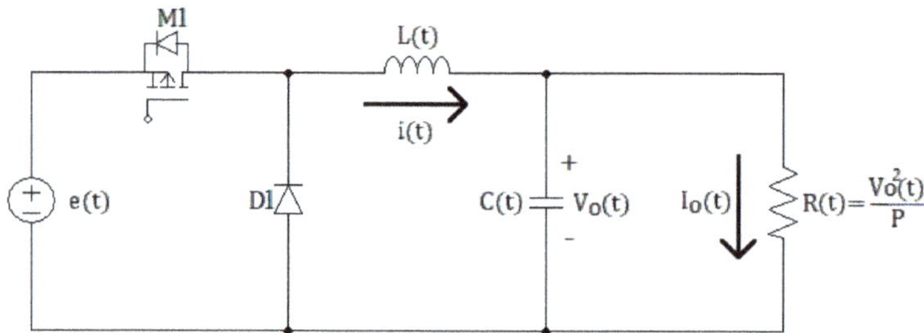

Figure 3. Schematic of the buck converter with a constant power load (CPL).

From [24], the above mathematical representation is valid also for boost and for buck-boost topologies depending on the u value, that is in a reconfigurable scheme. Three operation modes are allowed, and each one of them is selected by a proper switching activation and by a pulse width modulation (PWM) MOSFET activation:

1. S1, S2 closed, S3, M2 open and PWM switching on M1, buck.
2. M1, S2 closed, S1, S3 open and PWM switching on M2, boost.
3. M2, S3 closed, S1, S2 open and PWM switching on M1, buck-boost.

where the following is obtained:

- a buck PEC if $0 \leq u \leq 1$
- a boost PEC if $u > 1$
- a buck-boost PEC if $u < 0$

and with the PWM duty cycle for buck ($\breve{u}$), boost ($\hat{u}$) and buck-boost ($\tilde{u}$) calculated by:

- buck duty-cycle for $0 \leq u \leq 1$ is $u = \breve{u}$
- boost duty-cycle for $u > 1$ is $u = \frac{1}{1-\hat{u}}$
- buck-boost duty-cycle for $u < 0$ is $u = \frac{\tilde{u}}{\tilde{u}-1}$

Note that a few extra components are needed for the reconfigurable on-the-fly topology.

On the other hand, it is known that the instability property of a nonlinear system can be inferred from its linearization instability property (Theorem 4.15 of [26]). The linearization of (5) and (6) in $x_1 = V_x$, $x_2 = 0$ and $u = u_x$ is:

$$\dot{x}_1 = x_2 \tag{7}$$

$$\dot{x}_2 = \frac{e}{LC}u - \frac{1}{LC}x_1 + \frac{P}{CV_x^2}x_2 \tag{8}$$

and it has the following transfer function:

$$\frac{x_2(s)}{u(s)} = \frac{e}{LCs^2 - \frac{PL}{V_x^2}s + 1} \tag{9}$$

whose poles remain in the right-side complex semi-plane, even with no varying parameters, so that the instability property can be extended to the nonlinear system. Therefore, technically there exists the possibility of an output voltage drop for the open loop PEC with CPL even with time-invariant parameters. In the following section, a nonlinear control law that stabilizes the (7) and (8) system, even with varying parameters, is shown.

2.2. Current-Mode Modeling

From the dynamic system (1) and (2), it is easy to demonstrate that:

$$\frac{dI_o}{dt} = \frac{I_o^3 - iI_o}{CP} \tag{10}$$

$$\frac{di}{dt} = \frac{ue}{L} - \frac{P}{I_o L} \tag{11}$$

with u defined as in the previous section. The C^∞ differentiability property of the above system allows replacing Equations (10) in (11) to obtain the unified mathematical description:

$$\dot{y}_1 = y_2 \tag{12}$$

$$\dot{y}_2 = \frac{2y_1^2 y_2}{CP} + \frac{y_2^2}{y_1} - \frac{euy_1}{CPL} + \frac{1}{LC} \tag{13}$$

where $y_1 = I_o$ and $y_2 = \dot{I}_o$. The linearization of (12) and (13) in $y_1 = I_r$, $y_2 = 0$ and $u = u_y$ is:

$$\dot{y}_1 = y_2 \tag{14}$$

$$\dot{y}_2 = -\frac{eI_r}{CPL}u - \frac{eu_y}{CPL}y_1 + \frac{2I_r^2}{CP}y_2 + \frac{1}{LC} + \frac{eu_y I_r}{CPL} \tag{15}$$

and it has the following transfer function:

$$\frac{y_2(s)}{u(s)} = \frac{I_y}{CPLs^2 - 2LI_y^2 s + u_y} \tag{16}$$

whose poles remain in the right-side complex semi-plane, in a way that the instability property can be extended to the nonlinear system.

3. Controller Design and Stability

In the following, linearization is used in order to determine a linear parameter varying system for a robust stability analysis. Firstly, a feedback linearization is performed in voltage mode, and after, Taylor series linearizations are performed for voltage and current mode control. Once each linear system is obtained, a simplice polytopic system is constructed that allows a stability analysis by means of a common Lyapunov function, as well as appropriate values of the controllers' gains. Note that the controllers are easy to implement for linearizations in Taylor series.

3.1. Feedback Linearization for Voltage-Mode Controller

Consider the system (5) and (6). A feedback linearization is performed in order to obtain a linear polytopic system, with the main aim to demonstrate that it is robust against arbitrary parameter changes. The feedback proposed is as follows:

$$u = -k_1 x_1 - \left(\frac{PL}{ex_1^2} + k_2\right) x_2 \tag{17}$$

where k_1, k_2 are controller gains; the system under the above controller action is obtained substituting u in the system (5) and (6):

$$\dot{x}_1 = x_2 \tag{18}$$

$$\dot{x}_2 = -\alpha_1 x_1 - \alpha_2 x_2 \tag{19}$$

where:

$$\alpha_1 = \frac{k_1 e + 1}{LC} \tag{20}$$

$$\alpha_2 = \frac{k_2 e}{LC} \tag{21}$$

Selecting $k_1 > 0$ and $k_2 > 0$, one has that $\alpha_1 > 0$ and $\alpha_2 > 0$.

Consider parametric variation within known ranges, denoted by an underline for the minimum value and by an overline for the maximum value, as follows: $e(t) \in [\underline{e}, \overline{e}]$, $L(t) \in [\underline{L}, \overline{L}]$ and $C(t) \in [\underline{C}, \overline{C}]$. The linearized system (18) and (19) can be written as a polytopic, simplice system [27]:

$$\dot{x} = \theta_1 A_1 x + \theta_2 A_2 x + \theta_3 A_3 x + \theta_4 A_4 x \tag{22}$$

where $\Theta = \theta_1 + \theta_2 + \theta_3 + \theta_4 = 1, \theta_1 \geq 0, \theta_2 \geq 0, \theta_3 \geq 0, \theta_4 \geq 0$ (simplice) and:

$$A_1 = \begin{bmatrix} 0 & 1 \\ -\underline{\alpha}_1 & -\underline{\alpha}_2 \end{bmatrix}, A_2 = \begin{bmatrix} 0 & 1 \\ -\underline{\alpha}_1 & -\overline{\alpha}_2 \end{bmatrix}, A_3 = \begin{bmatrix} 0 & 1 \\ -\overline{\alpha}_1 & -\underline{\alpha}_2 \end{bmatrix}, \text{and } A_4 = \begin{bmatrix} 0 & 1 \\ -\overline{\alpha}_1 & -\overline{\alpha}_2 \end{bmatrix}. \tag{23}$$

That is, $\underline{\alpha}_1$ is the minimum value of α_1 for the defined parametric variation ranges and similarly for the other bound of α_1 and the boundaries of α_2. From [28], the quadratic stability of the system (22) is ensured if it is quadratically stable with $\theta_1 = 1$, with $\theta_2 = 1$, with $\theta_3 = 1$ and with $\theta_4 = 1$:

Proposition 1. *[28] The quadratic stability of the system (22) is equivalent to the existence of a $P \in \mathbb{R}^{2 \times 2}$ symmetric, positive definite matrix satisfying:*

$$PA_i + A_i^T P \prec 0, \ \forall i = 1, ..., 4 \tag{24}$$

where A_i denotes the i-th vertex and $\cdot \prec 0$ denotes a definite negative matrix.

That is, it is enough to demonstrate quadratic stability for all vertices; in fact, in such a case, a common Lyapunov function (CLF) can be constructed from one of the vertices. With that objective, from [26], every eigenvalue of the A_i-th matrix must take a negative real value in order to determine the quadratic stability of the vertex; it can be resolved that the second vertex conditions are the supreme, and it is enough to accomplish (see Appendix A for a proof):

$$k_1 > \frac{\overline{LC}}{4\underline{e}} \left(\frac{k_2 \overline{e}}{\underline{LC}} \right)^2 - \frac{1}{\underline{e}} \tag{25}$$

$$k_2 > 0 \tag{26}$$

to ensure the robust quadratic stability; so, it can be concluded that the proposed feedback linearization (17) stabilizes the nonlinear system (5) and (6) if the gains k_1, k_2 are selected so that (25) and (26) are met.

3.2. Taylor Series Linearization for Voltage-Mode Controller

Although a direct feedback linearization can be carried out in order to nullify the nonlinear term in (6), as described in the previous section, possibly time-varying and hardly-estimable terms are involved, for example $L(t)$. Instead of a direct feedback linearization, consider the Taylor-series linearization (7) and (8) and consider once again known ranges for parameter variation, additionally with $V_x(t) \in \left[\underline{V}_x, \overline{V}_x\right]$ as the output voltage operating point range, $P(t) \in \left[\underline{P}, \overline{P}\right]$ as the output power range and the linear feedback $u = -k_3 x_1 - k_4 x_2$. A polytopic, simplice system can be constructed as in Section 3.1, [27]:

$$\dot{x} = \theta_1 \mathscr{A}_1 x + \theta_2 \mathscr{A}_2 x + \theta_3 \mathscr{A}_3 x + \theta_4 \mathscr{A}_4 x \tag{27}$$

where:

$$\mathscr{A}_1 = \begin{bmatrix} 0 & 1 \\ -\underline{\sigma}_1 & -\underline{\sigma}_2 \end{bmatrix}, \mathscr{A}_2 = \begin{bmatrix} 0 & 1 \\ -\underline{\sigma}_1 & -\overline{\sigma}_2 \end{bmatrix}, \mathscr{A}_3 = \begin{bmatrix} 0 & 1 \\ -\overline{\sigma}_1 & -\underline{\sigma}_2 \end{bmatrix}, \text{ and } \mathscr{A}_4 = \begin{bmatrix} 0 & 1 \\ -\overline{\sigma}_1 & -\overline{\sigma}_2 \end{bmatrix}, \tag{28}$$

Following the same technique used in Section 3.1, it is easy to demonstrate that:

$$\sigma_1 = \frac{ek_3 + 1}{LC} \tag{29}$$

$$\sigma_2 = \frac{k_4 e}{LC} - \frac{P}{CV_x^2} \tag{30}$$

where the σ boundaries are constructed with the boundaries of the parameter variation ranges as in the previous section. The stability of all of the vertices is ensured if:

$$k_3 > 0 \tag{31}$$

$$0 < \frac{k_4 e}{\overline{LC}} - \frac{\overline{P}}{\underline{C V_x}^2} < 2\sqrt{\underline{\sigma}_1} \tag{32}$$

It is worth mentioning that in this analysis, there is no restriction on the rate of change in the parameters including the value of the constant power transferred to the load. Therefore, it can be concluded that the proposed feedback, stabilizes the polytopic system (27) if the gains k_3, k_4 are selected so that (31) and (32) are met. In order to accomplish these conditions, first select a k_3 and calculate σ_1 to determine the maximum value of k_4.

Note that several linearizations in different operating points can be performed to obtain a switched polytopic system, and the global stability can be demonstrated by a CLF; however, it is not the objective of this paper. Refer to [29] for a switched polytopic stability approach.

3.3. Taylor Series Linearization for Current-Mode Controller

In current-mode, a feedback linearization as in the case of voltage-mode is not possible due to the squared current change (y_2^2). In this case, a direct linearization of (12) and (13) in the operating point $y_1 = I_r, y_2 = 0, u = u_y$ and variable changes $z_1 = y_1 - I_r, z_2 = y_2$ are performed:

$$\dot{z}_1 = z_2 \tag{33}$$

$$\dot{z}_2 = \frac{e u_y}{CPL} z_1 + \frac{2 I_r^2}{CP} z_2 - \frac{e I_r}{CPL} u \tag{34}$$

Consider $u = k_5 z_1 + k_6 z_2$ where $k_3, k_4 > 0$ are controller gains; the system (33) and (34) is now:

$$\dot{z}_1 = z_2 \tag{35}$$

$$\dot{z}_2 = -\gamma_1 z_1 - \gamma_2 z_2 \tag{36}$$

where:

$$\gamma_1 = \frac{k_5 e I_r}{CPL} - \frac{e u_y}{CPL} \tag{37}$$

$$\gamma_2 = \frac{k_6 e I_r}{CPL} - \frac{2I_r^2}{CP} \tag{38}$$

Again, a polytopic system can be built as follows:

$$\dot{z} = \vartheta_1 \mathcal{A}_1 z + \vartheta_2 \mathcal{A}_2 z + \vartheta_3 \mathcal{A}_3 z + \vartheta_4 \mathcal{A}_4 z \tag{39}$$

where:

$$\mathcal{A}_1 = \begin{bmatrix} 0 & 1 \\ -\underline{\gamma}_1 & -\underline{\gamma}_2 \end{bmatrix}, \mathcal{A}_2 = \begin{bmatrix} 0 & 1 \\ -\underline{\gamma}_1 & -\overline{\gamma}_2 \end{bmatrix}, \mathcal{A}_3 = \begin{bmatrix} 0 & 1 \\ -\overline{\gamma}_1 & -\underline{\gamma}_2 \end{bmatrix}, \text{and } \mathcal{A}_4 = \begin{bmatrix} 0 & 1 \\ -\overline{\gamma}_1 & -\overline{\gamma}_2 \end{bmatrix}. \tag{40}$$

For the quadratic stability of the system (39), every eigenvalue of the $\mathcal{A}_i$-th matrix must take a negative real value in order to determine the quadratic stability of the vertex. Calculating such conditions for $\gamma_1, \gamma_2 > 0$, for all A_i and for all the eigenvalues results in the controller gains conditions:

$$k_5 > \frac{\overline{PLC}}{\underline{eI}_r} \left[\frac{1}{4} \left(\frac{k_6 \overline{e} \overline{I}_r}{\underline{PLC}} - \frac{2\overline{I}_r^2}{\overline{PC}} \right)^2 + \frac{\overline{e u}_y}{\underline{PLC}} \right] \tag{41}$$

$$k_6 > \frac{2\overline{I}_r^2 \overline{PLC}}{\underline{eI}_r \underline{CP}} \tag{42}$$

It is worth mentioning that, in this analysis, there is no restriction about the rate of change in the parameters including the value of the constant power transferred to the load. Once again, it can be concluded that the proposed feedback stabilizes the polytopic system (39) if the gains k_5, k_6 are selected in a way that (41) and (42) are met.

4. Experimental Behavior

In order to illustrate the robust stability of the proposed controller even under arbitrary (bounded) changes, a 100-W testbed was implemented (Figure 4) for a buck topology (although simulations were performed for a wide variety of tests including at high power ranges, such data are not presented here; the interested reader can refer to the Data Availability Section at the end of the paper for PSIM 11 schematics and data). At the top is a low voltage power source that feeds the low control board voltage at the bottom. In the middle are the power source and the electronic load from top to bottom.

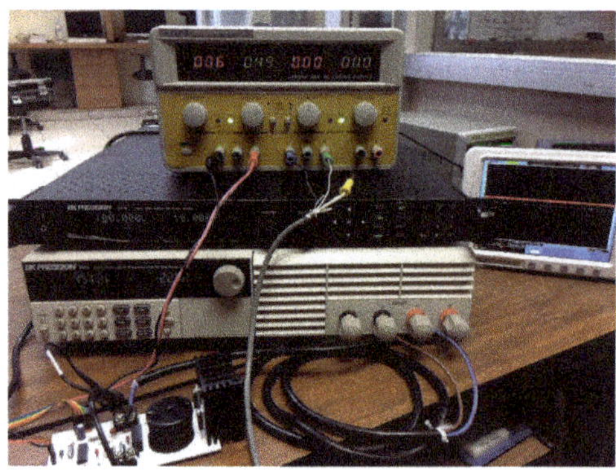

Figure 4. Experimental testbed.

The buck topology was selected due to lowering of the voltage, which is the most required case for the LVDCdistribution system for building/residential applications. Indeed, it is required to decrease the voltage from the main 375-V DC bus in the following cases: (a) loads with input rectifier (325 V DC); (b) loads with pure resistive loads (230 V DC); (c) loads without protection against indirect contacts (120 V DC); (d) light electronic appliance loads, based on the standard telecommunication industry (48 V DC); (e) EMerge Alliance Occupied Space Standard loads (24 V DC); and (f) automotive loads (12 V DC).

4.1. Equipment

In Figure 5, the configuration of the testbed used to perform the demonstrative experiments is shown. The PEC and PWM were built in a single PCB. The PCB that was built is shown in Figure 6. The low cost of the PEC, the 8-bit Microchip microcontroller running at 20 MHz, sampling at 625 kHz and the common 120-W P-MOSFET that were used, is worth highlighting. A programmable medium power DC power source (1200 W), Model 9116 from BK PRECISION, was used for the input; this source allows obtaining abrupt changes in output voltage and current. Additionally, a 600-W programmable DC electronic load, Model 8510 from BK PRECISION, was used as the output; this device allows setting a constant power (CPL) for some desired voltage or current and performs abrupt changes in the power level on-the-fly. Two mechanical switches SL and SC were laid to perform abrupt changes of the passive components of the PEC; that is, to connect/disconnect a parallel capacitor/inductor (CP/LP) with C and L. The output current is measured with a Tektronix A622 current probe with 100 mV/A (1 A) connected to a four-channel oscilloscope; the input and output voltages are measured by a direct connection to the oscilloscope.

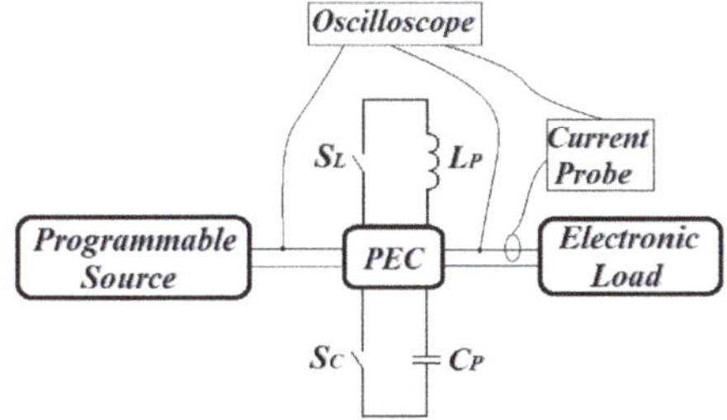

Figure 5. Experimental testbed configuration.

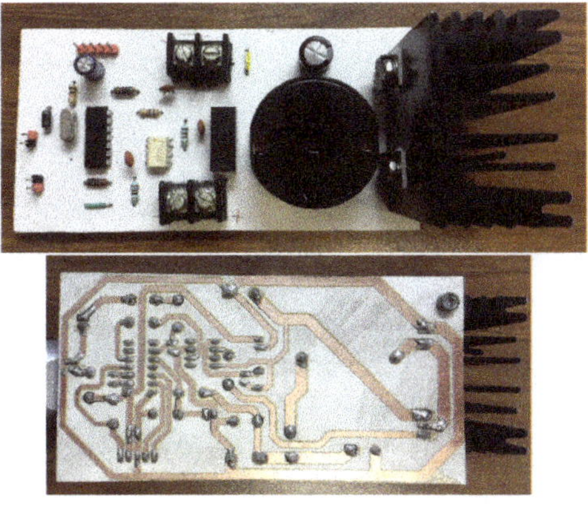

Figure 6. PCB of the DC-DC converter and robust CPL controller.

The parameter boundaries for this testbed were as follows: $L \in [2.2\,\text{mH}, 2.4\,\text{mH}]$, $C \in [0.9\,\mu\text{F}, 1.1\,\mu\text{F}]$, $P \in [45\,\text{W}, 100\,\text{W}]$, $e \in [80\,\text{V}, 100\,\text{V}]$. The voltage-mode robust controller (Section 3.2) was selected for the experiments due to its implementational simplicity (no current measure is performed to calculate the control output), such that controller gains that fulfill (31) and (32) were tuned with $V_x = 48$ V. That is, first, $k_3 = 3.5 > 0$ was tuned to obtain a desired behavior, and with this value, σ_1 was calculated to determine the value of k_4 that fulfills (32); in such a case, k_4 results in a very low positive value.

4.2. Procedure

In order to show the validity of the analysis of this paper, several tests were performed in the testbed described in the previous section. The overall tests were performed at normal temperature conditions (22 degrees for the room temperature and 40 degrees at the heat sink).

The first representative test consists of performing abrupt input voltage changes emulating a real microgrid scenario. For a 100-W CPL the following sequence of input voltage changes was used: 90 V, 80 V, 70 V, 80 V, 90 V and 100 V. The output voltage response is shown in Figure 7 in purple at the bottom, while the input voltage is shown in blue at the top and the output current in yellow in the middle. The CPL is maintained at 50 W at 48 V.

The second representative test consisted of CPL changes, that is the constant power load value was modified in the following sequence: 50 W, 1 W, 80 W and 1 W, in order to recreate a scenario with several CPLs in a changing ON-OFF condition with $e = 80$ V. No adverse effects were detected in this test, as shown in Figure 8; again, the output voltage response is shown in purple at the bottom, while the input voltage is shown in blue at the top and the output current in yellow in the middle.

The third representative test consisted of performing a capacitance abrupt change by connecting a series capacitor of 3.9 μF with C through a switch (mechanical). The first 5 s (approximately) shown in Figure 9 correspond to a 1 μF capacitance; then, the switch is turned on up to 10 sto increase the capacitance to 4.9 μF, which is out of the design bounds. Note that the current ripple in the CPL increases slightly since more current is needed to charge the capacitor; however, the CPL voltage remains stable without transients. This procedure is repeated one time, as shown in the figure.

The fourth test consisted of performing an inductor abrupt change by a switch (mechanical). A parallel inductor of the same nominal value (2.3 mH) is connected to L from zero to 5 s (approximately) till the total inductance is 1.15 mH, which is out of the design bounds. Then, the switch is turned off, and the inductance turns to 2.3 mH up to 10 s. Note that in spite of a CPL current ripple increase (with a small inductance), the CPL voltage remains stable without transients. This procedure is repeated as shown in the Figure 10.

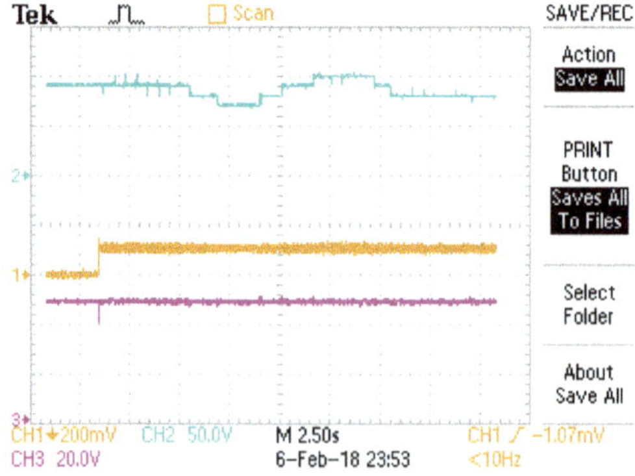

Figure 7. Output voltage (purple, bottom) and current responses (yellow, middle) for e (blue, top) abrupt changes.

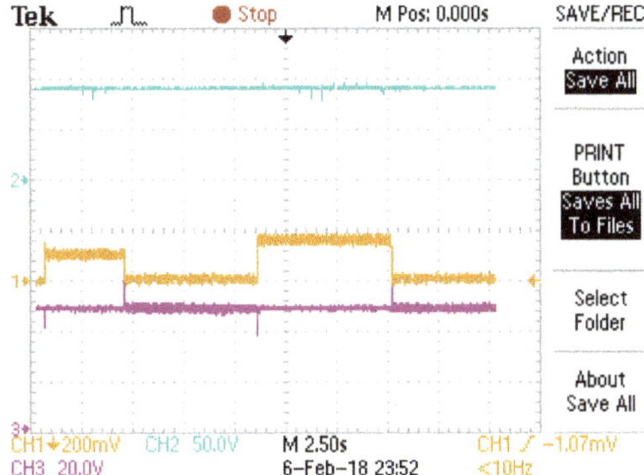

Figure 8. Output voltage (purple, bottom) and current responses (yellow, middle) for constant *e* (blue, top) and *P* abrupt changes.

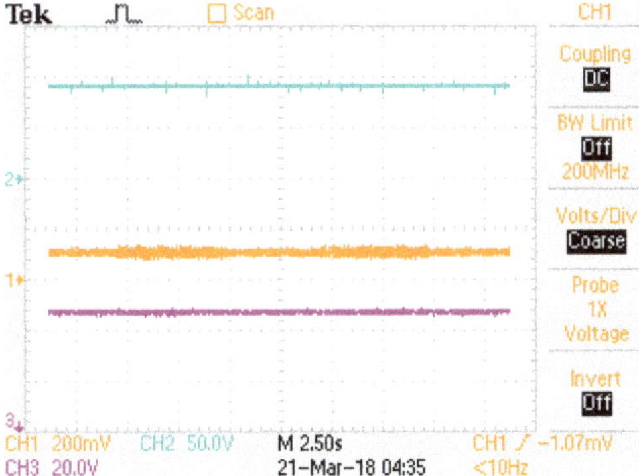

Figure 9. Output voltage (purple, bottom) and current responses (yellow, middle) for constant *e* (blue, top) and a capacitor change from 1 to 4.9 μF alternating every 5 s.

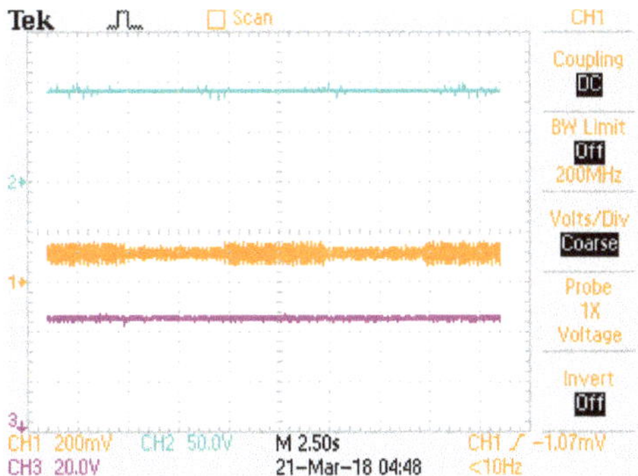

Figure 10. Output voltage (purple, bottom) and current responses (yellow, middle) for constant *e* (blue, top) and an inductor change from 1.15 to 2.3 mH alternating every 5 s.

The last representative test presented in this paper was performed demanding a power load wattage from 50 W to 110 W, which is out of the design boundary [45, 100] W; Figure 11 shows the obtained results; note that the controller was able to maintain the output voltage without signals of instability, and only a small voltage transient of a 15% was perceived during the CPL change.

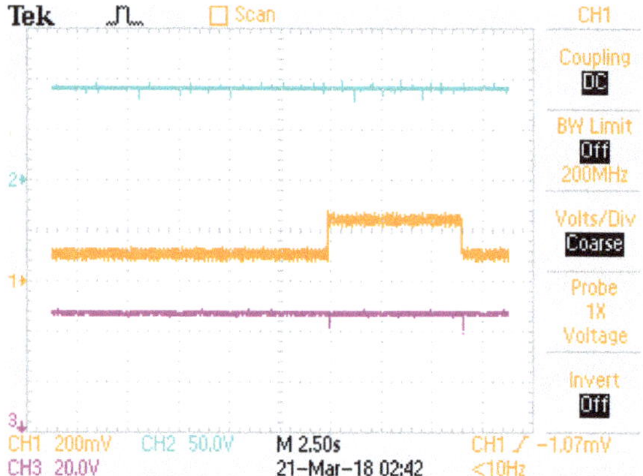

Figure 11. Output voltage (purple, bottom) and current responses (yellow, middle) for constant e (blue, top) and a large P change (out of the design bound $P(t) = 110 \notin [45, 100]$ W).

4.3. Discussion of the Results

For the first test, Figure 7 shows the results obtained for both the input and CPL voltage, as well as the CPL current, during abrupt input voltage changes. This is a normal situation in an LVDC microgrid since the voltage is provided from a non-regulated source. In spite of the abrupt changes in input voltage, it is easily seen that the CPL holds with a constant output voltage as expected from the analysis. Note that a small output transient is presented at the beginning of the test; this is reasonable since a derivative controller was used, and initially, the absolute value of the error is high. This can be diminished (if desired) by an ad hoc tuning.

One difficulty encountered during the experiment was that an abrupt increment in the CPL value can easily damage the MOSFET if the voltage is low because the current increases rapidly. The solution was to use a MOSFET, with an adequate working voltage ($V_{DS} = 200$ V).

For the second test, Figure 8 shows the results obtained for abrupt CPL changes. In this test, another LVDC microgrid real scenario is emulated since some loads in the microgrid can connect/disconnect from the bus unexpectedly. In spite of abrupt changes in CPL, it is easily seen that the CPL holds with a stable output voltage as expected from the analysis. Note that the CPL voltage ripple increases slightly when the CPL level is low (this is external to the boundaries of the design [45, 100] W).

For the third test, Figure 9 shows the results obtained for abrupt capacitance (C) changes. In this test, another real LVDC microgrid scenario is emulated since some loads in the microgrid can provide a parasitic impedance and/or the temperature or another variable can change the total capacitance of the PEC/RPEC. In spite of abrupt changes in capacitance, it is easily seen that the CPL holds with a stable output voltage as expected from the analysis. Note that the CPL voltage ripple increases slightly when the capacitance is high (as in the previous test; this is external to the boundaries of the design [0.9, 1.1] μF).

For the fourth test, Figure 10 shows the results obtained for abrupt inductance (L) changes. As in the previous test, another real LVDC microgrid scenario is emulated for parasitic impedances. In spite of the abrupt changes, it is easily seen that the CPL holds with a stable output voltage as expected from the analysis. Note that the CPL current ripple increases when the inductance is low (again, this is external to the boundaries of the design [2.2, 2.4] mH).

Finally, the fifth test emulates a CPL that is external to the boundaries of the design with a maximum of 110 W. Note in Figure 11 that a small output transient is present at every change due to the derivative action, but in spite of those, the CPL voltage remains stable.

4.4. Brief Comparison with Other Proposals

Though this proposed controller has an easy operation and dynamic response, its advantages have to be corroborated by comparing it with other proposals. In this paper, not only its complexity, theory and implementation, but also, the dynamic response is evaluated in order to fulfill this purpose. Because no experimental setup was built for additional proposals, the evaluation is performed based on data from existing literature on the subject. A brief comparison for the proposed controller with other schemes is given in Table 1. These are found to be classified in terms of the theory and implementational difficulty.

Table 1. Comparison with other proposals.

Type of Controller	Theory Difficulty	Implementation Complexity	Dynamic Response	Remarks
Active stabilizer [8]	High	High	Good	Good performance and a digital platform is used, in this case, a dSPACE.
Active damping [9]	Medium	Medium	Good	Good performance and a digital platform is used, in this case a pair of DSPs.
Linear [10]	High	High	Good	Good performance and only numerical results are reported.
Sum of squares [11]	High	High	Medium	Medium performance and only numerical results are reported.
Algebraic [12]	Medium	Medium	Medium	Medium performance and only numerical results are reported.
Sliding mode [13]	Medium	High	Good	Good performance and only numerical results are reported.
Linear damping [14]	Medium	High	Good	Good performance and a digital platform is used, in this case a DSP.
Robust stability [15]	High	High	Medium	Medium performance and only numerical results are reported.
Drop control [16]	High	High	Good	Good performance and only numerical results are reported.
Fractional order controller [17]	High	High	Medium	Medium performance and only numerical results are reported.
Robust stability [19]	High	High	Medium	Medium performance and only numerical results are reported.
Proposed approach	Medium	Low	High	Good performance and a low cost digital platform is used, in this case, a microcontroller.

It may be observed that the proposed approach presents a good compromise between theory implementation and dynamic response. Although most of the controllers that are based on non-linear models propose comparable characteristics, the active damping presents a good dynamic response [8]. However, it requires a more sophisticated digital platform for its implementation. Additionally, as seen in Table 1, several controllers are validated with only numerical simulations [10–13,15–17,19]. In contrast, this proposed controller offers a good dynamic response in comparison with the previous work. Additionally, as reported in Table 1, this proposal has a similar response to active damping; however, it does not require much time to cope with CPL, since no sophisticated digital platform is required.

Without a doubt, this proposal requires a natural digital implementation, however not necessarily a sophisticated one (an 8-bit microcontroller is used for the experimental tests). Besides, there are mathematical and experimental conditions where the controller may be operated properly, which do not occur with other proposals. For instance, abrupt changes in passive parameters. Although an in-depth practical comparison may be performed between references [8,9,14] and this proposal, this is beyond the purposes of this publication.

5. Conclusions

In the present work, the CPL stabilization problem was analyzed from a robust-polytopic point of view, for three basic converters even in a reconfigurable scheme while the activity is going on (on-the-fly).

It was also demonstrated that the design of a polytopic, robust controller is natural for the CPL problem since the ranges of power demand and other parameters are normally known. This controller is robust against reasonable (design) changes in the power source voltage/current level and in the values of the components.

It was analytically demonstrated and experimentally shown that a simple controller is enough to stabilize a CPL voltage/current even when all of the parameters are time-varying values if some conditions are met. Various scenarios of LVDC were emulated, and the validity of the analysis presented can be easily extended to other topologies of PEC, even for those reconfigurable ones while the activity is going on (on-the-fly).

The presented controller can be used in various real microgrid scenarios in which a stabilization of the output voltage with parameters that change with time and a piecewise constant power demand are present. The proposed controller and DC-DC converter can be extended for use in electric vehicles.

Author Contributions: M.-A.R.-L. conceived of and designed the experiments. M.-A.R.-L. performed the experiments. M.-A.R.-L., F.-J.P.-P., C.-A.H.-R. and J.-C.N.-P. analyzed the data. M.-A.R.-L., F.-J.P.-P., C.-A.H.-R. and J.-C.N.-P. contributed materials/analysis tools. M.-A.R.-L. and F.-J.P.-P. wrote the paper. Datasets of representative simulations in PSIM 11 and its simulation diagram, related to this article can be found at https://osf.io/d9m6t/download and https://osf.io/zbwku/download, respectively; an open-source online data repository hosted at the Open Science Framework (Brian Nosek, 2018).

Acknowledgments: The authors wish to thank the IPN for its support provided through the project SIP-20180020. In addition, the authors would like to express their gratitude to the COFAA for its financial support and to the CONACYT for Cátedra ID 4155.

Conflicts of Interest: The authors declare no conflict of interest for this paper.

Appendix A

Calculating conditions for A_1 (eigenvalues), one has:

$$|A_1 - \lambda I| = \left| \begin{bmatrix} -\lambda & 1 \\ -\underline{\alpha}_1 & -\underline{\alpha}_2 - \lambda \end{bmatrix} \right| = \lambda(\underline{\alpha}_2 + \lambda) + \underline{\alpha}_1 = \lambda^2 + \underline{\alpha}_2 \lambda + \underline{\alpha}_1 \tag{A1}$$

solving for λ:

$$\lambda_1 = \frac{-\underline{\alpha}_2 + \sqrt{\alpha_2^2 - 4\underline{\alpha}_1}}{2} \tag{A2}$$

$$\lambda_2 = \frac{-\underline{\alpha}_2 - \sqrt{\alpha_2^2 - 4\underline{\alpha}_1}}{2} \tag{A3}$$

It is enough for stability to get the negative real part of every eigenvalue, that is:

$$\underline{\alpha}_1 > \frac{\underline{\alpha}_2^2}{4} \tag{A4}$$

$$\underline{\alpha}_2 > 0 \tag{A5}$$

Calculating the values of $\underline{\alpha}_1$ and $\underline{\alpha}_2$ results in:

$$\underline{\alpha}_1 = \frac{k_1 \underline{e} + 1}{\overline{LC}} \tag{A6}$$

$$\underline{\alpha}_2 = \frac{k_2 e}{\overline{\overline{LC}}} \tag{A7}$$

Substituting (A6) and (A7) in (A4) and (A5), the controller gains conditions for quadratic stability can be determined:

$$k_1 > \frac{k_2^2 \underline{e}}{4 \overline{LC}} - \frac{1}{\underline{e}} \tag{A8}$$

$$k_2 > 0 \tag{A9}$$

For the second vertex, the same analysis can be performed, but now:

$$\underline{\alpha}_1 = \frac{k_1 \underline{e} + 1}{\overline{LC}} \tag{A10}$$

$$\overline{\alpha}_2 = \frac{k_2 \overline{e}}{LC} \tag{A11}$$

is used; and the controller gains are:

$$k_1 > \frac{\overline{LC}}{4\underline{e}} \left(\frac{k_2 \overline{e}}{LC} \right)^2 - \frac{1}{\underline{e}} \tag{A12}$$

$$k_2 > 0 \tag{A13}$$

For the third vertex, the same analysis can be performed, but now:

$$\overline{\alpha}_1 = \frac{k_1 \overline{e} + 1}{LC} \tag{A14}$$

$$\underline{\alpha}_2 = \frac{k_2 e}{\overline{\overline{LC}}} \tag{A15}$$

is used; and the controller gains are:

$$k_1 > \frac{LC}{4\overline{e}} \left(\frac{k_2 e}{\overline{\overline{LC}}} \right)^2 - \frac{1}{\overline{e}} \tag{A16}$$

$$k_2 > 0 \tag{A17}$$

For the fourth vertex, the same analysis can be performed, but now:

$$\overline{\alpha}_1 = \frac{k_1 \overline{e} + 1}{LC} \tag{A18}$$

$$\overline{\alpha}_2 = \frac{k_2 \overline{e}}{\underline{LC}} \tag{A19}$$

is used; and the controller gains are:

$$k_1 > \frac{k_2^2 \bar{e}}{4 \underline{L} C} - \frac{1}{\bar{e}} \tag{A20}$$

$$k_2 > 0 \tag{A21}$$

In order to obtain a single condition valid for the four vertices, it can be seen that selecting the greater of them will be enough, and quadratic stability will be achieved. The k_2 conditions are the same. For k_1, it is demonstrated that the right side of (A20) is greater than the right side of (A8):

$$\frac{k_2^2 \underline{e}}{4 \overline{L} \overline{C}} - \frac{1}{\underline{e}} < \frac{k_2^2 \bar{e}}{4 \underline{L} C} - \frac{1}{\bar{e}} \tag{A22}$$

Since:

$$\frac{k_2^2 \underline{e}}{4 \overline{L} C} - \frac{1}{\underline{e}} < \frac{k_2^2 \underline{e}}{4 \overline{L} C} - \frac{1}{\underline{e}} \tag{A23}$$

Therefore:

$$\frac{k_2^2 \underline{e}}{4 \overline{L} \overline{C}} - \frac{1}{\underline{e}} < \frac{k_2^2 \bar{e}}{4 \underline{L} C} - \frac{1}{\bar{e}} \tag{A24}$$

$$\frac{k_2^2 \underline{e}}{4 \overline{L} \overline{C}} < \frac{k_2^2 \bar{e}}{4 \underline{L} C} \tag{A25}$$

$$\frac{\underline{e}}{\overline{L} \overline{C}} < \frac{\bar{e}}{\underline{L} C} \tag{A26}$$

$$\frac{\underline{e} \underline{L} C}{\bar{e} \overline{L} \overline{C}} < 1 \tag{A27}$$

Since $\underline{e} < \bar{e}$, $\underline{C} < \overline{C}$ and $\underline{L} < \overline{L}$, the right side of (A20) is greater than the right side of (A8). Now, it is demonstrated that the right side of (A12) is greater than the right side of (A16):

$$\frac{\underline{L} C}{4 \bar{e}} \left(\frac{k_2 \underline{e}}{\overline{L} \overline{C}} \right)^2 - \frac{1}{\bar{e}} < \frac{\overline{L} \overline{C}}{4 \underline{e}} \left(\frac{k_2 \bar{e}}{\underline{L} C} \right)^2 - \frac{1}{\underline{e}} \tag{A28}$$

Since:

$$\frac{\underline{L} C}{4 \bar{e}} \left(\frac{k_2 \underline{e}}{\overline{L} \overline{C}} \right)^2 - \frac{1}{\underline{e}} < \frac{\underline{L} C}{4 \bar{e}} \left(\frac{k_2 \underline{e}}{\overline{L} \overline{C}} \right)^2 - \frac{1}{\bar{e}} \tag{A29}$$

it can be written out as:

$$\frac{\underline{L} C}{4 \bar{e}} \left(\frac{k_2 \underline{e}}{\overline{L} \overline{C}} \right)^2 - \frac{1}{\underline{e}} < \frac{\overline{L} \overline{C}}{4 \underline{e}} \left(\frac{k_2 \bar{e}}{\underline{L} C} \right)^2 - \frac{1}{\underline{e}} \tag{A30}$$

$$\frac{\underline{L} C}{4 \bar{e}} \left(\frac{k_2 \underline{e}}{\overline{L} \overline{C}} \right)^2 < \frac{\overline{L} \overline{C}}{4 \underline{e}} \left(\frac{k_2 \bar{e}}{\underline{L} C} \right)^2 \tag{A31}$$

$$1 < \frac{(\overline{L} \overline{C})^3 \bar{e}^2}{(\underline{L} C)^3 \underline{e}^2} \tag{A32}$$

Since $\underline{e} < \bar{e}$, $\underline{C} < \overline{C}$ and $\underline{L} < \overline{L}$, the right side of (A12) is greater than the right side of (A16). Now, it is demonstrated that the right side of (A12) is greater than the right side of (A20):

$$\frac{k_2^2 \bar{e}}{4 \underline{L} C} - \frac{1}{\bar{e}} < \frac{\overline{L} \overline{C}}{4 \underline{e}} \left(\frac{k_2 \bar{e}}{\underline{L} C} \right)^2 - \frac{1}{\underline{e}} \tag{A33}$$

Since:

$$\frac{\overline{LC}}{4\underline{e}}\left(\frac{k_2\overline{e}}{\underline{LC}}\right)^2 - \frac{1}{\underline{e}} < \frac{\overline{LC}}{4\underline{e}}\left(\frac{k_2\overline{e}}{\underline{LC}}\right)^2 - \frac{1}{\overline{e}} \tag{A34}$$

it can be written out as:

$$\frac{k_2^2\overline{e}}{4\underline{LC}} - \frac{1}{\overline{e}} < \frac{\overline{LC}}{4\underline{e}}\left(\frac{k_2\overline{e}}{\underline{LC}}\right)^2 - \frac{1}{\overline{e}} \tag{A35}$$

$$\frac{k_2^2\overline{e}}{4\underline{LC}} < \frac{\overline{LC}}{4\underline{e}}\left(\frac{k_2\overline{e}}{\underline{LC}}\right)^2 \tag{A36}$$

$$\frac{\underline{e}\,\underline{LC}}{\overline{e}\,\overline{LC}} < 1 \tag{A37}$$

Since $\underline{e} < \overline{e}$, $\underline{C} < \overline{C}$ and $\underline{L} < \overline{L}$, the right side of (A12) is greater than the right side of (A20). It can be resolved that the second vertex conditions are the supreme, and it is enough to accomplish:

$$k_1 > \frac{\overline{LC}}{4\underline{e}}\left(\frac{k_2\overline{e}}{\underline{LC}}\right)^2 - \frac{1}{\underline{e}} \tag{A38}$$

$$k_2 > 0 \tag{A39}$$

References

1. Wang, P.; Goel, L.; Liu, X.; Choo, F.H. Harmonizing AC and DC: A hybrid AC/DC future grid solution. *IEEE Power Energy Mag.* **2013**, *11*, 76–83. [CrossRef]
2. Kakigano, H.; Nomura, M.; Ise, T. Loss evaluation of DC distribution for residential houses compared with AC system. In Proceedings of the International Power Electronics Conference (IPEC), Sapporo, Japan, 21–24 June 2010; pp. 480–486.
3. Wang, B.; Sechilariu, M.; Locment, F. Intelligent DC microgrid with smart grid communications: Control strategy consideration and design. *IEEE Trans. Smart Grid* **2012**, *3*, 2148–2156. [CrossRef]
4. Hart, D.W. *Power Electronics*; Tata McGraw-Hill Education: New York, NY, USA, 2011.
5. Rodriguez-Diaz, E.; Chen, F.; Vasquez, J.C.; Guerrero, J.M.; Burgos, R.; Boroyevich, D. Voltage-level selection of future two-level LVdc distribution grids: A compromise between grid compatibiliy, safety, and efficiency. *IEEE Electr. Mag.* **2016**, *4*, 20–28. [CrossRef]
6. AL-Nussairi, M.K.; Bayindir, R.; Padmanaban, S.; Mihet-Popa, L.; Siano, P. Constant power loads (cpl) with microgrids: Problem definition, stability analysis and compensation techniques. *Energies* **2017**, *10*, 1656. [CrossRef]
7. Singh, S.; Gautam, A.R.; Fulwani, D. Constant power loads and their effects in DC distributed power systems: A review. *Renew. Sustain. Energy Rev.* **2017**, *72*, 407–421. [CrossRef]
8. Zadeh, M.K.; Gavagsaz-Ghoachani, R.; Martin, J.P.; Nahid-Mobarakeh, B.; Pierfederici, S.; Molinas, M. Discrete-time modeling, stability analysis, and active stabilization of dc distribution systems with multiple constant power loads. *IEEE Trans. Ind. Appl.* **2016**, *52*, 4888–4898. [CrossRef]
9. Chang, X.; Li, Y.; Li, X.; Chen, X. An active damping method based on a supercapacitor energy storage system to overcome the destabilizing effect of instantaneous constant power loads in dc microgrids. *IEEE Trans. Energy Convers.* **2017**, *32*, 36–47. [CrossRef]
10. Arocas-Pérez, J.; Griño, R. A local stability condition for dc grids with constant power loads. *IFAC-PapersOnLine* **2017**, *50*, 7–12. [CrossRef]
11. Herrera, L.; Yao, X. Computation of stability metrics in DC power systems using sum of squares programming. In Proceedings of the IEEE 18th Workshop on Control and Modeling for Power Electronics (COMPEL), Stanford, CA, USA, 9–12 July 2017; pp. 1–5.
12. van der Blij, N.H.; Ramirez-Elizondo, L.M.; Spaan, M.T.; Bauer, P. Stability of DC Distribution Systems: An Algebraic Derivation. *Energies* **2017**, *10*, 1412. [CrossRef]

13. Hossain, E.; Perez, R.; Padmanaban, S.; Siano, P. Investigation on the development of a sliding mode controller for constant power loads in microgrids. *Energies* **2017**, *10*, 1086. [CrossRef]

14. Srinivasan, M.; Kwasinski, A. European Conference on Power Electronics and Applications. In Proceedings of the International Power Electronics Conference (IPEC), Warsaw, Poland, 11–14 Sptember 2017; pp. 1–10.

15. Garces, A. Uniqueness of the power flow solutions in low voltage direct current grids. *Electr. Power Syst. Res.* **2017**, *151*, 149–153.

16. Zhao, Z.; Hu, J.; Xue, H.; Huang, R.; Li, X.; Zhang, X. Large signal stability analysis of DC microgrid under droop control with constant power load. In Proceedings of the Chinese Automation Congress, Jinan, China, 20–22 October 2017; pp. 1046–1051.

17. Asghar, M.; Khattak, A.; Rafiq, M.M. Comparison of integer and fractional order robust controllers for DC/DC converter feeding constant power load in a DC microgrid. *Sustain. Energy Grids Netw.* **2017**, *12*, 1–9.

18. Chen, Y.; Petras, I.; Xue, D. Fractional order control-a tutorial. In Proceedings of the American Control Conference (ACC'09), St. Louis, MO, USA, 10–12 June 2009; pp. 1397–1411.

19. Liu, J.; Zhang, W.; Rizzoni, G. Robust stability analysis of dc microgrids with constant power loads. *IEEE Trans. Power Syst.* **2018**, *33*, 851–860. [CrossRef]

20. Momayyezan, M.; Hredzak, B.; Agelidis, V.G. A Load-Sharing Strategy for the State of Charge Balancing Between the Battery Modules of Integrated Reconfigurable Converter. *IEEE Trans. Power Electron.* **2017**, *32*, 4056–4063. [CrossRef]

21. Momayyezan, M.; Hredzak, B.; Agelidis, V.G. Integrated Reconfigurable Converter Topology for High-Voltage Battery Systems. *IEEE Trans. Power Electron.* **2016**, *31*, 1968–1979. [CrossRef]

22. Rizzo, R.; Spina, I.; Tricoli, P. A single input dual buck-boost output reconfigurable converter for distributed generation. In Proceedings of the International Conference on Clean Electrical Power (ICCEP), Taormina, Italy, 16–18 June 2015; pp. 767–774.

23. Grigore, V.; Hatonen, J.; Kyyra, J.; Suntio, T. Dynamics of a buck converter with a constant power load. In Proceedings of the 29th Annual IEEE Power Electronics Specialists Conference, (PESC'98), Fukuoka, Japan, 22–22 May 1998; Volume 1, pp. 72–78.

24. Licea, M.A.R.; Pinal, F.J.P.; Gutiérrez, A.I.B.; Ramírez, C.A.H.; Perez, J.C.N. A Reconfigurable buck, boost, and buck-boost Converter: Unified Model and Robust Controller. *Math. Prob. Eng.* **2018**, *2018*, 6251787. [CrossRef]

25. Sira-Ramirez, H.J.; Silva-Ortigoza, R. *Control Design Techniques in Power Electronics Devices*; Springer Science & Business Media: Berlin, Germany, 2006.

26. Khalil, H. *Nonlinear Systems*; Prentice Hall: Upper Saddle River, NJ, USA, 2002.

27. Lay, S.R. Convex sets and their applications. In *Pure and Applied Mathematics*; John Wiley and Sons: Hoboken, NJ, USA, 1982.

28. Boyd, S.; El Ghaoui, L.; Feron, E.; Balakrishnan, V. *Linear Matrix Inequalities in System and Control Theory*; SIAM: Philadelphia, PA, USA, 1994.

29. Licea, M.R.; Cervantes, I. Robust indirect-defined envelope control for rollover and lateral skid prevention. *Control Eng. Pract.* **2017**, *61*, 149–162. [CrossRef]

 *energies*

Article

A Digital Hysteresis Current Control for Half-Bridge Inverters with Constrained Switching Frequency

Triet Nguyen-Van *, Rikiya Abe and Kenji Tanaka

Internet of Energy Laboratory, Department of Technology Management for Innovation, The University of Tokyo, Tokyo 113-8656, Japan; abe-r@tmi.t.u-tokyo.ac.jp (R.A.); tanaka@tmi.t.u-tokyo.ac.jp (K.T.)
* Correspondence: nguyen@tmi.t.u-tokyo.ac.jp

Received: 7 September 2017; Accepted: 12 October 2017; Published: 14 October 2017

Abstract: This paper proposes a new robustly adaptive hysteresis current digital control algorithm for half-bridge inverters, which plays an important role in electric power, and in various applications in electronic systems. The proposed control algorithm is assumed to be implemented on a high-speed Field Programmable Gate Array (FPGA) circuit, using measured data with high sampling frequency. The hysteresis current band is computed in each switching modulation period based on both the current error and the negative half switching period during the previous modulation period, in addition to the conventionally used voltages measured at computation instants. The proposed control algorithm is derived by solving the optimization problem—where the switching frequency is always constrained at below the desired constant frequency—which is not guaranteed by the conventional method. The optimization problem also keeps the output current stable around the reference, and minimizes power loss. Simulation results show good performances of the proposed algorithm compared with the conventional one.

Keywords: half-bridge inverters; digital control; hysteresis current control; switching frequency; optimization

---

1. Introduction

Nowadays, inverters and their controllers play an important role in the enabling of a wider proliferation of renewable energy generation [1,2], the realization of new grid concepts with high efficiency [3,4], and various applications in electronic systems [5,6]. Among various current control techniques developed over last few decades, there are three major classes of control scheme: sine-triangle Pulse Width Modulation (PWM), predictive dead-beat, and hysteresis current control. While the asynchronous sine-triangle PWM is a popular technique to modulate current error, it requires a PI (Proportional-Integral) regulator and often results in unavoidable delays. The predictive dead-beat control technique tends to give good performance in terms of response and accuracy. However, this technique is highly dependent on the accuracy of the predictive model and is complicated to implement [7]. Hysteresis current control, on the other hand, is simpler to implement and does not have such drawbacks. It has a fast dynamic response, and does not require any information about the system parameters, which enhances its robustness [5,8]. For this reason, it finds applications in a large variety of switching inverters.

The basic implementation of hysteresis current control is based on the switching signal that is derived by comparing the actual current and the reference current so that the current error is kept within the tolerance current band. In classical hysteresis current controllers, the hysteresis current band is normally fixed to a certain value, which makes the switching frequency vary in order to contain the current ripple within the band. This leads to unwanted heavy interference among the phases in the three-phase system [9]. In order to overcome this problem, an adaptive hysteresis current control technique has been developed and used in many applications [10–12]. In this technique, the hysteresis current band is not fixed, but controlled adaptively in each switching modulation period, and based on the measured output voltage values and the desired constant switching frequency [13].

Digital hysteresis current control usually requires a sufficiently high sampling frequency to control the switch devices with accurate switching time [14,15]. The ripple current varies during each switching modulation at a very fast rate, which is inversely proportional to the output inductance and proportional to the difference between the dc and output voltages. Because, during the sampling interval, the hysteresis current band could not be observed until the arrival of the next data sample, the low sampling frequency may lead to a large ripple current overshoot from the hysteresis current band. A high sampling frequency at the MHz level seems to be quite high for conventional Digital Signal Processors (DSP) and microcontrollers, which are employed in most power electronic applications at present. Additionally, such a high sampling frequency is beyond the scope of the Field Programmable Gate Array (FPGA) circuit, which has a high-speed clock and is becoming more and more popular in inverter control [16–18].

In many applications, the high switching frequency has several advantages including lower current ripple, faster transient response, and effective size reduction for the inverter circuit [19–21]. In some cases, the switching devices are operated near the highest possible switching frequency, for example when the controller is redesigned for an existing device, or in order to reduce the price of the inverters. For instance, when a motor is accelerated from a standstill to rated speed, an inverter is usually designed to operate near its highest possible switching frequency in order to achieve a low current distortion [22]. In these cases, if the switching devices are forced to change their on-off states at a frequency beyond their limit, especially under the influence of noise, short-circuit fault can occur and may lead to breakage of the switching device [23]. Therefore, it is important to properly control the hysteresis current band so that the frequency of the switching device does not exceed its highest possible switching frequency. However, the switching frequency fluctuation controlled by the conventional method is not guaranteed to never exceed this limit, especially with noise. The conventional method of computing a hysteresis current band uses only the measured voltage values at the instant when the switches start a new modulation period [13,24]. The present study proposes a new digital hysteresis current control algorithm that takes the feedback information from the preceding modulation period into account, keeping the switching frequency constant and always below the highest possible switching frequency of the switching device, even when noise is present. The controller is assumed to be using an FPGA while the sampling interval is at the MHz level.

This paper is organized as follows. Section 2 presents a conventional method of using the fixed band hysteresis current. In Section 3, a conventional adaptive hysteresis current control is described and its disadvantages are explained. Then, a new robustly-adaptive hysteresis current control algorithm is proposed in Section 4. In Section 5, the simulation results are shown to demonstrate that the proposed method has better performance than the conventional method. Conclusions are given in Section 6.

2. Classical Fixed Band Hysteresis Current Control

Consider a single-phase half-bridge inverter circuit, where the output is connected to a grid ac voltage v_g as in Figure 1. The dc voltage is supplied by two constant and balanced dc sources, each of which has a value of V_{dc}. Parameters L, C, and R present the output inductance, current ripple filter, and load respectively. The inverter output current i_L is controlled by the switching devices S_1 and S_2 to track a given reference current i_{ref}. The ripple component of the output current i_L is filtered by the current ripple filter, and the grid current i_o is derived without a ripple component.

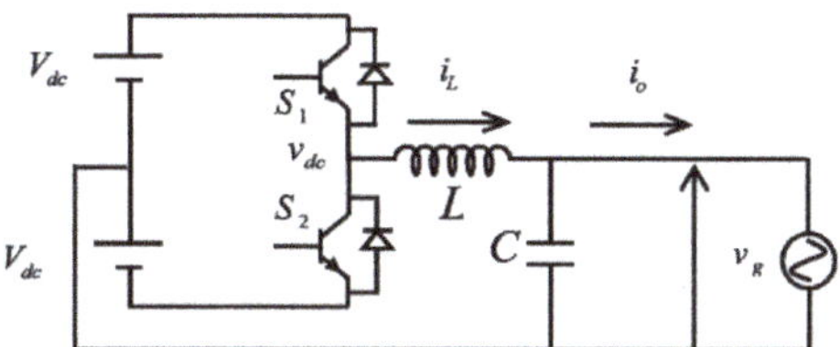

Figure 1. Single-phase half-bridge inverter circuit.

Figure 2 shows the structure of the digital control system. The measured output voltages, and currents are sampled by analog/digital converters, and used for computing on/off pulses of switches S_1 and S_2. The states of the switches and the corresponding inputs and outputs of the inverter are shown in Table 1.

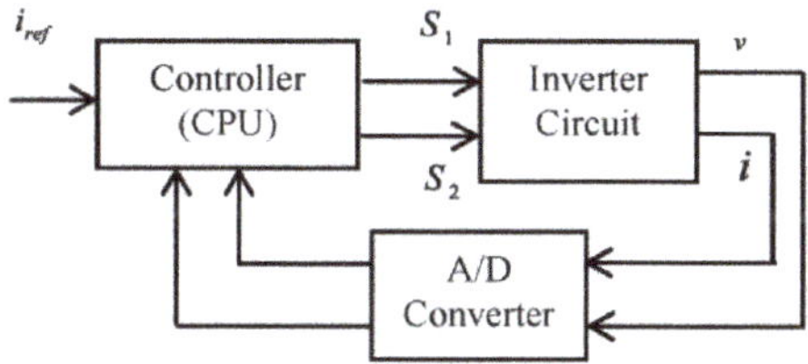

Figure 2. Structure of the digital control system.

Table 1. States of switches and corresponding outputs.

Half Period	S_1	S_2	Input Voltage v_{dc}	Output Current i_L
Positive	On	Off	V_{dc}	Increase
Negative	Off	On	$-V_{dc}$	Decrease

In the classical fixed band hysteresis current control, the tolerance current band is fixed to a certain value Δi_b [25]. While the measured output current i_L is between the upper and lower limits, no switching occurs. When the measured output current crosses above the upper limit of the hysteresis band $\left(i_{ref} + \Delta i_b\right)$, the switch S_1 is turned off, the switch S_2 is turned on and the current starts to decay. In contrast, when the measured current crosses below the lower limit of the hysteresis band $\left(i_{ref} - \Delta i_b\right)$, the switch S_1 is turned on, the switch S_2 is turned off and the current starts to increase (Figure 3).

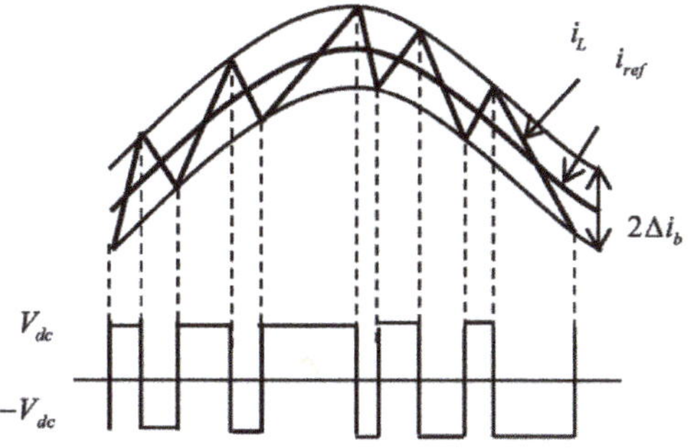

Figure 3. Fixed band hysteresis current control.

The hysteresis current band value is directly proportional to the current ripple and inversely proportional to the switching frequency. Thus, increasing the value of the hysteresis current band will increase the current ripple while a decrease in the band will increase the switching losses. In analog controllers, the current ripple is always kept exactly within the hysteresis band. However, in digital controllers, the hysteresis control is shown to be effective only if the current band is chosen to satisfy the following condition [5]:

$$\Delta i_b > \max\left(\frac{di_{ref}}{dt}\right)\frac{1}{f_{sp}},$$

(1)

where f_{sp} is the sampling frequency. The maximum switching frequency f_{sw_max} should be smaller than half the sampling frequency f_{sp}, i.e.,:

$$f_{sw\_max} \leq \frac{1}{2}f_{sp}.$$

(2)

3. Conventional Adaptive Hysteresis Current Control

3.1. Algorithm of the Convention Adaptive Hysteresis Current Control

The above mentioned fixed band hysteresis control has many advantages: its simplicity, its fast and stable response, and its independence from system parameters. However, a disadvantage is that the switching frequency needs to vary in order to keep the peak-to-peak current ripple controlled at all points on the fundamental frequency wave [10]. In order to solve this problem, the adaptive hysteresis current control technique has been presented in the literature [7,26]. In this technique, the hysteresis current band is not fixed, but controlled adaptively in each switching modulation period, based on the measured output voltage values and the desired constant switching frequency. The adaptive hysteresis current control is employed as shown below.

Define the current error $\Delta i(t)$ as:

$$\Delta i(t) = i_L(t) - i_{ref}(t),$$

(3)

where $i_L(t)$, and $i_{ref}(t)$ are the inverter output and the reference currents, respectively, at instant t. Consider the instant t_0, when the output current i_L tends to cross the lower hysteresis band, and the switch S_1 is switched on. The current error at t_0 is $\Delta i(t_0)$ (Figure 4). Assume that the switch S_1 is switched on during $[t_0, t_1)$, and is switched off during $[t_1, t_2)$ intervals. These intervals are called positive and negative half switching periods respectively.

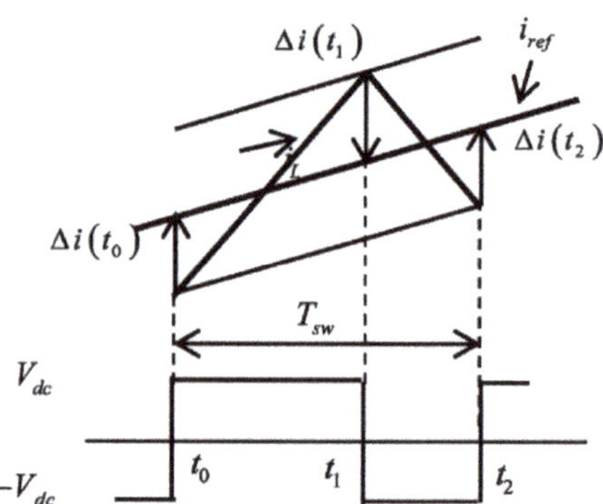

Figure 4. Conventional adaptive hysteresis current control.

The output current dynamic equation can be written as:

$$\frac{di_L(t)}{dt} = \frac{v_{dc}(t) - v_g(t)}{L} \tag{4}$$

for $t_0 \leq t \leq t_2$, where v_g is the instantaneous grid voltage, L is the output inductance, and $v_{dc}(t)$ is the inverter dc voltage, and can be elaborated as:

$$v_{dc}(t) = \begin{cases} V_{dc} & \text{if } S_1 \text{ is On} \\ -V_{dc} & \text{if } S_1 \text{ is Off} \end{cases}. \tag{5}$$

Define the output current slopes in the on and off periods by $\dot{I}_{on}$, and $\dot{I}_{off}$ respectively. Assuming that the output voltage is slowly varying during the switching modulation period $[t_0, t_2]$, the output current slopes (4) can be expressed as:

$$\dot{I}_{on} = \frac{di_L(t)}{dt} = \frac{V_{dc} - v_g(t_0)}{L} \tag{6}$$

for $t \in [t_0, t_1)$, and:

$$\dot{I}_{off} \triangleq \frac{di_L(t)}{dt} = \frac{-V_{dc} - v_g(t_0)}{L} \tag{7}$$

for $t \in [t_1, t_2)$.

The current errors in the positive and negative half switching periods are given as:

$$\begin{aligned} \Delta i(t_1) &= i_L(t_1) - i_{ref}(t_1) \\ &= i_{ref}(t_0) + \Delta i(t_0) + \dot{I}_{on}T_{on} - i_{ref}(t_1), \end{aligned} \tag{8}$$

$$\begin{aligned} \Delta i(t_2) &= i_L(t_2) - i_{ref}(t_2) \\ &= i_{ref}(t_0) + \Delta i(t_0) + \left(\dot{I}_{on}T_{on} + \dot{I}_{off}T_{off} \right) - i_{ref}(t_2). \end{aligned} \tag{9}$$

where T_{on} and T_{off} are the on and off periods of switch S_1, given as:

$$T_{on} = t_1 - t_0 \tag{10}$$

$$T_{off} = t_2 - t_1 \tag{11}$$

The reference current $i_{ref}(t)$ is slowly varying during the modulation period, such that it can be approximated as:

$$i_{ref}(t_1) = i_{ref}(t_0) + \dot{I}_{ref}(t_0)T_{on}, \tag{12}$$

$$i_{ref}(t_2) = i_{ref}(t_0) + \dot{I}_{ref}(t_0)\left(T_{on} + T_{off} \right), \tag{13}$$

where:

$$\dot{I}_{ref}(t_0) = \left. \frac{di_{ref}(t)}{dt} \right|_{t=t_0} \tag{14}$$

By substituting Equations (12) and (13) into Equations (8) and (9), one can write the current errors $\Delta i(t_1)$, and $\Delta i(t_2)$ as:

$$\Delta i(t_1) = \Delta i(t_0) + \dot{I}'_{on}T_{on} \tag{15}$$

$$\Delta i(t_2) = \Delta i(t_0) + \dot{I}'_{on}T_{on} + \dot{I}'_{off}T_{off} = \Delta i(t_1) + \dot{I}'_{off}T_{off} \tag{16}$$

where $\dot{I}'_{on}$ and $\dot{I}'_{off}$ are the current error slopes in positive and negative half switching periods, given as:

$$\dot{I}'_{on} = \dot{I}_{on} - \dot{I}_{ref}(t_0) = \frac{V_{dc} - v_g(t_0)}{L} - \dot{I}_{ref}(t_0),\tag{17}$$

$$\dot{I}'_{off} = \dot{I}_{off} - \dot{I}_{ref}(t_0) = \frac{-V_{dc} - v_g(t_0)}{L} - \dot{I}_{ref}(t_0)\tag{18}$$

Let f_{sw} be the desired constant switching frequency. In the conventional adaptive hysteresis current control method, the hysteresis current band $\Delta i_b(t_0)$ is derived by using the following conditions:

$$\Delta i(t_1) - \Delta i(t_0) = 2\Delta i_b(t_0)\tag{19}$$

$$\Delta i(t_2) - \Delta i(t_1) = -2\Delta i_b(t_0),\tag{20}$$

$$T_{on} + T_{off} = T_{sw},\tag{21}$$

where $T_{sw} = 1/f_{sw}$ is the desired constant switching period. Substituting Equations (19)–(21) into Equations (15) and (16), we can derive the hysteresis current band as:

$$\Delta i_b(t_0) = \frac{1}{2}\frac{\dot{I}_{on}\dot{I}_{off}}{\dot{I}_{off} - \dot{I}_{on}}T_{sw}\tag{22}$$

By substituting Equations (17) and (18) into Equation (22), the hysteresis band in Equation (22) can also be written in the form of [26] :

$$\Delta i_b(t_0) = \frac{V_{dc}T_{sw}}{4L}\left(1 - m^2(t_0)\right),\tag{23}$$

where:

$$m(t_0) = \frac{1}{V_{dc}}\left(v_g(t) + L\dot{I}_{ref}(t)\right)\tag{24}$$

3.2. Disadvantages of Conventional Adaptive Hysteresis Current Control

Consider the case where the inverter needs to be controlled so that the operating switching frequency is always strictly equal to or lower than a constant frequency, which may be the highest possible switching frequency of the switching devices.

The conventional adaptive hysteresis current control described in Section 3.1 has been shown to be able to improve on the disadvantage of the fixed band hysteresis current control of the varying switching frequency [8]. However, when the inverter is operated under the effect of noise or disturbances, this conventional method does not guarantee a stable response and constant switching frequency as, for example, in the following problems.

Problem 1. *If the negative half switching period of switch S_1 in the previous modulation period is T_{off_pre}, as shown in Figure 5, then:*

$$T_{off\_pre} = t_0 - t_{-1}\tag{25}$$

While the operating switching frequency can be defined from both the time periods $T_{on} + T_{off}$ and $T_{off\_pre} + T_{on}$, the calculation of the conventional hysteresis current band only guarantees that modulation period $T_{on} + T_{off}$ is equal to the desired constant switching period T_{sw}, as in Equation (21). If the noise or disturbances to voltage and current make the negative half switching period T_{off_pre} in the previous modulation shorter than the calculated value, then the operating switching frequency of the conventional method may exceed the highest possible switching frequency of the switching devices.

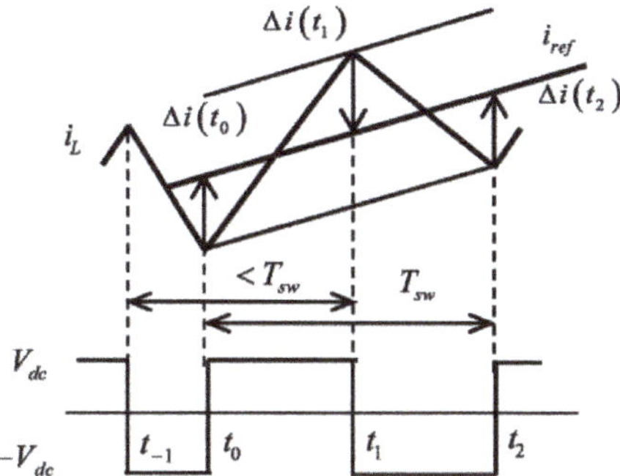

Figure 5. Conventional adaptive hysteresis current control does not guarantee that the operating switching frequency will be lower than the desired constant frequency.

Problem 2. *In order to keep the average value of the output current equal to the reference current in each switching modulation, the hysteresis current band should be computed so that:*

$$- \Delta i(t_1) = \Delta i(t_2) = -\Delta i_b \tag{26}$$

However, the hysteresis current band computed by the conventional method, as in Equations (19) and (20), in general, does not satisfy the Condition (26), except in the case of the ideal condition shown by the dashed line in Figure 6, where the current error at instant t_0 is identical to the computed hysteresis current band Δi_b, i.e.,:

$$\Delta i(t_0) = -\Delta i_b \tag{27}$$

When the Condition (27) is not satisfied, the average value of the output current during a switching modulation period deviates from the reference current, which may lead to an unstable response as shown in Figure 6 by the solid line.

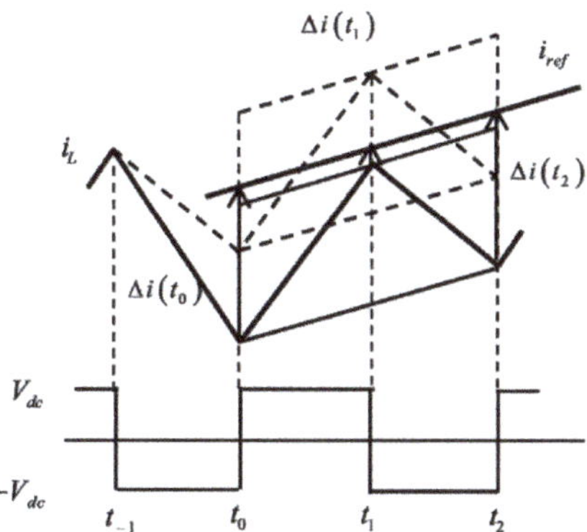

Figure 6. Conventional adaptive hysteresis current control does not guarantee the stability of the response current.

4. Proposed Robustly Adaptive Hysteresis Current Control

Consider the instant t_0, when the output current crosses to pass the lower limit of the hysteresis band $i_{ref} + \Delta i(t_0)$ and the switch S_1 starts to be switched on as shown in Figure 7. The negative half switching period of switch S_1 in the previous modulation period T_{off_pre} is:

$$T_{off\_pre} = t_0 - t_{-1}. \tag{28}$$

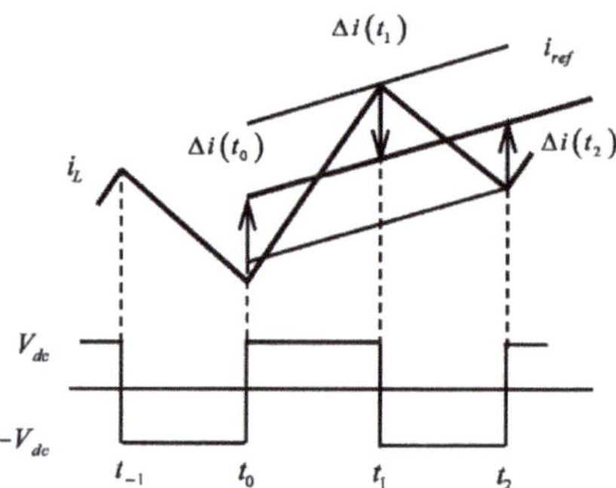

Figure 7. Proposed adaptive hysteresis current control.

In order to solve the problem of unstable switching frequency of the conventional adaptive hysteresis current control mentioned in Section 3, this study proposes a new hysteresis current band computation method by setting an optimization problem as:

$$
\begin{aligned}
&(i)\ T_{off-pre} + T_{on} \geq T_{sw}, \\
&(ii)\ T_{on} + T_{off} \geq T_{sw}, \\
&(iii)\ \Delta i_b = \Delta i(t_1) = -\Delta i(t_2), \\
&(iv)\ \Delta i_b \geq \Delta i_{conv}, \\
&\underset{\Delta i}{\text{minimize}}\ J = (\Delta i(t_1))^2 + (\Delta i(t_2))^2,
\end{aligned} \tag{29}
$$

where Δi_{conv} is the hysteresis current band computed by the conventional method as in Equation (22), and J is the objective function.

Constraint conditions (*i*) and (*ii*) in Equation (29) maintain the operating switching frequency to always be equal to or smaller than the desired instant switching frequency. Condition (*iii*) keeps the average value of the output current identical to the reference current during the switching modulation period. Condition (*iv*) keeps the output current from deviating from the reference current. The objective function J represents the power loss from the inverter in each switching modulation period. The hysteresis current band is computed such that the power loss J is minimized.

Equations (15) and (16) can be written as:

$$T_{on} = \frac{1}{\overset{.}{I}_{on}} (\Delta i(t_1) - \Delta i(t_0)), \tag{30}$$

$$T_{off} = \frac{1}{\overset{.}{I}_{off}} (\Delta i(t_2) - \Delta i(t_1)) \tag{31}$$

Substituting Equations (30) and (31) into Equation (29), the optimization problem (29) can be written as:

$$
\begin{aligned}
&(i)\ \Delta i(t_1) \geq \dot{i}'_{on}\left(T_{sw} - T_{off-pre}\right) + \Delta i(t_0),\\[4pt]
&(ii)\ \left(1 - \frac{\dot{i}_{on}}{\dot{i}_{off}}\right)\Delta i(t_1) + \frac{\dot{i}_{on}}{\dot{i}_{off}}\Delta i(t_2) \geq \dot{i}_{on}T_{sw} + \Delta i(t_0),\\[4pt]
&(iii)\ \Delta i(t_1) = -\Delta i(t_2) = \Delta i,\\[4pt]
&(iv)\ \Delta i \geq \Delta i_{conv},\\[4pt]
&\underset{\Delta i}{\text{minimize}}\ J = (\Delta i(t_1))^2 + (\Delta i(t_2))^2.
\end{aligned}
\tag{32}
$$

The constraints in the optimization problem given in Equation (32) are linear. They can be solved by using the graphical method for minimization problems [27], as below. The feasible region representing constraint conditions (*i*–*iv*) can be represented by the dark area on the phase-plane $(\Delta i(t_1), \Delta i(t_2))$ as in Figure 8. The objective function J is represented by the distance from the original phase-plane to the point $(\Delta i(t_1), \Delta i(t_2))$. Thus, the optimal solution for Problem (32) is the point $(\Delta i(t_1)_s, \Delta i(t_2)_s)$ on the feasible region, which is closest to the original point. The solution can be derived as:

$$
\Delta i_{opt} = \max(\Delta i_{conv}, \Delta i_A, \Delta i_B),
\tag{33}
$$

where $\Delta i_A, \Delta i_B, \Delta i_{conv}$ are the points which satisfy pair constraint conditions (*i, iii*), (*ii, iii*), (*iv, iii*), respectively, and are given by:

$$
\Delta i_A = \dot{i}_{on}\left(T_{sw} - T_{off-pre}\right) + \Delta i(t_0),
\tag{34}
$$

$$
\Delta i_B = \frac{\dot{i}_{on}T_{sw} + \Delta i(t_0)}{1 - 2\dot{i}_{on}/\dot{i}_{off}},
\tag{35}
$$

$$
\Delta i_{conv} = \frac{1}{2}\frac{\dot{i}_{on}\dot{i}_{off}}{\dot{i}_{off} - \dot{i}_{on}}T_{sw}
\tag{36}
$$

When the optimal solution is at Δi_A, it means that, for example under the effect of noise, the negative half switching period T_{off_pre} in the previous switching modulation was smaller than the regular value (Problem 1 in Section 3) and the hysteresis current band needs to be broadened to satisfy condition (i) in Equation (29). When the optimal solution is at Δi_B, it means that the average output current deviated from the reference current (Problem 2 in Section 3) and the hysteresis current band needs to be broadened to satisfy condition (*ii*) in Equation (29). When the optimal solution is at Δi_{conv}, it means that the conditions (*i*) and (*ii*) are satisfied and the hysteresis current band should be brought to the steady state as in the conventional method.

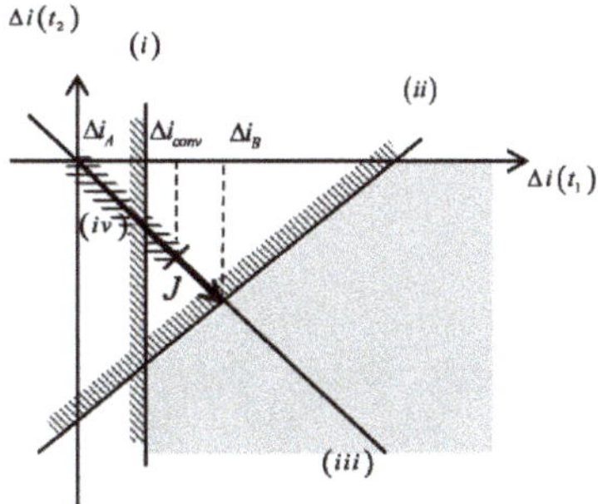

Figure 8. Domain and optimal solution for the hysteresis current band.

Remark 1. *In an ideal state, where the effect of noise is sufficiently small, the proposed method gives the same hysteresis current band as the conventional adaptive hysteresis current control method.*

5. Simulation Results

Simulations have been carried out to assess the performance of the proposed method as compared with the conventional method [13] using the Matlab-Simscape Power System. A single-phase half-bridge inverter was employed with the following parameters: an output inductance of $L = 1$ mH and a filter capacitor of $C = 10\,\mu\text{F}$. The switching transistors in the inverter circuit were modeled by the Insulated Gate Bipolar Transistor (IGBT) block in the Matlab-Simscape(R2016a, MathWorks, Natick, MA, USA) Power System. The IGBTs have internal and snubber resistances of 1 mΩ and 1 kΩ, respectively. The inverter was connected to the dc voltage source $V_{dc} = 175$ V, and the ac voltage of the grid was $v_g = 100\sqrt{2}\sin(100\pi t)$ V. The sampling frequency of the analog/digital converters was $f_{sp} = 2$ MHz. The reference current was set at $i_{ref} = 10\sin(100\pi t)$ A. The noise in the measured current, which may have come from the current sensor or the analog/digital converter, was assumed to be white noise with a variance of 0.01 A. Figures 9–11 show the operating switching frequencies and the output currents of the proposed and conventional methods compared with the reference current for the desired constant switching frequency f_{sw} at 40 kHz, 20 kHz, and 10 kHz respectively. The upper sub-figures show a part of the current hysteresis response. Regarding the desired constant switching frequencies, both methods yielded hysteresis output currents that were similar to the reference current. However, the operating switching frequency of the conventional method oscillated and went over the desired constant switching frequency. On the other hand, the proposed method yielded a switching frequency that was more stable than that of the conventional method and was always maintained at equal to or lower than the desired constant switching frequency. All the Figures also show that the hysteresis current band increased when the desired constant switching frequency was increased.

The power efficiency of the inverter has been calculated, taking into account hysteresis current loss. Table 2 shows the power efficiencies of the proposed and conventional methods with the tested switching frequencies. The efficiencies of both methods decreased when the switching frequency was decreased. It can be seen that the proposed method performed better than the conventional method in terms of power efficiency in all cases.

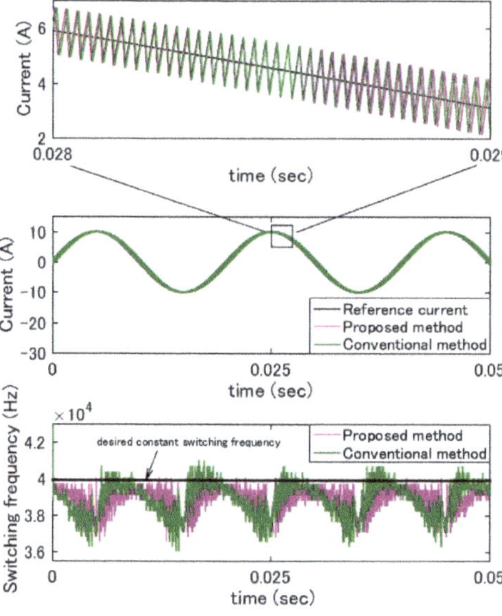

Figure 9. Current and switching period of the proposed method and conventional method for the desired switching frequency at 40 kHz.

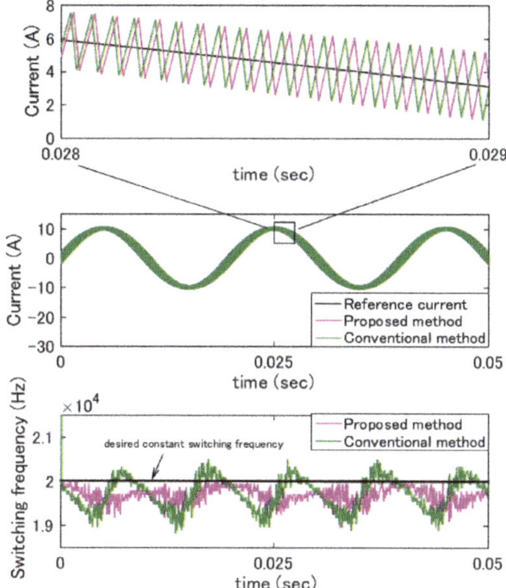

Figure 10. Current and switching period of the proposed method and conventional method for the desired switching frequency at 20 kHz.

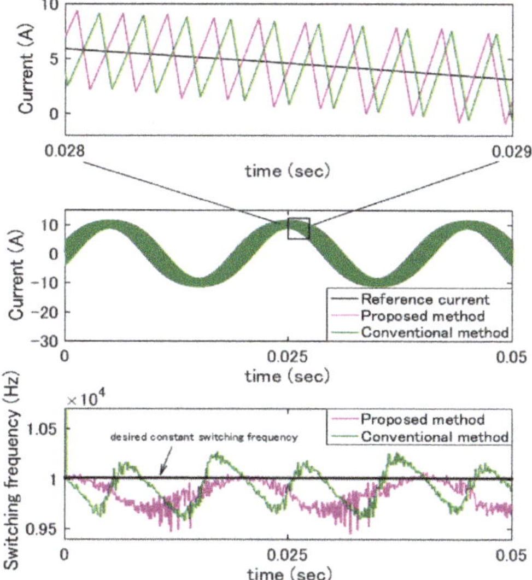

Figure 11. Current and switching period of the proposed method and conventional method for the desired switching frequency at 10 kHz.

Table 2. Power efficiency with various switching frequencies (%).

Switching frequency	$f_{sw} = 10$ kHz	$f_{sw} = 20$ kHz	$f_{sw} = 40$ kHz
Proposed method	97.1	98.4	99.3
Conventional method	95.2	97.2	98.5

6. Conclusions

A new digital hysteresis current control algorithm for single-phase half-bridge inverters has been proposed. By taking advantage of digital controls, the hysteresis current band is computed in each switching modulation period based on both the negative half switching period and the current error during the previous switching modulation period, in addition to the usual measured voltage value at the instant of computing. The hysteresis current control, with its simplicity and its fast and stable response, may be implemented on high-speed FPGA boards with a high sampling frequency. The proposed algorithm for computing the hysteresis current band is derived by solving the optimization problem where (i) the switching frequency is stable and always maintained at equal to or lower than the desired constant frequency; (ii) the output current is kept stable around the reference current; and (iii) the power loss is minimized. Although the proposed control method has not been evaluated in experiments, the theoretical numerical and simulation results show good performance of the proposed algorithm compared with those of the conventional method. The proposed method is expected to contribute to the control theory of high-speed FPGA-based hysteresis current control.

Author Contributions: Triet Nguyen-Van, Rikiya Abe, Kenji Tanaka performed and discussed the research; Triet Nguyen-Van carried out the simulations, analyzed the data, and wrote the paper.

Conflicts of Interest: The authors declare no conflict of interest.

References

1. Justo, J.J.; Mwasilu, F.; Lee, J.; Jung, J.-W. AC-microgrids versus DC-microgrids with distributed energy resources: A review. *Renew. Sustain. Energy Rev.* **2013**, *24*, 387–405. [CrossRef]
2. Khayyer, P.; Ozguner, U. Decentralized Control of Large-Scale Storage-Based Renewable Energy Systems. *IEEE Trans. Smart Grid* **2014**, *5*, 1300–1307. [CrossRef]
3. Abe, R.; Taoka, H.; McQuilkin, D. Digital Grid: Communicative Electrical Grids of the Future. *IEEE Trans. Smart Grid* **2011**, *2*, 399–410. [CrossRef]
4. Liu, C.-C.; McArthur, S.; Lee, S.-J. *Smart Grid Handbook*; Wiley: West Sussex, UK, 2016.
5. Marian, P.K.; Krishnan, R.; Blaabjerg, F.; Irwin, J.D. *Control in Power Electronics Selected Problems*; Academic Press: San Diego, CA, USA, 2002.
6. Séguier, G.; Labrique, F. *Power Electronic Converters: DC-AC Conversion*; Electric Energy Systems and Engineering Series; Springer: Berlin, Germany; New York, NY, USA, 1993.
7. Vazquez, G.; Rodriguez, P.; Ordonez, R.; Kerekes, T.; Teodorescu, R. Adaptive hysteresis band current control for transformerless single-phase PV inverters. In Proceedings of the 35th Annual Conference of IEEE. Industrial Electronics IECON, Porto, Portugal, 3–5 November 2009.
8. Buso, S.; Malesani, L.; Mattavelli, P. Comparison of current control techniques for active filter applications. *IEEE Trans. Ind. Electron.* **1998**, *45*, 722–729. [CrossRef]
9. Buso, S.; Fasolo, S.; Malesani, L.; Mattavelli, P. A dead-beat adaptive hysteresis current control. *IEEE Trans. Ind. Appl.* **2000**, *36*, 1174–1180. [CrossRef]
10. Kale, M.; Ozdemir, E. An adaptive hysteresis band current controller for shunt active power filter. *Electr. Power Syst. Res.* **2005**, *73*, 113–119. [CrossRef]
11. Bose, B.K. An adaptive hysteresis-band current control technique of a voltage-fed PWM inverter for machine drive system. *IEEE Trans. Ind. Electron.* **1990**, *37*, 402–408. [CrossRef]
12. Malesani, L.; Mattavelli, P.; Tomasin, P. Improved constant-frequency hysteresis current control of VSI inverters with simple feedforward bandwidth prediction. *IEEE Trans. Ind. Appl.* **1997**, *33*, 1194–1202. [CrossRef]

13. Poulsen, S.; Andersen, M.A.E. Hysteresis controller with constant switching frequency. *IEEE Trans. Consum. Electron.* **2005**, *51*, 688–693. [CrossRef]
14. Attaianese, C.; Monaco, M.D.; Tomasso, G. High Performance Digital Hysteresis Control for Single Source Cascaded Inverters. *IEEE Trans. Ind. Inform.* **2013**, *9*, 620–629. [CrossRef]
15. Hao, Y.; Fang, Z.; Feng, Z.; Zhenxiong, W. A Digital Hysteresis Current Controller for Three-Level Neural-Point-Clamped Inverter With Mixed-Levels and Prediction-Based Sampling. *IEEE Trans. Power Electron.* **2016**, *31*, 3945–3957.
16. Jens, O.K.; Markus, H.; Andreas, R.; Rolf, R. FPGA-based Control of Three-Level Inverters, in International Exhibition & Conference for Power Electronics. In Proceedings of the Intelligent Motion and Power Quality, Nuremberg, Germany, 17–19 May 2011; pp. 378–384.
17. Nguyen-Van, T.; Abe, R. An Indirect Hysteresis Voltage Digital Control for Half Bridge Inverters. In Proceedings of the 2016 IEEE Global Conference on Consumer Electronics, Kyoto, Japan, 11–14 October 2016.
18. Javier, C.; Domingo, B.; Francesc, G.; Alberto, P.; Francesc, M.; Eduard, A. FPGA-based design of a step-up photovoltaic array emulator for the test of PV grid-connected inverters. In Proceedings of the 2014 IEEE 23rd International Symposium on Industrial Electronics (ISIE), Istambul, Turkey, 1–4 June 2014; pp. 485–490.
19. Ang, S.S.; Oliva, A. *Power-Switching Converters*, 2nd ed.; Electrical and Computer Engineering; Taylor & Francis: Boca Raton, FL, USA, 2005.
20. Thiyagarajah, K.; Ranganathan, V.T.; Iyengar, B.S.R. A high switching frequency IGBT PWM rectifier/inverter system for AC motor drives operating from single phase supply. *IEEE Trans. Power Electron.* **1991**, *6*, 576–584. [CrossRef]
21. Stefan, H.; Eckart, H.; Oleg, Z.; Klaus-Dieter, L.; Gudrun, F. Reducing Inductor Size in High Frequency Grid Feeding Inverters. In Proceedings of the PCIM Europe 2015 International Exhibition and Conference for Power Electronics, Intelligent Motion, Renewable Energy and Energy Management, Nuremberg, Germany, 19–20 May 2015; pp. 196–203.
22. Tooley, M.H. *Plant and Process Engineering 360*, 1st ed.; Butterworth-Heinemann: Amsterdam, The Netherlands; Boston, MA, USA; London, UK, 2010.
23. Pei, X.; Kang, Y. Short-Circuit Fault Protection Strategy for High-Power Three-Phase Three-Wire Inverter. *IEEE Trans. Ind. Inform.* **2012**, *8*, 545–553. [CrossRef]
24. Nguyen-Van, T.; Abe, R. New hysteresis current band digital control for half bridge inverters. In Proceedings of the 2016 IEEE International Conference on Consumer Electronics-Taiwan (ICCE-TW), Nantou, Taiwan, 27–29 May 2016.
25. Zare, F.; Ledwich, G. A hysteresis current control for single-phase multilevel voltage source inverters: PLD implementation. *IEEE Trans. Power Electron.* **2002**, *17*, 731–738. [CrossRef]
26. Buso, S.; Caldognetto, T. A non-linear wide bandwidth digital current controller for DC-DC and DC-AC converters. *IEEE Trans. Ind. Electron.* **2015**, *62*, 7687–7695. [CrossRef]
27. Baker, K.R. *Optimization Modeling with Spreadsheets*, 3rd ed.; Wiley: West Sussex, UK, 2015.

Article

Design and Evaluation of an Efficient Three-Phase Four-Leg Voltage Source Inverter with Reduced IGBTs

Yixiao Luo [1,2], Chunhua Liu [1,2,*], Feng Yu [1,2] and Christopher H.T. Lee [3]

[1] Shenzhen Research Institute, City University of Hong Kong, Nanshan District, Shenzhen 518057, China; Yixiao.Luo@my.cityu.edu.hk (Y.L.); yufeng6575@sina.com (F.Y.)

[2] School of Energy and Environment, City University of Hong Kong, 83 Tat Chee Avenue, Kowloon Tong, Hong Kong, China

[3] Research Laboratory of Electronics, Massachusetts Institute of Technology, Cambridge, MA 02139, USA; chtlee@mit.edu

* Correspondence: chualiu@eee.hku.hk; Tel.: +852-3442-2885

Academic Editor: José Gabriel Oliveira Pinto
Received: 4 January 2017; Accepted: 10 April 2017; Published: 13 April 2017

Abstract: This paper presents a new three-phase four-leg voltage source inverter (VSI), which achieves a high cost effectiveness for mega-watt level system applications. The proposed four-leg inverter adopts the integrated topology with thyristors and insulated-gate bipolar transistors (IGBTs), which aims to reduce the number of IGBTs. In order to handle the zero sequence current, a neutral leg via incorporating IGBTs is artfully integrated with the regular phase legs. Furthermore, the modelling principles are elaborated and analyzed, which emphasizes switching states and voltage vectors in six segments based on the states of thyristors. Finally, by using the carrier-based pulse width modulation (PWM) method, the closed-loop current control of the proposed inverter is verified by both simulation and experimentation.

Keywords: power inverter; voltage source inverter; four-leg inverter; cost-effectiveness; current control; pulse width modulation

---

1. Introduction

Three-phase voltage source inverters (VSI) are widely used in industrial applications such as uninterruptable power supply (UPS) [1,2], motor drives [3–5], wireless power transfer [6] and distributed power generation system [7,8]. Unfortunately, this conventional topology is hard to keep the neutral point potential in a constant level under unbalanced load conditions, which leads to asymmetric output voltage [9]. To solve this problem, a split capacitor three-phase inverter is investigated in [10], but such a topology poses a problem of low DC voltage utilization. Alternatively, the three-phase four-leg VSI that connects the neutral point to the fourth leg is investigated by [11–18]. In fact, three-phase four-leg VSI is widely used in various industrial applications, such as distributed power generation, three phase UPS systems and series active filters, where the balanced three-phase voltage output is required during unbalanced load conditions [19–21]. However, the topology and control algorithm of the four-leg VSI are inherently more complex than the traditional three-phase type.

There has been a substantial increasing trend in control strategy investigations on three-phase four-leg inverters, including the carrier-based pulse width modulation (PWM) [22,23], space vector modulation algorithm [24,25], and digital predictive control [26,27]. Firstly, a minimally switched control algorithm is studied, which can minimize switching operations and monitor the current polarity for three-phase four-leg inverter [28]. Then, a selective harmonic elimination (SHE) control strategy on a three-phase four-leg inverter is reported in [29], where the lower order nontriplen harmonics are eliminated by using Fourier-based equations on line-to-line basis as conventional SHE technology to express the control signals of three legs. In addition, two methods that mitigate the load neutral point voltage (LNPV) for three-phase four-leg inverters are analyzed, which are based on the PWM strategy and the common mode filter, respectively [30]. Moreover, a decoupled sequence control strategy, employing positive sequence, negative sequence, and zero sequence controllers is studied in [31].

However, the available publications on three-phase four-leg inverters are limited to eight-insulated-gate bipolar transistors (IGBTs) topologies, as illustrated in Figure 1a. In mega-watt level applications, the IGBT with a rated current of thousands of ampere is not cost-effective and is not suitable for some cost-sensitive situations.

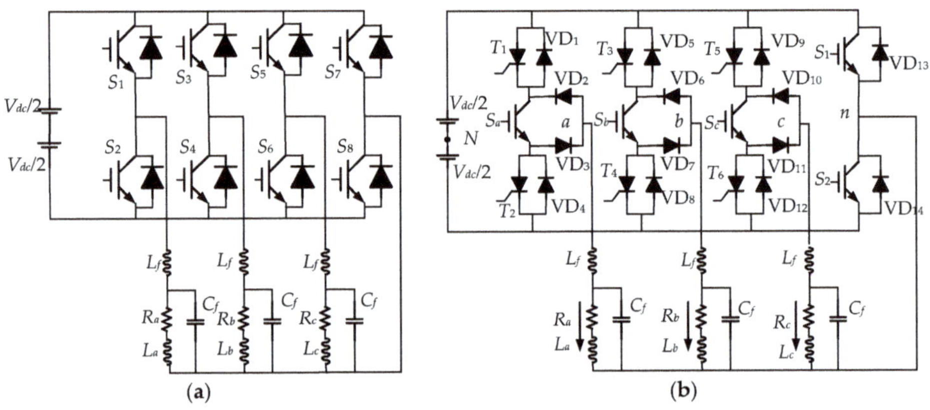

Figure 1. Configurations of a three-phase four-leg inverter, (**a**) Conventional topology with eight IGBTs; (**b**) Proposed topology with five IGBTs.

The purpose of this paper is to propose and evaluate an IGBT-reduced three-phase four-leg VSI that possesses the attractive feature of four-leg inverters. The proposed topology can reduce cost significantly, especially for mega-watt level applications such as electric ship propulsion systems, microgrid converters, motor drives, and so on. In Section 2, the configuration of the proposed inverter and the operation principle will be introduced. Section 3 will be devoted to deducing the representation of inverter output voltages as a three-dimensional space vector in a stator oriented $\alpha\beta\gamma$ rectangular coordinate. In Section 4, closed-loop current control strategy of the proposed inverter will be presented. Then, simulation and experiment results will be given to verify the validity of the proposed inverter in Section 5. Consequently, a quantitative comparison between the proposed inverter and the conventional three-phase four-leg inverter will be made in Section 6. Finally, conclusions will be drawn in Section 7.

2. Topology and Operation Principle of the Proposed Inverter

The proposed cost-effective three-phase four-leg VSI is shown in Figure 1b. In particular, the proposed inverter has the definite benefit of the reduction of three IGBTs. In addition, according to the current market price, with a rated current of one thousand ampere, the thyristor is much cheaper than the IGBT. Though the number of gating signals and control complexity increase inherently due to the existence of the fourth leg, it possesses the advantage of handling unbalanced loads over the three-leg inverter.

From Figure 1b, it also can be found that the thyristors are employed in phase legs, while the IGBTs and diodes are in the fourth leg. Upon the commutation of the phase current alone, the thyristors turn off naturally. Thus, the zero-crossing point of phase currents is the major factor which determines the switching time of thyristors. Also, it is worth mentioning that the directions of injected currents for the analysis of the converter operation modes are shown in Figure 1b. Normally, the thyristors in the same leg are not allowed to be switched on or off simultaneously. In particular, the upper one is switched on during the positive cycle, while the lower one is on during the negative cycle. Consequently, there are six combinations for upper thyristor switching states and each combination corresponds to a specific time interval, as shown in Figure 2.

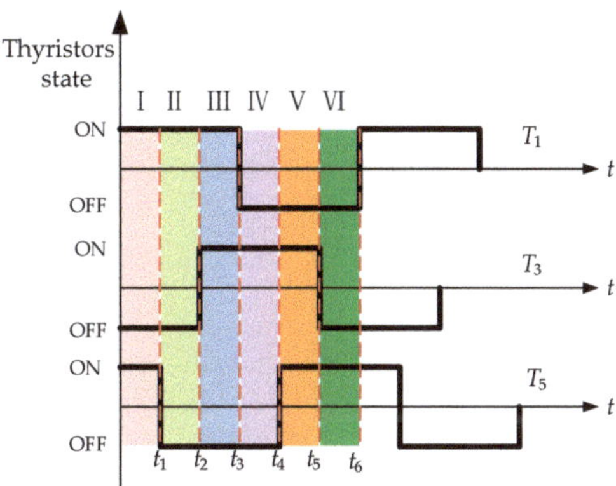

Figure 2. Six divided segments based on the thyristor states.

In Segment I, T_1 is switched on, T_3 is off and T_5 is on. For phase a, when S_a is turned on, current flows through T_1, S_a, and VD_3. When S_a is turned off, the freewheeling current flows through VD_4, as shown in Figure 3a. It should be noted the current direction in phase b is negative in Segment I. Thus, the current flows through VD_6, S_b, and T_4 when S_b is on, and the current passes through the freewheeling diode VD_5 when S_b is off. Meanwhile, phase c is in the same condition as phase a, while the fourth leg can be analyzed as a typical inverter leg. According to these situations, the thyristor-IGBT combined leg can be regarded as the same flowing path as a typical inverter leg. Therefore, based on the proposed topology, one IGBT can provide the flowing path for both positive and negative currents, whereas the current needs to pass through two different IGBTs in the conventional inverter leg.

The current flow paths for the other five segments can be analyzed in the same manner, while all individual paths are illustrated in Figure 3b–f.

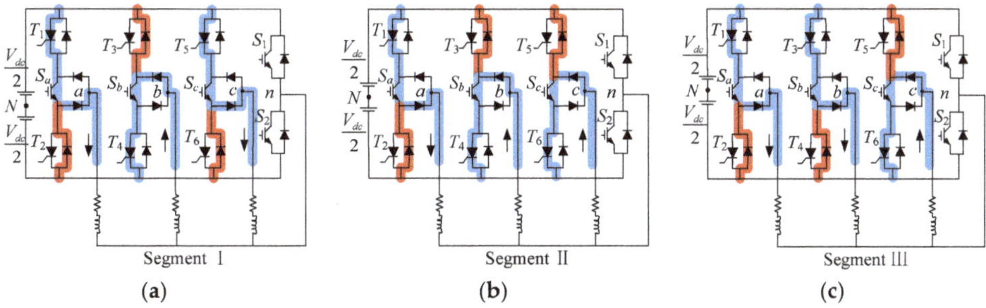

Segment I   Segment II   Segment III

(a)   (b)   (c)

Figure 3. *Cont.*

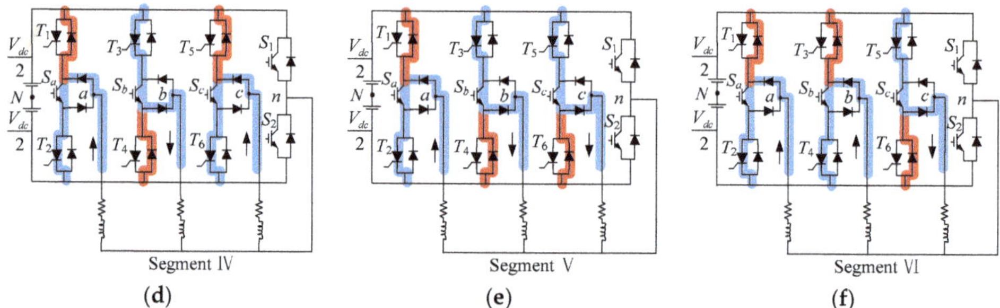

Segment IV   Segment V   Segment VI

(d)   (e)   (f)

Figure 3. Current flow path in six segments: (**a**) Segment I; (**b**) Segment II; (**c**) Segment III; (**d**) Segment IV; (**e**) Segment V; (**f**) Segment VI.

3. Mathematical Modelling

It should be noted that when the thyristors are involved, the voltage equations would differ from those produced by the conventional three-phase four-leg inverter. Referring to the middle point of the direct current (DC) link, the voltages in each leg are defined separately for the positive and negative cycles of the alternating current (AC) which is based on consideration of the thyristor states. The relationship can be described as,

$$\begin{bmatrix} v_{aN} \\ v_{bN} \\ v_{cN} \\ v_{nN} \end{bmatrix} = \begin{bmatrix} 2S_a - 1 \\ 2S_b - 1 \\ 2S_c - 1 \\ 2S_1 - 1 \end{bmatrix} \times \frac{V_{dc}}{2} \quad (T_1, T_3, T_5 \text{ is on}) \tag{1}$$

$$\begin{bmatrix} v_{aN} \\ v_{bN} \\ v_{cN} \\ v_{nN} \end{bmatrix} = \begin{bmatrix} 1 - 2S_a \\ 1 - 2S_b \\ 1 - 2S_c \\ 2S_1 - 1 \end{bmatrix} \times \frac{V_{dc}}{2} \quad (T_1, T_3, T_5 \text{ is off}) \tag{2}$$

where v_{aN}, v_{bN}, v_{cN}, and v_{nN} are the voltage potentials between points a, b, c, n and N, respectively.

$$S_i = \begin{cases} 1, & \text{when } S_i \text{ is on} \\ 0, & \text{when } S_i \text{ is off} \end{cases} \quad (i = a, b, c, 1) \tag{3}$$

Referring to Figure 1b, the phase voltage that is represented in the matrix can be governed by

$$\begin{bmatrix} v_{an} \\ v_{bn} \\ v_{cn} \\ v_n \end{bmatrix} = \begin{bmatrix} 1 & 0 & 0 & -1 \\ 0 & 1 & 0 & -1 \\ 0 & 0 & 1 & -1 \\ 0 & 0 & 0 & 0 \end{bmatrix} \begin{bmatrix} v_{aN} \\ v_{bN} \\ v_{cN} \\ v_{nN} \end{bmatrix} \tag{4}$$

where v_{an}, v_{bn}, and v_{cn} are the voltage potentials between points a, b, c, and n, respectively.

Normally, a four-leg inverter consists of $2^4 = 16$ possible switching states, while for the new topology, the states of thyristors have to be considered beside the states of IGBTs. The relationship between phase voltage and switching states of phase a are summarized in Table 1, where the UT stands for "Upper Thyristor". Based on the same rationale, the phase voltages of phase b and phase c can be analyzed in the same manner.

Table 1. Switching States and Phase Voltage of Phase *a*.

Switching States	Phase *a*							
S_1	1				0			
UT	1		0		1		0	
S_a	1	0	1	0	1	0	1	0
v_{an}	0	$-V_{dc}$	$-V_{dc}$	0	V_{dc}	0	0	V_{dc}

It can be seen that there are, in total, eight switching-state combinations in accordance to three voltages for each phase. When the states of UT are opposite, the corresponding phase voltage is inverse. This feature can be utilized to simplify the control algorithm of the proposed inverter. The typical converter's modes of voltage vectors are analyzed in the aforementioned six segments, also known as intervals, individually.

Since the state of the thyristor does not change within one segment, the voltage vectors can only be defined according to the IGBT state in each segment. Therefore, the basic voltage vectors of 16-switching states are expressed in a binary system as $0000, 0001, \ldots, 1111$, respectively, where the switching states are described in the order of S_a, S_b, S_c, and S_1. The distributions of the voltage vectors in all six segments can be obtained via Equations (1)–(4). The corresponding results are summarized in Table 2.

Table 2. Switching Combinations and Voltage Vectors.

Vectors	States	v_{an}, v_{bn}, v_{cn}					
		I	II	III	IV	V	VI
V_0	0000	$0, 1, 0$	$0, 1, 1$	$0, 0, 1$	$1, 0, 1$	$1, 0, 0$	$1, 1, 0$
V_1	0001	$-1, 0, -1$	$-1, 0, 0$	$-1, -1, 0$	$0, -1, 0$	$0, -1, -1$	$0, 0, -1$
V_2	0010	$0, 1, 1$	$0, 1, 0$	$0, 0, 0$	$1, 0, 0$	$1, 0, 1$	$1, 1, 1$
V_3	0011	$-1, 0, 0$	$-1, 0, -1$	$-1, -1, -1$	$0, -1, -1$	$0, -1, 0$	$0, 0, 0$
V_4	0100	$0, 0, 0$	$0, 0, 1$	$0, 1, 1$	$1, 1, 1$	$1, 1, 0$	$1, 0, 0$
V_5	0101	$-1, -1, -1$	$-1, -1, 0$	$-1, 0, 0$	$0, 0, 0$	$0, 0, -1$	$0, -1, -1$
V_6	0110	$0, 0, 1$	$0, 0, 0$	$0, 1, 0$	$1, 1, 0$	$1, 1, 1$	$1, 0, 1$
V_7	0111	$-1, -1, 0$	$-1, -1, -1$	$-1, 0, -1$	$0, 0, -1$	$0, 0, 0$	$0, -1, 0$
V_8	1000	$1, 1, 0$	$1, 1, 1$	$1, 0, 1$	$0, 0, 1$	$0, 0, 0$	$0, 1, 0$
V_9	1001	$0, 0, -1$	$0, 0, 0$	$0, -1, 0$	$-1, -1, 0$	$-1, -1, -1$	$-1, 0, -1$
V_{10}	1010	$1, 1, 1$	$1, 1, 0$	$1, 0, 0$	$0, 0, 0$	$0, 0, 1$	$0, 1, 1$
V_{11}	1011	$0, 0, 0$	$0, 0, -1$	$0, -1, -1$	$-1, -1, -1$	$-1, -1, 0$	$-1, 0, 0$
V_{12}	1100	$1, 0, 0$	$1, 0, 1$	$1, 1, 1$	$0, 1, 1$	$0, 1, 0$	$0, 0, 0$
V_{13}	1101	$0, -1, -1$	$0, -1, 0$	$0, 0, 0$	$-1, 0, 0$	$-1, 0, -1$	$-1, -1, -1$
V_{14}	1110	$1, 0, 1$	$1, 0, 0$	$1, 1, 0$	$0, 1, 0$	$0, 1, 1$	$0, 0, 1$
V_{15}	1111	$0, -1, 0$	$0, -1, -1$	$0, 0, -1$	$-1, 0, -1$	$-1, 0, 0$	$-1, -1, 0$

Note: The unit of the output voltage is V_{dc}.

The phase voltages can also be expressed regarding the load and current as

$$\begin{bmatrix} v_{an} \\ v_{bn} \\ v_{cn} \end{bmatrix} = L_f \begin{bmatrix} \frac{di_a}{dt} \\ \frac{di_b}{dt} \\ \frac{di_c}{dt} \end{bmatrix} + L_f C_f \begin{bmatrix} \frac{d^2 u_{sa}}{dt^2} \\ \frac{d^2 u_{sb}}{dt^2} \\ \frac{d^2 u_{sc}}{dt^2} \end{bmatrix} + \begin{bmatrix} u_{sa} \\ u_{sb} \\ u_{sc} \end{bmatrix} \tag{5}$$

where

$$\begin{bmatrix} u_{sa} \\ u_{sb} \\ u_{sc} \end{bmatrix} = \begin{bmatrix} L_a \frac{di_a}{dt} \\ L_b \frac{di_b}{dt} \\ L_c \frac{di_c}{dt} \end{bmatrix} + \begin{bmatrix} R_a i_a \\ R_b i_b \\ R_c i_c \end{bmatrix} \tag{6}$$

Moreover, (5) in $\alpha\beta\gamma$ coordinate can be further formulated as

$$\begin{bmatrix} v_\alpha \\ v_\beta \\ v_\gamma \end{bmatrix} = L_f \begin{bmatrix} \frac{di_\alpha}{dt} \\ \frac{di_\beta}{dt} \\ \frac{di_\gamma}{dt} \end{bmatrix} + C_f \begin{bmatrix} \frac{d^2 u_{s\alpha}}{dt^2} \\ \frac{d^2 u_{s\beta}}{dt^2} \\ \frac{d^2 u_{s\gamma}}{dt^2} \end{bmatrix} + \begin{bmatrix} u_{s\alpha} \\ u_{s\beta} \\ u_{s\gamma} \end{bmatrix} \tag{7}$$

where

$$\begin{bmatrix} u_{s\alpha} \\ u_{s\beta} \\ u_{s\gamma} \end{bmatrix} = \begin{bmatrix} L_a \frac{di_\alpha}{dt} \\ L_b \frac{di_\beta}{dt} \\ L_c \frac{di_\gamma}{dt} \end{bmatrix} + \begin{bmatrix} R_a i_\alpha \\ R_b i_\beta \\ R_c i_\gamma \end{bmatrix} \tag{8}$$

In particular, the quantities v_{an}, v_{bn}, v_{cn} and i_a, i_b, i_c in Equation (5) can be transformed to $\alpha\beta\gamma$ coordinate.

Through (1)–(8), the voltage vectors v_α, v_β and v_γ in $\alpha\beta\gamma$ coordinate can be obtained from Segment I to Segment VI, respectively. v_α and v_β form a hexagon in the $\alpha\beta$ plane as shown in Figure 4, where each vector corresponds to a combination of different switching states in different segments. Moreover, it should be noted that v_γ is determined by the state of the fourth leg. The graphical representation of all voltage vectors in the six segments in $\alpha\beta\gamma$ coordinate is shown in Figure 5. It can be seen that there are fourteen nonzero voltage vectors and two zero vectors in each segment. However, the same voltage vector in different segments is formed by different corresponding switching state combinations. For instance, 0110 in Segment I forms the same voltage vector as 0100 in Segment II. Moreover, there are seven $\alpha\beta$ planes in each coordinate. The distance between each of them is $1/3$.

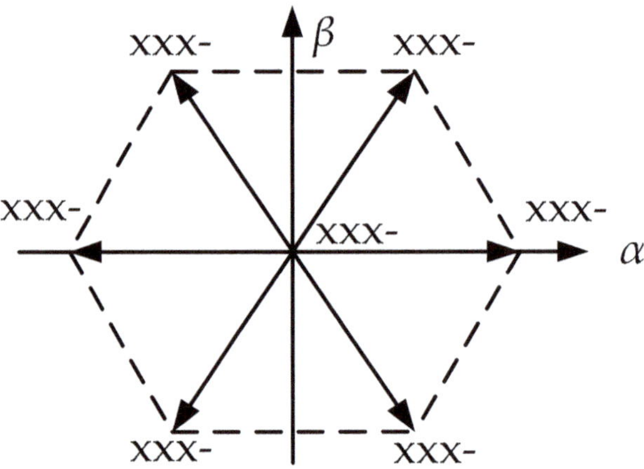

Figure 4. Switching projection of inverter vectors in $\alpha\beta$ plane.

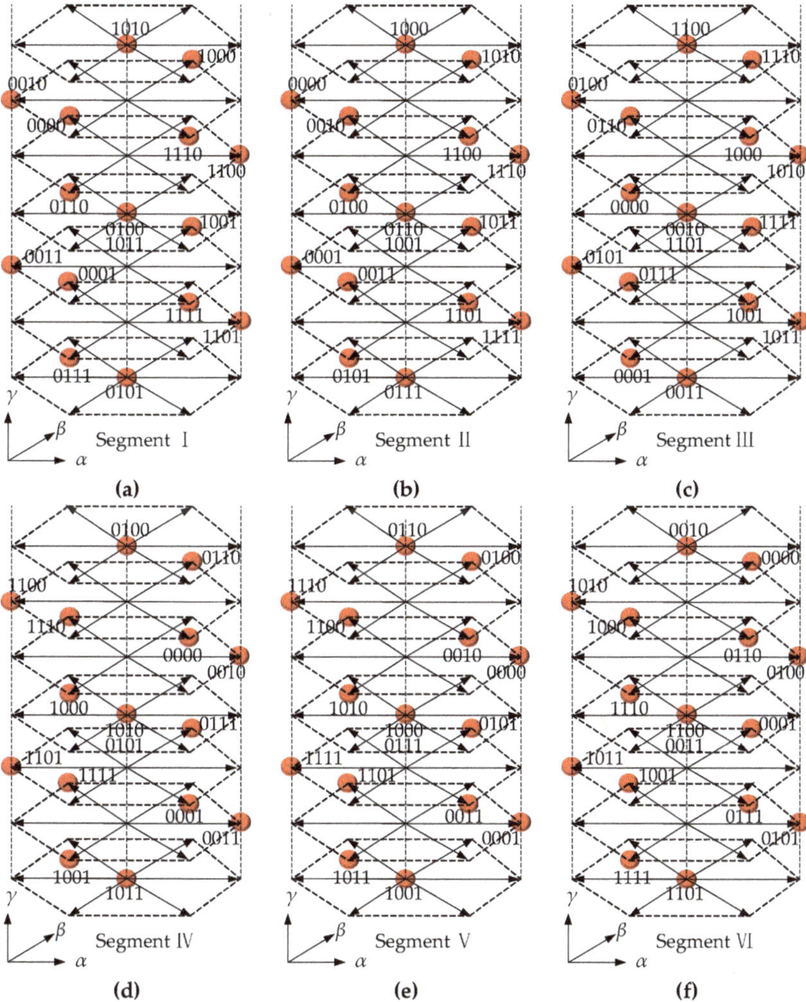

(a)   **(b)**   **(c)**

(d)   **(e)**   **(f)**

Figure 5. Switching projection of inverter vectors in $\alpha\beta\gamma$ plane: (**a**) Segment I; (**b**) Segment II; (**c**) Segment III; (**d**) Segment IV; (**e**) Segment V; (**f**) Segment VI.

4. Current Control Algorithm for Proposed Inverter

Carrier-based pulse width modulation (PWM) with proportional resonant (PR) controllers and a pulse computational module (PCM) are employed to regulate the output currents of the proposed inverter. The control algorithm is depicted in Figure 6.

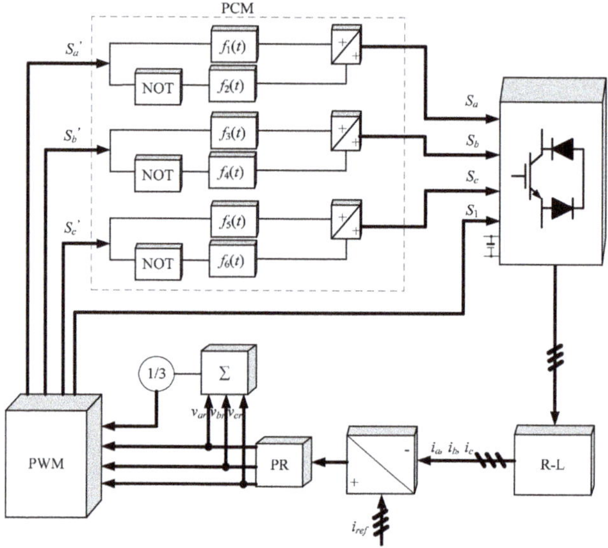

Figure 6. Current control diagram of the proposed inverter.

The PR regulator is used due to its capability of tracking the AC current with zero steady-state error. The practical PR controller is described by the transfer function [32]:

$$G(s) = k_p\left[1 + \frac{s}{T_i(s^2 + \omega_r s + \omega_0^2)}\right] \tag{9}$$

where k_p is the proportional gain, T_i is the integral gain, ω_r is the resonant cut off frequency, and ω_0 is equal to the inverter frequency.

The error between the measured current and the reference are sent to the PR controller to produce commands v_{ar}, v_{br} and v_{cr}. It is worth mentioning that the neutral leg is commanded by the average of the other three phase voltages,

$$v_M = \frac{1}{3}(v_{ar} + v_{br} + v_{cr}) \tag{10}$$

Each modulator generates a signal for one inverter leg when keeping with the principle of PWM. However, except for the neutral leg, the signals produced by PWM modulator have to be processed through a PCM before sending them to IGBTs. According to Table 1, it indicates that the opposite state thyristor corresponds to the inverse phase voltage. Therefore, the switching signal values generated from the PWM modulator for phase legs in the negative cycle should be flipped using a "NOT" function. This computation process can be expressed through the piecewise functions where, the pulse computation function for the positive cycle of phase *a* is shown as:

$$f_1(t) = \begin{cases} 1, & nT + \Delta t \leq t \leq nT + \frac{T}{2} + \Delta t, \ n = 0,1,2,3\cdots \\ 0, & nT + \frac{T}{2} + \Delta t \leq t \leq (n+1)T + \Delta t, \ n = 0,1,2,3\cdots \end{cases} \tag{11}$$

the pulse computation function for the negative cycle of phase *a* is shown as:

$$f_2(t) = \begin{cases} 0, & nT + \Delta t \leq t \leq nT + \frac{T}{2} + \Delta t, \ n = 0,1,2,3\cdots \\ 1, & nT + \frac{T}{2} + \Delta t \leq t \leq (n+1)T + \Delta t, \ n = 0,1,2,3\cdots \end{cases} \tag{12}$$

the pulse computation function for the positive cycle of phase b is shown as:

$$f_3(t) = \begin{cases} 1, & nT + \frac{T}{3} + \Delta t \le t \le nT + \frac{5T}{6} + \Delta t, \ n = 0,1,2,3 \cdots \\ 0, & nT + \frac{5T}{6} + \Delta t \le t \le (n+1)T + \frac{T}{3} + \Delta t, \ n = 0,1,2,3 \cdots \end{cases} \tag{13}$$

the pulse computation function for the negative cycle of phase b is shown as:

$$f_4(t) = \begin{cases} 0, & nT + \frac{T}{3} + \Delta t \le t \le nT + \frac{5T}{6} + \Delta t, \ n = 0,1,2,3 \cdots \\ 1, & nT + \frac{5T}{6} + \Delta t \le t \le (n+1)T + \frac{T}{3} + \Delta t, \ n = 0,1,2,3 \cdots \end{cases} \tag{14}$$

the pulse computation function for the positive cycle of phase c is shown as:

$$f_5(t) = \begin{cases} 1, & nT + \frac{2T}{3} + \Delta t \le t \le (n+1)T + \frac{T}{6} + \Delta t, \ n = 0,1,2,3 \cdots \\ 0, & (n+1)T + \frac{T}{6} + \Delta t \le t \le (n+1)T + \frac{2T}{3} + \Delta t, \ n = 0,1,2,3 \cdots \end{cases} \tag{15}$$

the pulse computation function for the negative cycle of phase c is shown as:

$$f_6(t) = \begin{cases} 0, & nT + \frac{2T}{3} + \Delta t \le t \le (n+1)T + \frac{T}{6} + \Delta t, \ n = 0,1,2,3 \cdots \\ 1, & (n+1)T + \frac{T}{6} + \Delta t \le t \le (n+1)T + \frac{2T}{3} + \Delta t, \ n = 0,1,2,3 \cdots \end{cases} \tag{16}$$

here, T is the period of the phase current.

Using the PCM, all IGBTs are controlled. Phase shift factor Δt is defined in piecewise functions. The optimum value of Δt is tuned to reduce the total harmonics distortion (THD) of the output current. To control the thyristors, a triggering pulse is sent to the upper thyristor at the start of each positive cycle and to the lower thyristor at the start of each negative cycle. When the current crosses zero, the thyristor is switched off accordingly.

5. Verification Results of Proposed Topology

In order to verify the proposed topology, simulations and experiments are conducted and analyzed. The system parameters for the simulation are listed in Table 3. The closed-loop current feedback control was implemented.

Table 3. Key Parameters of Simulation System.

Parameters	Values
DC link voltage	100 V
AC filter inductor for each phase	5 mH
Ac filter capacitor for each phase	1.5 μF
Switching frequency	5 kHz
Load inductor	1.5 mH
Load resistor	3-phase variable resistors

5.1. Simulation Results

Using the current feedback PWM scheme, the proposed inverter is operated with a balanced load, as shown in Figure 7. With the support of the PR controller loop, the static performance of the current is satisfactory. Moreover, the output current and phase voltage are well-balanced. Subsequently, the unbalanced load condition is investigated with the load resistors of 2 Ω, 1 Ω and 0.5 Ω, respectively. As shown in Figure 8a, the output currents stay balanced regardless of the unbalanced loads. In the meantime, the line to line voltages in balanced and unbalanced load conditions are shown in Figures 7c and 8c, respectively. It can be observed that under the unbalanced loads, the current THD is slightly increased, from 1.45% to 1.55%. Furthermore, when the inverter is operated as the controlled current source, the appropriate adjustments of the output voltages are shown in Figure 8b. In this way, the amplitudes of the output currents are kept constant under unbalanced loads.

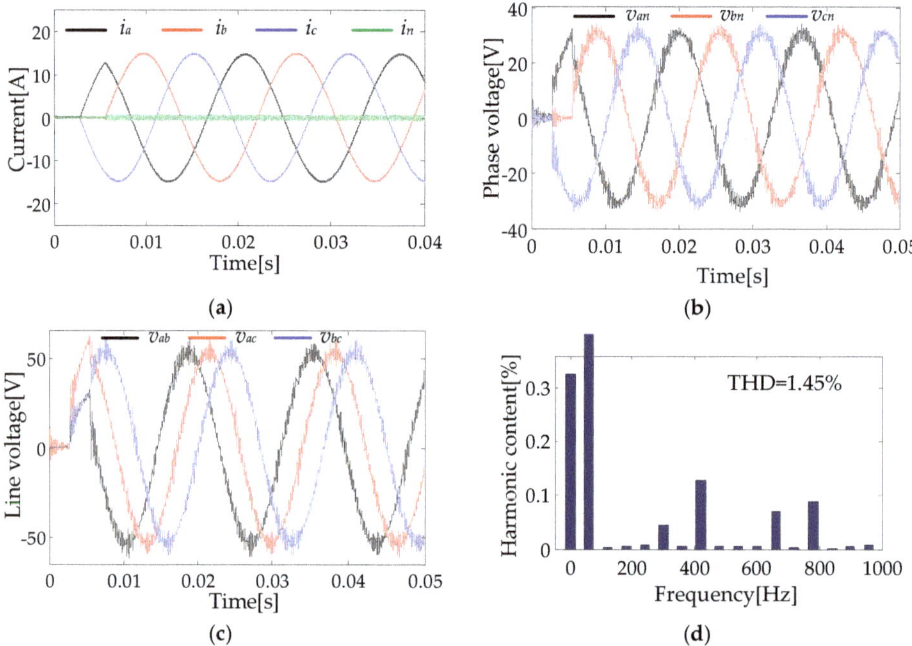

Figure 7. Simulated results of the closed-loop current control scheme in a balanced load condition: (**a**) Output currents; (**b**) Output phase voltages; (**c**) Output line to line voltages; (**d**) THD of phase *a* current.

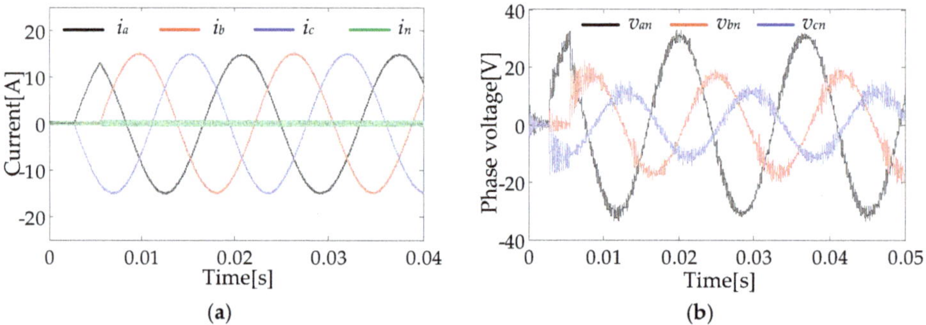

Figure 8. *Cont.*

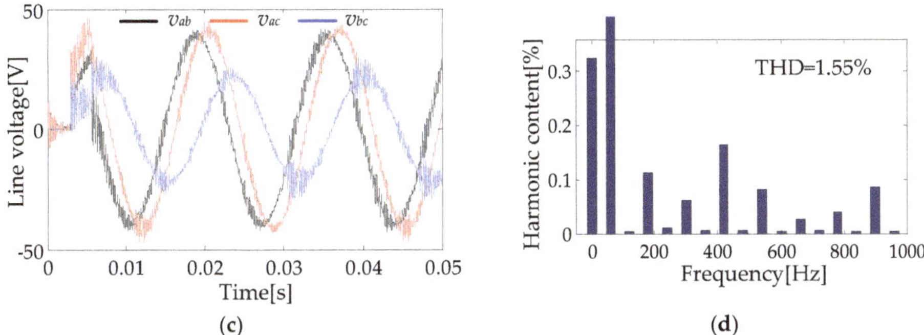

Figure 8. Simulated results of the closed-loop current control scheme in an unbalanced load condition: (**a**) Output currents; (**b**) Output phase voltages; (**c**) Output line to line voltages; (**d**) THD of phase *a* current.

5.2. Experimental Results

To further verify the performance of the proposed converter, a prototype of the new converter is built as shown in Figure 9. The experimental setup mainly includes the inverter prototype, drive circuit for IGBTs, drive circuit for thyristors, current sensors, R-L load, DC power supply and a dSPACE 1104 controller. The load parameters are: R = 5 Ω, 4 Ω, 3 Ω and L = 5 mH.

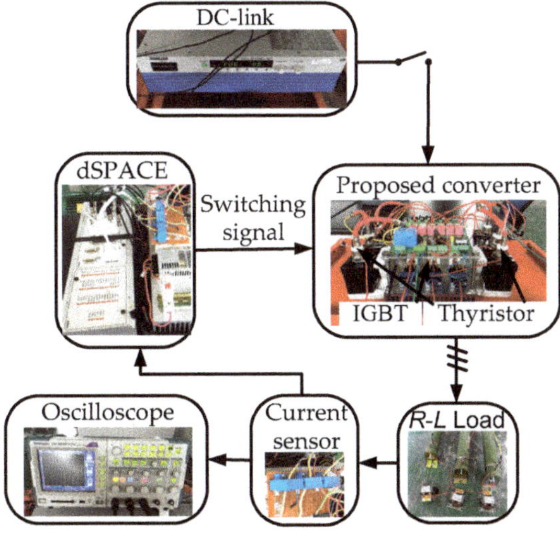

Figure 9. Experimental prototype.

The validity of the improved PWM-PR controller current control algorithm for the proposed three-phase four-leg converter is investigated. It can be observed from Figure 10a that the triggering pulse for the upper thyristor is generated when the current crosses zero from negative to positive, while the triggering pulse for the lower thyristor is generated when the current crosses zero from positive to negative. Figure 10b,c shows the phase current waveforms and THD of the proposed converter at a steady 2 A reference current with an unbalanced load. By importing the data from oscilloscope into Matlab, the THD of phase *a* current is obtained as 5.42%.

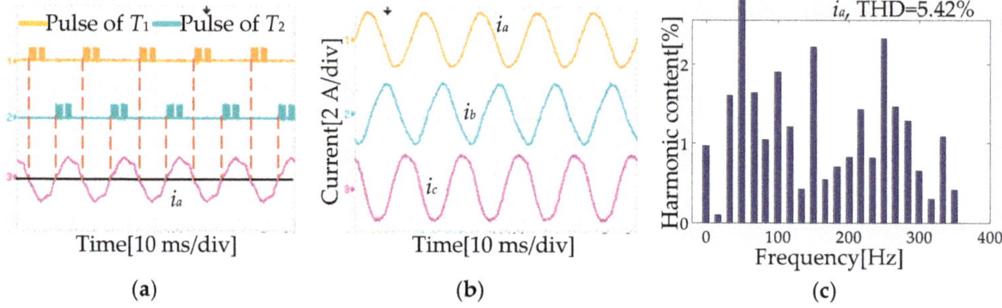

Figure 10. Experiment results of the proposed converter: (**a**) Triggering pulses for phase *a* thyristors; (**b**) Output three-phase currents; (**c**) THD analysis of phase *a* current.

Therefore, the characteristics of the proposed three-phase four-leg converter are verified by simulation and experimentation. It proves that the proposed cost-effective converter possesses a distinct merit for potential implementation in high power applications, such as electric ship propulsion systems, microgrid converters, motor drives, and so on.

6. Comparison

To further demonstrate the merit of the proposed three-phase four-leg inverter, a quantitative comparison with the conventional eight-IGBTs VSI is conducted and discussed. For a fair comparison, the converter ratings are unified as the power level of 2.88 MW, whereas the DC voltage and current are set to 1600 V and 1800 A. The specifications of IGBT, thyristor, and diode can be tabulated in Table 4.

Table 4. Specifications of Switches.

Item	IGBT Module (Single Switch)	IGBT Module (Dual Switch)	Thyristor	Diode
Model	FZ1800R16KF4	FF1800R17IP5	Y60KPE	ZP2000A
Cost (USD)	550	965	50	30
Size (H × W × D, cm)	3.81 × 19.0 × 14.0	3.8 × 23.4 × 8.9	6.0 × 15 × 5.0	3.2 × 9.9 × 9.9
Weight (kg)	2.31	1.4	0.85	0.52
I_r (Amps)	1800	1800	1800	2000
V_r (Volts)	1600	1700	1800	5000

First, the number of switches for the proposed converter and the conventional three-phase four-leg inverter can be determined according to Figure 1. The numbers of single switch IGBT module, dual switch IGBT module, thyristor and diode for the proposed converter and conventional converter are 0, 4, 0, 0 and 1, 1, 6, 12, respectively. Subsequently, size and cost comparison can be calculated. The size and weight are first compared with the specific data. Then, their cost is quantitatively compared and discussed.

The estimated data of the proposed and conventional inverters, namely, the size, weight and cost are summarized in Table 5. It can be observed that the proposed inverter has a definite advantage in terms of lower cost, although it suffers from its relatively complicated structure and larger size. Approximately USD 1094 can be saved by one three-phase four-leg inverter for the power level of a 2.88 MW system, which is appreciated for the demand side. Therefore, the proposed inverter has great potential for mega-watt level cost-sensitive applications.

Table 5. Comparison between Proposed and Conventional Inverter.

Item	Proposed Inverter	Conventional Inverter
Overall size (cm^3)	8268.4	3165.56
Overall weight (kg)	15.05	5.6
Cost (USD)	2766	3860

7. Conclusions

In this paper, a new integrated IGBT-thyristor three-phase four-leg VSI is proposed, analyzed and evaluated, specifically for mega-watt level applications. As compared with traditional four-leg inverters, the proposed inverter achieves a distinct advantage of higher cost-effectiveness. The operation principle of the proposed inverter is investigated and discussed. The inverter output voltage in $\alpha\beta\gamma$ coordinate of each segment is also presented. Moreover, the current control strategy based on carrier-based PWM with PR controllers and the PCM is developed and analyzed. Demonstration results of simulation and experimentation for closed-loop current control are accomplished, which verifies the functionality of the proposed inverter under balanced and unbalanced load conditions.

Acknowledgments: The work described in this paper was supported by a grant (Project No. 51677159) from the Natural Science Foundation of China (NSFC), China.

Author Contributions: The work presented in this paper is the output of research projects undertaken by Chunhua Liu. In specific, Yixiao Luo and Chunhua Liu developed the topic, designed the system, analyzed the results, and wrote this paper. Feng Yu and Christopher H.T. Lee helped review and improve the paper.

Conflicts of Interest: The authors declare no conflict of interest.

1. Kim, E.-K.; Mwasilu, F.; Choi, H.H.; Jung, J.-W. An observer-based optimal voltage control scheme for three-phase UPS systems. *IEEE Trans. Ind. Electron.* **2015**, *62*, 2073–2081. [CrossRef]
2. Hossain, R.; Lipo, T.; Kwon, B. A novel topology for a voltage source inverter with reduced transistor count and utilizing naturally commutated thyristors with simple commutation. In Proceedings of the 2014 International Symposium on Power Electronics Electrical Drives Automation and Motion (SPEEDAM), Ischia, Italy, 18–20 June 2014.
3. Liu, C.; Chau, K.T.; Zhang, X. An efficient Wind–Photovoltaic hybrid generation system using doubly excited permanent-magnet brushless machine. *IEEE Trans. Ind. Electron.* **2010**, *57*, 831–839.
4. Liu, C.; Chau, K.T.; Li, W. Loss analysis of permanent magnet hybrid brushless machines with and without HTS field windings. *IEEE Trans. Appl. Supercond.* **2010**, *20*, 1077–1080.
5. Liu, C.; Chau, K.T.; Lee, C.H.T.; Lin, F.; Li, F.; Ching, T.W. Magnetic vibration analysis of a new DC-excited multitoothed switched reluctance machine. *IEEE Trans. Magn.* **2014**, *50*, 1–4. [CrossRef]
6. Chen, W.; Liu, C.; Lee, C.H.T.; Shan, Z. Cost-effectiveness comparison of coupler designs of wireless power transfer for electric vehicle dynamic charging. *Energies* **2016**, *9*, 906. [CrossRef]
7. Bao, L.N.; Kim, K. An Improved Current Control Strategy for a Grid-Connected Inverter under Distorted Grid Conditions. *Energies* **2016**, *9*, 190. [CrossRef]
8. Wang, F.; Zhang, Z.; Ericsen, T.; Raju, R. Advances in Power Conversion and Drives for Shipboard Systems. *IEEE Proc.* **2015**, *103*, 2285–2311. [CrossRef]
9. Pei, X.; Zhou, W.; Kang, Y. Analysis and calculation of DC-link current and voltage ripples for three-phase inverter with unbalanced load. *IEEE Trans. Power Electron.* **2015**, *30*, 5401–5412. [CrossRef]
10. Jeong, C.Y.; Cho, J.G.; Kang, Y.; Rim, G.H.; Song, E.H. A 100 kVA power conditioner for three-phase four-wire emergency generators. In Proceedings of the IEEE PESC'98, Fukuoka, Japan, 22 May 1998; pp. 1906–1911.
11. Zhang, R.; Boroyevich, D.; Prasad, V.H. A three-phase inverter with a neutral leg with space vector modulation. *IEEE Proc.* **1997**, *2*, 857–863.
12. Ryan, M.J.; Lorenz, R.D.; de Doncker, R.W. Decoupled control of a four-leg inverter via a new 4×4 transformation matrix. *IEEE Trans. Power Electron.* **2001**, *16*, 694–701. [CrossRef]

13. Zhang, R.; Prasad, V.H.; Boroyevich, D.; Lee, F.C. Three-dimensional space vector modulation for four-leg voltage source converters. *IEEE Trans. Power Electron.* **2002**, *17*, 314–325. [CrossRef]

14. Perales, M.A.; Prats, M.M.; Portillo, R.; Mora, J.L.; Len, J.L.; Franquelo, L.G. Three-dimensional space vector modulation in abc coordinates for four-leg voltage source converters. *IEEE Power Electron. Lett.* **2003**, *1*, 104–109. [CrossRef]

15. Depenbrock, M.; Staudt, V.; Wrede, H. A theoretical investigation of original and modified instantaneous power theory applied to four-wire systems. *IEEE Trans. Ind. Appl.* **2003**, *39*, 1160–1168. [CrossRef]

16. Li, Y.W.; Vilathgamuwa, D.M.; Loh, P.C. Microgrid power quality enhancement using a three-phase four-wire grid-interfacing compensator. *IEEE Trans. Ind. Appl.* **2005**, *41*, 1707–1719. [CrossRef]

17. Dong, G.; Ojo, O. Current regulation in four-leg voltage-source converters. *IEEE Trans. Ind. Electron.* **2007**, *54*, 2095–2105. [CrossRef]

18. Escobar, G.; Valdez, A.A.; Torres-Olguin, R.E.; Martinez-Montejano, M.F. A model-based controller for a three-phase four-wire shunt active filter with compensation of the neutral line current. *IEEE Trans. Power Electron.* **2007**, *22*, 2261–2270. [CrossRef]

19. Teodorescu, R.; Liserre, M.; Rodriguez, P.; Blaabjerg, F. *Grid Converters for Photovoltaic and Wind Power Systems*; Wiley: Chichester, UK, 2011.

20. Dos Santos, E.; Jacobina, C.; Rocha, N.; Dias, J.; Correa, M. Single-phase to three-phase four-leg converter applied to distributed generation system. *IET Power Electron.* **2010**, *3*, 892–903. [CrossRef]

21. Wu, B.; Lang, Y.; Zargari, N.; Kouro, S. *Power Conversion and Control of Wind Energy Systems*, 2nd ed.; IEEE Press Series on Power Engineering; Wiley: Hoboken, NJ, USA, 2011.

22. Kim, J.-H.; Sul, S.-K. A carrier-based PWM method for three-phase four-leg voltage source converters. *IEEE Trans. Power Electron.* **2004**, *19*, 66–75. [CrossRef]

23. Dai, N.-Y.; Wong, M.-C.; Ng, F.; Han, Y.-D. A FPGA-based generalized pulse width modulator for three-leg center-split and four-leg voltage source inverters. *IEEE Trans. Power Electron.* **2008**, *23*, 1472–1484. [CrossRef]

24. Zhang, M.; Atkinson, D.J.; Ji, B.; Armstrong, M.; Ma, M. A near-state three-dimensional space vector modulation for a three-phase four-leg voltage source inverter. *IEEE Trans. Power Electron.* **2014**, *29*, 5715–5726. [CrossRef]

25. Li, X.; Deng, Z.; Chen, Z.; Fei, Q. Analysis and simplification of three-dimensional space vector PWM for three-phase four-leg inverters. *IEEE Trans. Ind. Electron.* **2011**, *58*, 450–464. [CrossRef]

26. Rivera, M.; Yaramasu, V.; Llor, A.; Rodriguez, J.; Wu, B.; Fadel, M. Digital predictive current control of a three-phase four-leg inverter. *IEEE Trans. Ind. Electron.* **2013**, *60*, 4903–4912. [CrossRef]

27. Bayhan, S.; Abu-Rub, H.; Balog, R. Model predictive control of quasi-z source four-leg inverter. *IEEE Trans. Ind. Electron.* **2016**, *63*, 4506–4516. [CrossRef]

28. Lohia, P.; Mishra, M.K.; Karthikeyan, K.; Vasudevan, K. A minimally switched control algorithm for three-phase four-leg VSI topology to compensate unbalanced and nonlinear load. *IEEE Trans. Power Electron.* **2008**, *23*, 1935–1944. [CrossRef]

29. Zhang, F.; Yan, Y. Selective harmonic elimination PWM control scheme on a three-phase four-leg voltage source inverter. *IEEE Trans. Power Electron.* **2009**, *24*, 1682–1689. [CrossRef]

30. Liu, Z.; Liu, J.; Li, J. Modeling, analysis and mitigation of load neutral point voltage for three-phase four-leg inverter. *IEEE Trans. Ind. Electron.* **2013**, *60*, 2010–2021. [CrossRef]

31. Li, B.; Jia, J.; Xue, S. Study on the Current-Limiting-Capable Control Strategy for Grid-Connected Three-Phase Four-Leg Inverter in Low-Voltage Network. *Energies* **2016**, *9*, 726. [CrossRef]

32. Zmood, D.N.; Holmes, D.G. Stationary Frame Current Regulation of PWM Inverters with Zero Steady-State Error. *IEEE Trans. Power Electron.* **2003**, *18*, 814–822. [CrossRef]

Article

Analysis of Voltage Variation in Silicon Carbide MOSFETs during Turn-On and Turn-Off

Hui Li, Xinglin Liao *, Yaogang Hu, Zhangjian Huang and Kun Wang

State Key Laboratory of Power Transmission Equipment & System Security and New Technology, School of Electrical Engineering, Chongqing University, No. 174, Shazhengjie Road, Shapingba, Chongqing 400044, China; cqulh@163.com (H.L.); huyaogang345@163.com (Y.H.); 20151102094t@cqu.edu.cn (Z.H.); 20161102076t@cqu.edu.cn (K.W.)
* Correspondence: lxl108381@126.com; Tel.: +86-023-6510-2437

Received: 5 July 2017; Accepted: 16 September 2017; Published: 21 September 2017

Abstract: Due to our limited knowledge about silicon carbide metal–oxide–semiconductor field-effect transistors (SiC MOSFETs), the theoretical analysis and change regularity in terms of the effects of temperature on their switching characteristics have not been fully characterized and understood. An analysis of variation in voltage (dV_{DS}/dt) for SiC MOSFET during turn-on and turn-off has been performed theoretically and experimentally in this paper. Turn-off variation in voltage is not a strong function of temperature, whereas the turn-on variation in voltage has a monotonic relationship with temperature. The temperature dependence is a result of the competing effects between the positive temperature coefficient of the intrinsic carrier concentration and the negative temperature coefficient of the effective mobility of the electrons in SiC MOSFETs. The relationship between variation in voltage and supply voltage, load current, and gate resistance are also discussed. A temperature-based analytical model of dV_{DS}/dt for SiC MOSFETs was derived in terms of internal parasitic capacitances during the charging and discharging processes at the voltage fall period during turn-on, and the rise period during turn-off. The calculation results were close to the experimental measurements. These results provide a potential junction temperature estimation approach for SiC MOSFETs. In SiC MOSFET-based practical applications, if the turn on dV_{DS}/dt is sensed, the device temperature can be estimated from the relationship curve of turn on dV_{DS}/dt versus temperature drawn in advance.

Keywords: power semiconductor device; temperature; switching transients; variation in voltage

1. Introduction

Although silicon power devices have developed rapidly in the past few decades, many of them are reaching their physical limits. In recent years, silicon carbide (SiC) power devices have offered a probable solution to this problem due to their wide bandgap and electrical and physical characteristics [1]. In comparison with Si devices, the superiority of SiC metal–oxide–semiconductor field-effect transistors (MOSFETs) has been demonstrated [2]. Since Si insulated-gate bipolar transistors (IGBTs) are widely used to construct power electronics converters in many industrial products, such as photovoltaic generation, motor drives, and uninterruptable power supplies, many efforts have been devoted to the promotion and expansion of the applications of SiC MOSFETs [3]. In the future, SiC MOSFETs may replace Si IGBTs in the voltage range of 1200 V and above, due to the fact they offer considerable performance and smaller switching time.

Due to our limited knowledge about SiC MOSFETs, the theoretical analysis and change regularity in terms of the effects of temperature on its switching characteristics have not been fully characterized and understood. Previous work by Zhu et al. [4] analyzed the temperature dependence of on-resistance. In [5,6], the threshold voltage has been investigated, which is not stable under different temperatures because of electron tunneling into and out of the oxide traps. The switching characteristics of SiC MOSFETs were investigated in [7,8], but the impact of temperature was not considered. The relationship between dV_{DS}/dt and temperature for SiC MOSFET can be observed in some published reports [9–14]. In [9] and [14], the characterization and comparison of three types of 1.2 kV SiC MOSFETs produced by different manufacturers is presented at 25 °C and 175 °C. Similar measurements have also been performed in [10,11]. In [12], a SiC Implantation and Epitaxial MOSFET (SiC-IEMOSFET) has been evaluated at the temperatures of 25 °C and 125 °C. In reference [13], a behavioral model of SiC MOSFET in Pspice over a wide temperature range is provided. The static and dynamic behavior is simulated using the presented model and compared to the measured waveforms. However, the effects of temperature on switching characteristics were only examined at two temperature conditions. From the aforementioned literature, we don't know if the dV_{DS}/dt varies linearly with temperature due to the absence of measurement data. Furthermore, many issues still remain unclear, such as the effects of supply voltage, load current and gate resistance on dV_{DS}/dt and the modeling of dV_{DS}/dt. However, these are important for SiC MOSFET-based practical applications. It is necessary to investigate the effects of temperature on the switching characteristics, which is useful for understanding how the variation of voltage varies with temperature. Alternatively, in future SiC MOSFET-based practical applications, the junction temperature measurement will become an important topic. However, for SiC MOSFETs there is less temperature sensitivity for the same electrical parameters as in Si IGBTs owing to their unipolar nature. Because of the wider bandgap, the lower intrinsic carrier concentration and the faster switching speed, some conventional indicators of junction temperature estimation for Si devices would fail for SiC MOSFETs [15,16]. Hence, it is also important to find a temperature-sensitive electrical parameter to evaluate device temperature in SiC MOSFETs.

In this paper, a thorough analysis was completed of the variation in voltage for SiC MOSFETs during turn-on and turn-off. The temperature dependency of turn-off variation in voltage and turn-on variation in voltage were found to be different. The turn-off variation in voltage was not strongly correlated with temperature, while the turn-on variation in voltage was a function of temperature. The relationship is nearly linear. From the relationship curve, the junction temperature of SiC MOSFET can be derived. Hence, turn-on variation in voltage is suitable as a temperature-sensitive electrical parameter for junction temperature measurement. In addition, a temperature-based analytical model of variation in voltage is presented and the effects of supply voltage, load current, and gate resistance on the temperature dependency of variation in voltage are analyzed. Finally, the implementations of the temperature dependence of variation in voltage are discussed.

2. Model

2.1. Overview of the Turn-On and Turn-Off Process

Figure 1 shows the typical structure of a SiC MOSFET consisting of three electrodes, namely drain, gate, and source, gate oxide, JFET region, and N-drift. It is a vertical device with a planar gate. Apart from that, the equivalent circuit of SiC MOSFET is also shown, including three internal parasitic capacitances: gate–drain (C_{GD}), gate–source (C_{GS}), and drain–source (C_{DS}).

Based on an inductive load circuit, the switching characteristic of a power MOSFET is illustrated in Figure 2, showing the four phases during turn-on and turn-off. Due the existence of switching loop stray inductances, the induced voltage across stray inductances will reshape the waveforms of a drain source voltage V_{DS}, causing a drop in turn-on and a peak in turn-off, as a result of the induced positive voltages and negative voltages, respectively [17].

The current distribution in the device is changed due to the effect of the drain–source capacitance C_{DS} of the SiC MOSFET, which usually cannot be observed outside the SiC MOSFET. During the turn-on process, the energy stored in C_{DS} discharges through MOS channel, which causes the channel current $I_{channel}$ to be larger than the drain current I_D measured outside. While at turn-off, the capacitance C_{DS} is charged as the drain-source voltage V_{DS} rises. In this process, a part of the load current flows to C_{DS}, which causes $I_{channel}$ to be smaller than I_D. As a result, the gate-source plateau voltage V_{GP} in turn-on is higher than that in turn-off.

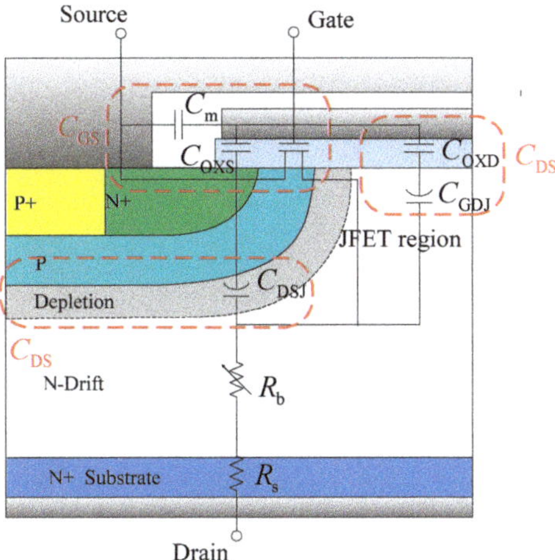

Figure 1. Cross structure of a silicon carbide (SiC) metal–oxide–semiconductor field-effect transistor (MOSFET).

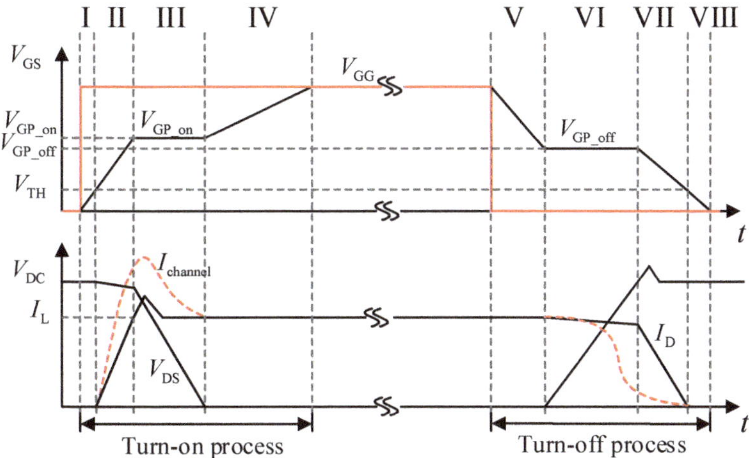

Figure 2. Swtiching characteristic of a power MOSFET with an inductive load.

2.2. Temperature-Based dV_{DS}/dt Model

The dynamic behavior of SiC MOSFETs is strongly dependent on their terminal capacitances: C_{GD}, C_{GS} and C_{DS}. The capacitances C_{GD} and C_{GS} govern the switching transient since they are charged and discharged during the turn-on and turn-off processes. The effect of capacitance C_{DS} can't also be neglected since it also charges and discharges. In Figure 2, the MOSFET is in the saturation region and V_{GS} remains almost unchanged at the voltage fall period during turn-on. Consequently, the gate current deviates from the gate source capacitance C_{GS} and the Miller capacitance C_{GD} to mainly charge the C_{GD}. In addition, the drain source capacitance C_{DS} is discharged in this phase and the discharging currents will inject into the channel of the MOSFET. Similarly, the C_{DS} is charged during the voltage rise period during turn-off, and a part of the load current will be shunt. Owing to the discharging/charging of terminal capacitances, the current flowing through MOSFET, named the channel current, is not equal to the current measured through the drain terminal, namely I_D, which is larger or smaller than I_D. Figure 3a shows the discharging of capacitances C_{GD} during turn-on, whereas the charging of capacitances C_{GD} is shown in Figure 3b. The discharging/charging process of Miller capacitance is also exhibited in this figure.

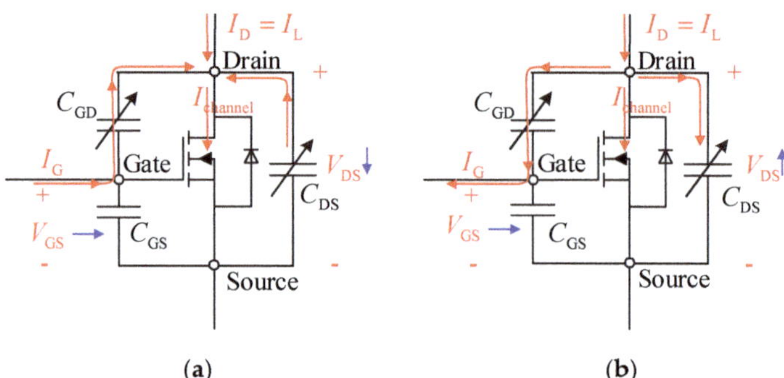

Figure 3. Terminal capacitances charging and discharging at: (**a**) the voltage fall phase and (**b**) the rise phase.

In the voltage rise phase during turn-off, the voltage across the inversion channel is higher than the saturation voltage, $V_{DS,sat}$, defined as $V_{GS} - V_{TH}$, and the SiC MOSFET is in the saturation region [18,19]. The channel current is a function of saturation voltage $V_{DS,sat}$ and can be given as :

$$I_{channel} = \frac{B}{2}V_{DS.sat}^2(1 + \lambda V_{DS}) = \frac{W\mu_0 C_{OX}}{2L}(V_{GS} - V_{TH})^2(1 + \lambda V_{DS}) \tag{1}$$

where $B = W\mu_0 C_{OX}/L$ is the transconductance parameter, W is the channel width, L is the channel length, V_{TH} and V_{GS} are the threshold voltage and gate-source voltage, respectively, λ is the channel length modulation parameter, μ_0 is the effective mobility of the electrons in the channel, and C_{OX} is the gate–oxide capacitor.

Since the drive voltage V_{GG} has become the low level V_{GG_L} in this phase, and the gate source voltage V_{GS} reaches its Miller plateau voltage that is equal to $V_{TH}+I_L/g_m$, the gate drive current I_G can be calculated as shown in Equation (2), where R_G is the gate drive resistance, I_L is the load current, and g_m is the device's transconductance:

$$I_G = \frac{V_{GS} - V_{GG\_L}}{R_G} = \frac{V_{TH} + \frac{I_{load}}{g_m} - V_{GG\_L}}{R_G} = \frac{V_{TH} + \sqrt{\frac{I_{load}L}{W\mu_0 C_{OX}}} - V_{GG\_L}}{R_G} \tag{2}$$

The capacitance C_{DS} is a drain–source voltage V_{DS} –sensitive parameter due to the variation in depletion region width of the drain–body junction with V_{DS}, which is given by [20,21]:

$$C_{DS} = A_{DS}\sqrt{\frac{qN_A\varepsilon_{SiC}}{2(V_{DS} + V_{bi})}} \tag{3}$$

where:

$$V_{bi} = \frac{kT}{q}\ln(\frac{N_A N_D}{n_i^2}) \tag{4}$$

while q is the fundamental electronic charge, ε_{SiC} is the dielectric constant in SiC, N_A is the p-well region doping, A_{DS} is the drain–source overlap area, V_{bi} is the junction potential in drain-body junction, N_D is the doping in drift region, and n_i is the intrinsic carrier concentration of SiC.

The capacitance C_{GD} is consisted of the gate oxide capacitance $C_{OX} = \varepsilon_{OX} A_{GD}/t_{ox}$ in series with the bias-dependent depletion capacitance under gate oxide $C_{GDJ} = A_{GD}(qN_A\varepsilon_{SiC}/2V_{DS})^{1/2}$, which is given by:

$$C_{GD} = \frac{C_{OX}C_{GDJ}}{C_{OX} + C_{GDJ}} \tag{5}$$

According to the charging and discharging processes of C_{GS}, C_{GD}, and C_{DS} as seen in Figure 3, the variation in voltage can be calculated for the voltage rise period during turn-off:

$$\frac{dV_{DS}}{dt} = \frac{I_D - I_{channel} + \frac{C_{GD}}{C_{GS}+C_{GD}}I_G}{C_{DS} + \frac{C_{GS}C_{GD}}{C_{GS}+C_{GD}}} \tag{6}$$

where C_{GS} is gate-source capacitance and I_D is drain current of the SiC MOSFET, which is equal to load current I_L. According to different operating conditions, Mc Nutt et al. [22] found that capacitance C_{GD} can be further simplified. Figure 4 shows how the capacitance C_{GD} varies under different operating conditions. When gate-source voltage V_{GS} is greater than threshold voltage, and drain-source voltage V_{DS} is greater than zero, capacitance C_{GD} is approximately equal to the gate oxide capacitance C_{OX} because depletion capacitance C_{GDJ} is far in excess of the C_{OX}. Hence, Equation (6) can be further simplified by using C_{OX} instead of C_{GD}. Additionally, during turn-on, the expression of the variation in voltage can also be derived is similar to Equation (6), where gate current is defined as $I_G = (V_{GG\_H} - (I_L/g_m + V_{TH}))/R_G$.

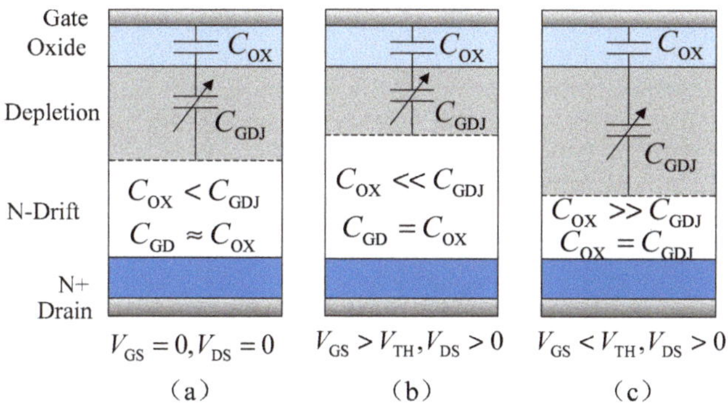

Figure 4. The gate–drain capacitance (C_{GD}) varies under different operating conditions. (**a**) $V_{GS} = 0$, $V_{DS} = 0$; (**b**) $V_{GS} > V_{TH}$, $V_{DS} > 0$; (**c**) $V_{GS} < V_{TH}$, $V_{DS} > 0$.

2.3. Dependency Analysis

Equation (6) shows that the variation in voltage is dependent on temperature, since the channel current and gate current are correlated with temperature. The load current, voltage, and gate drive resistance also have some influence on variation in voltage. The relationship between channel current and temperature depends on the temperature dependency of the threshold voltage and the effective mobility in SiC MOSFET. When the surface potential of the channel in the MOSFET is exactly twice the bulk potential, the value of the gate voltage is called the threshold voltage. This means that the gate potential has induced sufficient band bending for the intrinsic Fermi level in the p-type body of the device to be below the Fermi level. The electron concentration in the channel is exactly equal to the p-body doping and the channel is properly inverted, which is the minimum gate voltage to make the device work. Hence, the threshold voltage can be calculated as shown below in Equation (7), where ψ_B

is Fermi-potential, Q_f is the fixed oxide charges, and φ_{ms} is the work function difference between metal and semiconductor [23]:

$$V_{TH} = \phi_{ms} - \frac{Q_f}{C_{OX}} + 2\psi_B + \frac{\sqrt{4\varepsilon_{SiC} \cdot q \cdot N_A \cdot \psi_B}}{C_{OX}} \tag{7}$$

with:

$$\psi_B = \frac{kT}{q} \ln\left(\frac{N_A}{n_i}\right) \tag{8}$$

Due to the work-function difference φ_{ms} and fixed oxide charges Q_f being essentially independent of temperature, the threshold voltage temperature dependency can be given through differentiating Equation (7) with respect to temperature:

$$\frac{dV_{TH}}{dT} = \frac{d\psi_B}{dT}\left(2 + \frac{1}{C_{OX}}\sqrt{\frac{\varepsilon_{SiC} \cdot q \cdot N_A}{\psi_B}}\right) \tag{9}$$

From Equation (3), assuming a constant drain-source voltage V_{DS}, it is known that C_{DS} varying with temperature is dominated by the temperature dependency of V_{bi}. In the expression of V_{bi} in Equation (4), as temperature T increases, kT/q rises and $\ln(N_A N_D / n_i^2)$ decreases. Hence, the more dominant parameter will determine how C_{DS} changes with temperature. Theoretically, the $\ln(N_A N_D / n_i^2)$ term dominates the temperature dependency of C_{DS} since V_{bi} decreases with temperature shown in Figure 5, where the drift region doping N_D is 3.8×10^{15} cm^{-3}. From Figure 5, C_{DS} increases with temperature owing to its inverse proportional relationship with V_{bi}, according to Equation (4). As reported by Chen et al. [7], experimentally, the C-V characteristics of SiC MOSFET almost overlap under different temperatures. Hence, the temperature dependency of C_{DS} is neglected in the following analysis. Additionally, the C_{GS} and C_{OX} are generally considered to be constant.

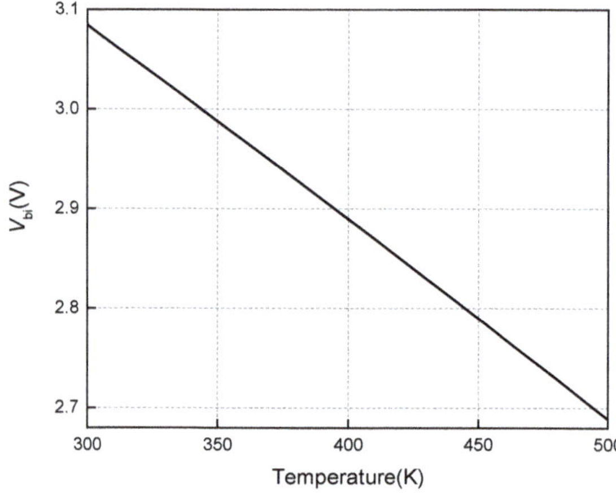

Figure 5. The junction potential in drain–body junction versus temperature.

From Equation (2), gate current is also correlated with threshold voltage and effective mobility, since the Miller plateau voltage is determined by the two parameters, resulting in the gate current varying under different temperatures. Hence, according to the temperature derivative of Equation (6), the temperature dependence of dV_{DS}/dt can be obtained, as shown in Equation (10):

$$\frac{d^2 V_{DS}}{dt\,dT} = \frac{-1}{C_{DS} + \frac{C_{GS}C_{OX}}{C_{GS}+C_{OX}}}\left(\frac{dI_{channel}}{dT} - \frac{C_{OX}}{C_{GS}+C_{OX}}\frac{dI_G}{dT}\right) \tag{10}$$

According to Equations (1) and (2), the temperature dependence of the channel and gate currents can be given by:

$$\frac{\mathrm{d}I_{\text{channel}}}{\mathrm{d}T} = \frac{1}{2}(1+\lambda V_{\text{DS}})(V_{\text{GS}}-V_{\text{TH}})[(V_{\text{GS}}-V_{\text{TH}})\frac{dB}{dT} - 2B\frac{dV_{\text{TH}}}{dT}]$$
$$= \frac{WC_{\text{OX}}}{2L}(1+\lambda V_{\text{DS}})(V_{\text{GS}}-V_{\text{TH}})[(V_{\text{GS}}-V_{\text{TH}})\frac{d\mu_0}{dT} - 2\mu_0\frac{dV_{\text{TH}}}{dT}] \tag{11}$$

$$\frac{\mathrm{d}I_{\text{G}}}{\mathrm{d}T} = \frac{dV_{\text{TH}}}{dT} - \frac{I_L}{g_m^2}\frac{dg_m}{dT} = \frac{dV_{\text{TH}}}{dT} - \frac{1}{2\mu_0}\sqrt{\frac{I_{\text{load}}L}{W\mu_0 C_{\text{OX}}}}\frac{d\mu_0}{dT} \tag{12}$$

The equation of mobility can be adopted from Mudholkar et al. [22]:

$$\mu_0 = \frac{947}{1+(\frac{N_D}{1.94\times10^{17}})^{0.61}}(\frac{T}{300})^{-2.15} \tag{13}$$

Substituting Equations (9), (11) and (12) into Equation (10), the temperature sensitivity of dV_{DS}/dt can be derived via the temperature dependency of threshold voltage and effective mobility. From Equation (13), mobility μ_0 possesses a negative temperature coefficient and decreases with temperature. However, for SiC MOSFET, $d\mu_0/dT$ is very low and can be neglected due to its wide band-gap characteristics as reported in Hasanuzzaman et al. [24]. Hence, the temperature sensitivity of dV_{DS}/dt is dominated by dV_{TH}/dT and is negative because the temperature coefficient of threshold voltage is negative. Hence, turn-off dV_{DS}/dt decreases as temperature increases. Similarly, the variation tendency of dV_{DS}/dt under different temperatures during the turn-on process can also be predicted.

From Equation (6), the variation in voltage is also affected by load current, supply voltage, and gate drive resistance. Since the derivative of dV_{DS}/dt with respect to the load current is positive, as shown in Equation (14), it has a positive impact on the variation in voltage, which means the variation in voltage increases with increasing load current:

$$\frac{\mathrm{d}^2 V_{\text{DS}}}{\mathrm{d}t\mathrm{d}I_L} = \frac{1}{C_{\text{DS}} + \frac{C_{\text{GS}}C_{\text{OX}}}{C_{\text{GS}}+C_{\text{OX}}}}(1 + \frac{C_{\text{OX}}}{R_G g_m(C_{\text{GS}}+C_{\text{OX}})}) \tag{14}$$

For the voltage, the impact on the variation in voltage is similar to that of the load current because the capacitance C_{DS} varies under different voltages, as described in Equation (3). As seen from Equation (2), the higher the gate resistance, the lower the gate current. Hence, the variation in voltage decreases with increasing gate resistance. Likewise, for the voltage fall period during turn-on, the variation in voltage also depends on the gate resistance, load current, voltage, and temperature, and their impacts on variation in voltage can be obtained through a similar analysis process as for turn-off. A flow diagram for the proposed temperature-based analytical model, described in function blocks, is given in Figure 6. The temperature dependency of variation in voltage is a result of the variation of the intrinsic carrier concentration n_i and the effective mobility of the electrons μ_0 with temperature. Owing to the positive temperature effect of the intrinsic carrier concentration n_i and the negative temperature effect of effective mobility of the electrons μ_0, dV_{DS}/dt decreases with temperature for turn-off and increases for turn-on. Since SiC MOSFET has wider band-gap energy, the temperature sensitivity of the effective mobility of the electrons μ_0 can be neglected and can be considered approximately constant in the model.

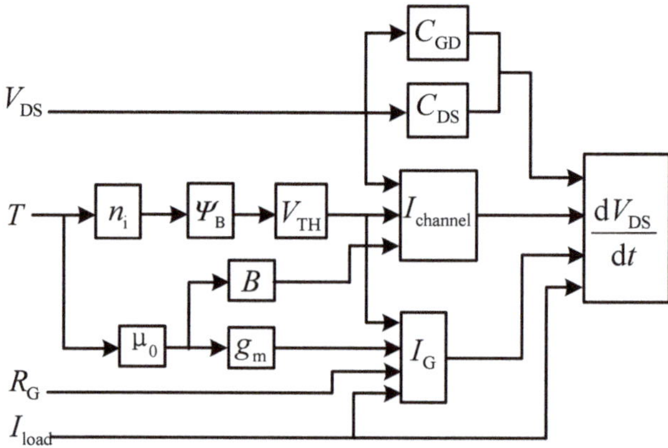

Figure 6. Summary of dependency of dV_{DS}/dt on operation condition.

3. Experiment Details

The static characteristics of SiC MOSFET were measured using a B1505A curve tracer (Agilent, Santa Clara, CA, USA) with the device placed in an environment chamber to control the temperature. The switching characteristics were obtained by the clamped inductive double-pulse test circuit shown in Figure 7.

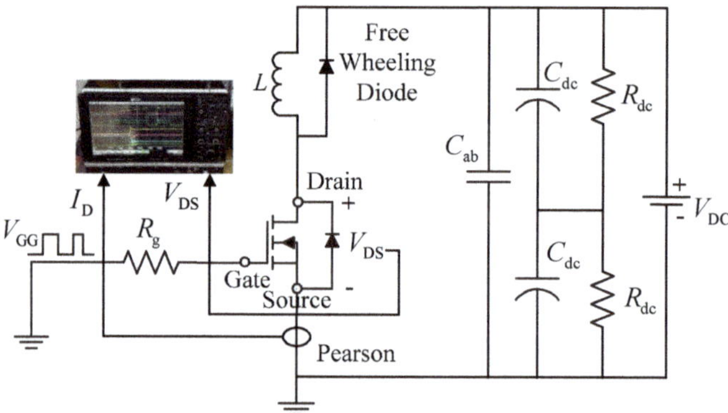

Figure 7. Schematics of the transient characteristics test.

The switching waveforms were captured using a 610Zi digital oscilloscope (Lecroy, New York, NY, USA) which has a bandwidth of 1 GHz and a sample rate of 20 GS/s. During the experiment, different ambient temperatures were simulated using a heater, and the SiC MOSFET was mounted at the bottom of the test circuit board, connected to the heater through an aluminum plate with some thermal paste for reliable heat transfer, shown in Figure 8. The heater can vary the temperature from room temperature to 450 °C. Theoretically, the SiC MOSFET chip can be normal operation above 300 °C due to its wider band gap and higher thermal conductivity. However, the maximum recommended operation temperature per the device manufacturer is 175 °C, owing to the considerations listed on the packaging and reliability issues. Hence, the maximum of 175 °C was selected for experimental measurements. A fan was used for the heat dissipation of the test circuit board to reduce the effects of temperature on the other components. In the test, the SiC MOSFET and diode were SCT2080KE (Rohm, Kyoto, Japan) and SCS220AM (Rohm, Kyoto, Japan) devices, respectively.

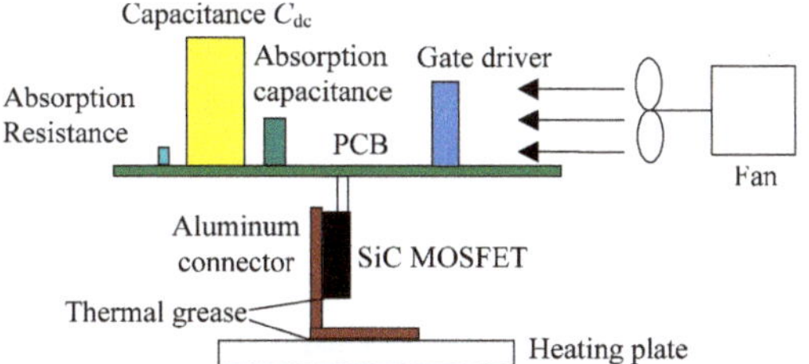

Figure 8. Schematic of the simulation of different ambient temperatures.

4. Experimental Results

4.1. Static Characteristics under Different Temperatures

Figures 9 and 10 show SiC MOSFET transfer and output characteristics at varying temperatures, respectively. The temperature dependence of the threshold voltage was obtained, as shown in Figure 11. The square represents the tested values under different temperatures, the solid line represents the fitted values, and the dashed line is representative of the calculation based on Equation (7). As seen, the threshold voltage significantly decreases with temperature increase, which is typical for 4H-SiC MOSFETs and has been observed in previous studies [13,14]. The effect is caused by the increase of intrinsic carrier concentration at higher temperature, seen in Equation (7), due to increased thermal generation of carriers across the band gap, which forms the channel easier.

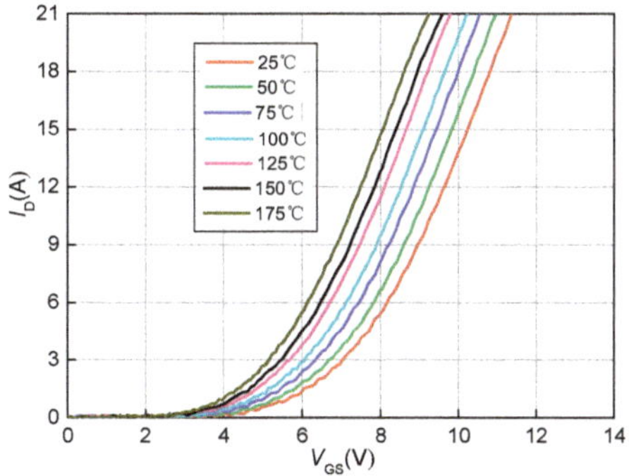

Figure 9. Temperature-dependent transfer characteristics.

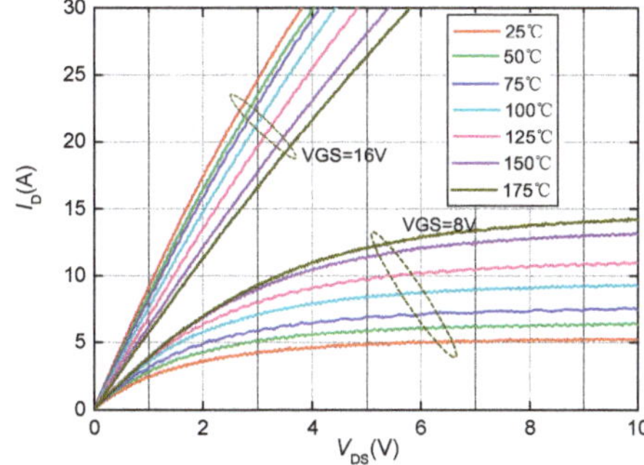

Figure 10. Temperature-dependent output characteristics at gate-source voltages of 8 V and 16 V.

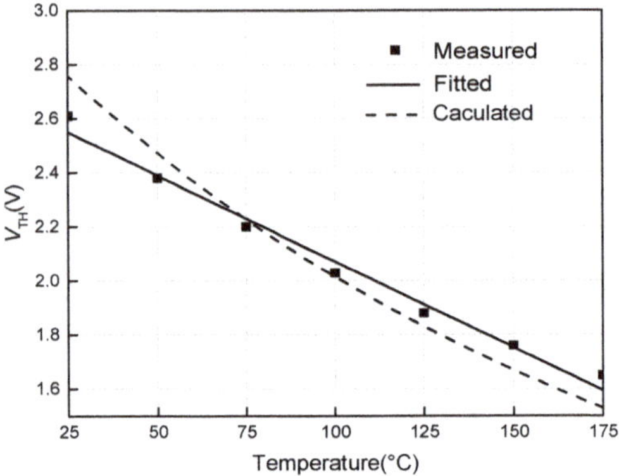

Figure 11. Threshold voltage of SiC MOSFET under different temperatures.

From Figure 11, the temperature dependency of threshold voltage is approximately linear. The temperature sensitivity coefficient k_{VT} is about -6.37 mV/$^\circ$C. If nominal threshold voltage V_{TH0} at room temperature is known, a simple expression can be used to describe threshold voltage at any measured temperature T:

$$V_{TH}(T) = V_{TH0} - k_{VT}(T - T_0) \tag{15}$$

Figure 11 also shows the calculated values of threshold voltage according to Equation (7). Reasonably good agreement exists between calculations and measurements. Nevertheless, the analytical model of threshold voltage, given by Equation (7), is often used for theoretical analysis since its parameters relate to the physics of the device. These parameters are usually hard to obtain. The linear fitting expression of the threshold voltage, given by Equation (15), is usually adopted in practical applications. In order to simplify the calculation, Equation (15) is used for the following calculations. The temperature dependency of transconductance gm is dominated by the effective mobility μ_0. As aforementioned analysis, the $d\mu_0/dT$ of SiC MOSFET is very low and can be neglected due to its wide band-gap characteristics. Hence, the temperature dependency of transconductance gm is also not considered.

4.2. Temperature Dependency of dV_{DS}/dt

From Equation (6), dV_{DS}/dt depends on device temperature, load current, gate resistors, and voltage. Here, the effects of these factors are investigated and the results are shown in Figures 12–19, where the turn-off waveforms of drain voltage are shown in Figures 12–15, and the turn-on waveforms are shown in Figures 16–19. The test conditions were set to voltages of 200, 400 and 600 V, and load currents of 10, 15 and 20 A. The gate resistor varied from 10 Ω to 150 Ω and the temperature ranged from 25 $^\circ$C to 175 $^\circ$C. An external drive system with 24/$-$5 voltage was used in the test and a double pulse signal was generated by a pulse generator. The current was measured using a Pearson current sensor. As seen, load current, voltage, temperature, and gate resistor have different impacts on dV_{DS}/dt during turn-off and turn-on. During turn-off, the relationship between dV_{DS}/dt and temperature is strongly dependent on the gate resistor. With larger values for the gate resistor, the curves nearly overlap under different temperatures. With smaller values for the gate resistor, the variation in voltage barely changed as temperature increased. However, the time at which the voltage rose obviously increased with temperature, which means the delay time varied as temperature increased.

In the voltage fall period during turn-on, at larger gate resistor values, the impact of temperature on the variation in voltage is obvious and its magnitudes increased as temperature increased. The time at which the voltage falls is different, and decreased with temperature. For smaller gate resistor values, the variation in voltage was nearly the same under different temperatures, since the waveforms overlapped. Moreover, load current and voltage were also important factors for the variation in voltage with temperature.

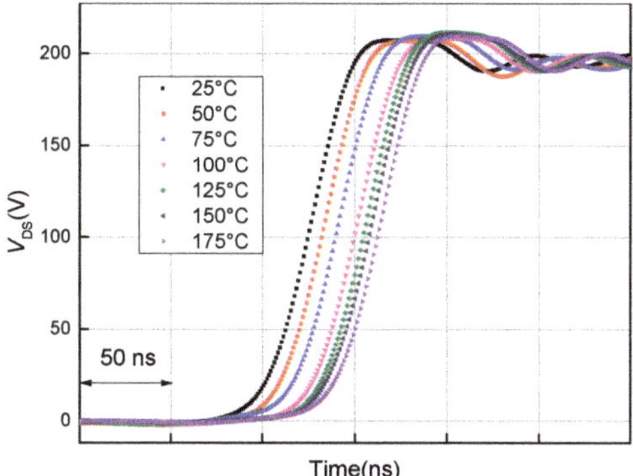

Figure 12. Turn-off waveforms for drain-source voltage (V_{DS}) at a voltage of 200 V, a load current 15 A, and at different temperatures with a gate drive resistance (R_G) = 10 Ω.

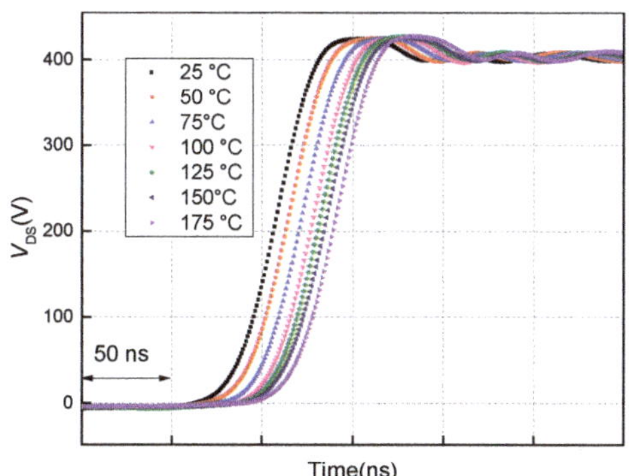

Figure 13. Turn-off waveforms for V_{DS} at a voltage of 400 V, a load current 15 A, and at different temperatures with R_G = 10 Ω.

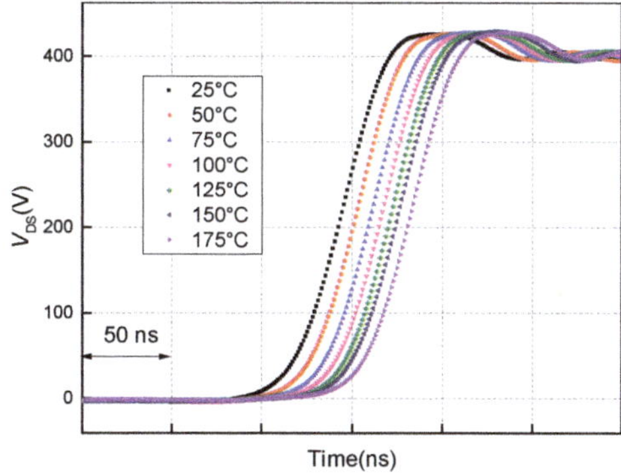

Figure 14. Turn-off waveforms for V_{DS} at a voltage of 400 V, a load current 20 A, and at different temperatures with R_G = 10 Ω.

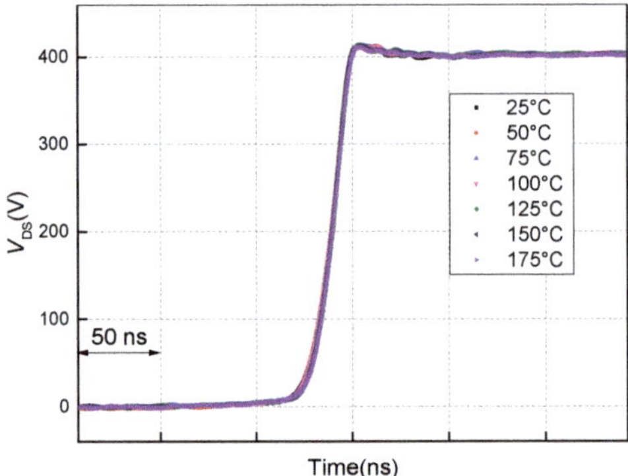

Figure 15. Turn-off waveforms for V_{DS} at a voltage of 400 V, a load current 20 A, and at different temperatures with $R_G = 150\ \Omega$.

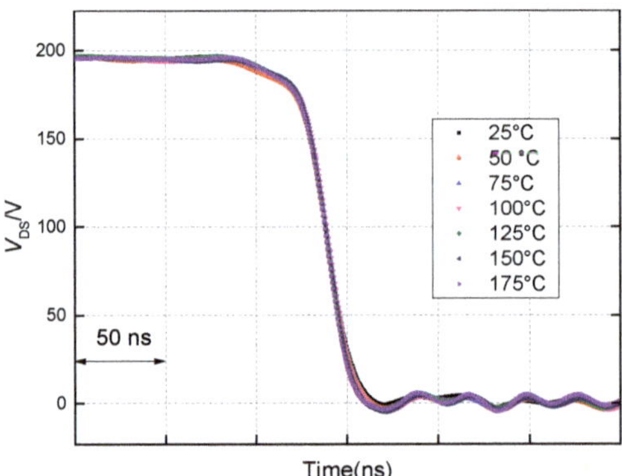

Figure 16. Turn-on waveforms for V_{DS} at a voltage of 200 V, a load current 15 A, and at different temperatures with $R_G = 10\ \Omega$.

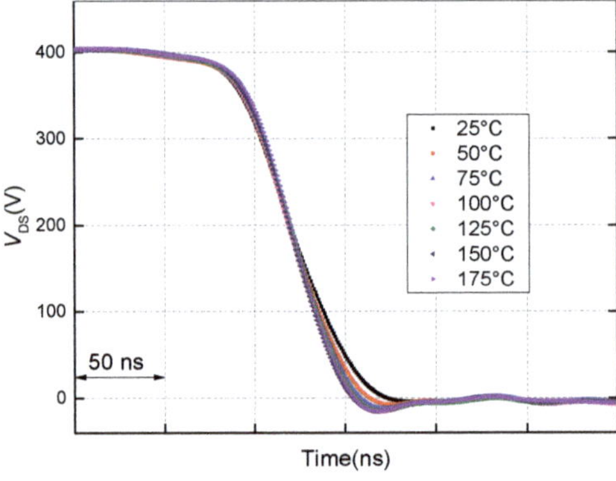

Figure 17. Turn-on waveforms for V_{DS} at a voltage of 400 V, a load current 15 A, and at different temperatures with $R_G = 10\ \Omega$.

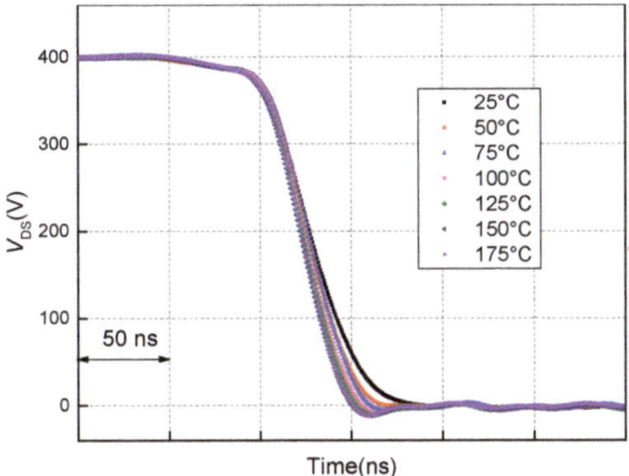

Figure 18. Turn-on waveforms for V_{DS} at a voltage of 400 V, a load current 20 A, and at different temperatures with $R_G = 10\ \Omega$.

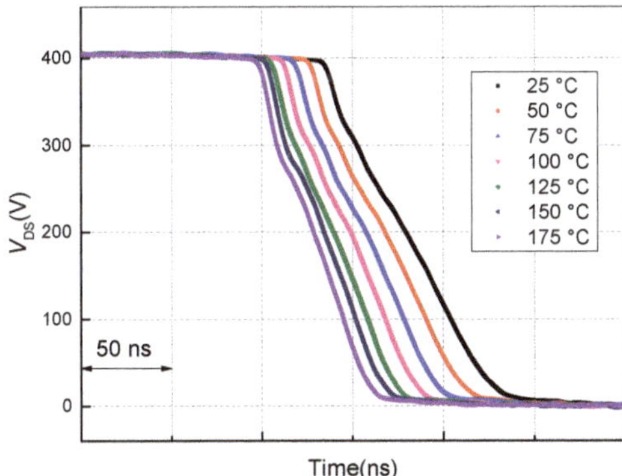

Figure 19. Turn-on waveforms for V_{DS} at a voltage of 400 V, a load current 20 A, and at different temperatures with $R_G = 150\ \Omega$.

From the above curves, the measured values of dV_{DS}/dt, which varies with temperature under different measurement conditions, were obtained. The results are shown from Figures 20–25, where Figures 20–22 are the results for turn-off, and Figures 23–25 are for turn-on. The calculated values are also shown in these figures. In the calculations, the values of C_{GS} and C_{OX} are 2.192 nF and 3.387 nF, respectively, extracted by a constant gate current circuit during turn-on. Indeed, for a SiC MOSFET the relative change in channel length is very small comparing with the long channel, λ can be treated as zero. According to some points in the device output characteristic saturation region, the values of W and L were obtained. The drain–source overlap area A_{DS} was 5.3676 mm^2. Equation (15) was adopted as the expression of threshold voltage under different temperatures.

For a fixed voltage of 400 V and gate resistance of 10 Ω, the relationship between dV_{DS}/dt and temperature under different load currents is presented in Figure 20. Figure 21 shows the temperature dependence of dV_{DS}/dt under different voltages at 15 A load current of and 10 Ω gate resistance. Note that dV_{DS}/dt has little fluctuation as the temperature increases, meaning that turn-off dV_{DS}/dt is not a strong function of temperature. The dV_{DS}/dt also increases with load current and supply voltage. The positive voltage coefficient of dV_{DS}/dt is due to the fact that drain-source capacitance and Miller capacitance reduce with increasing supply voltage. The Miller capacitance is consisted of a fixed oxide capacitance and a depletion capacitance that varies with voltage. The depletion width increases as voltage increases resulting in a small depletion capacitance and hence the Miller capacitance declines at a large voltage. While, drain-source capacitance is comprised of the depletion region capacitance of drain-body junction and the region width also increases with voltage. Hence drain-source capacitance decreases. The calculated results show dV_{DS}/dt is minimally temperature sensitive and decreases with temperature. Some discrepancies can be found between the measurements and calculations. The reason for this may be the temperature dependence of the effective mobility compensating for the temperature dependence of the threshold voltage in realistic experimental conditions. Furthermore, the impact of the gate resistors on the temperature dependence of dV_{DS}/dt was also measured at 400 V and 15 A. The results are exhibited in Figure 22. The dV_{DS}/dt varied with different resistors, and was large with smaller resistor values and was small with larger resistor values. The reason results from the variation of gate current at different resistors observed in aforementioned analysis.

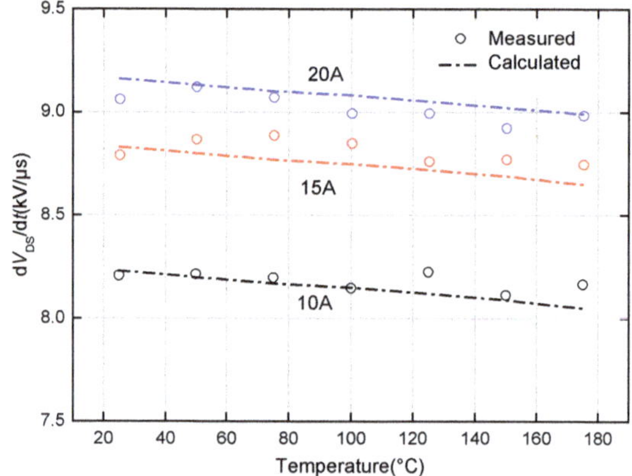

Figure 20. Measured and calculated dV_{DS}/dt as a function of temperature at a voltage of 400 V and a gate resistor of 10 Ω during turn-off, shown for different load currents.

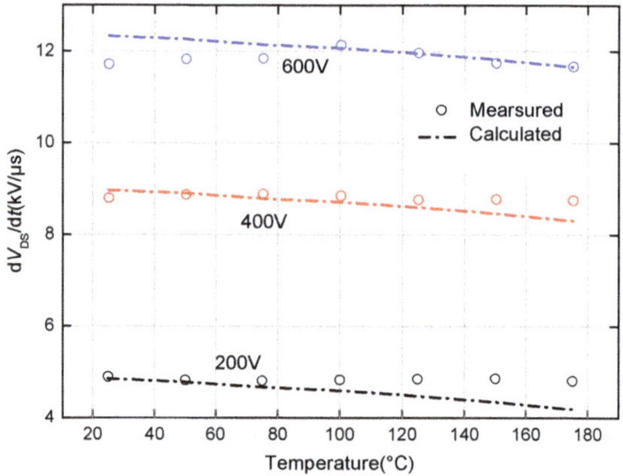

Figure 21. Measured and calculated dV_{DS}/dt as a function of temperature at a load current of 15 A and a gate resistor of 10 Ω during turn-off, shown for different voltages.

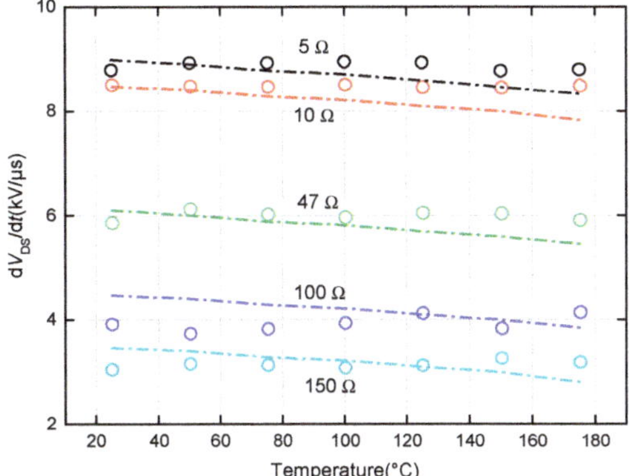

Figure 22. Measured dV_{DS}/dt as a function of temperature at a voltage of 400 V and a load current 15 A during turn-off, shown for different gate resistances.

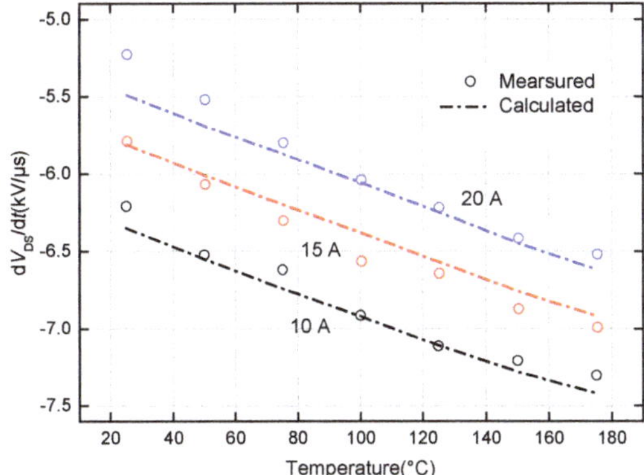

Figure 23. Measured and calculated dV_{DS}/dt as a function of temperature at a voltage of 400 V and a gate resistor of 10 Ω during turn-on, shown for different load currents.

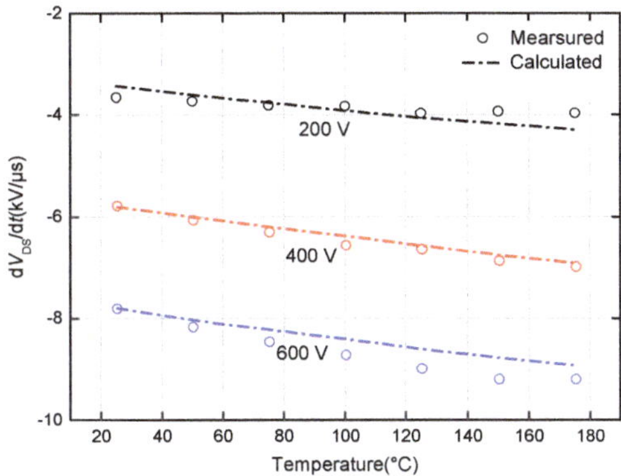

Figure 24. Measured and calculated dV_{DS}/dt as a function of temperature at a voltage of 15 A and a gate resistor of 10 Ω during turn-on, shown for different voltages.

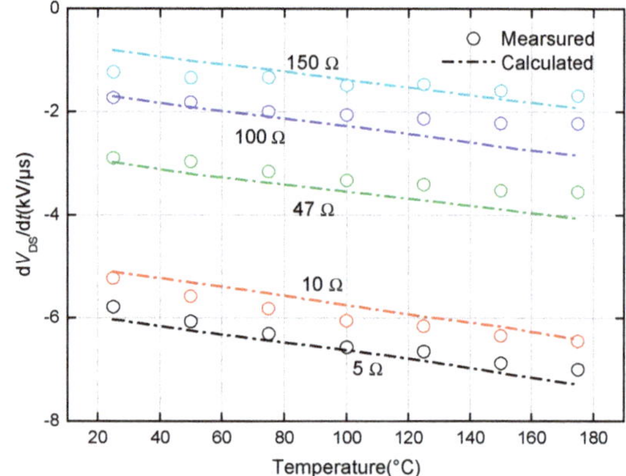

Figure 25. Measured and calculated dV_{DS}/dt as a function of temperature at a voltage of 400 V and a load current 15 A during turn-on, shown for different gate resistances.

During the period of voltage fall during turn-on, the relationship between dV_{DS}/dt and temperature, under different load currents, at 400 V and 10 Ω, is presented in Figure 23. For different voltages, the relationship at 15 A and 10 Ω is shown in Figure 24. Notably, dV_{DS}/dt is less than zero and its magnitude increases with temperature, as expected. Moreover, the temperature dependency of dV_{DS}/dt exhibits near-linear characteristics. The results are interesting since the monotonic relationship may be a potential indictor for junction potential measurement for SiC MOSFET. At the same temperature, a larger load current results in a smaller dV_{DS}/dt. For voltage, the magnitude of dV_{DS}/dt is larger for a larger voltage. To evaluate the effects of gate resistors on the temperature dependency of turn-on dV_{DS}/dt, the temperature dependency of turn-on dV_{DS}/dt was measured at 400 V and 15 A under different gate resistor values. The results are shown in Figure 25. A variation in dV_{DS}/dt can be observed in Figure 25 for different gate resistors at the same temperature. A smaller gate resistor results in a larger magnitude of dV_{DS}/dt. Moreover, in the above figures, the calculated and measured values show good agreement at the evaluated temperature range.

5. Discussion

The variation trend between dV_{DS}/dt and temperature for SiC MOSFET can be also seen in previous studies. In Chen et al. [9] and DiMarino et al. [14], the measurements were performed at a supply voltage of 600 V and load current of 10 A with 10 Ω gate resistance. The impact of temperature is clear and the magnitude of dV_{DS}/dt decreases as temperature increases for turn-off, but increases with increasing temperature for turn-on. Othman et al. [11] experimentally showed the temperature-sensitivity of turn-off dV_{DS}/dt is very low and approximately constant under 400 V, 15 A, and 28 Ω gate resistance test conditions. When temperature ranges from 25 °C up to 175 °C, the value of dV_{DS}/dt is approximately 10.44 V/ns. However, for turn-on dV_{DS}/dt, the magnitude increases from 4.6 V/ns at 25 °C to 6.62 V/ns at 175 °C. As reported by Takao et al. [12], the same conclusions were drawn from the switching waveforms of SiC MOSFETs at 25 °C and 175 °C at 600 V, 10 A, with 11.36 Ω gate resistance. The magnitude of turn-on dV_{DS}/dt is approximately 12 V/ns at 25 °C and 15 V/ns for 175 °C, whereas the values of turn-off dV_{DS}/dt are all approximately 29.11 V/ns for both 25 °C and 175 °C. These provide a potential solution for SiC MOSFET junction temperature estimation.

In the case of Si IGBTs, the turn-off dV_{CE}/dt under different temperatures has been reported by Bryant et al. [25]. The dV_{CE}/dt possesses a negative temperature coefficient and decreases with temperature decreases. For different load currents, the slope is the same and its value is 6.75 V/(μs°C). Because of this fixed sensitivity and linearity, turn-off dV_{CE}/dt can be used as an effective indicator for junction temperature measurement of an IGBT. However, for a SiC MOSFET, turn-off dV_{DS}/dt is not strongly correlated with temperature. Hence, monitoring turn-off dV_{DS}/dt seems to be infeasible for junction temperature measurement of a SiC MOSFET.

Equation (10) shows the two temperature-sensitive parameters that dominate the temperature dependency of dV_{DS}/dt are the threshold voltage and the effective mobility. For Si MOSFET, the effective mobility decreases with temperature and its temperature dependency is negative. But for SiC MOSFETs, as reported in [24], the temperature dependency of effective mobility can be neglected and be considered approximately constant due to its wide band-gap characteristics. Hence, it can be concluded theoretically that the temperature dependency of dV_{DS}/dt in Si MOSFETs is less than that in SiC MOSFETs owing to the compensating effects of the temperature characteristic of the effective mobility to the temperature characteristic of the threshold voltage in Si MOSFETs. However, it is hard to find the experimental measurements on the temperature dependency of dV_{DS}/dt for Si MOSFETs in previous literature. A comprehensive comparison between Si MOSFETs and SiC MOSFETs, regarding the dependency of the dV_{DS}/dt on temperature, load current and gate resistor, will be performed experimentally in the next step.

From the measurements, the turn–on dV_{DS}/dt is shown to have good temperature sensitivity and is approximately linear. The results are interesting and significant because it may be a potential indicator for junction temperature measurement of a SiC MOSFET. Others factors can also effect turn-on dV_{DS}/dt, as shown in Figures 23–25, but this is not an issue. Because the type of device and system parameters, such as voltage, load current, and gate resistance, are usually fixed in practical SiC MOSFET-based applications, and turn-on dV_{DS}/dt is only dominated by temperature. Before the turn-on dV_{DS}/dt–based method is used for temperature assessment, the calibration curves between dV_{DS}/dt and temperature should be drawn experimentally and used as a lookup table. Indeed, the lookup table is easily implemented by digital signal processing (DSP). The dV_{DS}/dt can also be easily measured by a RC high-pass connected to the drain and source terminals of device from the switching waveforms. After the dV_{DS}/dt is measured, the corresponding junction temperature can be estimated from the lookup table.

6. Conclusions

A thorough analysis of variation in voltage for SiC MOSFETs during turn-on and turn-off was completed. It can be concluded that turn on variation of voltage is a function of temperature and has near-linear dependency with temperature. The relationship between turn-off variations in voltage and temperature was also investigated. Turn-off variation in voltage was approximately invariant with temperature with fixed supply voltage, load current, and gate resistance test conditions. Using a temperature-based analytical model for variations in voltage, we demonstrated that the temperature dependency of variation in voltage results from the positive temperature dependency of the intrinsic carrier concentration and negative temperature dependency of the effective mobility of the electrons in SiC MOSFETs. The analytical model also demonstrated the ability to correctly predict how variation in voltage varies with temperature. Additionally, the effects of supply voltage, load current, and gate resistance on the temperature dependency of variations in voltage were discussed. Due to good linearity, turn on variation in voltage may be considered as a practical temperature-sensitive electrical parameter for junction temperature estimation in SiC MOSFETs. With a database of turn-on variations in voltage with temperature, supply voltage, load current, and gate resistance, the junction temperature of SiC MOSFETs can be derived from the calibration curve. Future work includes a thorough comparison with other models, a verification of whether the turn-on variation in the voltage–based temperature measurement method represents correctly the junction temperature at realistic operation conditions, and a comprehensive assessment in terms of selectivity, linearity, generality, and possibility for online measurement.

Acknowledgments: This research work is supported by the Fundamental Research Funds for the Central Universities (106112016CDJZR158802), Graduate scientific research and innovation foundation of Chongqing,

China(CYB16020), National Natural Science Foundation of China (51377184, 51607016), and Chongqing Scientific &Technological Talents Program (KJXX2017009).

Author Contributions: Hui Li and Xinglin Liao put forward to the main idea and designed the entire structure of this paper. Yaogang Hu made some suggestions in the article writing. Zhangjian Huang and Kun Wang provided some help in the experiments.

Conflicts of Interest: The authors declare no conflicts of interest.

References

1. Hudgins, J. Power electronic devices in the future. *IEEE J. Emerg. Sel. Top. Power Electron.* **2013**, *1*, 11–17. [CrossRef]

2. Jiang, D.; Burgos, R.; Wang, F.; Boroyevich, D. Temperature dependent characteristics of SiC devices: Performance evaluation and loss calculation. *IEEE Trans. Power Electron.* **2012**, *27*, 1013–1024. [CrossRef]

3. Millan, J.; Godignon, P.; Perpina, X.; Perez-Tomas, A.; Rebollo, J. A survey of wide bandgap power semiconductor devices. *IEEE Trans. Power Electron.* **2014**, *29*, 2155–2163. [CrossRef]

4. Zhu, P.; Wang, L.; Ruan, L.G.; Zhang, J.F. Temperature Effects on Performance of SiC Power Transistors (SiC JFET and SiC MOSFET). In Proceedings of the 2015 IEEE European Conference on Power Electronics and Applications (EPE'15 ECCE-Europe), Geneva, Switzerland, 8–10 September 2015; pp. 449–454.

5. Lelis, A.J.; Habersat, D.; Green, R.; Ogunniyi, A.; Gurfinkel, M.; Suehle, J.; Goldsman, N. Time dependence of bias-stress-induced SiC MOSFET threshold voltage instability measurements. *IEEE Trans. Power Electron.* **2008**, *55*, 1835–1840. [CrossRef]

6. Gurfinkel, M.; Xiong, H.D.; Cheung, K.P.; Suehle, J.S.; Bernstein, J.B.; Shapira, Y.; Lelis, A.J.; Habersat, D.; Goldsman, N. Characterization of transient gate oxide trapping in SiC MOSFETs using fast I-V techniques. *IEEE Trans. Power Electron.* **2008**, *55*, 2004–2012. [CrossRef]

7. Chen, Z.; Boroyevich, D.; Burgos, R.; Wang, F. Characterization and modeling of 1.2 kV, 20 A SiC MOSFETs. In Proceedings of the 2009 IEEE Energy Conversion Congress and Exposition (ECCE), San Jose, CA, USA, 20–24 September 2009; pp. 1480–1487.

8. Hull, B.; Das, M.; Husna, F.; Callanan, R.; Agarwal, A.; Palmour, J. 20 A, 1200 V 4H-SiC DMOSFETs for Energy Conversion Systems. In Proceedings of the 2009 IEEE Energy Conversion Congress and Exposition (ECCE), San Jose, CA, USA, 20–24 September 2009; pp. 112–119.

9. Chen, Z.; Yao, Y.Y.; Danilovic, M.; Boroyevich, D. Performance Evaluation of SiC Power MOSFETs for High-Temperature Applications. In Proceedings of the 2012 IEEE Power Electronics and Motion Control Conference (EPE/PEMC), Novi Sad, Serbia, 4–5 September 2012; pp. DS1a.8-1–DS1a.8-9.

10. Chen, Z.; Yao, Y.Y.; Boroyevich, D.; Ngo, K.D.T.; Mattavelli, P.; Rajashekara, K. A 1200-V, 60-A SiC MOSFET Multichip Phase-Leg Module for High-Temperature, High-Frequency Applications. *IEEE Trans. Power Electron.* **2014**, *29*, 2307–2320. [CrossRef]

11. Othman, D.; Berkani, M.; Lefebvre, S.; Ibrahim, A.; Khatir, Z.; Bouzourene, A. Comparison study on performances and robustness between SiC MOSFET & JFET devices—Abilities for aeronautics application. *Microelectron. Reliab.* **2012**, *52*, 1859–1864.

12. Takao, K.; Harada, S.; Shinohe, T.; Ohashi, H. Performance evaluation of all SiC power converters for realizing high power density of 50 W/cm^3. In Proceedings of the 2010 IEEE International Power Electronics Conference (IPEC), Sapporo, Japan, 21–24 June 2010; pp. 2128–2134.

13. Cui, Y.T.; Chinthavali, M.; Tolbert, L.M. Temperature Dependent Pspice Model of Silicon Carbide Power MOSFET. In Proceedings of the 2012 IEEE Applied Power Electronics Conference and Exposition (APEC), Orlando, FL, USA, 5–9 February 2012; pp. 1698–1704.

14. DiMarino, C.; Chen, Z.; Danilovic, M.; Boroyevich, D.; Burgos, R.; Mattavelli, P. High-Temperature Characterization and Comparison of 1.2 kV SiC Power MOSFETs. In Proceedings of the 2013 IEEE Energy Conversion Congress and Exposition (ECCE), Denver, CO, USA, 15–19 September 2013; pp. 3235–3242.

15. Avenas, Y.; Dupont, L.; Khatir, Z. Temperature measurement of power semiconductor devices by thermo-sensitive electrical parameters—A review. *IEEE Trans. Power Electron.* **2012**, *27*, 2081–3092. [CrossRef]

16. Sundaramoorthy, V.K.; Bianda, E.; Bloch, R.; Angelosante, D.; Nistor, I.; Riedel, G.J.; Zurfluh, F.; Knapp, G.; Heinemann, A. A study on IGBT junction temperature (Tj) online estimation using gate-emitter voltage (Vge) at turn-off. *Microelectron. Reliab.* **2014**, *54*, 2423–2431. [CrossRef]

17. Wang, J.J.; Chung, H.S.-h.; Li, R.T.-h. Characterization and Experimental Assessment of the Effects of Parasitic Elements on the MOSFET Switching Performance. *IEEE Trans. Power Electron.* **2013**, *28*, 573–589. [CrossRef]
18. Yin, S.; Tu, P.F.; Wang, P.; Tseng, K.J.; Qi, C.; Hu, X.L.; Zagrodnik, M.; Simanjorang, R. An Accurate Subcircuit Model of SiC Half-Bridge Module for Switching-Loss Optimization. *IEEE Trans. Ind. Appl.* **2017**, *53*, 3840–3849. [CrossRef]
19. Arribas, A.P.; Shang, F.; Krishnamurthy, M.; Shenai, K. Simple and Accurate Circuit Simulation Model for SiC Power MOSFETs. *IEEE Trans. Electron. Devices* **2015**, *62*, 449–457. [CrossRef]
20. Chen, K.; Zhao, Z.G.; Yuan, L.Q.; Lu, T.; He, F. The Impact of Nonlinear Junction Capacitance on Switching Transient and Its Modeling for SiC MOSFET. *IEEE Trans. Electron Devices* **2015**, *62*, 333–338. [CrossRef]
21. McNutt, T.R.; Hefner, A.R.; Mantooth, H.A.; Berning, D.; Ryu, S.H. Silicon carbide power MOSFET model and parameter extraction sequence. *IEEE Trans. Power Electron.* **2007**, *22*, 353–363. [CrossRef]
22. Mudholkar, M.; Ahmed, S.; Ericson, M.N.; Frank, S.S.; Britton, C.L.; Mantooth, H.A. Datasheet Driven Silicon Carbide Power MOSFET Model. *IEEE Trans. Power Electron.* **2014**, *29*, 2220–2228. [CrossRef]
23. Baliga, B.J. *Fundamentals of Power Semiconductor Devices*; Spring-Verlag: New York, NY, USA, 2008.
24. Hasanuzzaman, M.; Islam, S.K.; Tolbert, L.M.; Alam, M.T. Temperature dependency of MOSFET device characteristics in 4H- and 6H-silicon carbide (SiC). *Solid-State Electron.* **2004**, *48*, 1877–1881. [CrossRef]
25. Bryant, A.; Yang, S.Y.; Mawby, P.; Xiang, D.; Ran, L.; Tavner, P.; Palmer, P.R. Investigation Into IGBT dV/dt During Turn-Off and Its Temperature Dependence. *IEEE Trans. Power Electron.* **2011**, *26*, 3019–3031. [CrossRef]

Article

Development of Propulsion Inverter Control System for High-Speed Maglev based on Long Stator Linear Synchronous Motor

Jeong-Min Jo *, Jin-Ho Lee, Young-Jae Han, Chang-Young Lee and Kwan-Sup Lee

Korea Railroad Research Institute, Uiwang-si, Gyeonggi-do 16105, Korea; jinholee@krri.re.kr (J.-H.L.);
yjhan@krri.re.kr (Y.-J.H.); cylee@krri.re.kr (C.-Y.L.); kslee@krri.re.kr (K.-S.L.)
* Correspondence: jmjo@krri.re.kr; Tel.: +82-31-460-5619

Academic Editor: José Gabriel Oliveira Pinto
Received: 19 September 2016; Accepted: 20 January 2017; Published: 3 February 2017

Abstract: In the case of a long-stator linear drive, unlike rotative drives for which speed or position sensors are a single unit attached to the shaft, these sensors extend along the guideway. The position signals transmitted from a maglev vehicle cannot meet the need of the real-time propulsion control in the on-ground inverter power substations. In this paper the design of the propulsion inverter control system with a position estimator for driving a long-stator synchronous motor in a high-speed maglev train is proposed. The experiments have been carried out at the 150 m long guideway at the O-song test track. To investigate the performance of the position estimator, the propulsion control system with, and without, the position estimator are compared. The result confirms that the proposed strategy can meet the dynamic property needs of the propulsion inverter control system for driving long-stator linear synchronous motors.

Keywords: high speed maglev; long-stator synchronous motor; propulsion inverter control system; position estimator

---

1. Introduction

A maglev train system is a new transportation system technology that replaces the mechanical components by electronic devices and, thus, can help to overcome the technical limitations that wheels have in railway technology. In comparison with the existing wheel-on-rail systems, the maglev train system ensures high-speed operation, high safety, low pollution, and low energy loss for mass transport [1]. A long stator linear synchronous motor (LS-LSM) has the following special features as compared to a rotary-type synchronous motor: Since the LS-LSM is made through a combination of levitation magnets and the stator of propulsion motor, the magnetic flux density of the motor is determined by the weight of the vehicle, and the thrust force is affected by the phase angle and the magnitude of the stator current. Therefore, the position information of the mover is essential for the LS-LSM drive inverter [2–5].

In the case of the Transrapid maglev system, as the position signal detection system PRW (coming from the German word: polradwinkelverfahren) is mounted on the head portion and rear portion of the train and the position information flag is installed in the guideway, with the absolute position signals of the vehicle being transmitted to the propulsion system on the ground. The position information transmission method is to transmit the position signals using radio frequency at intervals of 20 ms with the use of the RS-485 communication protocol, and the transmission rate is 512 kbps. However, the vehicle position information obtained through this method does not meet the requirements of the propulsion inverter control system [6].

This paper seeks to develop a propulsion inverter control system technology for the SUMA550-1, a high-speed maglev train system based on an LS-LSM. The test vehicle driven by an LS-LSM was fabricated with one-half the length of the final target model for this project and a LS-LSM guideway of 150 m in length was constructed to test and evaluate the developed vehicle. The differences of propulsion control between a rotating motor and a linear motor are analyzed and the propulsion system with the vehicle magnetic flux position estimator is proposed in order to solve the problem of linear motors. A difference in thrust force control between the rotative motor and linear unit was analyzed, and the propulsion control system that has the vehicle flux position observer to solve the problem which may occur in the linear unit is presented.

For performance verification of the propulsion inverter control system of the SUMA550-1, the response characteristics of the current controller and speed controller, and the stability on the control system were analyzed through a simulation model to which the system parameters and propulsion controller gain parameters are applied. The control performance by the maximum torque, the characteristic of position estimator, and the speed profile was confirmed through a propulsion control experiment at the test track with 150 m in length.

2. Development of Propulsion Inverter Control System Based on Long Stator Linear Synchronous Motor

2.1. The Thrust Force Characteristics of Propulsion System for a High-Speed Maglev Train

The final model to be developed through the research and development of high-speed maglev train system is the SUMA550, whose configuration consists of three cars/one organization. The specifications of the SUMA550-1, which is the first prototype model of the high-speed maglev train system developed in this research project, are summarized in Table 1.

Table 1. High-speed maglev train specifications. LS-LSM: long stator linear synchronous motor.

Parameter	Description	SUMA550	SUMA550-1
Vehicles	Train configuration	3 cars	½ cars
	Seats	156	25
	Gross mass (ton)	150	27
	Primary suspension	Magnetic Suspension	
	Secondary suspension	Passive Suspension	
	Air gap (mm)	10 (landing gap: 20)	
Propulsion System	Motor type	LS-LSM	
	Maximum speed	550 km/h	8 m/s@150 m
	Maximum acceleration	1.1 m/s^2	
Guideway	Gauge	2600	
	Static deflection	L (=15)/5600	At-grade
	Dynamic deflection	L (=15)/4000	

The entire vehicle running resistance F of the SUMA550 is expressed by the sum of all resistances to vehicle movement. Equation (1) represents the entire vehicle running resistances of high speed maglev system:

$$F = F_a + F_m + F_B \tag{1}$$

where F_a is aerodynamic train running resistance; F_m is eddy current running resistance; and F_B is running resistance caused by linear generator.

Each of equations for running resistance are described in Equation (2):

$$F_a = 2.8 \times v^2 \times (0.265 \times N + 0.3)\,[\text{N}]$$
$$F_m = N \times (0.1 \times v^{0.5} + 0.02 \times v^{0.7}) \times 10^{-3}\,[\text{N}]$$
$$F_B = \begin{cases} 0\,[\text{N}] \ (\text{for } 0 \leq v \leq 20\,\text{km/h}) \\ N \times 7.3 \times 10^3\,[\text{N}] \ (\text{for } 20 \leq v \leq 70\,\text{km/h}) \\ N \times (146/v - 0.2) \times 10^3\,[\text{N}] \ (\text{for } 70 \leq v \leq 500\,\text{km/h}) \end{cases} \tag{2}$$

where N is the number of vehicles and v is the velocity of the vehicle.

Figure 1 shows the required thrust force curve and electromagnet arrangement of the SUMA550, and the electromagnet arrangement of the SUMA550-1.

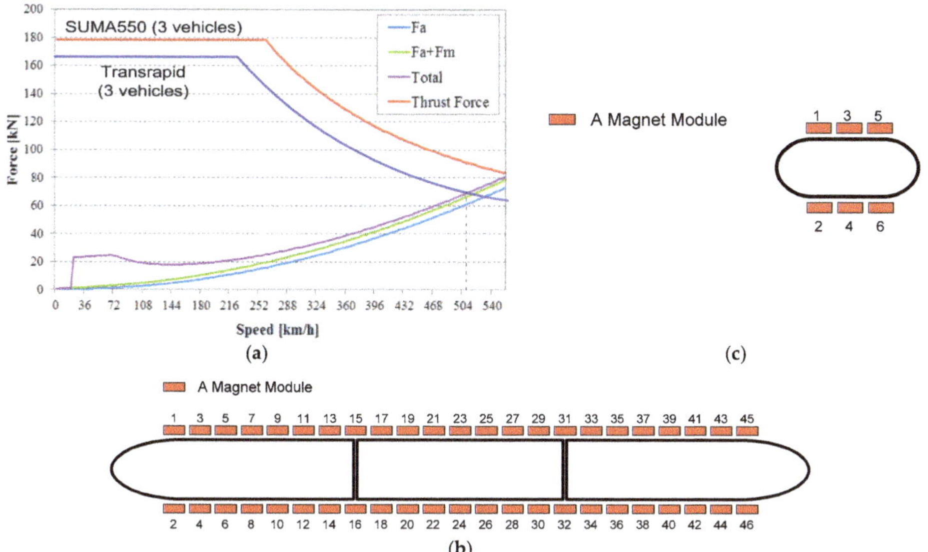

Figure 1. Required thrust force curve and electromagnet arrangement of the SUMA550, and the electromagnet arrangement of the SUMA550-1. (**a**) Running resistance curve according to the gradient slope of the maglev train; (**b**) arrangement of electromagnets in the final design model of the SUMA550; and (**c**) the arrangement of electromagnets for the developed SUMA550-1 model.

2.2. Linear Synchronous Motor Propulsion Control System Configuration

2.2.1. Positioning System Configuration

For a propulsion system based on an LS-LSM, the phase angle error with respect to the magnetic flux angle of the vehicle levitation electromagnet (secondary side of the motor) should be maintained at less than 15° in order to maintain a high thrust force, and if it exceeds this range, the thrust force is sharply reduced to less than 80%. Since the pole pitch of LS-LSM is 240 mm, the position error between the levitation magnet on the vehicle and the LS-LSM on the guideway should be maintained within 10 mm.

The positioning system of the SUMA550-1 uses a combination of the absolute positioning system and relative positioning system to detect the absolute information of the vehicle through the absolute positioning system and the relative positioning system installed on both sides along the guideway. The absolute position information of the vehicle is detected by the position reader of the vehicle.

The absolute position information detected from vehicle is transmitted to the propulsion control system on the ground via a radio communication system.

2.2.2. Design of the Propulsion Control System

The rotator flux orientated vector control has widely been used for the thrust force of propulsion systems in trains. The synchronous rotating axis in the rotator flux orientated vector control should be aligned to the flux angle of the rotator. Therefore, detecting precise position signals of the mover flux is important to the thrust force control of the synchronous motor. To this end, a resolver or an encoder is mainly utilized as a train speed and rotor position detection device, and the rotor position information with high resolution is provided to the propulsion control system [7]. However, in the propulsion system on the ground, such as high speed maglev train systems, the current position is measured in the moving vehicle, and the absolute position information of the vehicle is transmitted to the inverter on the ground in real-time. This method has three problems, as follows: First, it cannot provide the absolute position information of the 50 μs cycle required by the propulsion inverter due to the limits of the radio communication speed. Second, a failure in the radio communication makes the absolute position information update cycle variable. Third, a delay in the communication time for the absolute position information occurs when the vehicle runs at high speed.

In this paper, rotor flux oriented vector control was performed as a basic algorithm to control the thrust force of the vehicle and the flux position observer of the vehicle was utilized to align the flux of the vehicle with the moving magnetic flux of the LS-LSM. Since the propulsion inverter of the SUMA550-1 has the current control loop of a 500 μs cycle, it needs the precise magnetic flux position information of the 500 μs cycle to perform accurate phase current control in real-time. This system receives the position signal delayed by approximately 5 ms from the vehicle in a 2 ms cycle and, therefore, it requires the magnetic flux position observer that compensates the delay of the position signal [8–11]. In this paper, the closed loop full-order position observer based on the DC motor numerical model that has the position information from the vehicle as an input is proposed in order to serve the high-resolution position information required in the propulsion inverter in real-time. The detail on the design of the position observer are shown in reference [12].

$$
\begin{aligned}
\theta_{comp} &= \theta + 2\pi \left\{ \frac{v \times T_d}{1000} \right\} \\
\frac{d\theta}{dt} &= \hat{v} + \omega_n (2\zeta + 1)(\theta_{comp} - \hat{\theta}) \\
\frac{d\hat{v}}{dt} &= -\frac{1}{M}\hat{\tau}_d + \frac{1}{M}F_t + \omega_n^2 (2\zeta + 1)(\theta_{comp} - \hat{\theta}) \\
\frac{d\hat{\tau}_d}{dt} &= \omega_n^3 (\theta_{comp} - \hat{\theta})
\end{aligned}
\tag{3}
$$

where v represents the vehicle speed, $\hat{v}$ represents a state estimate of V, T_d represents the communication delay time constant, θ represents the phase angle measured in the vehicle, θ_{comp} represents the communication delay compensation phase angle, $\hat{\theta}$ represents an state estimate of θ, M represents vehicle mass, F_t represents the propulsion torque of the vehicle, τ_d represents the disturbance torque, $\hat{\tau}_d$ represents a state estimate of τ_d, ω_n represents the observer natural frequency, and ζ represents the observer damping ratio.

2.3. Experiments and Results Analysis

For the propulsion controller experiment of the SUMA550-1, a 150 m long guideway was constructed. An LS-LSM was installed under the both sides of the guideway, and barcodes and markers were also installed in combination with the LS-LSM. The vehicle is levitated up to 10 mm by the mover of the LS-LSM and propulsion force is generated in accordance with the phase angle and coil current of the ground LS-LSM.

The outward appearance of the SUMA550-1 is shown in Figure 2. The empty vehicle weight is 24 tons, the gross weight is 27 tons, the maximum acceleration is 1.1 m/s^2, the maximum jerk is

0.5 m/s^3, the maximum speed at a short-distance test line is 8 m/s, and there is no gradient of the 150 m-long guideway. The key parameters of the LS-LSM are summarized in Table 2.

Figure 2. High-speed maglev based on the LSM.

Table 2. The key parameters of LS-LSM.

Symbol	Value	Parameters
M	27,000 kg	Gross mass
R_W	0.36 ohm	Phase resistance
L_d	4.41 mH	d-axis stator self-inductance
L_q	1.85 mH	q-axis stator self-inductance
τ	0.24 m	Magnet pole pitch

2.4. Experiment Results and Analysis

For the performance verification of the propulsion control system of SUMA550-1, the maximum thrust force experiment, performance test of the position estimator, and automatic operation test by the speed profile were conducted in the test track. As described in Section 2.1, since the maximum thrust force of the SUMA550 generated by 46 magnet modules is 180 kN, the maximum design thrust force of the SUMA550-1 equipped with six magnet modules is 23.49 kN.

Figure 3 shows the maximum thrust force by applying the rated current $I_{qs} = 500[\text{A}]$ to the LS-LSM, and it is 24.88 kN, which is slightly more than the design value. This is because as the mover magnetic flux of the LS-LSM increases, the thrust force slightly exceeds the maximum design value under the full weight condition of the vehicle.

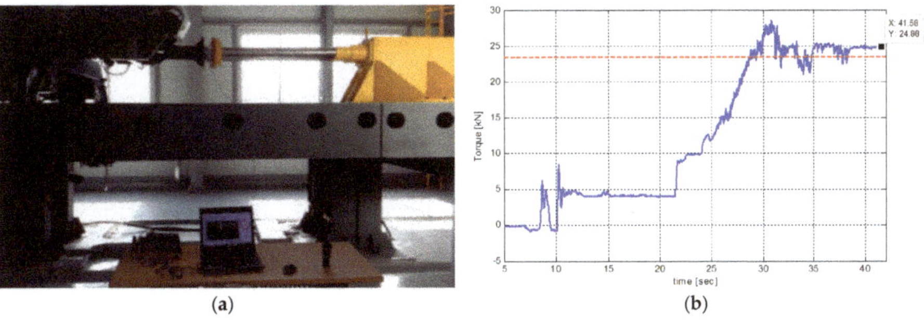

(a)   (b)

Figure 3. The thrust force experiment and the result waveforms of the SUMA550-1. (**a**) Complete view of the maximum thrust force experiment; and (**b**) the thrust force experiment result waveforms (the maximum thrust force of 24.88 kN).

Figures 4 and 5 show the experimental result waveforms when the vehicle is operated manually at speeds of up to 7 m/s. Figure 4a shows the results of the experiment performed by applying the absolute position signals received from the vehicle, and each waveform represents the speed, vehicle flux phase angle and estimated phase angle, and phase-U current of the LS-LSM from the top. In order to analyze the distortion of the current by the discontinuous signals of the absolute position signal and the subsequent current distortion, the expansion results of the 18.5–19.5 s section are represented in Figure 4b. Figure 4 shows that there is no problem in the overall operation of the propulsion control system to which the absolute position signal is applied, but the distortion of the phase-U current of the LS-LSM stands out due to the discontinuous signal of the absolute position signal as shown in Figure 4b. This current distortion can lead to the decrease in the thrust force of the propulsion control system, instability of the communication system by the harmonic current, and increased loss of the power converter system.

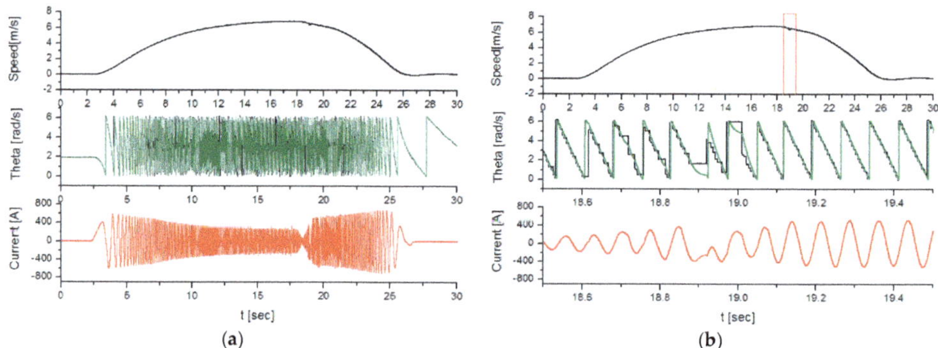

Figure 4. Waveforms of the propulsion control system in the application of the absolute position signal. (**a**) Speed, phase angle, and estimated phase angle phase-U current; and (**b**) the expanded waveforms of the 18.5–19.5 s section.

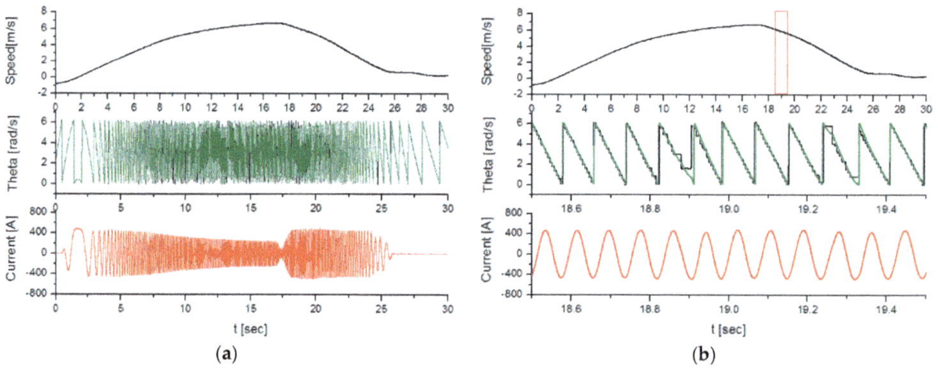

Figure 5. Waveforms of the propulsion control system in the application of the vehicle position estimator. (**a**) Speed, phase angle, and estimated angle, phase-U current; and (**b**) the expanded waveforms of the 18.5–19.5 s section.

Figure 5a shows the results of experiment performed by applying the vehicle position estimator of the propulsion control system, and each waveform represents the speed, vehicle flux phase angle and estimated phase angle, and phase-U current of the LS-LSM from the top. In order to analyze the discontinuous signal of the absolute position and the subsequent performance of the position estimator, the expansion results of the 18.5–19.5 s section were represented in Figure 5b.

Figure 5 shows that the propulsion control system to which the vehicle position estimator is applied is generally stable, and the phase angle of the vehicle position estimator is linearly estimated with respect to the discontinuous signal of the absolute position as shown in Figure 5b.

The deviating behavior is shown in Figures 4 and 5 when the ground false position packets are received. This is because the position estimator cannot distinguish whether the absolute position information received from the vehicle is right or not. Thus, the position estimator tries to follow the wrong information.

Nevertheless, this suggests that the phase-U current of the LS-LSM maintains a sine wave with the output current controlled by the linear phase angle.

Figure 6 shows the fast fourier transform (FFT) analysis results of the current distortion depending on the presence of the vehicle position estimator in the entire running section. Figure 6a shows the FFT analysis results on the phase-U current in the case where the vehicle position estimator is not applied, and Figure 6b shows the FFT analysis results on the phase-U current of the LS-LSM in case the vehicle position estimator is applied. The comparison of the results revealed that the application of the vehicle position estimator contributed to improving the running current harmonic wave, as well as the running current fundamental wave, and the improvement rate was particularly high in the 100–200 Hz sections among running current harmonic components.

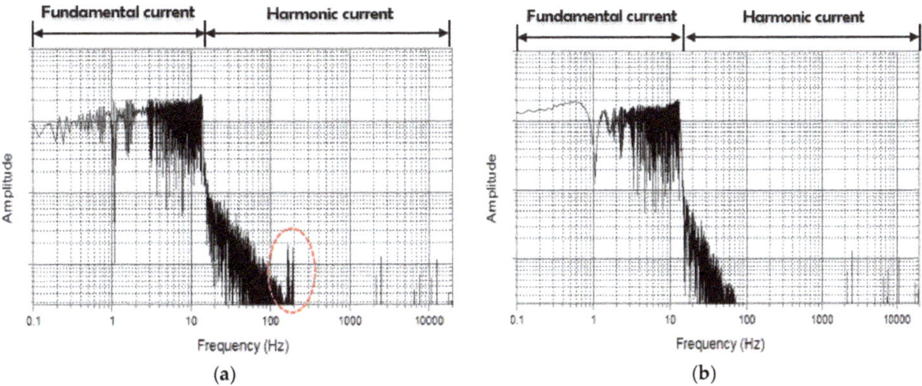

Figure 6. Phase-U current fast fourier transform (FFT) analysis results of the LSM. (**a**) LSM current FFT analysis results (without the estimator); and (**b**) the LSM current FFT analysis results (with the estimator).

Figure 7 represents the vehicle position estimation capability in the position information update cycle of 30 ms, which is 15 times later than the current position information update cycle of 2 ms. In order to identify the characteristics of the vehicle position estimator in case the absolute position information is transmitted from the vehicle in a 30 ms cycle, which is the same as that of the Transrapid. The results of Figure 7 confirmed that the phase angle of the vehicle position estimator is linearly estimated even in the position information update information of two to three times per cycle within the propulsion current and, therefore, the output current appears sinusoidal.

The operation parameters for generation of the speed profile were entered to analyze the performance of the automatic operation. The set operation parameters include the constant operation time of 2 s, maximum jerk limit of 0.5 m/s^3, maximum acceleration of 0.5 m/s^2, deceleration of 0.5 m/s^2, and maximum speed of 4.2 m/s, and the generated speed profile and automatic operation test results are shown in Figure 8.

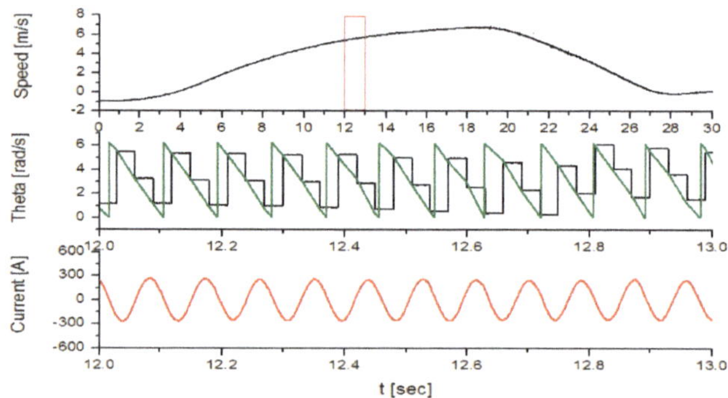

Figure 7. The experimental results of the position estimator with the position information update cycle of 30 ms.

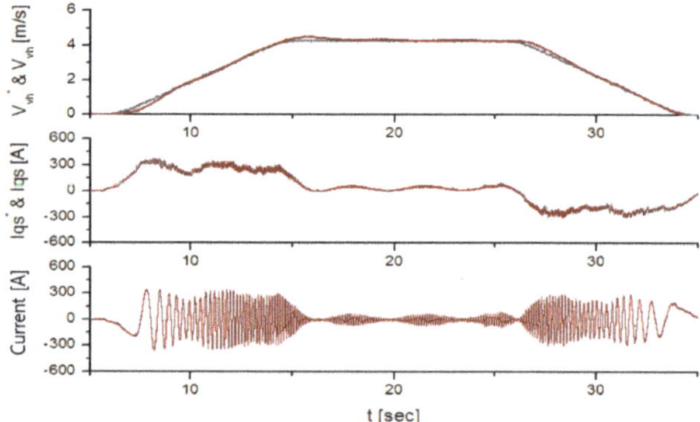

Figure 8. Automatic operation results according to the speed profile (top subplot: speed command and vehicle speed; middle subplot: i_{qs}* and i_{qs}; bottom subplot: i_{us}).

The experimental results showed that the speed tracking according to the speed profile is made smoothly when the q-axis components of stator current i_{qs} is tracked without error with respect to the q-axis components of the stator reference current i_{qs}*. In addition, the command value of the actual travel distance and the actual travel distance were calculated by integrating the command velocity and vehicle speed, respectively, in order to obtain the error rate of the automatic operation. According to the calculation results, the travel distance by the speed command profile is 84.75 m, the actual travel distance is 85.22 m and, therefore, the error rate of the travel distance is about 0.52%, if converted into the error rate of the automatic operation.

3. Conclusions

This paper aims to develop the propulsion control system technology for a high-speed maglev train system based on an LS-LSM. To this end, a test vehicle was fabricated with one-half the length of the final model, and a 150 m long LSM guideway was constructed to conduct the test and evaluation of the developed vehicle. The vehicle flux position observer was designed to analyze the difference in thrust force control between the rotating machine and the linear machine and, thus, to solve the problem that may occur in the liner machine. In addition, the characteristics of the current controller and speed controller were analyzed to design the propulsion controller of the LS-LSM.

For verification of the propulsion control performance, experiments on the automatic operation according to the speed profile and manual operation, and a comparison of the propulsion power quality enhancement on the position observer and torque required by the vehicle were performed, and the results confirmed that the design objectives were archived. For future work, a system for position detection within 5 mm, even at the maximum speed of 550 km/h, is planned to be developed for high-speed operation.

Acknowledgments: This research was supported by a grant from Research and Development Program of the Korea Railroad Research Institute, republic of Korea.

Author Contributions: Jeong-Min Jo designed the propulsion system and programmed the algorithms. Jin-Ho Lee designed the positioning system. Young-Jae Han organized the paper. Chang-Young Lee designed the LS-LSM. Kwan-Sup Lee helped to respond to the comments and revise this paper. All authors contributed to the writing of the manuscript, and approved the final manuscript.

Conflicts of Interest: The authors declare no conflict of interest.

1. Sadrnia, M.A.; Jafari, A.H. Robust control design for maglev train with parametric uncertainties using μ–systhesis. In Proceedings of the World Congress on Engineering 2007, London, UK, 2–4 July 2007.
2. Leidhold, R.; Mutshler, P. Speed sensorless control of a long-stator linear synchronous motor arranged in multiple segments. *IEEE Trans. Ind. Electr.* **2008**, *54*, 3246–3254. [CrossRef]
3. Qian, C.; Han, Z. Research on absolute positioning system for high-speed maglev train. In Proceedings of the 2007 IEEE International Conference on Mechatronics and Automation, Takamatsu, Japan, 5–8 August 2007.
4. Wu, X.M. *Maglev Train*; Science and Technology Press: Shanghai, China, 2003.
5. Yang, S.; Ke, S. Performance evaluation of a velocity observer for accurate velocity estimation of servo motor drives. *IEEE Trans. Ind. Appl.* **2000**, *36*, 98–104. [CrossRef]
6. Qian, C.; Wei, R.; Wang, X.; Ge, Q.; Li, Y. Analysis the Position signal problem in propulsion system with long stator synchronous motor. In Proceedings of the Maglev 2011 Conference, Daejeon, Korea, 10–13 October 2011.
7. Su, G.; McKeever, J.W.; Samons, L.S. Modular PM motor drives for automotive traction applications. In Proceedings of the 27th Annual Conference of the IEEE Industrial Electronics Society, Denver, CO, USA, 29 Novermber–2 December 2001.
8. Bado, A.; Bologani, S.; Zigliotto, M. Effective estimation of speed and rotor position of a PM synchronous motor drive by a Kalman filtering technique. In Proceedings of the 23rd Annual IEEE Power Electronics Specialists Conference, Toledo, Spain, 29 June–3 July 1992.
9. Zhu, G.; Kaddouri, A.; Dessiant, L.; Akhrif, O. A nonlinear state observer for the sensorless control of a permanent-magnet AC machine. *IEEE Trans. Ind. Electr.* **2001**, *48*, 1098–1108.
10. Hamada, D.; Uchida, K.; Yusivar, L.F.; Wakeo, S.; Onuki, T. Sensorless control of PMSM using a linear reduced order observer including disturbance torque estimation. In Proceedings of the 8th European Conference on Power Electronics and Applications, Lausanne, Switzerland, 7–9 September 1999.
11. Bhangu, B.; Williams, C.; Bingham, C.M.; Coles, J. EKFs and other nonlinear state-estimation techniques for sensorless control of automotive PMSMs. In Proceedings of the Speedam Conference 2002, Ravello, Italy, 11–14 June 2002.
12. Jo, J.; Lee, J.; Han, Y.; Lee, C. Design of position estimator for propulsion inverter driving long stator LSM in high speed maglev. *J. Int. Conf. Electr. Mach. Syst.* **2014**, *3*, 252–255. [CrossRef]

MDPI

St. Alban-Anlage 66

4052 Basel

Switzerland

Tel. +41 61 683 77 34

Fax +41 61 302 89 18

www.mdpi.com

Energies Editorial Office

E-mail: energies@mdpi.com

www.mdpi.com/journal/energies